ERRATA

The paper "Chemical Ionization Mass Spectrometry of Hydrocarbons and Halohydrocarbons" by Dr. A. G. Harrison contains a number of errors which were introduced as a result of the paper being retyped, an editorial policy adopted for all papers. Because the retyped paper was not submitted to the author for proofreading and was not properly proofread by me, I accept full responsibility for these errors and sincerely regret the embarrassment and inconvenience which this has caused Dr. Harrison.

D. Mackay

page 265, line 4:	for *quantitaton* read *quantitation*
page 265, line 26:	for *addition ion* read *additive ion*
page 266, eq. (6):	$R^{+} + M \rightarrow M^{+} + R$
page 269, eq. (10):	$CH_5^{+} + n\text{-}C_6H_{14} \rightarrow C_6H_{13}^{+} + H_2 + CH_4$
page 270, line 23:	for $C_nH_{2n-1}^{+}$ read $C_nH_{2n+1}^{+}$
page 271, line 17:	for *4-decane* read *4-decene*
page 272, Fig. 1:	for *4-decane* read *4-decene*
page 272, line 3:	for CH_5 read CH_5^{+}
page 272, line 4:	for *MH* read MH^{+}
page 273, line 6:	for *reaction* read *reactant*
page 274, line 21:	for *mistures* read *mixtures*
page 274, line 39:	for $[MH^{+}\text{-}HX]^4$ read $[MH^{+}\text{-}HX]$
page 275, line 25:	for *with* read *which*
page 276, line 14:	for $(MH^{+}M[M\text{-}H]^{+})$ read $(MH^{+}, [M\text{-}H]^{+})$
page 276, line 28:	for $[M^{+}\text{-}CH1]$ read $[M^{+}\text{-}HC1]$
page 276, line 38:	for $(ClC_6H_4)CH\text{-}R$ read $(ClC_6H_4)_2CH\text{-}R$
page 278, eq. (14):	$X^{-} + HY \rightarrow HX + Y^{-}$
page 279, line 22:	for $D(H\text{-}C)$ read $D(H\text{-}X)$
page 278, line 27:	for *produce* read *product*
page 279, line 13:	for *fromed* read *formed*

HYDROCARBONS AND HALOGENATED HYDROCARBONS IN THE AQUATIC ENVIRONMENT
Edited by B. K. Afghan and D. Mackay 0-306-40329-3

Hydrocarbons and Halogenated Hydrocarbons in the Aquatic Environment

Environmental Science Research

Recent Volumes in this Series

Volume 7 – ECOLOGICAL TOXICOLOGY RESEARCH: Effects of Heavy Metal and Organohalogen Compounds
Edited by A. D. McIntyre and C. F. Mills

Volume 8 – HARVESTING POLLUTED WATERS: Waste Heat and Nutrient-Loaded Effluents in the Aquaculture
Edited by O. Devik

Volume 9 – PERCEIVING ENVIRONMENTAL QUALITY: Research and Applications
Edited by Kenneth H. Craik and Ervin H. Zube

Volume 10 – PESTICIDES IN AQUATIC ENVIRONMENTS
Edited by Mohammed Abdul Quddus Khan

Volume 11 – BIOLOGICAL CONTROL BY AUGMENTATION OF NATURAL ENEMIES: Insect and Mite Control with Parasites and Predators
Edited by R. L. Ridway and S. B. Vinson

Volume 12 – PENTACHLOROPHENOL: Chemistry, Pharmacology, and Environmental Toxicology
Edited by K. Rango Rao

Volume 13 – ENVIRONMENTAL POLLUTANTS: Detection and Measurement
Edited by Taft Y. Toribara, James R. Coleman, Barton E. Dahneke, and Isaac Feldman

Volume 14 – THE BIOSALINE CONCEPT: An Approach to the Utilization of Underexploited Resources
Edited by Alexander Hollaender, James C. Aller, Emanuel Epstein, Anthony San Pietro, and Oskar R. Zaborsky

Volume 15 – APPLICATION OF SHORT-TERM BIOASSAYS IN THE FRACTIONATION AND ANALYSIS OF COMPLEX ENVIRONMENTAL MIXTURES
Edited by Michael D. Waters, Stephen Nesnow, Joellen L. Huisingh, Shahbeg S. Sandhu, and Larry Claxton

Volume 16 – HYDROCARBONS AND HALOGENATED HYDROCARBONS IN THE AQUATIC ENVIRONMENT
Edited by B. K. Afghan and D. Mackay

Hydrocarbons and Halogenated Hydrocarbons in the Aquatic Environment

Edited by
B. K. Afghan
National Water Research Institute
Burlington, Ontario, Canada

and

D. Mackay
University of Toronto
Toronto, Ontario, Canada

Associate Editors

H. E. Braun A. S. Y. Chau J. Lawrence D. R. S. Lean
O. Meresz J. R. W. Miles R. C. Pierce G. A. V. Rees
R. E. White D. M. Whittle D. T. Williams

PLENUM PRESS · NEW YORK AND LONDON

Library of Congress Cataloging in Publication Data

International Symposium on the Analysis of Hydrocarbons and Halogenated Hydrocarbons in the Aquatic Environment, McMaster University, 1978.
Hydrocarbons and halogenated hydrocarbons in the aquatic environment.

(Environmental science research; v. 16)
Proceedings of the symposium, held May 23–25, 1978, and organized by the National Water Research Institute, Canada Centre for Inland Waters, and the Institute for Environmental Studies, University of Toronto.
Includes index.
1. Hydrocarbons–Environmental aspects–Congresses. 2. Halocarbons–Environmental aspects–Congresses. 3. Hydrocarbons–Analysis–Congresses. 4. Halocarbons–Analysis–Congresses. 5. Water–Pollution–Environmental aspects–Congresses. I. Afgha, B. K. II. Mackay, D. III. Canada. National Water Research Institute. IV. University of Toronto. Institute for Environmental Studies. V. Title.
QH545.H92I54 1978 574.5'263 79-26462
ISBN 0-306-40329-3

Proceedings of the International Symposium on the Analysis of Hydrocarbons and Halogenated Hydrocarbons in the Aquatic Environment, held in Ontario, Canada, May 23–25, 1978.

A Division of Plenum Publishing Corporation
227 West 17th Street, New York, N. Y. 10011

Printed in the United States of America

INTERNATIONAL SYMPOSIUM ON THE ANALYSIS
OF
HYDROCARBONS AND HALOGENATED HYDROCARBONS

Program Committee

B.K. AFGHAN
General Chairman

D. MACKAY
Co-Chairman

B.F. Scott
Secretary and Budget Director

R. Knechtel
Arrangements

H.E. Braun
O. Meresz
J.R.W. Miles
R. Pierce
G.A.V. Rees
G. Rosenblatt
R.E. White
D.T. Williams

A.S.Y. Chau
J. Lawrence
D.R.S. Lean
F.I. Onuska
D.M. Whittle
J.J. Lichtenberg
T. Meiggs
J. McGuire

ORGANIZED
BY

National Water Research Institute
Canada Centre for Inland Waters
Burlington, Ontario

and

Institute of Environmental Studies
University of Toronto
Toronto, Ontario

CO-SPONSORS

National Health and Welfare
Ottawa, Ontario

Water Quality Branch (Ontario Region)
Burlington, Ontario

Great Lakes Biolimnology Laboratories
Burlington, Ontario

Contaminants Control Branch, EPS
Ottawa, Ontario

Laboratory Services Branch, Ontario Ministry of Environment
Rexdale, Ontario

SUPPORTERS

Agriculture Canada
Ottawa, Ontario

Wastewater Technology Centre, EPS
Burlington, Ontario

International Joint Commission
Windsor, Ontario

National Research Council
Ottawa, Ontario

Ontario Ministry of Agriculture and Food

International Standards Organization

Chemical Institute of Canada, Analytical Chemistry Division

Great Lakes Fisheries Laboratory
Ann Arbor, Michigan

PREFACE

This volume contains papers selected from those presented at the International Symposium on the Analysis of Hydrocarbons and Halogenated Hydrocarbons in the Aquatic Environment, May 23-25, 1978. The Symposium was organized by the National Water Research Institute of Environment Canada and the Institute for Environmental Studies of the University of Toronto.

The purpose of the Symposium and of this volume was to bring together information on the analyses, behaviour and effects of hydrocarbons and halogenated hydrocarbons on the aquatic environment. This class of contaminants presents many difficult analytical problems, and to a large extent our ability to identify environmental problems and assess their severity, depends on the availability of proved analytical techniques. By exposing workers in this field to the techniques and results of others we hope that progress can be made towards solving the many problems caused by these substances.

The papers in this volume are divided into five categories: two plenary session addresses given by J.P. Bruce and O. Hutzinger, 11 papers describing quantitative analytical results, 12 papers on analytical methods, 9 papers on incidence, monitoring and pathways; and finally, 13 papers on drinking water, health and biological effects.

We were fortunate in having the support of the many organizations and individuals listed after this preface. In particular we thank B.F. Scott and R. Knechtel for their help in organizing the Symposium, S. Paterson, C. Straka, B. Reuber, A. Bobra, R. Wan, S. Austin, G.E. Toll and B. Missen for their help in publishing the proceedings, and the session chairmen and associate editors for their help at the Symposium and later in reviewing and assembling these proceedings. The success of the Symposium and this volume is largely attributable to the enthusiasm and efforts of these individuals. We are deeply indebted to them.

B. Afghan and D. Mackay

CONTENTS

KEYNOTE ADDRESS

J.P. Bruce

Assistant Deputy Minister

Environmental Management Service, Environment Canada

Ottawa, Ontario, K1A 0E7

As many of you are only too aware, I am not an expert in chemistry but now simply a government bureaucrat. In order to speak to this learned group without too much fear of serious scientific criticism and contradiction, I decided to talk first about the Mackenzie River system and secondly to deal with the halogenated hydrocarbons and hydrocarbons from a policy perspective.

The Mackenzie River you say! What has this to do with analysis of hydrocarbons, halogenated or not? Well, you'll have to be patient - I'll make the connection a little later. Who outside Canada has ever heard of the Mackenzie River; Mackenzie Valley Pipeline maybe, but the river itself? And it is not exactly as if the Mackenzie were just a piddling little stream. In fact, it is the largest river whose basin is entirely within Canada and is well up in the top ten in the world in terms of discharge to the sea. A flow exceeding 1 million cubic feet per second was recorded there last summer - the largest flow ever recorded in Canada. On average its total flow is roughly ten percent greater than that of the St. Lawrence and two percent less than that of the Mississippi, though in some years it may discharge more water to the Arctic Ocean than the Mississippi does to the Gulf of Mexico. Only the St. Lawrence and the Nile Rivers drain such extensive lakes as the Mackenzie. Great Bear Lake and Great Slave Lake are each considerably larger than Lake Erie, while Lake Athabasca and Lesser Slave Lake are both of them bigger than Lake Ontario. Until a few years ago, the Mackenzie, running through Sub Arctic and Arctic, supported no industry and little human settlement so that its enormous discharge to the sea was essentially unaffected by human activity. The small amount of human settlement in the catchment area of the MacKenzie, a few score thousand native people, Dene and Inuit, made little or no impact upon this enormous discharge of water.

And yet the waters of the Mackenzie have at times been so tainted with hydrocarbons as to be undrinkable.

Different parts of the main river flowing from one lake to another have different names, often because they were discovered independently in the eighteenth and early nineteenth century and were not known to be parts of the same river system. The Athabasca River, for instance, was well known by the middle of the eighteenth century and in 1788 Alexander Mackenzie was sent by the Montreal Fur Company to take charge of the fur trade in this area. He established Fort Chipewyan on the southern short of Lake Athabasca. It has since been moved to the northern shore. The following year he set out, in an endeavour to follow the Athabasca River to its outlet. He avoided the deadly "Rapids of the Drowned", named for two canoe loads of adventurers who had perished in 1786, and during June of 1789 he groped his way for days through the fog and ice of Great Slave Lake to find the exit which was about eight or ten miles wide, but choked with islands.

Now here's where we're coming to the hydrocarbons. Drifting down the river he noted one place on the bank where an exposed bitumen seam was on fire. It is still burning today. On a shore excursion his mocassins became mired in oozing petroleum, an episode which he noted in his journal. Finally he made his way down the Mackenzie to the Beaufort Sea and the Arctic Ocean.

He seems to have had no thrill from this achievement, but instead a sense of bitter disappointment. He had hoped that the river would turn westward to provide a new route to the Pacific and to the fabled Indies. He notes in his journal: "It is evident that thse waters empty themselves into the Hypoborean Sea" and he bestowed upon the river the name "River of Disappointment". Later his own name was substituted.

That section of the Mackenzie Basin which is called the Athabasca River basin contains one of the richest deposits of fossil hydrocarbons in the world. David Laird and Charles Mair who negotiated Treaty No. 8, by which the greater part of the basins of the Athabasca River, Peace River, Lake Athabasca and Laird River were ceded to Canada by the Dene Tribes in 1900, noted the potential economic value of all this. Mair wrote: "In the neighbourhood of McMurray there are several tar wells, so called, and there, if a hole is scraped in the bank, it slowly fills in with tar mingled with sand. This is separated by boiling, and is used, in its native state, for gumming the canoes and boats. Farther up are immense towering banks, the tar oozing at every pore, and underlaid by great overlapping dykes of disintegrated limestone, alternating with lofty clay exposures, crowned with poplar, spruce and pine." He noted giant clay escarpments everywhere streaked oozing tar and "smelling like an old ship." "These tar cliffs are here hundreds of feet high ... the impregnated clay appears to be constantly falling off the almost sheer face of the slate-brown cliffs, in great sheets, which plunge into the river's edge in broken masses." Further downstream, further north that is, near the juncture of the Great Bear River, asphalt and coal seams are evident in a number of localities. Near Fort Norman, Norman Wells was first brought into production in 1920 and by the Second World War three other wells were producing in this area.

Hydrocarbons represent such a chemical store of energy that we may wonder how they have avoided decomposition by biological processes. One thing is clear at any rate, that fossil hydrocarbons must represent that fraction of the biogenic chemical store which has, for some reason, been least available to the biota. There is general agreement among scientists that fossil hydrocarbons are of biogenic origin and that the compositions of different crude oils reflects in some way the precise origin of the organisms whose remains have gone to make up the petroleum deposits. Thus a high wax content in crude oil is thought to indicate a depositional environment where land plants have been a major source of organic material. The dominant alkane of algae is comparitively short - C_{17} (heptadecane). In contrast, land plants synthesize alkanes greater than C_{17}, usually around C_{31}. The presence of these C_{31} waxes may thus be taken to indicate a major contribution from land plants. Similarly, polycyclic tri-terpenoids are usually taken as evidence of non-marine origin or at least of a non-marine contribution to the petroleum. Tri-terpenoids are a major constituent of many terrestrial plants but are minor constituents of marine microorganisms. In Chile, the Daniel crude oil and the Dicky crude oil, both from the Cretaceous era, are both believed to have formed in the delta zone of major rivers, and show high wax and tri-terpenoid content characteristics of a major contribution by land plants. By contrast, the Pembina crude of the Cretaceous of Canada appears to have formed in a marine environment and lacks these components in any major degree.

There is also evidence that besides the formation of petroleum deposits in the deltaic zone, they may also form a lacustrine regime, particularly under anoxic conditions. Such an oil would be characterized by a high level of waxes and a high ratio of pristane (tri-terpenoid) to heptadecane.

In the marine environment it appears that the sedimentation rate is a major factor in the formation of petroleum or rather of pre-petroleum. Bacteria are concentrated at the sediment water interface and a rapid sedimentation rate will reduce the residence time of organic matter in the surface zone. A low sedimentation rate would mean that much of the organic matter would be consumed by the bacteria while, with a rapid sedimentation rate, the organic matter would be bedded below the zone of bacterial activity before it could all be consumed.

At the present day the Northwestern Indian Ocean represents such an environment where anaerobic sediments are formed in pockets along the continental margins.

The Cariaco Trench on the continental shelf of Venezuela and the bottom waters of the Black Sea are also well known areas receiving a heavy sedimentation rich in organic matter and where the first stages of petroleum formation under anoxic conditions may be noted.

In the Saanich Inlet in British Columbia there are intermittently anoxic conditions where the bottom waters become anoxic in late summer. The sediments here have a high content of organic matter mainly due to input from the land and, therefore, are high in waxes. The sediments contain nearly 5 percent by dry weight organic carbon and the pristane to phytane ratio is greater than 3. By contrast the pristane to phytane ratios of the permanently anoxic recent sediments are generally less than one.

In more general terms we can say that the wax content, the pristane to phytane ratio, and the sulfur and porphyrin contents of crude oils and tars reflect to a large extent the particular conditions under which the sediments were laid down. By comparison with recent sediments we can surmise that a petroleum deposit high in waxes is either lacustrine or deltaic in origin, while one high in porphyrins is almost certainly marine. High pristane to phytane ratios indicate at least occasional oxic conditions, while low values of this ratio indicate anoxic conditions.

By these criteria, and by other more subtle criteria, such as the molybdenum, cobalt, uranium and lead content, it is clear that the Athabasca Tar Sands of Northern Alberta and the Northwest Territories represent an essentially marine deposit. These Tar Sands, representing the largest reserve of petroleum deposits in the world, larger than the crude oil reserves of Saudi Arabia, are, as we know, only now beginning to be exploited because of the great technological difficulties in extraction.

But even before any major exploitation, the contribution to the water quality of the Mackenzie River system has been enormous. Natural seepage from the Tar Sands has led to levels of hydrocarbons and other constituents of the tar which make the waters unacceptable for many purposes. The Inland Waters Directorate has been monitoring the concentrations of the vanadium in the waters of the Mackenzie, since the vanadium porphyrins are particularly abundant in the Tar Sands. The levels found are in excess of the standards set by the World Health Organization for Drinking Water, and this before there has been any significant exploitation of the Tar Sands. These levels of vanadium represent primarily natural seepage from the Sands. At last, you are likely saying, he's beginning to come to questions of analysis of hydrocarbons and closely associated compounds like the vanadium porphyrins. It is noteworthy too that the levels of vanadium which have been recorded and which are already unsatisfactorily high are probably an underestimate. Only in the last couple of years have our scientists been able to develop a method of analysis of vanadium which will adequately quantify that part of the vanadium which is present in the porphyrins. Some of these porphyrins are so stable that they may be detected in the exhaust of a diesel engine after exposure to temperatures of 2500°C, while vanadium oxides are extremely refractory. As a result of these two factors it appears that the standard method of atomic absorption spectrometry records only about one-third of the vanadium actually present. Most of the figures we have for vanadium in the Mackenzie River may be as much as three times too small.

The public at large tends to think of the environmental impact of a tanker spill where any analysis might conceivably be done on a bucket-and-spade wet chemistry basis. But in the inland waters of Canada and the United States, we are more often concerned not with large spills of this kind, but rather with the effects of the continuous input to the environment both of naturally occurring substances (whether of fossil hydrocarbons or of recent contributions from the flora, such as the waxes of land plants) and of man-made non-point sources. We must be constantly on the alert for the deterioration of our environment from such causes. We must also be in a position to distinguish man-made effects from those of natural processes. And this, of course, is why analyses of the pristane:phytane ratio may enable us to distinguish a man-made spill from the background of naturally occurring hydrocarbons. The Department of Environment is supporting development of new and sensitive methods for the detection and quantification of microscopic amounts of these compounds. Classical methods no longer suffice.

A recent unpublished study by David Carlisle on hydrocarbons in the Trent Canal in Ontario has indicated that during the summer the contribution to the waters of the system by the exhaust of outboard motor boats amounts to about 7 micrograms per litre (PPB). About the same amount again is contributed by various weeds, especially Eurasian water milfoil, growing in the system. Conventional methods of analysis will not quantify these amounts, much less separate the origin of these hydrocarbons. Some of these components have also been shown to undergo biological accumulation. Thus, the livers of crayfish from the Trent System may contain as much as 5 miligrams per kilogram (5ppm) of waxes derived from these two sources. Biological accumulation of this kind is of course much better known in relation to the chlorinated hydrocarbons. In any case, studies of this kind and our experience in the Mackenzie provides an incentive for the improvement of analytical methods for both hydrocarbons and chlorinated hydrocarbons.

Of course, the fossil hydrocarbons and plant waxes are only one aspect of the analytical problems you will be addressing in the next three days. The environmental headlines much more frequently focus on the halogenated hydrocarbons such as PCB's, PPB's, PCT's, Mirex and so on and analyses for low concentrations of these and other persistent toxic compounds is becoming increasingly important.

Over the past decade, we have begun to realize the enormity of the problems we face with toxic persistent environmental contaminants - many of them halogenated hydrocarbons. We have learned that they have very harmful effects on creatures at the higher levels of the food chain, particularly fish-eating birds and mammals. We have learned that they pose a threat to man - some are carcinogenic or teratogenic. We have discovered that they are spread far and wide on the globe. PCB's are present in loons of northern lakes and even in polar bears. These and many other compounds are spread around the globe in the long range transport of air pollutants. We have discovered that some of these substances are very persistent in the environment and can accumulate over long periods of time at higher levels in the food chain.

We have learned all of these things through the art and science of the analytical chemists who have been able to find ways of detecting very low concentrations in many different media - from the organs of mammals, to the eggs of herring gulls, in water, sediments, in precipitation and in the air.

But the challenge is only beginning. There are many compounds we have only begun to learn about - both in terms of their biological effects and their pathways in the environment. Control measures for both the well-known and newer contaminants will require increasingly sensitive and measured techniques.

For example, we expect that the revised Great Lakes Water Quality Agreement between Canada and the United States will be signed this summer. It will undoubtedly contain a number of specific environmental objectives recommended by the International Joint Commission. The International Joint Commission has recommended, for example, that PCB's in fish should not exceed 1 mg/kilogram to protect birds and animals which consume fish. This is estimated to mean that concentrations in water shouldn't exceed

1 part per trillion. Such low concentrations are at present impossible to monitor - yet to protect the biota, including man, we will have to continue to push back the threshold of sensitivity of analytical methods.

The revised Great Lakes Water Quality Agreement will lay major stress on control of persistent contaminants. The goal of Canada and the United States will be to have halogenated hydrocarbons "substantially absent" from the water environment. To these ends, domestically, Environment Canada uses the Fisheries Act and the Environmental Contaminants Act.

The Environmental Contaminants Act gives the Minister of the Environment the power to investigate and control the use of chemical compounds that pose a danger to human health or the environment.

The implementation of the Act has concentrated on the evaluation of information on the following organohalogan compounds: polychlorinated biphenyls, mirex, polybrominated biphenyls and polychlorinated terphenyls. A regulation has been promulgated for PCB's and regulations are being written for control of the others. Increasingly in the future the emphasis will be on the assessment of new compounds as opposed to these existing problem chemicals. This will pose increasing and new challenges to the analytical chemist.

The Fisheries Act prohibits the deposit of "deleterious substances" of any type into waters frequented by fish. Regulations under the Fisheries Act have been promulgated for a number of industrial sectors, the one of most interest to this conference being Petroleum Refining.

In protecting human health, a matter of major controversy currently exists as to the desirability of wastewater disinfection by chlorination in the Great Lakes and elsewhere in Canada. Those concerned with bacteriological contamination of beaches and drinking waters suggest that needless hazards would be introduced by not chlorinating. Those responsible for protection of aquatic life have demonstrated the deleterious effect of chlorine residuals and compounds. Health officials too are concerned about chlorinated hydrocarbons that may form in lakes and rivers or water supplies through chlorine uses - are these a health hazard? Clearly, a disinfection policy must address these divergent views and provide for protection of water supplies as well as the other beneficial uses of the water resource. I am delighted to see that a number of papers related to this controversy are on the program of this Conference.

I have tried to demonstrate, with a number of examples, why environmental agencies must place a high priority on development, standardization and use of micro-analytical methods for hydrocarbons and halogenated hydrocarbons. This Conference is an important and timely gathering and we are most grateful to the organizers. The outcome of your discussions over the next few days will make an important contribution to our understanding of contamination of the environment. In addition, your work will lead to the development and use of the analytical tools essential to controlling toxic contaminants. This is important to all of us in our efforts to sustain the ecosystems of which we are a part - and so protect the health and well-being of both this and future generations of mankind.

HYDROCARBONS AND HALOGENATED HYDROCARBONS IN THE AQUATIC ENVIRONMENT: SOME THOUGHTS ON THE PHILOSOPHY AND PRACTICE OF ENVIRONMENTAL ANALYTICAL CHEMISTRY

O. Hutzinger and A.A.M. Roof

Laboratory of Environmental and Toxicological Chemistry
University of Amsterdam, Nieuwe Achtergracht 166
1018 WV Amsterdam, The Netherlands

ABSTRACT

Sophisticated analytical techniques are now available for the analysis of hydrocarbons and halogenated hydrocarbons, and the number of compounds found in waters of different origins is now well over one thousand.

The limited resources and manpower available for environmental analysis make it imperative to avoid unnecessary determinations.

In this paper an attempt is made to outline the important tasks of environmental analysis. It is concluded that there should be more than just determination of levels in various environmental sub-compartments. Rather, programs should be designed to contribute to the overall picture of importance of the compounds analysed in the environment. This implies that information on transport and transformation including data on transformation products be obtained.

INTRODUCTION

Throughout recorded history we find much greater preoccupation with air pollution than with water pollution. The earliest laws for regulating air pollution in England, for instance, date from the first part of the fourteenth century. These laws were introduced as a reaction to the burning of bituminous coal which began in London during the 13th century (Papetti and Gilmore, 1971).

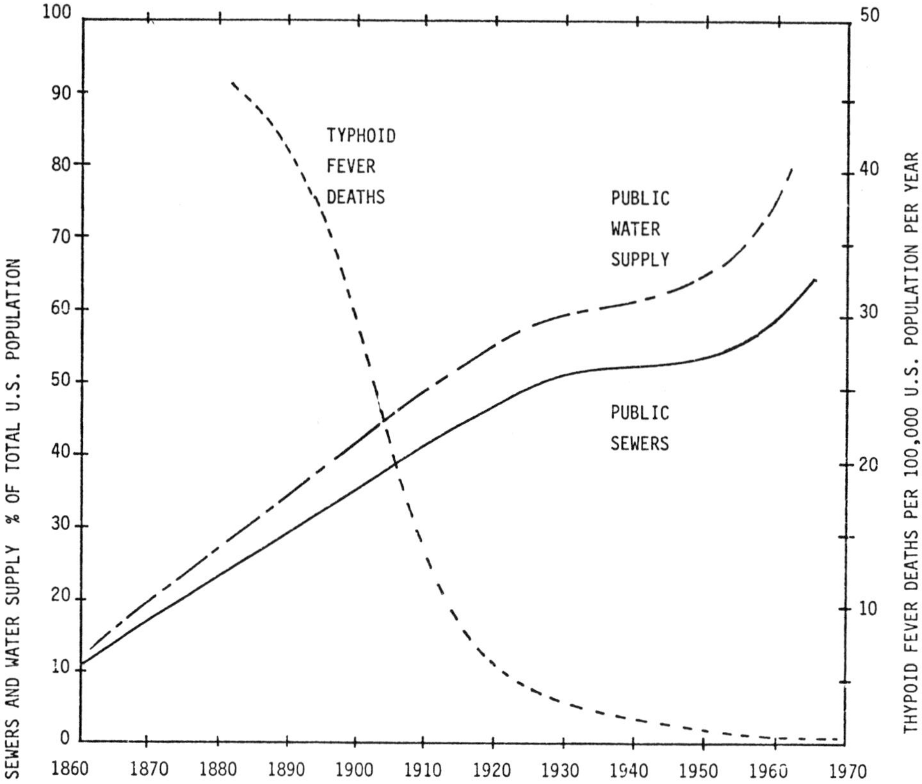

Figure 1. The relationship between typhoid fever deaths and availability of public water supplies and public sewers in the U.S.

From the design of sewer facilities and drinking water sources of cities in the middle ages we know, on the other hand, that people were rather unconcerned about the quality of their water during this time period and even later (Zajic, 1971).

This lack of concern about water quality was completely unjustified as the well known relation between water hygiene and certain infectious diseases indicates (Figure 1). Without proper water treatment procedures, contamination of drinking water with pathogenic microorganisms would still be the most serious health problem today - even in our age of chemical contaminants.

Experts consider the following water pollution problems of more recent history most important (Mount, 1978):

to 1950	domestic sewage and oxygen demanding waste
1950 - 1970	insecticides in surface water
1970 - 1975	electric power generation
from 1975	highly toxic substances from daily commerce.

Many of the chemicals of daily use are the most important source of the so-called micropollutants in our water. A large percentage of the micropollutants, in turn, are hydrocarbons and halogenated hydrocarbons, the compounds of concern in this symposium.

Chemical compounds in water. Grob (1978) has recently proposed an outline of organic compounds in water for consideration by analytical chemists:

Classification of organic compounds in water:
- All substances from the atmosphere
- Man-made organics
- By-products of chlorination
- Biological metabolites and breakdown products

A more complete outline which considers the source of compounds in water systematically is given below:

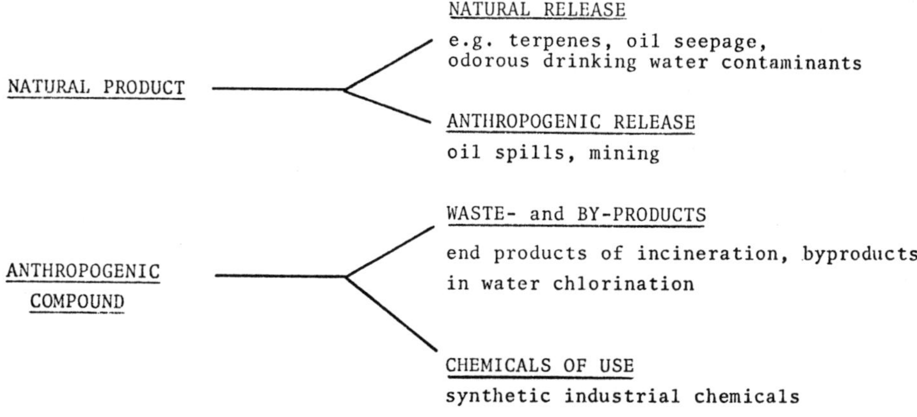

Among the many products released by natural processes are all the breakdown products and substances of biological origin which account for the largest quantity of organic carbon in natural water. Oil spills and related releases of petroleum are the most important sources of hydrocarbons in the second category.

Unintentionally formed products during the chlorination or ozonisation of (waste-) water are important examples of compounds in the third category and the millions of synthetic chemicals are a well-defined class in category four.

ENVIRONMENTAL ANALYSIS: WHY?

Numerous chemicals from all the classes of compounds mentioned above can be present in different waters which are found in the environment. From the point of view of the analytical chemist, this presence of chemicals in water and possible associated problems can be viewed in several ways.

In the aquatic environments, like in any other area of analytical chemistry, some basic quastions have to be asked:

which chemical species need to be analysed	WHAT?
by what methods are they best analysed	HOW?
why do they need to be analysed	WHY?

The most obvious approach would be answer and satisfy the first two questions and try to find out what compounds are actually present in any given natural body of water (qualitative analysis) and then attempt to find out how much there is (quantitative analysis). For the purpose of this symposium this would mean: what are the best analytical methods to determine hydrocarbons and halogenated hydrocarbons? This necessary and valid approach will be dealt with in the many contributions to this symposium. For the purpose of this article we will ask the third question: what does it mean to have these compounds in the water; where do they come from, what happens to them in the aquatic environment (and the environment generally) and what type of effects do they have. The question to the analytical chemist is: what can analysis do to clarify this situation, what can it do to help understand the fate and effect of chemical compounds in the aquatic environment.

Compounds in water are of interest only if they are toxic to organisms including man, or if they have some other undesirable effect (e.g. if the taste of drinking water is affected by them). Compounds without effects are compounds without interest. The great problem in this simple statement lies in the difficulty of making this assignment in a sound and scientific manner.

In an attempt to define and find compounds which have an effect, a systematic look at water pollution is desirable and the following course of events should be considered:

CHEMICAL SUBSTANCE	RELEASE → INTO ENVIRONMENT	PRESENCE AND → FATE IN ENVIRONMENT	CONTACE WITH (TARGET) ORGANISM	EFFECT
1	2	3	4	5

1. The chemical substance in this sequence can be any structure of natural or synthetic origin.

2. The release of a substance into the environment is the "act of polluting".

3. In the environment the presence of the substance, and the levels in various environmental compartments can be observed but, in a more dynamic outlook, the behaviour (fate) of a compound during exposure to the complex interplay of forces within the environment can be observed.

4. The compound, or transformation product thereof, can contact a living organism which may be any biotic component in the environment, including man.

5. The chemical or its environmental transformation product may have a toxic or other undesirable effect on the organism.

This toxic or other effect on an organism is the real and final reason for concern about a chemical pollutant. The toxic effect which may be observed depends, from the aquatic pollutant's point of view on:

a. the chemical species (structure) and
b. the quantity ("dose") which is present at the "target" organism.

Questions related to the nature (quality) and amount (quantity) of a chemical contaminant can obviously be answered by analytical investigations.

The exposure of a target organism by a chemical compound, the "what" and "how much" of a substance at any given location, depends to a large extent on what had happened to a compound released into the environment before.

For an understanding of the contributing factors which determine the concentration of a chemical compound in an environmental subcompartment (thus the potential exposure concentrations to organism) the following have to be considered:

	CONCERN	PRINCIPAL QUESTION ASKED
1.	Nature of chemical type of structure	What is the compound?
2.	Factors contributing to possible release origin, source, production method, storage, shipment, use pattern, disposal.	Where does the compound come from?
3.	Entry into environment environmental compartment, geographic area, pattern (point source, diffuse, continuous, discontinuous), release mechanisms.	Where does it enter, what are the release mechanisms?
4.	Physical processes in the environment responsible for distribution transport processes based on solubility, volatility, adsorption and sedimentation processes	Where is the preferred residence compartment?
5.	Chemical processes within the environment responsible for transformation chemcial, photochemical, enzymatic (mainly microbial)	What are the breakdown routes and mechanisms; what potentially toxic transformation are formed?

When the environment is treated as a simple one-compartment model and simplified rate constants are applied to the movement of a chemical (Figure 2) (Hutzinger et al, 1978), it becomes obvious that points 2 and 3 of the chemical scheme above are responsible for k_r and points 4 and 5 for the behaviour of the compound in the environment, and all of them for the "exposure level concentration to target organisms".

The "behaviour" or "fate" of chemicals in the environment, often referred to as environmental dynamics of chemicals, chemodynamics or ecokinetics, has been the subject of investigation during the last years (National Academy of Sciences, 1972; de Freitas et al, 1974; Haque and Freed, 1974; National Academy of Sciences, 1975; Haque and Freed, 1975; Freed et al, 1977; Suffet, 1977; Butler, 1978) and is also considered in relation to natural cycles in the environment (Nriagu, 1976; Stumm, 1977).

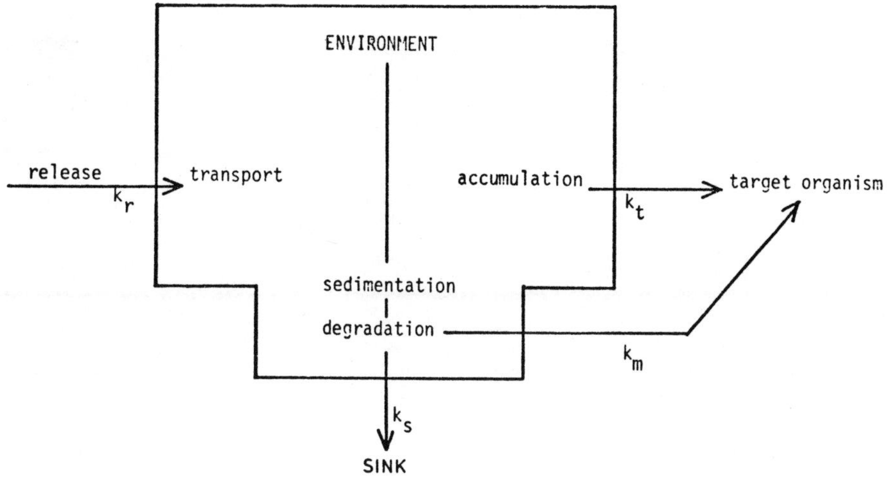

Figure 2. Representation of environmental chemodynamics (ecokinetics)
k_r = rate of release of chemical
k_t = rate at which chemical (toxicant) becomes available to organism
k_m = same for metabolite
k_s = rate of disappearance of chemical (sink)

An attempt can now be made to define environmental analytical chemistry. It should do more than just report the presence and levels of certain compounds, it should give information on the fate of compounds in the environment by studying trends in different environmental compartments and by following the structures and their concentration levels in degradative pathways.

A definition of Environmental Analytical Chemistry can now be considered which is more dynamic than the determination of the nature and quantities of chemical contaminants in different environmental compartments: FOLLOW IMPORTANT CHEMICAL SPECIES and their TRANSFORMATION PRODUCTS IN THE ENVIRONMENT.

The argument for the scope of environmental analysis can be carried further. An _effect_ (on a given organism) does not automatically follow exposure but depends on uptake, distribution, excretion and availability at a specific receptor within the organism which is sensitive to the toxicant. The behaviour of drugs and similar compounds within mammalian organisms is well studied in the science of pharmacodynamics, but much less is known on the fate of pollutants within aquatic organisms. In addition, possible

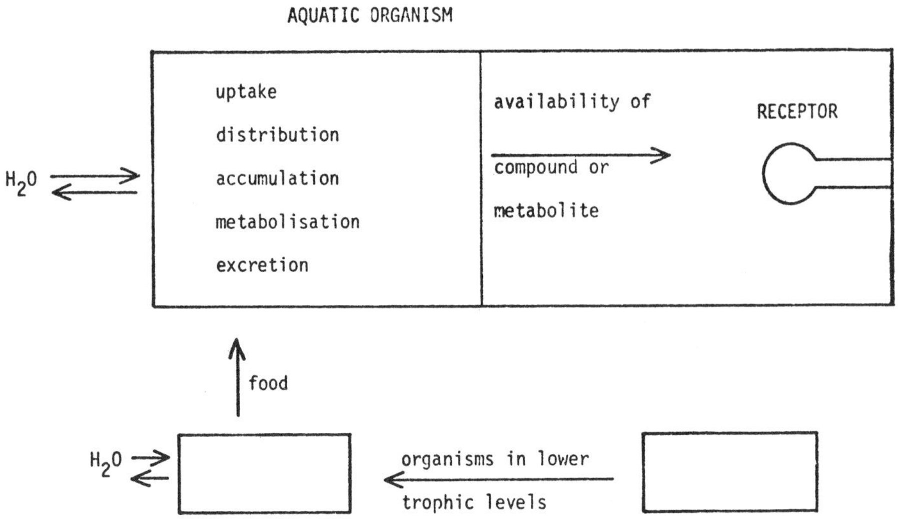

Figure 3. Schematic model of important pharmacodynamic parameters for aquatic organisms.

trophic level effects have to be considered where concentrations much higher than in the ambient water are available to an organism via its food. Both effects are shown schematically in Figure 3.

A vision and ideal of environmental analytical chemistry in its widest sense could be:

FOLLOW IMPORTANT CHEMICAL SPECIES AND THEIR TRANSFORMATION PRODUCTS THROUGHOUT THE ENVIRONMENT AND WITHIN ORGANISMS TO THE RECEPTOR OF TOXIC ACTION OR SITE OF OTHER EFFECT

ENVIRONMENTAL CHEMICALS

More than four million chemicals are known today (Maugh, 1978); over 60,000 of these are in common use and about one thousand new chemicals are brought into commerce every year (UNEP, 1978; EPA Office of Toxic Substances, 1978 - anon.) (Table 1).

It is estimated (Harris, 1976) that over 10,000 chemicals are produced world-wide in amounts exceeding 500 kg.

The production of all halogenated organic compounds in the U.S. was more than 5 x 10^6 tons in 1974 (Tariff Commission Reports, 1974). Total world production of chlorinated aliphatic hydrocarbons

Table 1 Chemical Compounds

Registered in Chemical Abstracts, November 1977	4 million
Weekly addition	~ 6,000
Active pesticide ingredients, pharmaceutical and food additives	~ 11,000
Compounds likely to be in common use	~ 63,000
First list of compounds recommended for testing (EPA)	~ 30,000
EPA priority list	50 - 200

probably exceeds 3 x 10^6 tons (Goldberg, 1971). U.S. production of synthetic chemicals increased 36% from 106 x 10^9 lbs in 1967 to 143 x 10^9 lbs in 1971 (U.S. Tariff Commission, 1973). Total world production in 1977 was probably 1.5 x 10^8 tons. These figures reflect only those more or less well defined compounds which are described in the literature and do not include most components of complex natural mixtures (e.g. oil) or incidentally formed products such as hydrocarbons from combustion processes.

Routes into the environment. The source and route of entry of chemical compounds vary widely and range from natural product release (e.g. IC_4) to totally synthetic compounds released into the environment accidentally. The mode of release ranges from incidental to continuous and from specific point sources (e.g. stacks) to large surface area diffuse release (e.g. highways, cities).

In principle, all chemicals ever produced, synthesized or formed by other mechanisms can enter the environment. The significance of release from a quantitative point of view depends on the amounts produced, the method of shipping and storage, the use of the compound and the disposal methods.

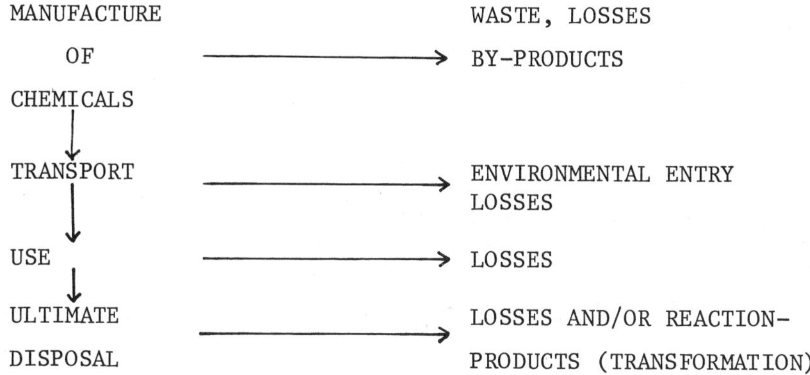

A pesticide, for instance, which is designed to be deliberately applied to the environment will be an environmental contaminant for essentially 100% of its amount of production, whereas only a small portion of a chemical in a closed system will reach the environment during proper use. In the production and use of chemicals, release of material into the environment cannot be avoided. While there is a difference between "open" and "closed" (i.e. controlled) manufacture and use, escape by leaking, evaporation, etc. is unavoidable.

The methods of final disposal also have a strong influence on the total amdount of release for different compounds. Non-biodegradable material may pass biological water purification procedures and may remain unchanged during composting; volatile and water-soluble compounds may escape from landfills and thermally stable compounds may survive incineration processes. Effluents from waste water treatment plants and gaseous industrial emissions are thus a potential source of environmental contamination.

Apart from the initial geographic distribution, the environmental compartment of "first entry" is important, e.g.

water	-	river (waste water) oceans (oil spill)
air	-	lower atmosphere (exhaust gases) stratosphere (supersonic aircraft)
soil	-	(pesticide spraying, land fill)
sediment	-	(sludge dumping)

Environmental behaviour of chemicals. The main questions asked are the following:

- where do they go?
  - transport
- how much remains?
  - persistence
  - biodegradation
  - sedimentation
- are important transformation products formed?
  - chemical, photochemical reactions
  - metabolism
- are dangerous concentrations observed in the biosphere?
  - bioaccumulation

For clarity and to aid the development of concepts, it is convenient to look at transport (physical processes), transformation (change of chemical structure) and bioaccumulation separately as if they were stepwise processes, although in practice they occur concurrently.

Transport and related physical processes. Transport and adsorptive behaviour depends on physical properties of a chemical compound, e.g. vapour pressure, solubility and partition coefficients and the existence of natural transport processes (water cycle, wind).

Transformations (alteration of chemical structure). While many compounds are completely degraded in the environment (mineralized to CO_2 and mineral elements) the formation of intermediate, stable transformation products of possible toxicological importance is and important feature. Significant in the case of degradation (lack of degradation = persistence) is the structure of the compound (chemical, biochemical stability, "robustness"), availability of the compound (e.g. adsorbed vs. in solution) and the presence of "reactants" in the aquatic environment. Such reactants are short wavelength sunlight (ca. 300 - 350 nm), OH^- or H^+ ions (pH), catalytic surfaces, nucleophiles, radicals and other reactive species and a variety of enzymes usually available from microbial sources.

Bioaccumulation. The most important factors in determining bioaccumulations are lipophilicity (high partition factor), a molar size suitable for membrane passage, and biochemical stability (not easily metabolized).

Figure 4 presents an outline of the most important features in the behaviour of an aquatic pollutant (Branson, 1976).

Biological effects. The following types of biological effects of aquatic pollutants can be distinguished:

1. Effect on Humans. Because of the relatively low concentration of organic contaminants in drinking water, the main concern is about low-level, long-term effects such as carcinogenesis. Much higher concentrations of certain compounds (e.g. PCB) can be found in food from aquatic sources.

As more surface water has to be processed for drinking water, more, and a higher incidence of, organic compounds from this category can be expected.

2. Effect on Commercial Aquatic Species. The difference between 2 and 3 is only one of human viewpoint. The importance of economic aspects in relation to human food expressed in availability of certain species and yields is considered here.

3. Effect on Aquatic Ecosystems. The importance here is on species which play an important role in the community of an ecosystem. The emphasis is on the health of the ecosystem and not on human health or human economic interest.

Certain non-toxic effects of aquatic pollutants are considered under interference with chemical communication in an aquatic ecosystem such as chemotaxis and a series of chemically mediated relationships such as food markers, sex attractants and defence substances. Most of these relationships are not fully understood but are potentially of great significance.

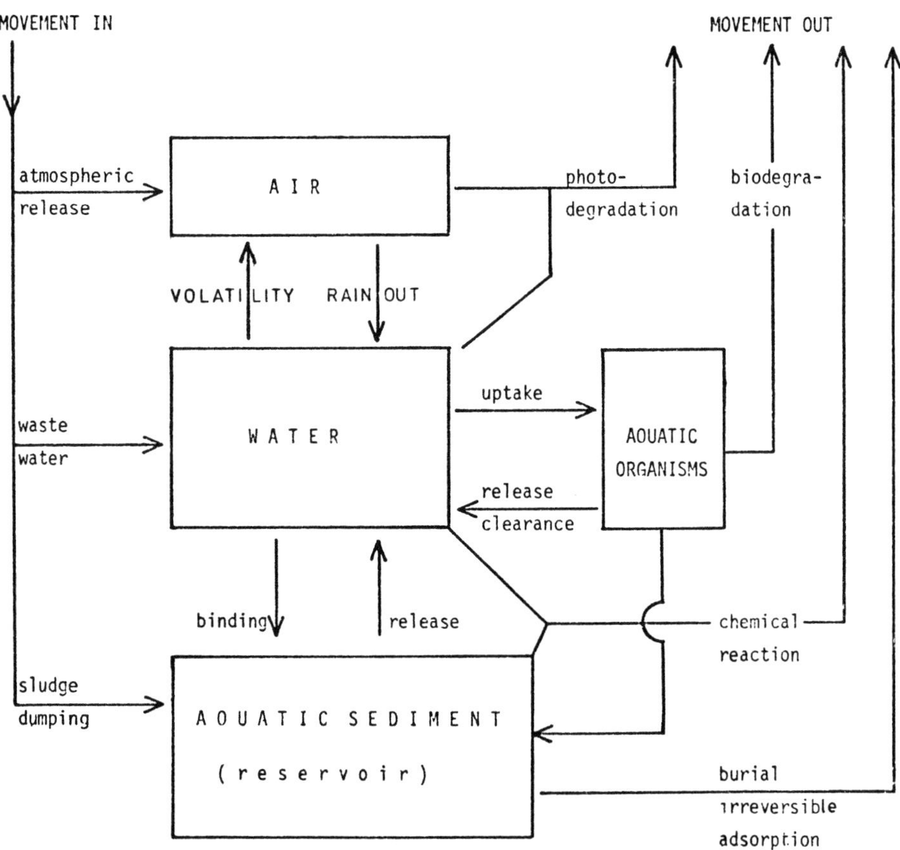

Figure 4. Schematic model for the behaviour of a chemical in the aquatic environment.

Table 2 Survey Results on Carcinogenic and Mutagenic Activities of Aquatic Pollutants

Classification	No. of Chemicals
recognized carcinogens	22
suspected carcinogens	42
tumor promotors/cocarcinogens	27
chemicals - inadequate test results	314
chemicals not tested	1323
mutagens	50
suspected mutagens	15
chemicals tested - no evidence of activity	101
chemicals - no test data found	1562

Two comments on human carcinogenesis may now be in order. The information in Table 2 is related to carcinogenic and mutagenic properties of compounds found in water and illustrates the state of ignorance in this area (Kraybill et al, 1978).

Figure 5 is concerned with carcinogens generally, but has a more direct meaning for analytical chemistry since it relates carcinogenic properties of certain chemicals with the quantity necessary to produce the effect (Fox, 1977). The difference between the most active and the least active carcinogen on this list is a factor of greater than 10 million. This indicates that any statement of occurrence of a carcinogen in water (or any commodity for that matter) without quantitative information is incomplete.

It is difficult to evaluate the toxicity of an aquatic pollutant but it is even more difficult to evaluate the overall risk which is associated, for example, with contaminants in drinking water.

Recently, there has been much concern and activity about risk assessment and risk-benefit evaluation. From statistical information human acts and conditions are compared to their "danger content" expressed as a reduced life expectancy (Cohen, 1977; Cohen, 1978; Kletz, 1977).

While the list in which "equidangerous" entities are listed cannot be taken altogether seriously, it illustrates some of the difficulties in this area and possible avenues of approach.

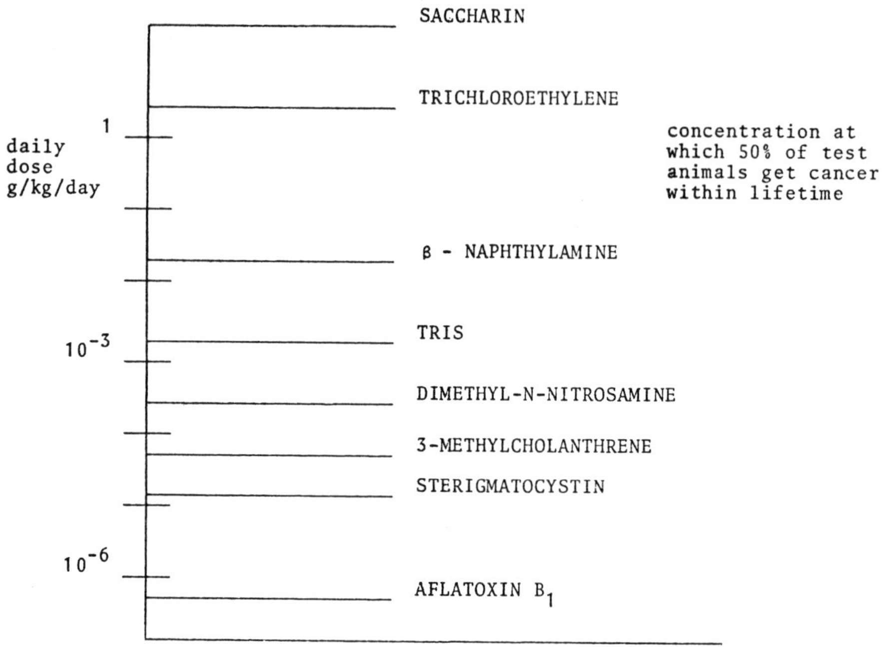

Figure 5. Effective dose of carcinogenic chemicals, concentrations at which 50% of test animals get cancer within lifetime.

Human Activity or Condition Leading to a Reduced Life Expectancy of Approximately 10 Minutes

Smoking one single cigarette

Drinking 60 diet soft drinks (saccharin) per lifetime

Taking 1 km car ride

Being 100 g overweight

Living in U.S.A. with 65 nuclear reactors

Drinking x litres tap water in Y city.

ANALYTICAL OUTLOOK

A number of analytical approaches in the environmental area are important.

Routine analysis - monitoring. (Measurement of known compounds).

need:
- cheap field methods;
- more accurate, selective and sensitive methods;
- methods for unambiguous identification;
- automation.

Analysis of complex environmental samples. (Detection, identification and quantification of the "unknowns").

need:
- methods for separation and detection of new (e.g. polar and less volatile) compounds;
- identification of new compounds, metabolites and breakdown products including use of sophisticated methods such as computer-assisted structure elucidation and fourier IR;
- storage of findings in data banks.

Good advances have been made in the detection and identification of organic compounds in water and well over a thousand entries can be found in the collection of data from different types of water in the environment (COST Project 64b, Commission of the European Communities, 1976; Garrison, 1976; Scackelford and Keith, 1976; Garrison et al, 1978).

It is well publicized (e.g. Cohen, 1978; Donaldson, 1977) that our present method of identification of organic water pollutants, GC-MS, fails to detect 80-90% of organic compounds in water since the gas chromatograph introduces a bias against non-volatile chemicals, such as polar or large molecular weight compounds. There is now a danger that a second, additional bias is introduced by the exclusive use of computer matching of mass spectra from data banks. Generally, a compound for which a suitable reference mass spectrum is not found in the files remains unidentified. Since it is extremely difficult to do proper strucutre identification on very small quantities of organic compounds there is a growing danger of rapidly approaching the limit of detectable compounds simply because of lack of suitable reference spectra in the computer files.

Environmental analysis with toxicological feedback. (Concentrate on compounds with proven or suspected toxic effect).

need:
- in cooperation with toxicologist, the most significant com-

pounds have to be selected for the monitoring and detailed studies among the large number of proven and potential water pollutants, one possibility is testing of crude fractions of environmental samples and investigating only the active ones.

Even if thousands of compounds in water were properly identified this information is of little value without knowing about their toxicological significance. This information can only be obtained by expensive and time-consuming procedures and a monitoring system for significant compounds can therefore not readily be set up.

An example of analysis with toxicilogical feedback is the popular Ames test. In practice a battery of tests including some which indicate effects to aquatic organisms will have to be used. It is crucial to obtain a biologically active fraction from, for instance, a water extract and to get on the right track (toxicologically important compounds) early in the analytical investigatons.

Analytical methods in laboratory studies. (Analysis as aid in laboratory studies designed to predict environmental behaviour and study toxicity).

need:
- analytical methods in bioaccumulation and biodegradation studies; determination of log P;
- analytical methods in pharmocokinetics, elucidation of metabolic pathways and identification of toxic intermediates.

As an example of contribution of analytical methodology in this area, the determination of the partition coefficient log P by reverse phase HPLC can be given. The partition coefficients for highly lipophilic compounds (log P > ca. 4) cannot easily be obtained since measurements by the classical n-octanol/water partitioning technique are impossible because the low concentrations remaining in the aqueous phase are often below the detection limit of most convenient methods of analysis, and calculations often yield unreliable results (Tulp and Hutzinger, 1978). Veith (personal communication), has recently used reverse phase HPLC with aqueous methanol as the eluant. A good correlation between log P and log retention time was obtained.

Final comments - the environmental analytical chemist. Environmental analysis will remain or perhaps increase its role as a most valuable and necessary tool in environmental problem-solving. Special care should be taken, however, that large numbers of indiscriminate and eventually unnecessary analyses not be carried out. Behind each analysis should be a plan.

The analytical chemist shares part of the responsibility for deciding how resources should be used, i.e. how existing instrumentation and manpower should be available for specific analytical tasks. In the field of environmental analysis the analytical chemist is, indeed, often the only person qualified for such judgement and information on environmental fate and toxicity need to be taken into consideration.

In the list below various activities related to analysis are correlated to the type of person who can carry out such activity and the order is in increasing level of sophistication.

Analytical Activity	Carried out by
Generate samples	almost anyone
Find time for analysis	anyone competent that is organized
Do good analysis by known methods	properly trained technician
Evaluate methods, new techniques ("how")	analytical chemist
Judgement of importance, relevance and need of analysis ("why")	ENLIGHTENED analytical chemist

It is easiest to generate samples (this does not refer to proper sampling techniques but to the fact of "bringing samples to be analyzed"), in fact, it is too easy and hundreds of samples of water, sediment and biota can be taken in a short period of time. It is more difficult to find the time for analysis and it takes a trained technician to do a good analysis. A properly educated analytical chemist is necessary to evaluate methods and to develop new techniques - i.e. to understand the HOW of analysis. The highest level of analytical understanding is reserved for the ENLIGHTENED ANALYTICAL CHEMIST. This is someone who has the capacity to judge the importance and need of analysis, the master of the wisdom of WHY.

In many cases, unfortunately, the course of events is as follows: samples are delivered, time if found to do the analysis by a known method first, if this does not work other methods are evaluated and then - sometimes - the question is asked why was it done?

The ideal situation exists if we ask the critical question why first, then evaluate methods, get prepared and finally ask for a sample taken in the proper manner.

Finally, it should be pointed out that Analytical Chemistry is a branch of chemistry with a long and proud tradition. Analytical Chemistry is, in fact, very closely connected with the beginning of scientific chemistry. This is reflected linguistically in some Germanic languages where chemistry and analysis are, or were, used almost synonymously. The old German word for Chemistry "Scheidekunde" and the "Scheikunde" which is still used in Dutch for chemistry mean literally "separation science" and thus analysis.

Analytical chemistry, however, does not only have a proud tradition but also a bright future. We live in a time where science, at least part of it, has to be applied to the needs and problems of society. Environmental analytical chemistry is ideally suited to make contributions in this area and analytical chemistry is for these and other reasons more important now than ever before.

REFERENCES

Branson, D.R. 1976. ASTM Symposium on Aquatic Toxicology, Memphis, Tenn., Oct. 25-26, p.130.

Butler, G.C. (ed.) 1978. "Principles of Ecotoxicology", John Wiley, (in press).

Cohen, B.L. 1977. American Scientist, 64, 550.

Cohen, B.L. 1978. Science, 199.

Commoner, B. 1976. "Chemical Carcinogens in the Environment" in "Identification and Analysis of Organic Pollutants in Water", L.H. Keith, ed., Ann Arbor Science Publishers, Ann Arbor.

de Freitas, A.S.W., D.J. Kushner, and S.U. Quadri (eds.) 1974. Proceedings of the International Conference on Transport and Persistent Chemicals in Aquatic Ecosystems, National Research Council, Ottawa.

Donaldson, W.T. 1977. Environ. Sci. Technol., 11, 348.

Freed, V.H., C.T. Chiou and R. Haque. 1977. Environ. Health Perspect. 20, 55.

Fox, J.L. 1977. Chem. Eng. News, Dec. 12. p.34.

Garrison, A.W. 1976. "Analysis of Organic Compounds in Water to Support Health Effect Studies", WHO International Reference Centre for Community Water Supply, Technical Paper Series 9, Leidschendam.

Garrison, A.W., L.H. Keith and W.M. Shackelford. 1978. "Occurrence, Registry and Classification of Organic Pollutants in Water, with Development of a Master Scheme for their Analysis" in "Aquatic Pollutants, Transformation and Biological Effects" (O. Hutzinger, L.H. van Lelyveld, and B.C.J. Zoeteman, eds.), Pergamon Press, Oxford.

Goldberg, E.D. 1971. In Impingement of Man on the Oceans, D.W. Hood (ed.), Wiley Interscience, New York.

Grob, K. 1978. Plenary Lecture, 8th Annual Symposium on the Analytical Chemistry of Pollutants, April 5-7, Geneva.

Haque, R., and V.H. Freed (eds.) 1975. Environmental Dynamics of Pesticides, Plenum Publ. Co., New York.

Harris, R.C. 1976. Suggestion for the Development of a Hazard Procedure for Potentially Toxic Chemicals, MARC Report, Nr. 3, The Monitoring and Assessment Research Centre, Chelsea College, University of London.

Hutzinger, O., M.Th.M. Tulp, and V. Zitko. 1978. "Chemicals with Pollution Potential", in "Aquatic Pollutants, Transformation and Biological Effects", O. Hutzinger, L.H. van Lelyveld, and B.C.J. Zoeteman (eds.), Pergamon Press, Oxford.

Kletz, T.A. 1977. New Scientist, May 12, p.320.

Kraybill, J.F., C. Tucker Helmes, and C.C. Sigman. 1978. "Biomedical Aspects of Biorefractories in Water", in "Aquatic Pollutants, Transformations and Biological Effects" (O. Hutzinger, L.H. van Lelyveld and B.C.J. Zoeteman, eds.), Pergamon Press, Oxford.

Maugh II, T.H. 1978. Science, 199, 162.

Mount, D.J. 1978. Quarterly Report, Environmental Research Laboratory, Duluth.

National Academy of Sciences. 1972. Degradation of Synthetic Organic Molecules in the Biosphere, Proceedings of a Conference in San Francisco, Printing and Publishing Offices, National Academy of Sciences, Washington.

National Academy of Sciences. 1975. "Principles for Evaluating Chemicals in the Environment", Printing and Publishing Office, National Academy of Sciences, Washington.

J.O. Nriagu (ed.) 1976. "Environmental Biogeochemistry", Vol.1 and 2, Ann Arbor Science Publishers.

Papetti, R.A. and F.R. Gilmore. 1971. Endeavour, 30, 107.

Shackelford, W.M., and L.H. Keith. 1976. "Frequency of Organic Compounds Identified in Water", EPA-600/4-76-062, Washington.

Stumm, W. (ed.) 1976. "Global Chemical Cycles and their Alterations by Man", Report of the Dahlem Workshop, Berlin, Abakon Verlagsgesellschaft, Berlin.

Suffet, I.H. (ed.) 1977. Fate of Pollutants in the Air and Water Environments, Part I and II, John Wiley and Sons, New York.

Tariff Commission Reports. 1974. Miscellaneous Chemicals, Washington, D.C.

Tulp, M.Th.M., and O. Hutzinger. 1978. Chemosphere, in press.

UNEP. 1978. International Register of Potentially TOXIC Chemicals (RPTC) Manuscript, July 1.

U.S., EPA. 1978. Anonymous, Implementing the Toxic Substances Control Act: Where we Stand, manuscript, Office of Toxic Substances, Washington, July 20.

U.S. Tariff Commission. 1973. Government Printing Office, Washington, D.C., 20402.

Veith, D.G. personal communication (unpublished manuscript).

Water Research Centre Stevenage. 1976. "A Comprehensive List of Polluting Substances which have been Identified in various Fresh Waters, Effluent Discharges, Aquatic Animals and Plants, and Bottom Sediments", COST Project 64b, Commission of the European Communities.

Zajic,E.C. 1971. "Water Pollution", Vol 1. Marcel Dekker, New York.

THE SAMPLING AND MEASUREMENT OF HYDROCARBONS IN NATURAL WATERS

B.W. de Lappe, R.W. Risebrough, A.M. Springer, T.T. Schmidt, J.C. Shropshire, E.F. Letterman and J.R. Payne

University of California, Bodega Marine Laboratory
Bodega Bay, California 94923

A considerable volume of recent research has demonstrated that the chemistry of natural waters, including the world's oceans, is changing as a consequence of technological activities of man. In part these changes are associated with the global dissemination of organic compounds and elements that are synthesized and/or mobilized by man. In addition, global energy demands have increased both the rate of exploitation of submarine petroleum deposits and the volume of tanker traffic; these activities have increased the incidental input of petroleum into the marine environment.

We might anticipate that almost all exogenous substances will effect changes in an aquatic system when concentrations exceed a certain threshold. For this reason the measurement and study of exogenous substances, as well as naturally-occurring biogenic compounds, in natural waters is becoming increasingly important. In this paper we consider hydrocarbons and related compounds with boiling points encompassing the range between those of the alkanes n-C14 and n-C34; they are furthermore defined by the extraction and analytical methodologies employed. Generally they are sufficiently nonpolar to be detected without chemical derivatization by the gas chromatographic techniques currently in use. They consist primarily of a considerable variety of biogenic compounds; those of special interest are the synthetic organochlorine compounds and compounds that are present in petroleum or that are derived from petroleum.

The history of the determination of hydrocarbons in seawater appears to parallel the history of the determinations in seawater of heavy metals. Prior to about 1975, almost all of the determinations of lead in seawater appear to have been too high because of contamination during collections (Participants of the Lead in

Seawater Workshop, 1974; Patterson et al., 1976). Similarly, previous determinations of zinc in seawater were anomalously high because of sample contamination (Bruland et al., 1978); extreme precautions must be taken during sampling of seawater for cadmium measurements (Martin et al., 1979). As a consequence, virtually all of the measurements made before 1975 of metals in seawater by a considerable number of investigators should now be considered as part of the learning process.

Previously (Risebrough *et al.*, 1976) we have suggested that many recent determinations of chlorinated hydrocarbons, particularly polychlorinated biphenyls, in seawater were too high, perhaps by an order of magnitude. A number of investigators have reported PCB levels in both the Atlantic and Pacific Oceans, and in the Mediterranean Sea, in surface waters and throughout the water column, that exceeded 1 ng/1 (Williams and Robertson, 1975; Scura and McClure, 1975; Harvey et al., 1973; Harvey et al., 1974; Elder, 1976). Usually, the PCB detected approximated in composition that of Aroclor 1254, a mixture consisting principally of tetra-, penta-, and hexachlorobiphenyls. If, however, the mean concentration of polychlorinated biphenyls in the North Atlantic between the equator and 65°N, (an area of approximately 47 x 10^6 km^2 and a mean depth of 3900 meters; Sverdrup et al., 1942), were 1 ng/1, the PCB content would be in the order of 180,000 tonnes, higher than the total U.S. domestic use of these PCB's between 1957 and 1974. Since global production might be assumed to have been approximately twice that of the U.S., concentrations of this magnitude would imply that most of the PCB of this chlorine content that had ever been manufactured was in the waters of the North Atlantic. Contamination of the samples during some stage of the sampling would appear to be the most likely explanation for these reported concentrations that were clearly inconsistent with global mass balance considerations. Alternatively, even with the sensitivity of the electron capture detectors, sample volumes may not have been sufficient to provide adequate resolution and signal above background.

Similarly, many of the reported values in seawater of total hydrocarbons, usually considered to be a mixture of petroleum and biogenic compounds, may also be too high, perhaps by an order of magnitude. The diversity of sampling methodologies used for these determinations has been reviewed by de Lappe et al. (1978). Almost all values obtained before 1975 with a variety of analytical techniques, have been in excess of 1 microgram/1 (Levy, 1971; Keizer and Gordon, 1973; Gordon and Keizer, 1974; Zsolnay, 1972; Zsolnay, 1973; Barbier et al., 1973; Iliffe and Calder, 1974; Koons and Brandon, 1975; Brown and Huffman, 1976; Brown and Searl, 1976). From recent studies carried out in the Pacific, ranging from the Aleutians westward to Japan and Singapore, southwards to New Zealand and the Antarctic, and southeastwards of Tahiti, Brown and

Searl (1976) have suggested that the "natural" background level in surface seawater of the Pacific is in the order of 1.5 μg/l. Brown and Huffman (1976) have reported the results of comparable studies in the Atlantic and Indian Oceans: the median concentration of non-volatile hydrocarbons was reported to be 4 μg/l in surface waters of the Atlantic.

A mean concentration of 1 μg/l in the top hundred meters over the area of the world's oceans (361 x 10^6 km^2; Sverdrup et al., 1942) would be equivalent to a total hydrocarbon content in the top 100 meter layer in the order of 40 million tonnes. Estimated recent yearly input of petroleum into the global marine environment from all sources is in the order of 6 million tonnes (National Academy of Sciences, 1975); estimates of the yearly input of biogenic hydrocarbons into the marine environment are of the same order of magnitude (National Academy of Sciences, 1975; Koons and Monaghan, 1976; Farrington and Meyers, 1975). Since both petrogenic and biogenic hydrocarbons are found throughout the water column and are not restricted to the top 100 meters (this paper), and since residence times of several years would appear to be too long for many hydrocarbons, it is likely that either the input estimates are too low or the estimated mean concentration of hydrocarbons in surface waters is too high.

In a candid review of both their own work and that of others, Gordon et al. (1974) stressed the difficulties encountered in obtaining uncontaminated seawater samples and concluded that many "of the published data, including our own, are probably erroneously high". Jensen et al. (1972), Farrington (1974), Wong et al. (1976), Risebrough et al. (1976), Dawson and Riley (1977), and Zsolnay (1978) have further discussed contamination problems associated with hydrocarbon sampling.

In this paper we describe two approaches to the problem of obtaining seawater samples with reduced or known levels of contamination: the development of an *in situ* sampling system which is a modification of one previously described (Risebrough et al., 1976), and the use of the Bodega-Bodman sampler, designed to obtain samples of 90 liters at any depth throughout the water column down to 10 meters above the bottom.

The Development of *in situ* Sampling Systems

The use of sampling devices which permit *in situ* extraction of organic compounds from water overcomes many of the problems associated with other samplers and offers the following advantages:

Size: Since water is extracted at depth rather than collected, *in situ* samplers are considerably smaller than conventional large volume samplers. This limited size:

- facilitates shipboard handling and eliminates the need for most, if not all, special ship preparations. Minimal winch and crane requirements permit the use of smaller, less expensive and more readily available research vessels;

- reduces the cost of materials and construction;

- reduces personnel required for sampling operation to two.

Volume: in situ systems offer greater flexibility in the volume of water sampled, since the volume of water extracted is a function of pumping time and is not limited by sampler capacity. The ability to extract organic compounds from water in situ permits the collection of the large volumes desirable for detailed chemical characterization, particularly of open ocean seawater which contains only trace levels of compounds of interest.

Contamination control: in situ samplers offer considerable advantages in respect to this essential requirement:

- The extraction of water may occur directly at depth, with no intermediate sources of contamination (collection vessel, transfer lines, storage containers, etc.);

- in situ samplers are small enough to be hand carried into shipboard cleanrooms for sampling preparations and thorough solvent cleaning between successive casts;

Previously (Risebrough et al., 1976) we have descibed the construction of the Slocum sampler and its application to determine levels of chlorinated hydrocarbons in Pacific coastal waters. The Slocum Sampler consisted primarily of a stainless steel adaptor with Viton O rings used to deploy modular columns packed with high density open cell polyurethane foam. Polyurethane foam offers the following advantages for trace organic sampling:

- It is easily handled and packed within sampling columns. The microstructure of polyurethane foam is relatively uniform and will not deteriorate as a result of drying;

- As with other adsorbents, polyurethane foam must be thoroughly cleaned prior to use. Soxhlet extraction will, however, routinely produce background blanks suitable for pg/l determinations;

- Polyurethane foam will extract a wide range of non-polar organic compounds from dilute solutions. Unlike activated carbon, which has been reported to produce inconsistent recoveries due to irreversible adsorption, compounds extracted by polyurethane foam may be efficiently recovered without substantial chemical alteration or deterioration;

- Sampling flow rates may be significantly increased by a proportional increase in the diameter of the adsorbent.

Studies which have reported on the use of polyurethane foam adsorbents for the extraction of organics from water include those of Bowen (1970), Gesser _et al._ (1971), Uthe _et al._ (1972), Uthe _et al._ (1974), Bieri _et al._ (1974), Bedford (1974), Musty and Nickless (1974), Musty and Nickless (1976), Bidleman and Olney (1974), Bidleman and Olney (1976), Webb (1975), Carmignani and Bennett (1976), Saxena and Basu (1976), and Basu and Saxena (1978a; 1978b).

The Modified Slocum _in situ_ Sampler

The Modified Slocum Sampler (Figure 1) provides collection of a separate particulate fraction which is retained on a glass fiber filter mounted inline with a column that extracts the "dissolved" components. It incorporates the main configuration of the original version, but through the use of an additional adaptor connects directly to a standard stainless steel Millipore filter holder (C, Figure 1; Model YY 22 293 02; 293 mm diameter; Millipore Corp., Bedford, Mass.). The large diameter reduces the pressure drop and minimizes clogging when sampling water with a high particulate load. Glass fiber filters contain no organic binder (Gelman A/E glass fiber) and are rated to retain 99.7% of particles greater than or equal to 0.3 μ, (Gelman Instrument Co., Ann Arbor, Michigan).

The Modified Slocum Sampler is supported by three 3/32 " (2.38 mm) stainless steel cables (B) which are swaged to support tabs bolted to the top plate of the filter holder. A vaned intake tube (F) is attached to the filter holder with a Swagelok full flow quick-connect fitting (G) which causes the intake tube to orient into the ambient current during sampling. A small teflon cap is force-fit over the intake to provide protection from surface layer contamination during the deployment of the sampler. The cap is pulled off once the sampler is 2-3 meters below the surface.

A machined stainless steel column adaptor (H) is connected to the outlet of the filter holder using a 3/4 " (1.90 cm) FPT (Female Pipe Thread) to 1/2" (1.27 cm) MPT (Male Pipe Thread) reducing adaptor (D). Each column adaptor is fitted with two 1/8 " (3.2 mm) viton O rings (A-230). Three 1/2 " stainless steel rods (E) are bolted to the filter holder and to a nylon collar (J) to provide structural support for the attachment of teflon adsorptive columns. A one-way check valve (L) with a viton O ring and a 1/3 lb (150 g) spring is located in-line between the lower column adaptor and the hose bib (N), used for the connection of vacuum tubing. The check valve is intended to prevent the back flow of seawater and possible contamination from the vacuum tubing.

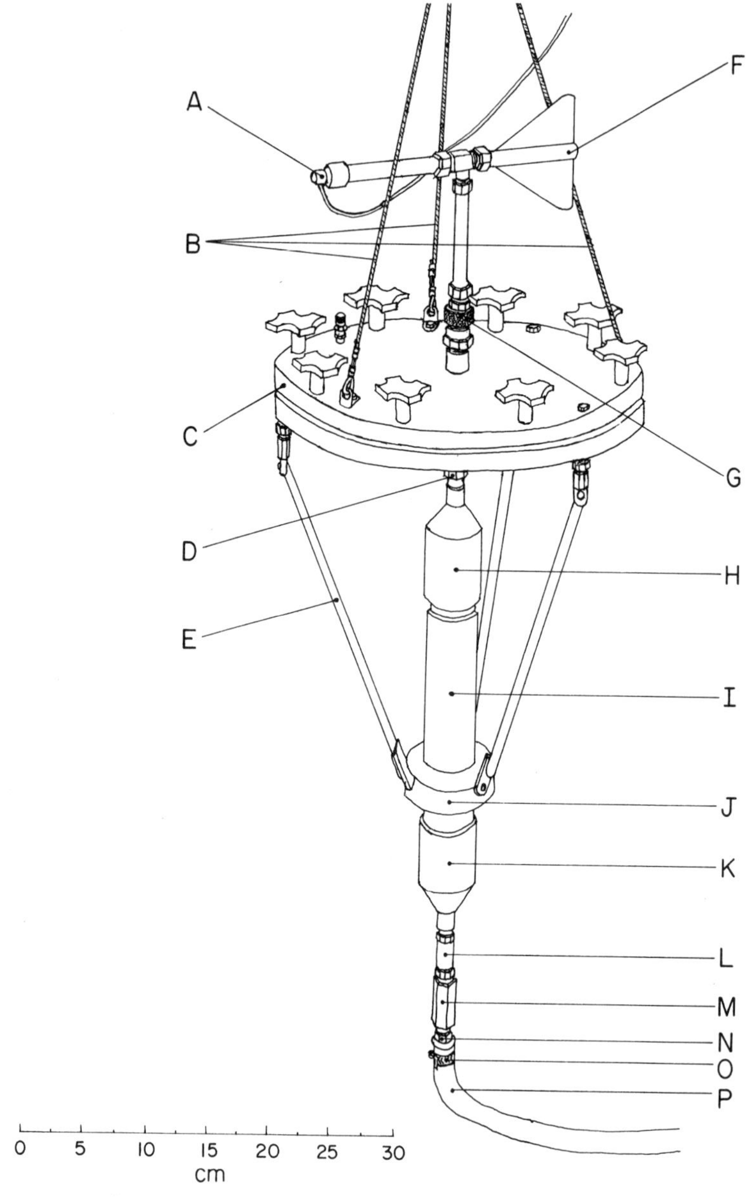

Figure 1. The Modified Slocum *in situ* Sampler.
Caption continued on following page.

Figure 1, continued.

A. Teflon cap (force fit, remove below air-sea interface)
B. 3/32" (2.38 mm) stainless steel support cables
C. Stainless steel Millipore(R) filter holder (Model YY 2229302, 293 mm diameter
D. Reducing adaptor; 3/4" (1.90 cm) Female Pipe Thread to 1/2" Male Pipe Thread (Cajon(R) SS-12-RA-8)
E. Support frame; 1/2" (1.27 cm) stainless steel rods
F. Vaned intake tube; 1.90 cm stainless steel tubing and rod with Swagelok fittings
G. Full flow quick-connect; Swagelok SS-QF12-B-1210,SS-QF12-S-12 PM
H. Column adaptor; machined from stainless steel rod with 2 viton O rings
I. Teflon column; packed with high density polyurethane foam (30.5 cm long, 6.35 cm O.D., 5.08 cm I.D.)
J. Nylon collar
K. Column adaptor
L. One-way valve; Nupro(R) SS-8CP2-1/3(viton O ring)
M. 1.27 cm Female Pipe Thread stainless steel coupling; Cajon(R) SS-8-HCG
N. Stainless steel hose bib; Cajon(R) SS-8-HC-1-8
O. Stainless steel hose clamp
P. Rubber vacuum tubing; 1.27 cm I.D., 2.86 cm O.D.

Weight: 23 kg

The molded teflon tubing used for the adsorptive columns requires only minimal precision machining of the outer surface at each end to provide the appropriate vacuum-tight force fit with the viton O rings of the stainless steel adaptors. Additionally, a shallow groove is cut on the inner surface at each end of the column in order that stainless steel spring clips in the form of a "W" may be inserted to retain the foam plugs packed within the column. The standard column packing consists of five polyurethane foam plugs (66 mm diameter, 57 mm thick) providing a total adsorbtive volume of 975 cm^3. The composition of the high density polyurethane foam (0.036 g/cm^3) is approximately 19% polyether glycol and 79% toluene diisocyanate (United Foam Co., UF-2064).

The Bodega-Bodman Sampler

To meet requirements for the collection of seawater volumes in the order of 90 liters throughout the water column and to reduce the potential for sample contamination, the Bodega Marine Laboratory modified the design of the original Bodman sampler (Bodman *et al.*, 1961) in collaboration with R. Hamblin, Oceanic Industries, Osterville, Mass. The resulting sampler, which we have named the Bodega-Bodman Sampler, is shown in Figure 2. The modifications included:

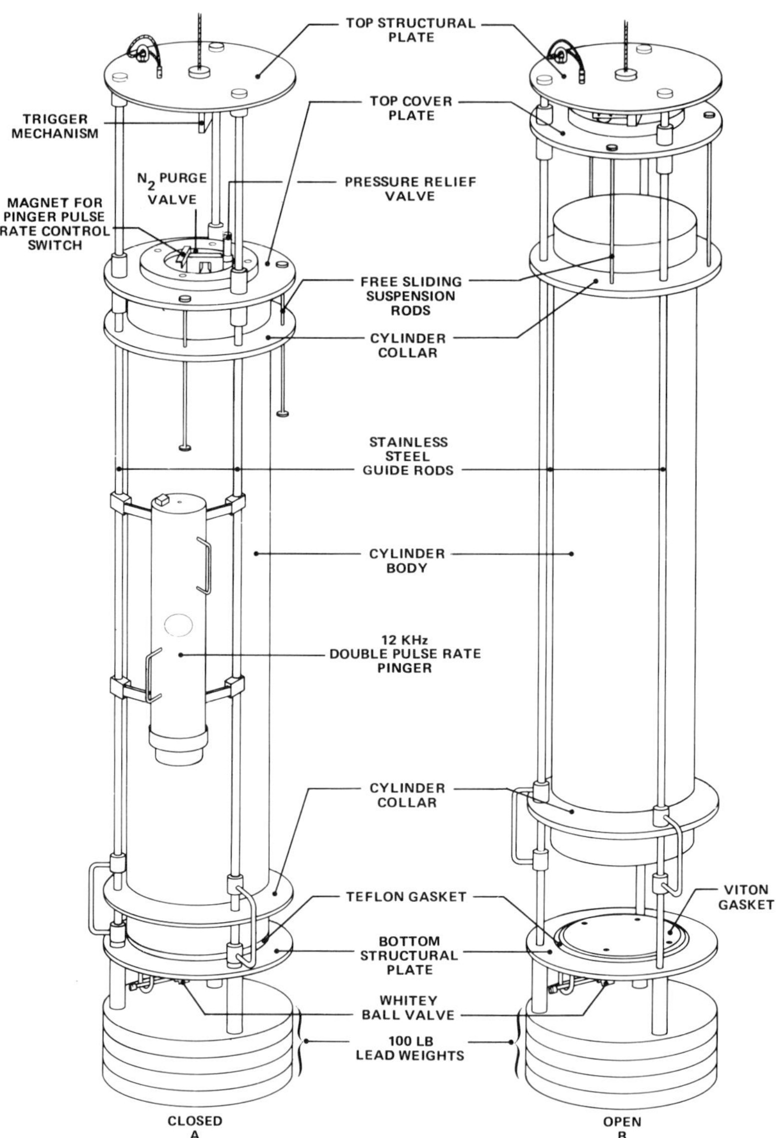

Figure 2. The Bodega-Bodman Samplers.

1) construction with non-contaminating materials, - anodized aluminum, stainless steel, teflon and viton; 2) 100 lb disk lead weights bolted directly to the bottom of the sampler and a one-way relief valve in the lid to permit subsurface cocking, such that the sampler could pass closed through the surface film, thereby minimizing a potential source of contamination; 3) the cylinder, constructed of aluminum, was maintained at its full diameter at both ends, facilitating cleaning and enhancing flushing as it descends through the water column; 4) the use of a Kullenberg piston core release mechanism allows sampling at precise distances from the bottom; 5) a magnetic switch on the upper plate of the sampler switches an integrally mounted 12 KHZ pinger to double pulse rate at the time of the trip, thus permitting continuous shipboard monitoring of the position and condition of the sampler; 6) a Swagelok Quick-connect purge valve in the lid for the attachment of a nitrogen gas line and a stainless steel ball valve with standard Swagelok connections at the bottom of the sampler allows the contents of the Bodega-Bodman to be transferred directly to a cleanroom for processing with limited exposure to the shipboard atmosphere; 7) positioning of the outlet valve in the center of the flat bottom plate reduces the potential for loss of larger settling particulates between sample collection and processing, which has been reported as a deficiency in similar samplers (Gardner, 1977).

METHODS

The polyurethane plugs were initially cleaned by a minimum of 72-hour reflux extraction with acetone in a large Soxhlet apparatus. The plugs were then packed into the teflon columns, cold acetone and hexane were successively passed through the column, and the hexane reduced to a small volume for GC analysis until a satisfactory blank was obtained. The columns were then sealed in teflon bags for transport to the field. The glass fiber filters were kiln-fired overnight at 500°, and packed in solvent-rinsed aluminum foil for transport to the field. All components of the sampler were rinsed with acetone and hexane in the field prior to sampling. Blanks were obtained by carrying columns into the field, either opening them but not exposing them to the water, or were lowered into the water and a minimum volume (less than three liters) was passed through them. Background levels ranged from < 6 to < 60 picograms/liter of PCB in an equivalent volume of 40 liters. The cleanup and gas chromatographic conditions employed were equivalent to those previously described (Risebrough et al., 1976).

After pressurization with pre-purified nitrogen, the contents of the Bodega-Bodman samplers were transferred directly to a shipboard cleanroom through stainless steel tubing. The air supply entering the cleanroom passed through a filter which removed particulates; air within the cleanroom was recirculated through a charcoal filter. The seawater passed through a glass fiber filter equivalent to those used in the Modified Slocum Sampler; the filtrates were stored in five gallon (19 liter) glass carboys which had been solvent rinsed and checked for background values. The receiving carboy was evacuated using an oil-less vacuum pump (ITT Corp.) with in-line dessicant and charcoal traps. After each carboy was filled, 100 ml chloroform was added to prevent bacterial growth, the carboy was stoppered with an aluminum foil-lined cork, and stored in the dark until analysis. Filters were wrapped in kiln-fired, chloroform-washed aluminum foil and frozen until analysis.

Filters were reflux-extracted in 250 ml round bottom flasks with successive 150 ml volumes of chloroform and hexane. Filtrates were extracted in the original carboys with three successive 100 ml volumes of chloroform. The seawater and chloroform were agitated using an electric mixer with a stainless steel shaft and folding blades which allowed passage through the narrow neck of the carboy. After settling, the chloroform was transferred through 1/8" (3.2 mm) stainless steel tubing to a 500 ml separatory funnel by pressurizing the carboy with pre-purified nitrogen. The stopper of the transfer apparatus was constructed of teflon with a viton O ring. After partitioning from residual water, the chloroform was collected and reduced to near dryness by rotary evaporator prior to column chromatography.

All the solvents used were glass-distilled non-spectro grade (Burdick and Jackson). Column chromatography separations utilized Alcoa alumina F-20 80/100 mesh and Hi-Flosil silica gel 60/200 mesh (Applied Science Laboratories). The silica and alumina were precleaned with methanol rinses and activated overnight at 240° C. Chromatography columns were prepared in a length diameter ratio of 20:1 with a minimum 200 parts SiO_2 and 100 parts of Al_2O_3 to one part water extract by weight. All sample transfers were accomplished with glass syringes with stainless steel needles. Extracts were applied to chromatography columns with 2-3 mls of hexane. Two fractions were collected by eluting with two bed volumes of hexane followed by two bed volumes of benzene. Sample reductions were performed using a rotary evaporator to volumes of approximately 3 mls and by a stream of pre-purified nitrogen to final volumes of 10-20 μl prior to GC analysis.

The flame ionization gas chromatography of the Bodega-Bodman samples was performed using Hewlett-Packard 5840A gas chromatographs equipped with 7176A automatic liquid samplers and 18835A capillary inlet systems. Injection volumes were 1 or 2 μl in the splitless mode. Inlet and detector temperatures were maintained at 250° and 350°, respectively. Saturate (hexane) fractions were analysed using a 30 meter, 0.25 mm I.D., SP-2100 WCOT glass capillary column (J & W Scientific). Aromatic (unsaturate) fractions were run on a column of the same dimensions but with a more polar SE-54 coating. Helium was used as the carrier gas. Benzene fractions were injected in toluene. Initial temperatures of the SP-2100 and SE-54 columns were 65° and 75°, respectively. All other conditions were identical for each column: initial hold, 2 minutes; rate, 3.5^{o}/minute, final temperature, 270°. Instrument attenuation ranged between 2^3 and 2^5. Each gas chromatograph was calibrated daily with a standard containing all the normal alkanes between n-C12 and n-C32, pristane and phytane.

Both Hewlett-Packard gas chromatographs were equipped with 18846A digital communications interface boards which permitted the digitized data summaries of each chromatogram to be transmitted to a Texas Instruments 733 ASR data terminal to be recorded on cassette tapes. Data summaries were subsequently played into a computer system where an interactive program converted the retention time of each resolved peak to a Kovat index and converted the integrated area of each peak into concentrations by comparison with the response factors of the calibration standard. Unresolved areas were measured by planimetry and quantified on the basis of the average response factor of the n-alkanes in the boiling point range.

Solvent rinses of the carboys used to transport and store seawater samples from the summer cruise in the Southern California Bight were checked for contamination by electron-capturing substances to permit subsequent analysis for chlorinated hydrocarbons. When such contamination was found, rinses were repeated until satisfactory blanks were obtained.

Sampling Sites

Waters of the Cape Cod and New Bedford areas in Massachusetts were sampled with the Modified Slocum Sampler 30 July - 3 August, 1976. Collection sites are shown in Figure 3. Water of western Lake Ontario in the vicinity of Burlington was sampled on 13 August and on 19 October, 1976, during a collaborative program with the Canada Centre for Inland Waters. The sampler was suspended at a depth of approximately three meters from a boom mounted on a small vessel, and water was pumped through it at a rate of approximately 60 liters/hour using a shipboard vacuum system powered by an onboard generator.

Waters of the Southern California Bight were sampled in February and in August-September, 1977, with the Bodega-Bodman samplers at 10 stations (Figure 4) on the R/V THOMAS THOMPSON. Samples were taken three or five meters below the surface, at an intermediate depth below the thermocline if present, and ten meters above the bottom.

RESULTS AND DISCUSSION

Concentrations of PCB and total DDT (the sum of p,p'-DDT and p,p'-DDE) in the "dissolved" phase of seawater samples from the Southern California Bight obtained with the Bodega Bodman Samplers in the summer of 1977 are presented in Table 1. They are compared with the concentrations of pristane, phytane and squalene measured in the same samples, and with concentrations of these compounds in samples obtained in the winter of 1977. The PCB and DDT values presented are considered maximum concentrations, since it was not possible to obtain a representative blank. The PCB values, ranging between 0.1 and 0.9 ng/l, are comparable to those recorded in the same area two years previously with the Slocum Sampler (Risebrough et al., 1976). Maximum PCB values at that time at stations that were not in the vicinity of wastewater outfalls, particularly the outfall of the Los Angeles County Sanitation Districts, were in the order of 0.1-0.7 ng/l; the limiting factor in determining maximum concentrations was the presence of interfering substances in the extracts. Measured values near the outfall were in the order of 1-2 ng/l; several samples from other stations adjacent to the Los Angeles and San Diego areas had sufficiently low levels of interfering materials to permit determinations of PCB in the range of 0.04-0.3 ng/l.

The maximum DDT values recorded in 1977 in waters of the Bight, in the range of 0.1-0.6 ng/l, were generally comparable to those recorded in 1975 with the Slocum Sampler, which were in the order of 0.05-0.3 ng/l. Total DDT concentrations in 1975 exceeded 1 ng/l only in the vicinity of the outfall of the Los Angeles County Sanitation Districts (Risebrough et al., 1976).

The data of Table 1 permit a preliminary answer to the question: in a relatively polluted area of the marine environment, such as the Southern California Bight, how do levels of anthropogenic compounds compare with levels of naturally occurring biogenic compounds and of petrogenic chemicals? Among the biogenic compounds measured in waters of the Bight in 1977, pristane and squalene were among those present in highest concentrations. Levels of squalene generally exceeded those of other biogenic compounds. It is, however, a ubiquitous laboratory contaminant, and the levels represent maximum values. Concentrations of squalene in the dissolved phase were found to range between 1 and 10 ng/l, one to three orders of magnitude higher than those of individual DDT and PCB compounds. Concentrations of individual PCB compounds are considered to be an order of magnitude lower than the concentrations of the total PCB mixture. Pristane concentrations in the dissolved phase were generally betwen 0.1 and 1.0 ng/l, equivalent to the concentrations of total DDT and PCB, and an order of magnitude higher than concentrations of individual PCB compounds. Since the majority of biogenic

Table 1. Concentrations of pristane, phytane, squalene, PCB* and Total DDT** in the dissolved phase of seawater of the Southern California Bight in 1977. Nanograms/liter. N=3 for replicate samples, one standard deviation.

Station Depth (m)	Pristane Winter	Pristane Summer	Phytane Winter	Phytane Summer	Squalene Winter	Squalene Summer	PCB Summer	DDT Summer
Coal Oil Point								
3-5	0.6±0.2	0.3	0.5±0.2	0.2	6.7±5.5	1.1	0.9	0.2
13-33	1.7	0.3	1.7	0.2	3.5	0.5	0.4	0.2
56-57	0.4±0.2	0.05	0.3±0.1	0.09	1.4±0.9	0.8	0.2	0.6
Santa Barbara Basin								
3.5	0.7±0.5	0.5	0.6±0.4	0.1	10±13	1.4	0.1	0.1
17-40	0.9±0.4	0.4	0.9±0.4	0.2	2.4±0.3	6.6	0.3	0.2
558-564	0.9±0.4	0.2	0.8±0.3	0.2	15±13	1.5	0.3	0.3
San Miguel								
3-5	0.8	0.9	0.7	0.3	15	2	0.3	0.2
10-34	0.6	0.9	0.8	0.3	1.8	5.6	---	---
196-237	0.1	0.4	0.8	0.3	1.2	4.7	0.9	0.3
Santa Rosa								
3-5	0.2	0.3	0.1	0.1	4.1	4.7	0.4	0.2
40-50	0.6	0.4	0.6	0.3	2.1	0.8	0.9	0.2
95-108	0.6	0.2	0.5	0.2	5.4	0.1	0.2	0.1
Santa Cruz Basin								
3-5	0.2	0.5	0.2	0.2	0.3	0.8	0.4	0.2
19-34	0.6	0.2	0.4	0.1	1.0	0.5	0.2	0.1
1839-1857	0.2	0.2	0.4	0.2	0.1	3.2	---	---

continued on next page

Table 1, continued.

Station Depth (m)	Pristane		Phytane		Squalene		PCB	DDT
	Winter	Summer	Winter	Summer	Winter	Summer	Summer	
San Nicholas Basin								
3-5	0.7±0.5	0.2	0.8±0.5	0.1	2.9±2.5	10	0.4	0.2
21-34	0.8	0.1	1.0	0.1	1.4	2.0	0.2	0.4
1720-1762	0.6±0.3	0.1	0.8±0.7	0.1	0.7±0.8	0.6	0.2	0.3
Tanner Bank								
3-5	0.4±0.1	0.2	0.4±0.2	0.2	2.1±2.9	7.4	0.7	0.3
19-32	0.7	0.4	0.9	0.3	2.2	1.0	0.1	0.3
62-73	0.5±0.2	0.1	0.5±0.3	0.1	1.7±1.4	4.6	0.2	0.2
San Pedro Basin								
3-5	0.5	0.7	0.3	0.3	1.6	3.3	---	---
435-440	0.2	0.3	0.2	0.2	3.7	1.5	0.3	---
863-871	0.3	0.4	0.3	0.3	0.5	0.8	---	---
Huntington								
3-5	0.6	0.3	0.9	0.2	1.0	13	---	0.2
12-17	0.5	0.4	0.7	0.2	2.0	5.6	0.8	0.3
20-34	1.4	0.6	1.3	0.4	1.3	9.5	0.3	0.4
Santa Monica Basin								
3-5	0.7	0.3	1.2	0.1	6.6	3.7	0.4	0.2
37-450	0.5	0.2	0.5	0.2	1.4	3.8	0.9	0.3
874-885	1.4	0.3	1.8	0.2	3.7	0.7	0.2	0.3

hydrocarbons are present in substantially lower concentrations than is pristane, concentrations of individual PCB and DDT compounds in the Bight would have exceeded those of most biogenic compounds.

Phytane is of particular interest since its presence and relative abundance have been used as an indicator of the presence of petroleum. Attempts to isolate phytane from organisms, including marine algae (Clark and Blumer, 1967), zooplankton and the liver of a basking shark (Blumer and Snyder, 1967) were not successful. Blumer (1970) stated that the "...source of phytane in our water samples remains obscure. It may be derived from plants or animals which have not yet been analysed or from the rather abundant phytanic acid." Although the highest value was recorded at Coal Oil Point, in the vicinity of a natural oil seep, that concentration, 1.7 ng/l, was of the same order of magnitude as many of the other values recorded throughout the Bight and reported in Table 1. The extent to which phytane can be used as an indicator of the presence of petroleum is not yet resolved.

Generally, concentrations of phytane were of the same order as those of pristane in the dissolved phase. Pristane concentrations, however, were frequently higher in particulates. In the Santa Barbara Basin, phytane concentrations in winter in the particulate phase on a whole water basis were 0.1±0.03 (N=3), 0.1±0.01 (N=3), and 0.1 ng/l in surface, mid-depth, and bottom samples, all substantially lower than the corresponding concentrations in the dissolved phase (see Table 1). Pristane concentrations in the particulates at this station were 1.6±0.6 (N=3), 4.3±3.0 (N=3), and 0.2±0.2 (N=3), respectively, higher than or equivalent to the corresponding dissolved values. Additional data on particulate concentrations of pristane and phytane, showing the same relationships as the data presented here, have been reported by de Lappe et al. (1979). Figures 5 and 7 (see below) show, for example, prominant pristane peaks in chromatograms of particulate phases and the virtual absence of phytane.

Concentrations of PCB measured in seawater in the vicinity of Cape Cod, Massachusetts, in 1976 are presented in Table 2. Exceptionally high concentrations were recorded in New Bedford Harbor, strongly suggesting a point source of local contamination. The majority of the PCB in New Bedford Harbor consisted of the tri- and tetra-chloro isomers that are found in Aroclor 1242, or its equivalent, Aroclor 1016. The separation made in Table 2 among the three classes of PCB is, of necessity, arbitrary. Since the gas chromatography was performed with packed columns, it was not possible to distinguish individual PCB isomers; the application of glass capillary techniques has since permitted much higher resolution of PCB mixtures (Onuska and Comba, in press). When the observed mixture predominantly ressembled one particular commercial preparation

Table 2. PCB concentrations in seawater in the vicinity of Cape Cod, Massachusetts, 1976. Concentrations in nanograms/liter.

Site	Replicate Number	Volume (1)	Phase	PCB 1242	PCB 1248	PCB 1254
Monomoy Is.	1	197	Dissolved	---	.31	.12
			Particulate	---	---	.27
	2	197	Dissolved	---	.45	.17
			Particulate	---	---	.32
Ram Is. Buzzards Bay	3	99.0	Dissolved	---	4.1	1.2
			Particulate	---	---	1.2
	4	99.5	Dissolved	---	3.5	1.6
			Particulate	---	---	.85
Widows Cove Buzzards Bay	5	98.5	Dissolved	---	---	.33
			Particulate	---	---	.42
	6	99.5	Dissolved	---	---	.20
			Particulate	---	---	.48
Bird Is. Buzzards Bay	7	99.0	Dissolved	---	---	.38
			Particulate	---	---	.54
	8	100	Dissolved	---	---	.72
			Particulate	---	---	1.90
New Bedford	10	99.5	Dissolved	320	---	---
			Particulate	150	---	90
	11	97.0	Dissolved	340	---	---
			Particulate	172	---	70

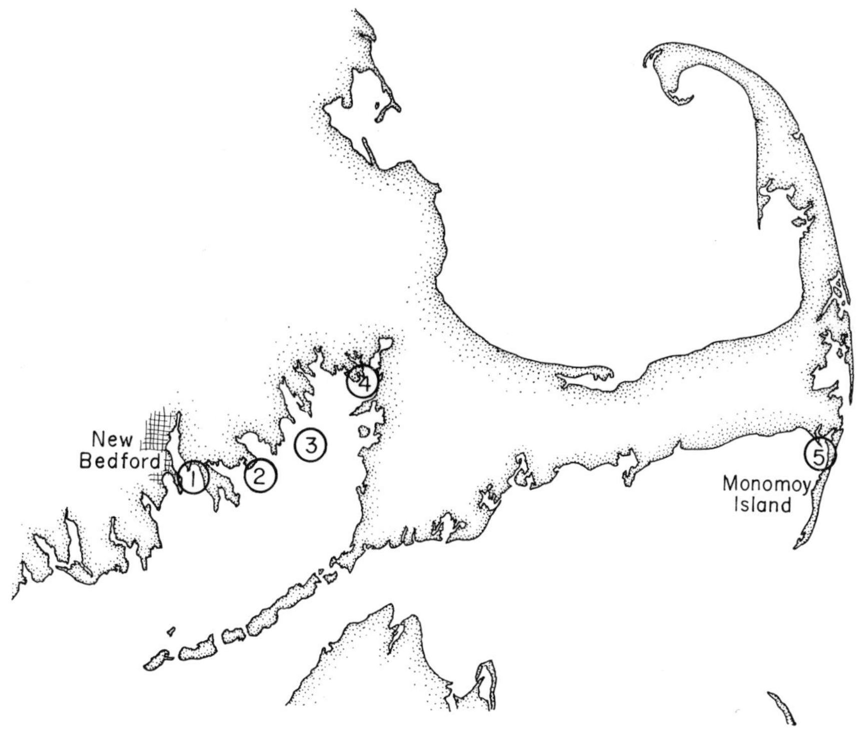

Figure 3. Sampling sites in coastal Massachusetts, 30 July - 3 August, 1976. 1) New Bedford Harbor; 2) Ram Island 70°48'W, 41°38'N; 3) Bird Island, 70°43'W, 41°40'N; 4) Widow's Cove, 70°39'W, 41°43'N; 5) Monomoy Island west of Chatham.

of PCB, it was quantified by comparison with that mixture, and no attempt was made to quantify the other components with fewer or greater numbers of chlorine atoms on the biphenyl molecules. When the observed mixture had a greater variation in chlorine content, it was quantified as though it consisted of two or more commercial preparations. Such a distinction permits the conclusion that the PCB composition in the particulate phase generally had a greater proportion of the higher-chlorinated isomers than were present in the "dissolved" phase retained in the polyurethane foam column. In several samples, PCB concentrations in the dissolved phase exceeded those in the particulates, the latter determined on a whole water basis.

The coastal zone of Cape Cod, including Buzzard's Bay, like the Southern California Bight, is a relatively polluted area. The majority of PCB values recorded in 1976 were below 1 ng/l (Table 2). They may be compared with concentrations reported from the Hano Bight off the south Swedish coast in 1975 (Osterroht, 1977) and in British coastal waters in 1974 (Dawson and Riley, 1977). Among eight values reported from seawater in the area south of Sweden, six ranged between 0.3 and 0.9 ng/l; the other two were 1.0 and 3.0 ng/l. In British coastal waters, concentrations ranged up to 1.5 ng/l in the more heavily polluted areas, and were in the order of 0.2 ng/l in areas least subject to pollution. There is, therefore, a general agreement among PCB determinations in relatively polluted coastal waters, including the Southern California Bight, between 1974 and 1977; the majority of values, obtained with a diversity of sampling methodologies, were in the range 0.1-1.0 ng.l.

Data on PCB levels in water of western Lake Ontario are presented in Table 3. Concentrations were up to an order of magnitude higher than those recorded in seawater of the Southern California Bight and of coastal Massachusetts. In all samples, the majority of the PCB was in the "dissolved" phase and the PCB composition differed significantly between the two phases, with lower-chlorinated isomers predominating in the "dissolved" phase. Previously, the only recorded PCB values from Lake Ontario appear to be those reported by Haile et. al. (1975) who obtained water in Van Dorn samplers at 8 sites in 1972. Volumes of 10 liters were extracted at the collection sites by passage through polyurethane plugs according to procedures of Uthe et al. (1972). The PCB's were quantified by perchlorination of the extracts and measuring the amounts of the resulting decachlorobiphenyl. PCB values ranged from 35 ng/l to 97 ng/l, with a mean of 55 ng/l, substantially higher than the values we report here. Since there appears to be no evidence that PCB levels in Lake Ontario declined by over an order of magnitude between 1972 and 1976, such apparent discrepancies would indicate the need for frequent intercalibration of sampling and well as analytical methodologies. Three water samples from Lake Superior obtained in 1972-73, extracted with polyurethane foam plugs and analysed for PCB by perchlorination yielded a mean value of 0.8 ng/l of PCB as Aroclor 1254 (Veith et al., 1977). This value would appear to be consistent with those we report from Lake Ontario; a program undertaken by the Canadian Wildlife Service in collaboration with the Canada Centre for Inland Waters using eggs of the Herring Gull, *Larus argentatus*, as indicators of levels of organochlorine contamination in the Great Lakes, has shown that PCB levels in eggs obtained from the Lake Superior region in 1975 were approximately half those recorded in eggs from Lake Ontario (Fox et al., 1975).

Data on the levels of total hydrocarbons in the dissolved phase of seawater from the Southern California Bight in 1977 are presented in Table 4. Generally, concentrations of total hydrocarbons were below 1 μg/l. Since the area has high biological productivity,

Table 3. PCB levels in water of Lake Ontario, two miles east of the Burlington Causeway and in Burlington Harbour, August and October, 1976. Concentrations in nanograms/liter.

Date	Volume (liters)	Phase	PCB 1242	PCB 1248	PCB 1254	PCB 1260
13 August '76	93[a]	Dissolved	---	2.5	---	---
		Particulate	---	---	0.45	---
	93[a]	Dissolved	---	2.9	---	---
		Particulate	---	---	0.50	---
	47.5[b]	Dissolved	9.1	---	---	---
		Particulate	---	---	---	4.4
19 October '76	39.5[a]	Dissolved	---	1.1	---	---
		Particulate	---	---	0.16	---
	93[a]	Dissolved	---	1.5	---	---
		Particulate	---	---	0.41	---
19 October '76	184[a]	Dissolved	---	1.1	---	---
		Particulate	---	---	0.19	---

[a]Two miles east of Burlington Causeway

[b]Center of Burlington Harbour

Table 4. Hydrocarbons in waters of the Southern California Bight in 1977. Nanograms/liter of the dissolved phase.

Station Depth (m)	Winter, 18-24 February				Summer, 28 August-1 September			
	Total Unresolved	Total Resolved	Total Alkanes	HMW* Olefins	Total Unresolved	Total Resolved	Total Alkanes	HMW* Olefins
Coal Oil Point								
3-5	320±74**	45±6**	12±3**	6.5±4.6**	110	42	1.9	5.5
13-33	360	45	21	3.0	79	15	2.5	2.4
56-57	160±65	34±19	8.4±4.9	1.3±0.2	50	21	1.6	7.3
Santa Barbara Basin								
3-5	280±77**	74±26	15±8	26±29	41	14	1.3	<1
17-40	380±130	64±16	28±13	7.9±3.5	145	1100	2.6	1100
558-564	360±88	70±10	15±5	12±8	140	24	2.9	1.6
San Miguel Island								
3-5	290	63	13	9.5	99	110	6.9	92
10-34	240	22	8.1	1.9	97	120	5.9	88
196-237	160	68	2.1	24				
Santa Rosa Island								
3-5	110	14	3.9	0.8	53	17	1.9	8.5
40-50	390	100	23	12	110	36	3.6	24
95-108	380	110	23	15	67	17	3.0	1.7
Santa Cruz Basin								
3-5	150	36	4.6	3.1	69	34	4.2	11
19-34	200	62	13	2.9	50	23	2.0	17
1839-1857	330	97	11	1.6	53	35	2.7	11

**Means of three replicate samples, with one standard deviation

continued on next page

Table 4, continued

Station Depth (m)	Winter, 18-24 February Total Unresolved	Total Resolved	Total Alkanes	HMW* Olefins	Summer, 28 August-1 September Total Unresolved	Total Resolved	Total Alkanes	HMW* Olefins
San Nicholas Basin								
3-5	250±110	48±15	16±8	>1	87	34	2.6	11
21-34	470	110	63	2.1	120	29	2.0	1.5
1720-1762	250±130	80±100	44±62	4.2±5	97	12	1.5	<1
Tanner Bank								
3-5	450±380	44±25	12±4	<1	140	29	2.9	4.5
19-32	330	35	12	<1	110	72	4.8	53
62-73	260±110	72±44	16±13	4.4±3.8	140	17	1.5	3.5
San Pedro Basin								
3-5	180	15	9.3	<1	110	42	4.5	15
435-440	94	15	4.5	<1	94	12	3.9	2.2
863-871	150	17	6.5	<1	96	12	4.5	1.4
Huntington Beach								
3-5	430	120	23	3	180	80	5.7	<1
12-17	220	49	8.1	6.7	91	22	4.3	2.5
20-34	740	300	27	3	100	56	4.7	24
Santa Monica Basin								
3-5	610	110	17	16	73	38	1.7	13
37-450	380	58	19	9.3	64	28	3.0	7.5
874-885	550	71	18	10	150	13	3.2	<1

*Kovat indices of HMW olefins 2757, 2774, 2938, 2956, 2971, 3143, 3160, 3174, 3394, 3423, 3442, 2781, 3814, 3835.

Figure 4. Water column sampling stations in the Southern California Bight in 1977. Coal Oil Point, 119°54.1'W, 34°23.1'N; Santa Barbara Basin, 120°01.2'W, 34°16.2'N; San Miguel, 120°26.3'W, 33°57.4'N; Santa Rosa, 120°00.0'W, 33°48.9'N; Santa Cruz Basin, 119°36.2'W, 33°46.1'N; San Nicholas Basin, 119°02.0'W, 33°01.1'N; Tanner Bank, 119°08.7'W, 32°41.4'N; Huntington Beach, 118°02.1W, 33°36.0'N; San Pedro Basin, 118°25.9'W,33°36.0'N; Santa Monica Basin, 118°50.1'W, 33°45.4'N.

receives substantial amounts of human wastes, in the order of 4 billion liters of wastewaters per day from the treatment plants of southern California (Young et al., in press), contains active submarine oil seeps, and supports a high density of both recreational and commercial shipping, hydrocarbon levels in the water, both biogenic and anthropogenic, would be expected to be higher than in the majority of oceanic areas. On this basis we conclude that a value of 1 μg/l or higher of total hydrocarbons in surface marine waters may generally be considered as too high, and that many of the earlier values reported in the literature, cited above, likely represent some degree of sample contamination. At this time, however, we are unable to provide a satisfactory estimate of the contaminant component among the data we present in Table 4, particularly for the unresolved mixtures and the alkanes. Corrections were made for procedural blanks which showed that one batch of hydrochloric acid used to acidify the extracts contributed substantially to the unresolved component of the unsaturate (aromatic) fraction (de Lappe et al., 1979). Contamination introduced during the analyses was, however, relatively easy to account for. It proved impossible to obtain an adequate blank for the sampling and shipboard filtration procedure. Exposure of organic solvents and the interior of samplers to the highly polluted shipboard atmosphere invariably introduced a component of contamination which would generally be limited during normal sampling and shipboard operations. Subsequent work may therefore record successively lower values of many hydrocarbons in marine waters, paralleling the work with lead and other heavy metals cited above.

The data presented in Table 4 indicate that the normal alkanes constitute a significant proportion of the resolved components. The ratio of odd to even alkanes has been used as one of several parameters to distinguish mixtures of petroleum hydrocarbons from mixtures which that are predominantly biogenic: in petroleum, ratios of adjacent normal alkanes are usually 1:1; in biogenic mixtures, the odd-carbon alkanes frequently predominate (Farrington and Meyers, 1975). Among the samples obtained in 1977, a series of alkanes with a maximum in the n-C25 to n-C29 range was occasionally observed. A comparable series was observed by Calder (1977) in both dissolved and particulate phases of seawater samples from the Gulf of Mexico. Figure 5 shows chromatograms of the dissolved phase obtained at 17 meters and of particulates obtained at the surface (3 meters) in the Santa Barbara Basin in February, 1977. In the dissolved phase the later alkanes appear to be superimposed upon a series with an odd:even ratio of approximately unity. In the particulate sample shown in Figure 5, the absence of appreciable amounts of n-C18, n-C19, and n-C20 suggests that the alkanes were not petrogenic. A comparable series of alkanes has been observed in extracts of tissues of the mussel, *Mytilus californianus*, obtained from the Southern California Bight area, that had remained at room temperature for several hours before analysis. The absence of a comparable pattern in extracts of whole animals that were analysed immediately after removal from their shells suggests that these alkanes were of bacterial origin (Risebrough et al., 1979). Since the seawater samples were poisoned with chloroform immediately after collection and filtration, bacterial growth would not appear likely under those conditions. The detection of this series in water extracts suggests the presence of bacteria containing these alkanes in the water column; either the bacteria pass through the glass fiber filter, or the alkanes are present in a truly "dissolved" state in the water column.

The major early-eluting peak in the chromatogram of the saturate

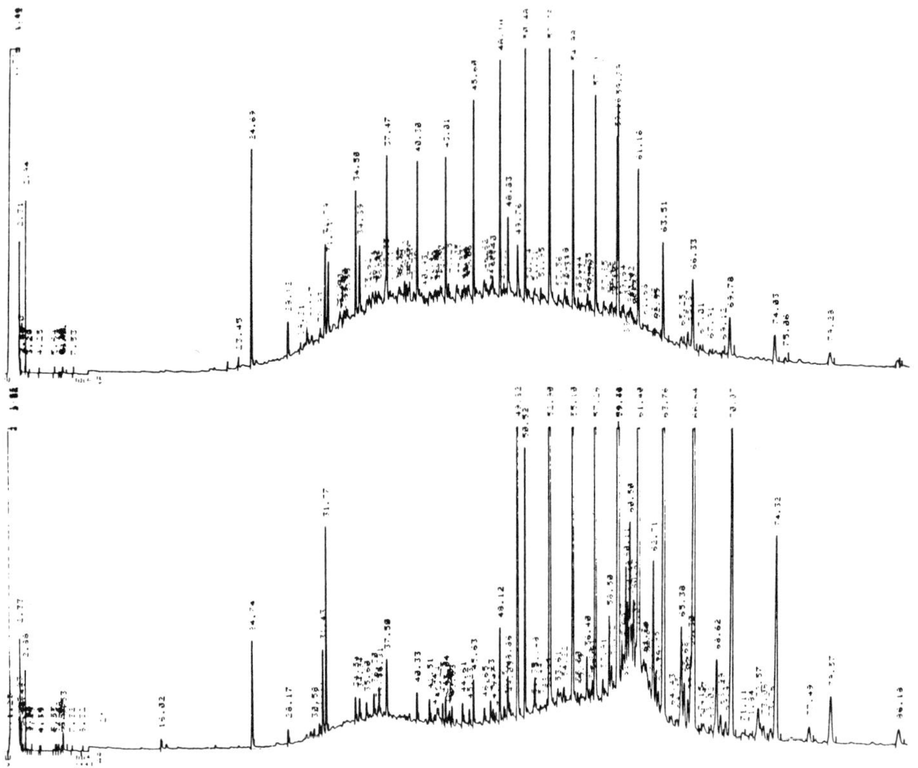

Figure 5. Above: dissolved phase of seawater, saturate fraction, 17 meters, Santa Barbara Basin, 23 February, 1977. 2.0 μl injection volume, 21 μl extract volume. Below: particulate phase from 5 meters at this station. Saturate fraction. 2.0 μl injection, 16 μl extract volume. Hewlett Packard 5840 gas chromatograph, 30 meter SP-2100 glass capillary column.

(hexane) fraction of the dissolved phase shown in Figure 5 is n-C15. In this sample it was present in substantially higher concentrations than the neighbouring alkanes n-C16 and n-C17. A comparable distribution is shown in Figures 6 and 7, which present chromatograms of seawater extracts from Tanner Banks and from the vicinity of San Miguel Island. In the majority of samples obtained in 1977, the relative concentrations of n-C15 and adjacent alkanes suggested that the origin of n-C15 differs from that of the other alkanes, or if there is a common origin, that the synthesis of n-C15 is favored. The distribution of n-C15 among the seawater samples obtained from the Bight in 1977 is presented in Table 5. Only relatively few of the particulate samples were analysed. In almost all cases, when

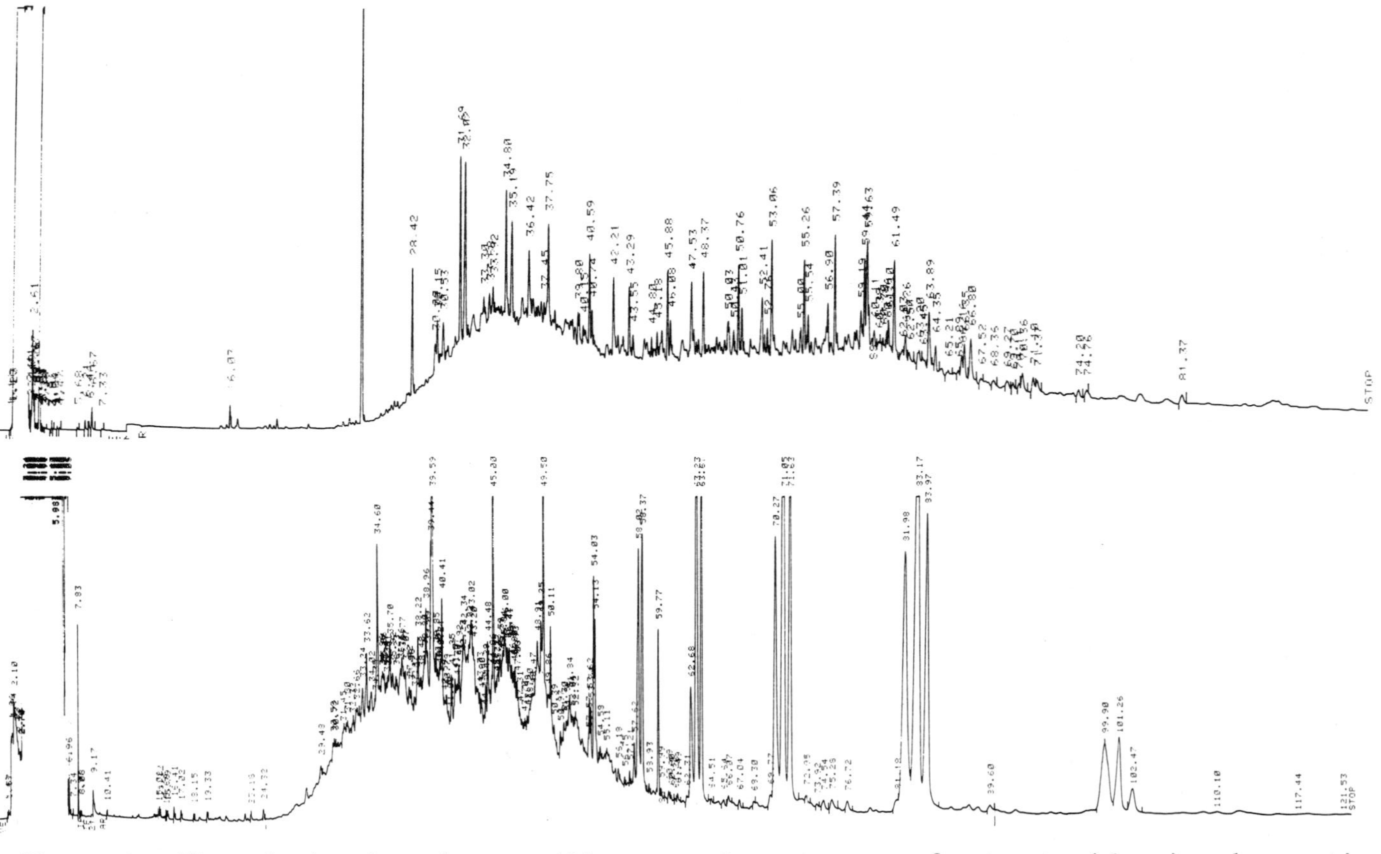

Figure 6. Flame ionization glass capillary gas chromatograms of saturate (above) and aromatic (below) fractions of the dissolved phase of seawater from 19 meters at Tanner Bank, 1 Sept., 1977. Top: 2.0 µl injection, 9 µl extract volume, 30 meter SP-2100 column; bottom: 2 µl injection, 10 µl extract volume, 30 meter SE-54 column. Hewlett Packard gas chromatograph.

Table 5. Distribution of n-C_{15} in dissolved and particulate phases of seawater of the Southern California Bight in 1977, concentrations in ng/l. N=3 for replicates, one std. deviation.

Station Depth (m)	Winter		Summer	
	Dissol.	Partic.	Dissol.	Partic.
Coal Oil Point				
3-5	2.5±0.5	---	.04	<.3
13-33	1.4	---	0.4	.3
56-57	1.4±1.0	---	0.1	.1
Santa Barbara Basin				
3-5	3.3±2.8	.45±.19	.3	---
17-40	2.0±0.9	.26±.13	.4	---
558-564	1.0±1.0	<.2	.2	---
San Miguel Island				
3-5	1.0	.1	2.6	.6
10-34	0.8	<.05	1.9	.25
196-237	0.4	<.05	.3	<.05
Tanner Bank				
3-5	1.8±1.1	---	.4	---
19-32	1.6	---	1.0	---
52-68	2.8±2.1	---	<.05	---
San Pedro Basin				
3-5	4.2	---	.6	---
435-440	.4	---	.2	---
863-871	.07	---	.3	---
Huntington Beach				
3-5	.2	.06	.8	---
12-17	.9	.5	2.1	---
20-34	3.0	.2	1.6	---
Santa Monica Basin				
3-5	4.0	---	.1	---
37-450	7.6	---	.02	---
874-885	0.9	---	.09	<.05
Santa Rosa				
3-5	1.1	---	.05	---
40-50	6.9	---	.23	---
95-108	1.3	---	.05	---
Santa Cruz Basin				
3-5	<.05	---	.7	---
19-34	3.9	---	<.05	---
1839-1857	<.05	---	.03	---
San Nicholas Basin				
3-5	3.0±2.8	---	<.05	---
21-34	1.1	---	<.05	---
1720-1762	0.6±0.4	---	<.05	---

both dissolved and particulate phases of the same sample were analysed, the levels of n-C15 were higher in the dissolved phase, frequently by as much as an order of magnitude. This distribution, however, may be an artifact of the analytical procedure. Particulates were reflux-extracted with boiling solvents (see methods section above) which may have resulted in relatively higher losses of the more volatile components of the hydrocarbon mixtures, including n-C15. Whatever the extent of loss during the extraction process, however, significant concentrations of n-C15 occured in the dissolved phase of seawater. Either it is present in bacteria that pass through the filter, or is excreted, secreted, or otherwise lost from the cells in which it was synthesized. In reviewing the phenomenon of apparent excretion of organic matter by marine phytoplankton, Sharp (1977) concluded that the balance of the evidence suggested that such excretion is not widespread. To assess the possibility that its observed distribution might be an artifact resulting from cell disruption during filtration, we also examined the distribution of a compound with Kovat Index 2035 on the SP-2100 column and 2057 on the SE-54 column assumed to be the hexa-olefin cis-heneicosa-3,6,9,12,15,18-hexaene, which has been shown to be a component of many species of marine algae (Blumer et al., 1970). Its distribution in southern California Bight waters in 1977 is presented in Table 6. In the majority of samples in which both the dissolved and particulate phases were analysed, most of the heneicosahexaene was retained on the glass fiber filters, indicating that disruption of the cells containing it was not occurring to a significant extent. This, however, would not prove a comparable stability of cells containing n-C15.

The chromatogram of the aromatic fraction shown in Figure 6 contains a series of triplets of later-eluting compounds. Total concentrations of these compounds are presented in Table 4 as higher-molecular weight olefins. They frequently constituted a majority of the resolved components, if not of the total hydrocarbons. Highest concentrations were present in the mid-depth waters (40m) in the Santa Barbara Basin in summer, where concentrations exceeded 1 μg/l. Concentrations were generally higher in the summer than in winter. Elsewhere, (de Lappe et al., 1979) we have reviewed the mass spectral evidence that these are long chain olefins of biological origin. At the present time, however, we are not able to associate them with any particular species or species group. The observed concentrations in the particulate phase were invariably lower than in the dissolved phase; also, the relative proportions of the several compounds were significantly different. We can not exclude, however, partial degradation during extraction resulting from the use of boiling solvents; the particulate data are consequently not presented. For the reasons discussed above, we can not exclude the possibility that more fragile cells were being disrupted during filtration, yet it appears probable that large quantities of these compounds were being secreted, excreted or otherwise discharged into seawater by organisms as yet uni-

Table 6. Distribution of heneicosahexaene in dissolved and particulate phases of seawater of the Southern California Bight in 1977. Concentrations in ng/l, N=3 for replicates, one std. dev.

Station Depth (m)	Winter		Summer	
	Dissol.	Partic.	Dissol.	Partic.
Coal Oil Point				
3-5	1.1±0.3	---	.05	3.3
13-33	0.45	---	.23	1.2
56-57	0.27±.08	---	.09	.09
Santa Barbara Basin				
3-5	.58±.17	3.1±0.3	2.1	---
17-40	.58±.09	5.5±1.2	1.7	---
558-564	.06±.05	<.05	.48	---
San Miguel Island				
3-5	0.94	13	.52	5.7
10-34	0.24	.58	1.2	14
196-237	<.05	<.05	.07	<.05
Santa Rosa Island				
3-5	.16	---	.25	---
40-50	.54	---	<.05	---
95-108	.06	---	.03	---
Santa Cruz Basin				
3-5	1.3	---	.21	---
19-34	1.3	---	.03	---
1839-1857	0.26	---	.06	---
San Nicholas Basin				
3-5	0.24±.12	---	.24	---
21-34	.62	---	.25	---
1720-1762	.03±.04	---	.05	---
Tanner Bank				
3-5	.20±.22	---	.36	---
19-32	.47	---	.17	---
52-68	.26±.08	---	.11	---
San Pedro Basin				
3-5	.16	---	.24	---
435-440	<.05	---	<.05	---
863-871	.09	---	<.05	---
Huntington				
3-5	.63	<.05	.58	---
12-17	.96	.13	.34	---
20-34	1.4	.45	.82	---
Santa Monica Basin				
3-5	.24	---	.39	---
37-450	.16	---	.08	---
874-885	1.4	---	<.05	---

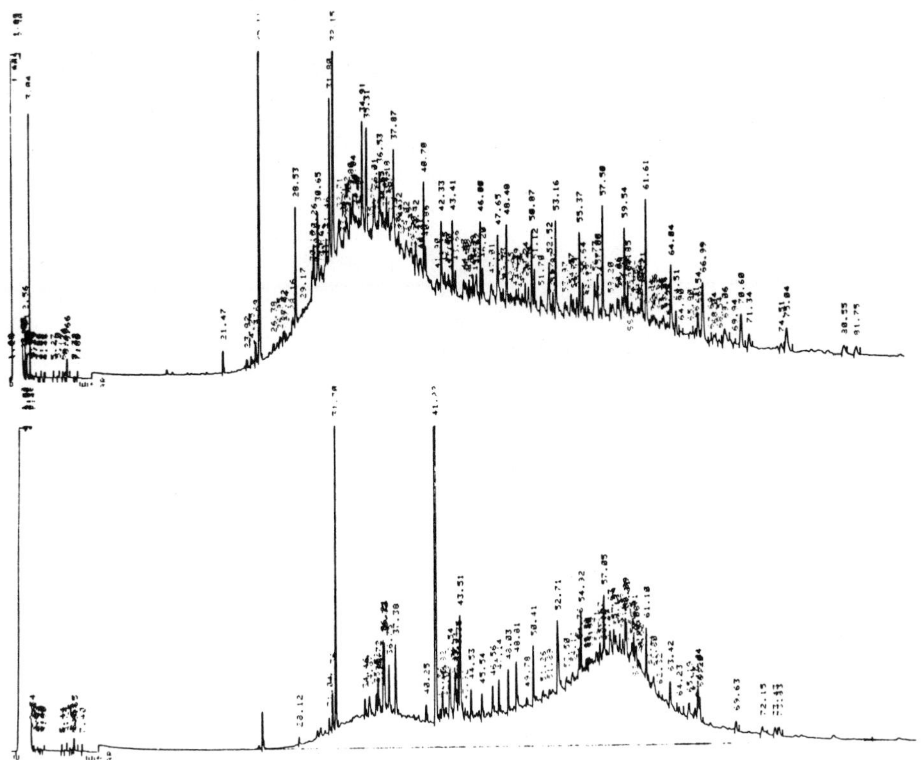

Figure 7. Dissolved (above) and particulate phases of seawater, saturate fractions, from 10 meters, San Miguel Island, 30 August, 1977. 30-meter SP-2100 glass capillary column. Top: 2.0 μl injection, 8 μl extract volume; bottom; 2.0 μl injection, 10 μl extract volume. Prominent peaks in the dissolved phase are n-C15, n-C17 and pristane (Kovat 1710); prominent peaks in the particulate phase are pristane and heneicosahexaene.

dentified. The biological significance of this phenomenon reamins to be elucidated.

Figure 7 shows chromatograms of the dissolved and particulate phases of the saturate fractions of a mid-depth sample (10 meters) obtained in the summer from the vicinity of San Miguel Island. The distribution of the unresolved component differs significanty between the two phases: the particulate fraction contains proportionately more of the higher-boiling components than does the dissolved phase. The maxima of the unresolved mixtures are at Kovats 2800 and 1850, respectively. This appears to be consistent with the partitioning of the PCB compounds between the dissolved and particulate phases that we have presented in Tables 2 and 3.

We have presented evidence above that two of the criteria used to determine whether petroleum hydrocarbons might be present in a sample, - the presence of phytane and the odd:even ratio of the alkanes, provide at best an ambiguous interpretation of the Southern California Bight data in determining whether petroleum is present in a particular sample. Elsewhere (de Lappe et al., 1979) we have presented mass spectral evidence that the ratios of phenanthrene, methyl and dimethyl phenanthrenes were significantly different at Coal Oil Point, the site of an active oil seep, than at several other stations in the Bight in 1977. At Coal Oil Point, the unsubstituted phenanthrene was less abundant than the methyl phenanthrenes and dimethyl phenanthrenes; at other stations, however, phenanthrene was more abundant than the total of both methyl and dimethyl phenanthrenes, suggesting a combustion source (Youngblood and Blumer, 1975). Gas chromatographic techniques alone did not permit distinguishing between unresolved mixtures of petrogenic origin from mixtures of combustion origin or from other sources.

The large-ship operations required for the deployment of samplers such as the Bodega-Bodmans, the increasing cost of such operations, the extreme difficulties encountered in reducing background contamination, the difficulty in obtaining representative blanks, and the necessity, in many cases, for sample volumes significantly larger than 90 liters for adequate chemical characterization argue for the continued development of *in situ* sampling systems.

To evaluate the extraction efficiency of petroleum hydrocarbons by polyurethane foam, laboratory recovery studies were conducted using filtered seawater that had been spiked with fuel oil or with a mixture of aliphatic and aromatic compounds. A spike of 5 g of No. 2 Fuel Oil was diluted with 20 ml of acetone, and injected with a glass syringe into 1000 liters of filtered seawater in a stainless steel tank. The tank contents were mixed with an electric stirrer for 24 hours prior to the onset of sampling. The fuel oil-spiked seawater was removed from the test tank through a stainless steel siphon tube, and passed by gravity flow through the foam columns or into 2-liter separatory funnels for 3-stage batchwise extraction with chloroform. Sample volumes were 12 liters for the chloroform extraction, and 41, 13.5 and 10.4 liters respectively in the foam column extraction. Recovery data for each technique are presented in Table 7. Chromatograms of the saturate (hexane) and unsaturate (benzene) fractions of both techniques are presented in Figures 8 and 9. Recovery was both qualitatively and quantitatively similar for both resolved and unresolved components.

A test mixture of aliphatic and aromatic standards was used in the second experiment. The foam and batchwise solvent extraction techniques used previously with the fuel oil were repeated. In addition the spiked seawater was also extracted with chloroform in the

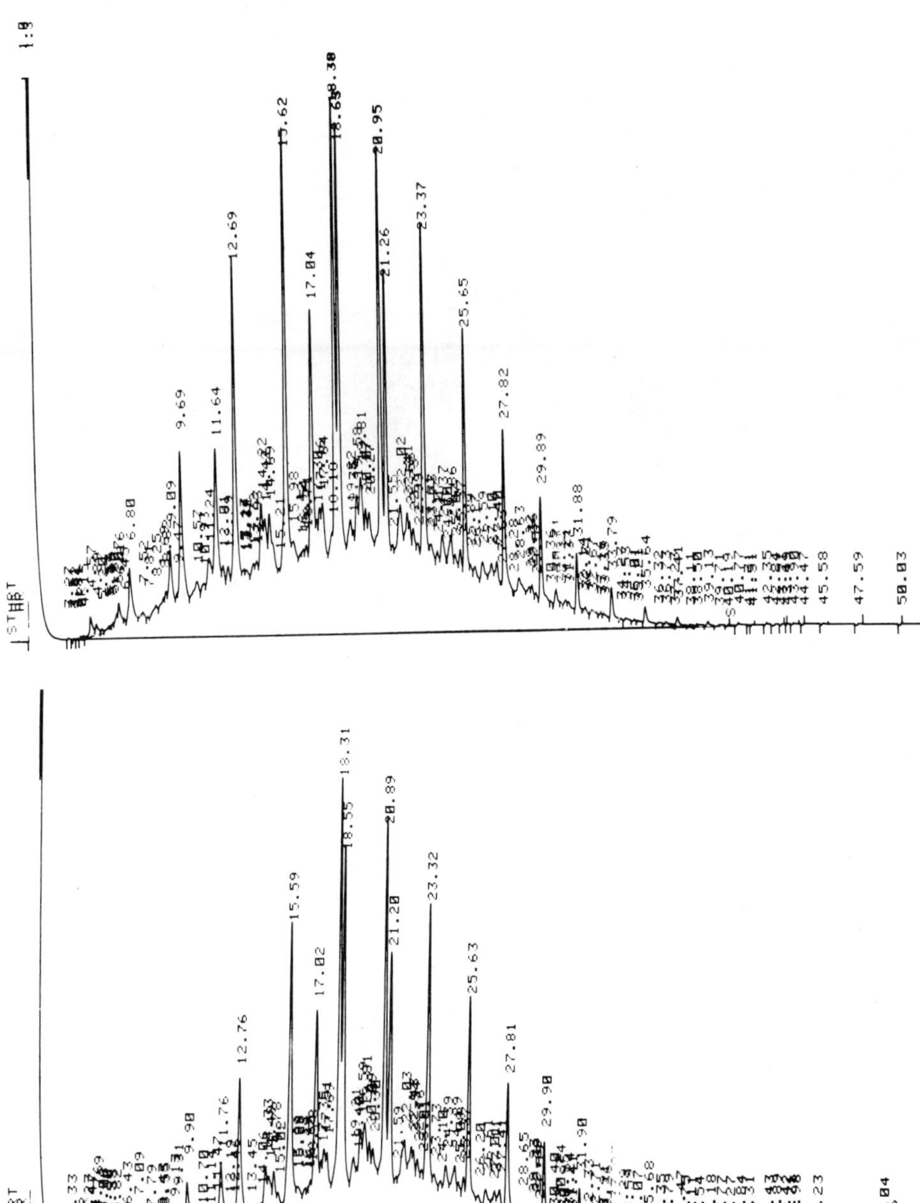

Figure 8. Recovery of fuel oil from filtered seawater, spiked with No. 2 Fuel Oil. Top: separatory funnel extraction with chloroform; bottom: polyurethane foam column extraction. Saturate fractions. Hewlett Packard 5830A gas chromatograph, 15.2 meter stainless steel SCOT OV-101 column (Perkin-Elmer); 0.4 μl injection, splitless mode. Initial temperature 130°, held 6 minutes, rate 4°/min., final temperature 280°. Helium carrier gas.

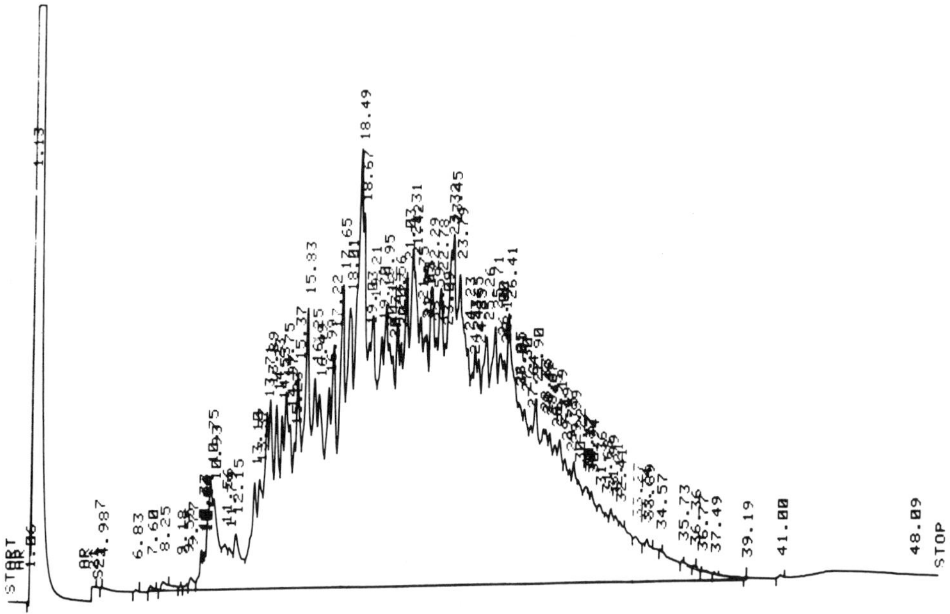

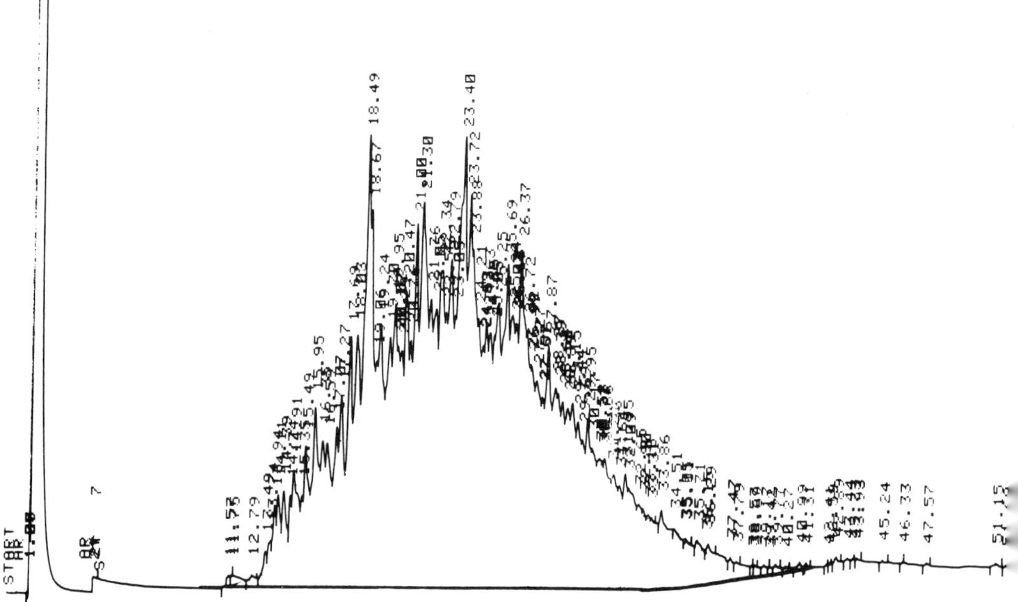

Figure 9. Recovery of No. 2 Fuel Oil from spiked seawater, aromatic fractions. Top: separatory funnel extraction with chloroform; bottom: polyurethane foam column. GC conditions as described in Figure 8.

Table 7. Recovery of No. 2 Fuel Oil from spiked, filtered seawater by polyurethane foam and by solvent extraction with chloroform. Concentrations in micrograms/liter. Experimental conditions described in the text. Means with one standard deviation, N = 3.

	Saturate Fraction			Aromatic Fraction	
	Total Resolved	Total Alkanes	Total Unres.	Total Resolved	Total Unres.
Separatory Funnel	1,700 ±810	560 ±250	710 ±380	170 ±19	100 ±14
Polyurethane Foam	1,300 ±0	460 ±11	620 ±35	170 ±22	100 ±22

continuous flow liquid-liquid apparatus developed by Ahnoff and Josefsson (Ahnoff and Josefsson, 1974), and in 19 l glass carboys using the mixing/settling transfer technique described earlier for extraction of the Bodega-Bodman samples. An additional group of three samples was acidified to pH 2 with HCl. Comparative recovery data are presented in Table 8. Estimates of absolute recovery are not presented because losses through volatilization or through adsorption to the walls of the test tank could not be determined. Recovery of the aliphatic compounds by the foam was generally comparable to recovery by chloroform, and the foam appears to produce higher recoveries of the aromatic compounds. A high efficiency of recovery of aromatic compounds by polyurethane foam has also been reported by Basu and Saxena (1978a; 1978b).

The polyurethane foam appears therefore to be a suitable adsorbent for use in continued development of *in situ* extraction techniques for the recovery of hydrocarbons from natural waters. The highly artificial environment of the laboratory test tank is not, however, equivalent to the state of natural waters which may contain high levels of a diversity of particulate material and of dissolved organics. Extensive intercalibrations in the field would therefore be required in future work.

SUMMARY

Levels of total hydrocarbons, constituting mixtures of petrogenic and biogenic compounds, in seawater of the Southern California Bight in 1977 were generally less than 1 μg/liter, lower than the majority of literature values of total hydrocarbons in seawater that have been reported in the past. Concentrations in seawater of the

Table 8. Extraction of aliphatic and aromatic compounds from spiked, filtered seawater by various methodologies. Concentrations in ng/liter. Means and one standard deviation.

Method	N	n-C18	n-C19	n-C21	n-C22	n-C23	n-C26	n-C30	n-C32	Phen.*	Pyrene
Separatory funnel, chloroform	3	21 ±36	26 ±45	63 ±73	97 ±77	102 ±69	140 ±99	240 ±160	330 ±230	15 ±27	83 ±110
Polyurethane foam columns	4	36 ±6	50 ±2	110 ±9	140 ±13	130 ±12	180 ±19	260 ±29	370 ±42	180 ±15	330 ±8
Tellusond liquid-liquid, chloroform	2	49 ±4	66 ±6	140 ±14	170 ±22	170 ±21	220 ±26	350 ±34	510 ±46	150 ±47	240 ±15
Carboy, normal pH chloroform	3	29 ±1	43 ±2	85 ±28	130 ±7	120 ±6	180 ±5	260 ±38	410 ±11	140 ±12	170 ±23
Carboy, pH 2 Chloroform	3	33 ±21	46 ±25	96 ±41	120 ±56	110 ±54	150 ±81	220 ±130	330 ±200	120 ±88	170 ±130

* Phenanthrene

open ocean might therefore be expected to be substantially lower. Gas chromatographic techniques alone, without GC/MS, were not sufficient to distinguish mixtures of petroleum compounds in seawater from mixtures derived from combustion or from other sources. Substantial levels of biogenic compounds were found in the "dissolved" or microparticulate phase of seawater; they frequently consitituted the majority of the total hydrocarbons. The question whether phytoplankton secrete, excrete, or otherwise loose significant quantities of higher-molecular weight organic compounds into the ambient seawater does not therefore appear to have been resolved.

Maximum PCB and total DDT levels in the Southern California Bight in the summer of 1977 were comparable to those obtained two years previously with the Slocum Sampler; the PCB levels, ranging between 0.1 and 0.9 ng/l, were equivalent to the majority of PCB concentrations measured with the Modified Slocum Sampler in waters of the vicinity of Cape Cod in 1976, and to values reported from coastal Sweden and Great Britain. A range of PCB concentrations between 0.1 and 1.0 ng/liter appears therefore to have been representative of realitively polluted coastal waters for the period 1974-1977. Estimated concentrations of individual PCB and DDT compounds exceeded concentrations of all but several biogenic hydrocarbons in Southern California Bight waters. PCB levels in western Lake Ontario in 1976 were approximately an order of magnitude higher than levels in coastal marine waters. Particulate phases were found to contain relatively more of the higher-molecular weight biogenic/petrogenic hydrocarbons and of PCB in marine waters and in Lake Ontario. Concentrations of PCB in the "dissolved" or microparticulate phase were generally higher than in the particulates, expressed on a whole water basis.

Experimental recovery studies in the laboratory have demonstrated the feasibility of using polyurethane foam in continuing development of *in situ* sampling systems for hydrocarbons in natural waters.

ACKNOWLEDGMENTS

Research was supported by the U.S. Bureau of Land Management, the International Decade of Ocean Exploration Program of the National Science Foundation, grant number IDO72-06412-A02, the Bodega Bay Institute of Pollution Ecology, and the National Audubon Society. We thank the personnel of the Canada Centre for Inland Waters, Burlington, for their assistance and collaboration.

LITERATURE CITED

Ahnoff, M. and B. Josefsson. 1974. Anal. Chem. 46:658-663.

Barbier, M., D. Joly, A. Saliot and D. Tourres. 1973. Deep-Sea Research 20:305-314.

Basu, D.K. and J. Saxena. 1978a. Environ. Sci. Technol. 12(7):

Basu, D.K. and J. Saxena. 1978b. Environ. Sci. Technol. 12(7):

Bedford, J. W. 1974. Bull. Environ. Contam. Toxicol. 12(5): 622-625.

Bidleman, T. F. and C. E. Olney. 1974. Bull. Environ. Contam. Toxicol. 11(5): 442-450.

Bidleman, T. F., C. P. Rice and C. E. Olney. 1976. pp. 323-351 in H. L. Windom and R. A. Duce, eds., Marine Pollutant Transfer D. C. Heath and Company, Lexington, Massachusetts. 391 pp.

Bieri, R. H., A. L. Walker, B. W. Lewis, G. Losser and R. J. Huggett. 1974. National Bureau of Standards Spec. Publ. 409, Pollution Monitoring (Petroleum), Proceedings of a Symposium and Workshop held at NBS, Gaithersburg, Maryland, May 13-17, 1974.

Blumer, M. and D. W. Snyder. 1965. Science 150:1588-1589.

Blumer, M. 1970. pp. 153-167 in D.W.Hood, ed. Symposium of Organic Matter in Natural Waters. University of Alaska Press.

Blumer, M., M. M. Mullin and R. R. L. Guillard. 1970. Marine Biology 6:226-235.

Bodman, R. H., L. V. Slabaugh and V. T. Bowen. 1961. J. Mar. Res. 19(3):141-148.

Bowen, H. J. M. 1970. J. Chem. Soc.(A), 1082-1085.

Brown, R. A. and H. L. Huffman, Jr. 1976. Science 191:847-849.

Brown, R. A. and T. D. Searl. 1976. pp. 240-255 in Sources, Effects and Sinks of Hydrocarbons in the Aquatic Environment, Proceedings of the Symposium, American University, Washington, D.C., 9-11 August 1976. The American Institute of Biological Sciences.

Bruland, K. W., G. A. Knauer and J. H. Martin. 1978. Nature 271 (5647): 741-743.

Calder, J. A. 1977. pp. 432-441 in D. A. Wolfe, ed. Fate and Effects of Petroleum Hydrocarbons in Marine Organisms and Ecosystems. Pergamon Press, New York.

Carmignani, G. M. and J. Bennett. 1976. Aquaculture 8:291-294.

Clark, R.C.,Jr. and M. Blumer. 1967. Limnol. Oceanogr. 12:79-87.

Dawson, R. and J. P. Riley. 1977. Esutar. Coast. Mar. Sci. 5: 55-69.

de Lappe, B. W., R. W. Risebrough, J. R. Clayton, P. L. Millikin, R. K. Okazaki, J. S. Parkin and J. R. Payne. 1978. Development of methodologies for the in situ extraction of petroleum compounds from seawater. Southern California Baseline Study, Vol. III, Report 3.2.4., pp. 1-164. Final Report to the Bureau of Land Management, Washington, D.C. by the Bodega Marine Laboratory, University of California, Bodega Bay, CA and Science Applications, Inc., La Jolla, CA.

de Lappe, B. W., R. W. Risebrough, J. C. Shropshire, E. F. Letterman, W. Sistek and W. Walker II. 1979. Hydrocarbons in the waters of the Southern California Bight in 1977. Report of the Bodega Marine Laboratory to Science Applications, Inc., La Jolla, California, and the U.S. Bureau of Land Management, Wash., D.C.

Elder, D. 1976. Mar. Pollut. Bull. 7: 63-64.

Farrington, J. W. 1974. Some problems associated with the collection of marine samples and analyses of hydrocarbons. Woods Hole Oceanographic Institution Tech. Rep., WHOI-74-23.

Farrington, J. W. and P.A. Meyers. 1975. Hydrocarbons in the Marine Environment. in Environmental Chemistry Vol. 1. Burlington House, London.

Fox, G.A., A.P.Gilman, D.J. Hallett, R.J. Norstrom, F.I. Onuska and D.B. Peakall. 1975. Canadian Wildlife Service Manuscript Reports 34. Canadian Wildlife Service, Ottawa. 35 pp.

Gardner, W. D. 1977. Limnol. Oceanogr. 22: 764-767.

Gesser, H. D., A. Chow and F. C. Davis. 1971. Anal. Letters 4 (12): 883-886.

Gordon, D. C., Jr. and P. D. Keizer. 1974. National Bureau of Standards Spec. Publ. 409. Marine Pollution Monitoring (Petroleum). Proceedings of a Symposium and Workshop held at NBS, Gaithersburg, Maryland, May 13-17, 1974.

Gordon, D. C., P. D. Keizer and J. Dale. 1974. Mar. Chem. 2: 251-261.

Haile, C. L., G.D. Veith, W.C. Boyle and G.F. Lee. 1975. Natl. Tech. Inform. Serv. PB-243.364.

Harvey, G. R., W. G. Steinhauer and J. M. Teal. 1973. Science 180:643-644.

Harvey, G. R., W. G. Steinhauer and H. P. Miklas. 1974. Nature 252: 387-388.

Iliffe, T. M. and J. A. Calder. 1974. Deep Sea Research 21: 431-488.

Jensen, S., L. Renberg and M. Olsson. 1972. Nature 240: 358-360.

Keizer, P.D. and D.C. Gordon. 1973. J. Fish. Res. Bd. Canada 30 (8): 1039-1046.

Koons, C.B. and D.E. Brandon. 1975. Proceedings of the 7th Annual Offshore Technology Conference, Vol. III.

Koons, C. B. and P. H. Monaghan. 1976. pp. 85-107 in Sources, Effects and Sinks of Hydrocarbons in the Aquatic Environment. Proceedings of the Symposium, American University, Wash., D.C., 9-11 August 1976. The American Institute of Biological Sciences.

Levy, E. M. 1971. Water Research 5: 723-733.

Martin, J. H., K. W. Bruland and W. W. Broenkow. 1976. pp. 159-184 in H. L. Windom and R. A. Duce, eds., Marine Pollutant Transfer. D. C. Heath and Company, Lexington, Mass. 391 pp.

Musty, P. R., and G. Nickless. 1974. J. Chromatogr. 89:185-190.

Musty, P. R. and G. Nickless. 1976. J. Chromatogr. 120: 369-378.

National Academy of Sciences - National Research Council. 1975. Petroleum in the marine environment. Wash., D.C. 105 pp.

Osterroht, C. 1977. Mar. Chem. 5: 113-121.

Participants of the Lead in Seawater Workshop. 1974. Mar. Chem. 2: 69-84.

Patterson, C., D. Settle, B. Schaule and M. Burnett. 1976. pp. 23-38 in H. L. Windom and R. A. Duce, eds., Marine Pollutant Transfer. D. C. Heath and Co., Lexington, Massachusetts, and Toronto. 391 pp.

Risebrough, R. W., B. W. de Lappe and W. Walker II. pp. 261-321 in H. L. Windom and R. A. Duce, eds., Marine Pollutant Transfer. D. C. Heath and Company, Lexington, Mass. and Toronto. 391 pp.

Risebrough, R.W.. B.W. de Lappe, E. F. Letterman, J.L. Lane A.M. Springer and W. Walker II. Report of the Bodega Marine Laboratory to Science Applications, Inc., La Jolla, CA and the Bureau of Land Management, Washington, D.C.

Saxena, J., J. Kozuchowski and D.K. Basu. 1977. in Methods and Standards for Environmental Measurement. NBS Special Publication 464. U.S. Department of Commerce. pp. 101-104.

Scura, E. D. and V. E. McClure. 1975. Mar. Chem. 3: 337-346.

Sharp, J.H. 1977. Limnol. Oceanogr. 22:381-399.

Sverdrup, H. O., M. W. Johnson and R. H. Fleming. 1942. The Oceans: Their Physics, Chemistry and General Biology. Prentice Hall, New York, 1942. 1087 pp.

Uthe, J. F., J. Reinke and H. Gesser. 1972. Environ. Letters 3(2): 117-135.

Uthe, J. F., J. Reinke and H. O'Brodovich. 1974. Environ. Letters 6(2): 103-115.

Veith, G.D., D.W. Kuehl, F.A. Puglisi, G.E. Glass and J.G. Eaton. 1977. Arch. Environ. Contam. Toxicol. 5(4):487-499.

Webb, R. G. 1975. Environmental Monitoring Series EPA-660/4-75-003, 30 pp. National Environmental Research Center, Office of Research and Development, U.S. Environmental Protection Agency, Corvallis, Oregon 97330.

Williams, P. M. and K. J. Robertson. 1975. Fish. Bull. 73:445-447.

Wong, C. S., W. J. Cretney, J. Piuze and P. Christensen. 1977. in Methods and Standards for Environmental Measurement. NBS Special Publication 464. U.S. Department of Commerce. pp. 249-258.

Young, D.R., R.W. Risebrough, B.W. de Lappe, T.C. Heesen and D. McDermott-Ehrlich. Pestic. Monit. J. (in press).

Youngblood, W.W. and M. Blumer. 1975. Geochim. Cosmochim. Acta 139: 1303-1314.

Zsolnay, A. 1972. Kieler Meeresforsch. 27: 129-134.

Zsolnay, A. 1973. Deep Sea Research 20: 923-925.

Zsolnay, A. 1978. Mar. Pollut. Bull. 9:23-24.

FAST QUANTITATIVE ANALYSIS OF A WIDE VARIETY OF HALOGENATED COMPOUNDS IN SURFACE-, DRINKING-, AND GROUND-WATER

G.J. Piet, P. Slingerland, G.H. Bijlsma, and C. Morra

National Institute for Water Supply

P.O. Box 150, 2260 AD Leidschendam, The Netherlands

ABSTRACT

A fast quantitative analysis of a wide range of halogenated compounds in water below the microgram/litre level is performed with a glass capillary column coupled to an electron capture detector on line with a flame ionization detector. Specific parameters which are used for the confirmation of individual organo-halogen compounds in complex mixtures are very accurate retention indices and the signal ratio numbers of the E.C.D. and F.I.D. detector. A split-less injection system and a technique to avoid base line drift during temperature programming make it possible to analyze a wide variety of halogen compounds from chlorobenzene up to PCBs in one straight run.

When this technique is combined with a direct head-space procedure, compounds such as trihalomethanes, trichloroethane etc. can be analyzed too.

Results of the analyses of water of the river Rhine, and tapwater derived from surface water are presented.

INTRODUCTION

The quantitative analysis of a wide variety of gaschromatographable organo-chlorine compounds in water is an important aspect in the evaluation of water quality (1).

Not only severe taste and odour impairment of drinking water can be induced by industrial chemicals with low odour threshold concentrations such as chloroanilines, chlorotoluidines, chlorinated dienes which easily pass a drinking water treatment system (2), but long term human health effects might be involved too when chemicals such as vinylchloride, chloroform, tetrachloromethane, hexachlorobutadiene, bis(chlorethyl)ether and trichloroethylene are present in tapwater (3,4).

The discharge of such compounds in water sources of industrial countries takes place on a relatively large scale and often effluent treatment systems are not adequate to remove them sufficiently. They enter the environment particularly at those places where large scale production of synthetic organic compounds takes place (5).
In this way the drinking water supply can be threatened when surface water is used as a raw water source. Compounds such as e.g. tetrachloromethane and bis(2-chloriisopropyl)ether appear to be rather persistent particularly when they are introduced into the soil as it is the case with bankfiltration.

Another important threat is groundwater pollution by trichloroethene, tetrachloromethane and tetrachloroethene, which compounds have a greater density than water so that they easily reach underground water bodies. Water chlorination as applied in disinfection and drinking water treatment techniques such as breakpoint chlorination and transport chlorination can be other sources of organo-chlorine compounds (6). A compound such as hexachlorobutadiene, which tends to be bioaccumulated like some pesticides is of importance to analyze regularly.
For these reasons an adequate monitoring system for a wide variety of organo-chlorine compounds, was developed to analyse surface water used as a raw water source for drinking water production, polluted rain water in the neighbourhood of open water storage reservoirs, groundwater destined for human consumption and finished potable water.

EXPERIMENTAL

A monitoring system should be equipped with a sample preparation-, sample introduction- and analysis system coupled to a data assimilation system.

Automated gas-chromatography with high resolution glass columns is an adequate system for multicomponent analysis of complex mixtures of organic compounds. But even when the conditions of the GC are extremely stable and reproducible, no real guarantee exists that the identity of the eluting compounds is not confused as long as the retention index is the only parameter to confirm the identity of each compound.

When a second parameter is introduced such as the signal-response ratio of two different detectors measuring a compound simultaneously, not only more certainty about the identity is obtained but in some cases chemicals with identical indices can be measured quantitatively if their response ratios differ sufficiently.

The compounds to be analyzed are concentrated by extracting a litre sample with 20 ml of a cyclohexane-diethylether (1:1) solvent. The recoveries of the individual compounds is emperically determined by adding calibration substances (standard-addition method) to a sample and perform the extraction according to the standard procedure.

In this way extraction recoveries can be calculated.
The recoveries, depending on the type of water to be analyzed, vary. Chemicals such as detergents can affect the extraction-efficiency. To avoid emulsification during the vigorous shaking in a three dimensional shaking machine, the extraction is performed in an almost completely filled 1 litre sample bottle. The extracts, of which 1 µl is brought on the capillary column with a splitless injection system are stored at -20 °C before analysis. The very narrow peaks of the separated compounds which elute from the capillary column are not broadened by the detectors when all dead volumes in the system are eliminated and when right amounts of suppletion gases are added at a constant temperature of 225 °C just before the detectors. In this way base line drift of the electron capture detector during temperature programming is reduced (fig.1).

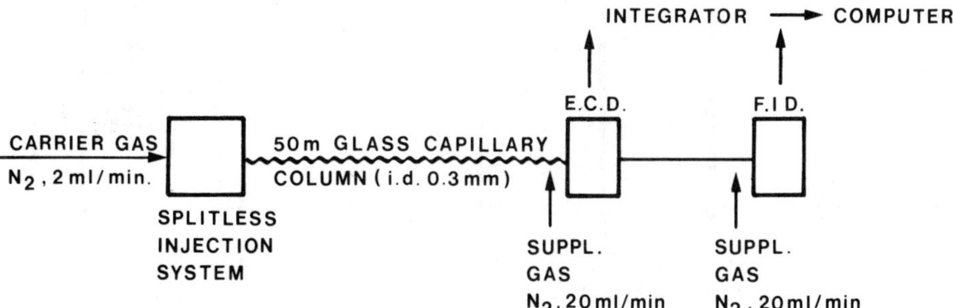

Figure 1. Instrumental set-up for the analysis of organo halogen compounds with a double detector system

In a modified Varian 2800 gaschromatograph equipped with a splitless Grob injection system temperature programming ($4^{\circ}C$/min) is applied from 30-225°C using a 50 m, 0,3 mm i.d. capillary column for the analysis of non-polar and slighly polar compounds. A 50 m UCON HB 5100 capillary column is used to analyse semi-polar compounds.

A chromatogram which is produced by the described system is shown in figure 2. It presents the analysis of water of the river Rhine at Lobith where the river Rhine crosses the border of The Netherlands.
Normal alkanes from 6-26 C atoms are added to an extract to determine very accurately the retention index of a compound with a variation of 1/1000.

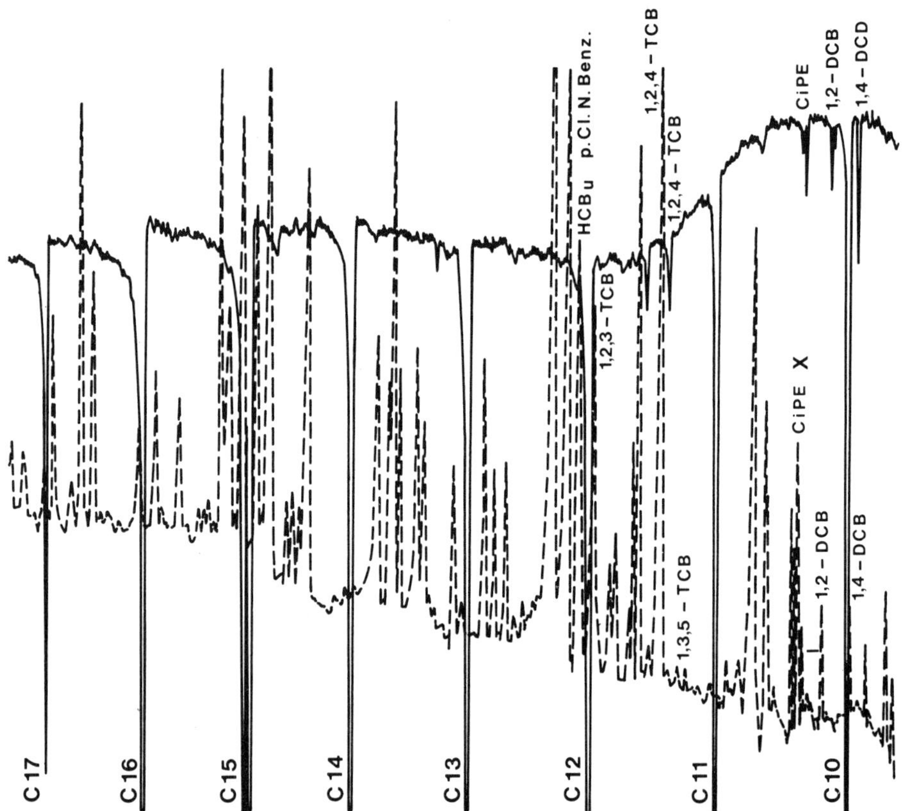

Figure 2. Analysis of Rhine water at Lobith (Determination of accurate retention indices)
DCB = dichlorobenzene
CIPE = bis(2-chloroisopropyl)ether
TCB = trichlorobenzene
HCB = hexachlorobutadiene

The E.C.D. and F.I.D. signals, which are registered by an automated integrator during the routinely applied procedure can be read from the presented chromatogram and the E.C.D.-F.I.D.signal response ratios for the individual compounds are calculated.

For the confirmation of each compound calibration mixtures have to be analyzed once a day to determine the individual retention index and the E.C.D.-F.I.D.response ratio (see figure 3).

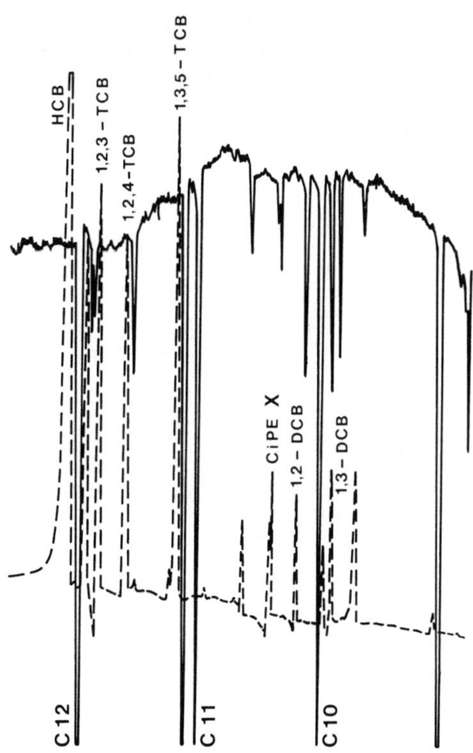

Figure 3. Analysis of a calibration mixture of organo-halogen compounds present in surface water

In this way important industrial pollutants such as chloroanilines, chlorotoluidines, chlorinated benzenes and toluenes, chlorinated ethers and dienes can be analyzed without confusion of the identity when the internal standard of n-alkanes is added. Phtalates can also be measured because of their electron-negative character.

Very volatile pollutants are measured separately with a direct head space method (9) in which the gas phase of a water vapour equilibrium at 30°C is injected on a glass capillary OV-225 column coupled to a ^{63}Ni E.C.D. with a temperature program from 0 - 110°C. Detection limits from 0.1 - 0.01 microgram/litre water are obtained for pollutants such as trichloroethene, tetrachloromethane, tetrachloroethene, 1,1,1-trichloroethane, 1,1-dichloroethene, 1,2-dichloroethane and trihalomethanes. For the trace analysis of vinyl-chloride the sensitivity of a flame ionisation detector is necessary because of the relatively low E.C.D. signal response.

For the analysis of these and other very volatile organic chlorine compounds, a double detector system is not really necessary, as the separation power of a wall coated whisker capillary column (10) is adequate to eliminate confusion of component identity in almost all cases.

RESULTS

Table 1

Organo-chlorine compounds in bankfiltered Rhine water (retention time in the soil more than 1 year)

bis(2-chloroethyl)ether
bis(2-chloroisopropyl)ether
chloroaniline (o,p)
chlorotoluidine (o,p)
chloroform
chloronitrobenzene (o,m,p)
dichlorobenzene(o,m,p)
3,4-dichloroaniline
1,2-dichloroethane
1,2-dichloropropane
hexachlorobutadiene
monochlorobenzene
tetrachloromethane
tetrachloroethene
tri(2-chloroethyl)phosphate
trichlorobenzenes
trichloroethene

When tapwater is derived from polluted surface water, quite a few organic chemicals pass into the drinking water supply. Some compounds cause taste- and odour impairment (table 2).

Table 2

Taste impairment by organo-chlorine compounds of tapwater from bankfiltered Rhine water

Compounds *	O.T.C. (μg/l)
o- and p-dichlorobenzene	10 and 0.3
1,2,4-trichlorobenzene	5
hexachlorobutadiene	6
bis(2-chloroisopropyl)ether	200
2-chloroaniline	3
3,4-dichloroaniline	3
5-chloro-o-toluidine	5

* compounds which are present at concentrations above 1 % of their Odour Threshold Concentration

Some of these chemicals have to be reduced by water sanitary programs. One of the substances present in Rhine water is hexachlorobutadiene. The monitoring of this substance in the Rhine where this river crosses the border of The Netherlands gave the following results (fig. 4).

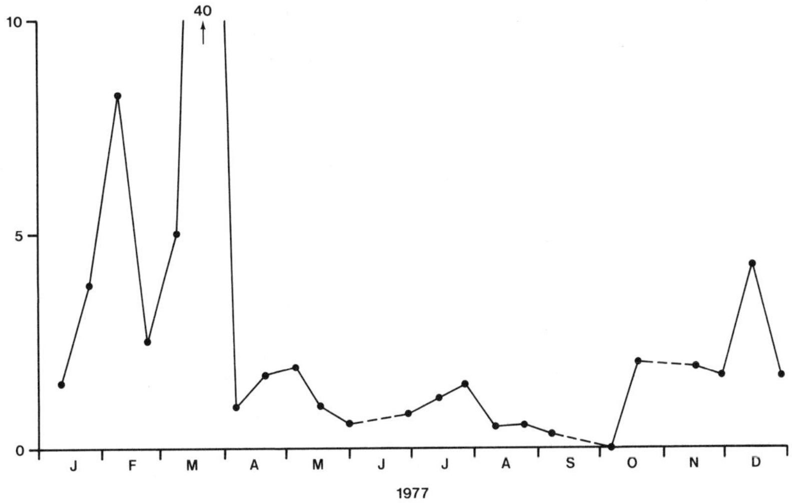

Figure 4. Hexachlorobutadiene, River Rhine, Lobith, The Netherlands 1977 (conc. μg/l)

Other odour intensive compounds are presented in figure 5 and 6.

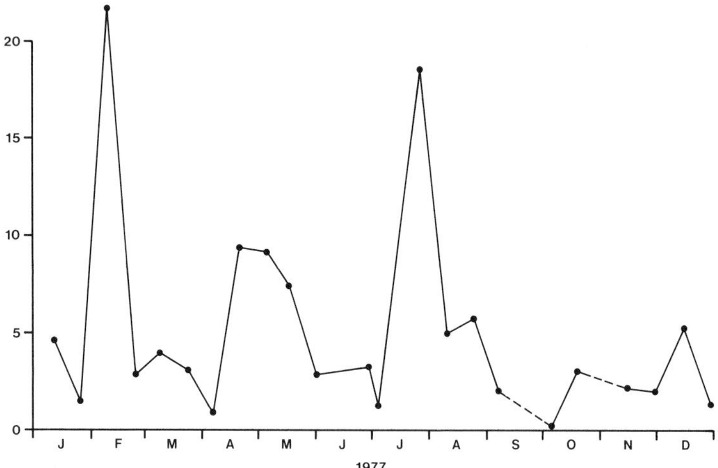

Figure 5. Σ Dichlorobenzenes, River Rhine, Lobith, The Netherlands. 1977 (conc. μg/l)

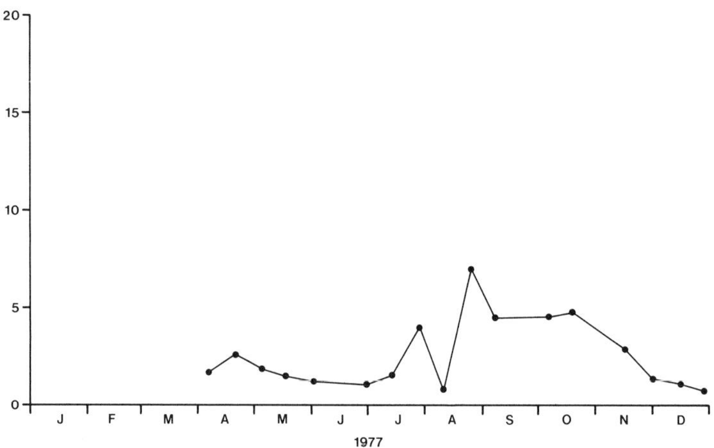

Figure 6. Σ Trichlorobenzenes, River Rhine, Lobith, The Netherlands. 1977 (conc. μg/l)

The concentrations are calculated by adding the concentrations of the individual compounds. The measurement of the concentration of the individual compounds makes it possible to evaluate their contribution to the odour impairment of water.

One of the substances which is rather persistent in water treatment systems is bis(2-chloroisopropyl)ether. Its O.T.C. is probably much lower than the reported value of 200 μg/l. (8). Its concentration in the river Rhine is reported in figure 7.

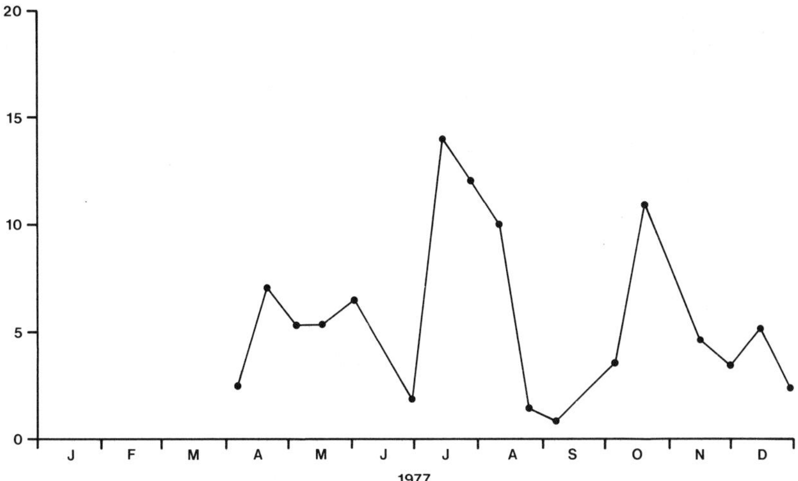

Figure 7. Bis(2-chloroisopropyl)ether, River Rhine, Lobith, The Netherlands. 1977 (conc. μg/l)

More volatile organo-chlorine compounds such as tetrachloromethane, an animal carcinogen, and chemicals such as chloroform trichloroethene, 1,2 dichloroethene etc. are also present in surface water. An example of tetrachloromethane pollution of the river Rhine is shown in figure 8.

The analysis of halogenated compounds in water sources is important because they also affect the quality of finished water of the waterwork.When water sanitary programs and improved water treatment systems are installed the measurement of a wide variety of organo halogen compounds is of great assistance to measure the quality of raw water sources and the drinking water derived from these sources.

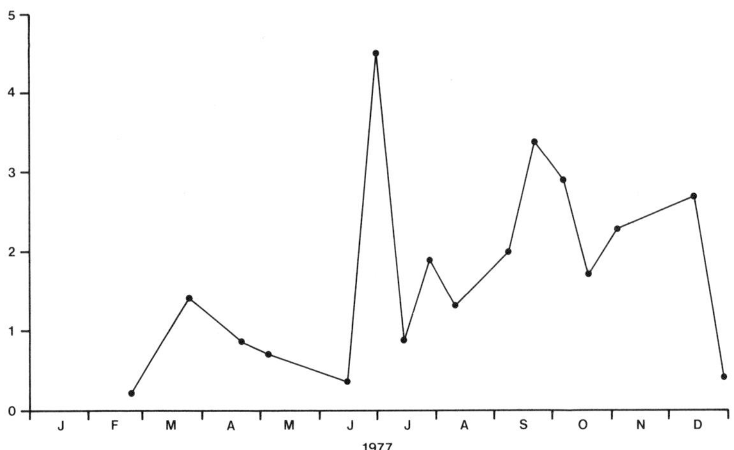

Figure 8. Tetrachloromethane, River Rhine, Lobith, The Netherlands. 1977 (conc. µg/l)

CONCLUSIONS

- The analysis of several organo-chlorine compounds is of great importance for the drinking water supply to evaluate the effectiveness of processes such as bankfiltration to remove specific substances which impair drinking water quality. Though the effectiveness of infiltration techniques can be determined by the measurement of sum-parameters such as chemical oxygen demand, biological oxygen demand and total organic chlorine content, the knowledge of the fate of individual suspected substances is of great importance to protect the drinking water supply by well-chosen additional treatment techniques.
- As organochlorine compounds suspected in relation to long term human health effects are formed during chlorination of water destined for human consumption, a monitoring program to analyze a wide range of organo-chlorine compounds in finished water is necessary.
- These measurements will also give information concerning the effect of treatment techniques of different types of water.
- As groundwater pollution in Europe and other countries is also caused by organo chlorine compounds which contaminate the environment and which are introduced into the groundwater by careless users, an analysis of these compounds in groundwater is necessary to protect groundwater against further pollution.
- The pollution of rivers and surface water by organo-chlorine compounds is evident. In several countries water sanitation programs have to be introduced and the concentration of

specific suspected substances has to be decreased, particularly when surface water is used as a raw water source for the drinking water supply.

REFERENCES

Analysis of organic compounds in water to support health effect studies.
Voorburg,WHO International Reference Centre for Community Water Supply, 1976.
W.H.O. Technical Paper Series No. 9. (1)

Zoeteman, B.C.J.
Sensory assessment and chemical composition of drinking water.
Thesis, State University Utrecht.
Voorburg, National Institute for Water Supply, 1978. (2)

The National Research Council.
Drinking water and health.
Washington, National Academy of Sciences, 1977. (3)

Health effects relating to direct and indirect re-use of waste water for human consumption.
Voorburg, WHO International Reference Centre for Community Water Supply, 1975.
W.H.O. Technical Paper Series No. 7. (4)

Keith, L.H.
Identification and analysis of organic pollutants in water.
Ann Arbor, Ann Arbor Science, 1976. (5)

Piet, G.J., Zoeteman, B.C.J., Slingerland, P.
The effect of chlorination on drinking water quality.
Presented at the seminar:
"Gesundheitliche Probleme des Wasserchlorung und Bewertung des dabei gebildeten organischen Verbindungen" at December 12th, 1977, Bundes Gesundheitsambt, WaBoLu, Berlin. (6)

Slooff, W., Zoeteman, B.C.J.
Toxicological aspects of some frequently detected organic compounds in drinking water.
Voorburg, National Institute for Water Supply, 1976.
RID-Report 76-15 prepared for the Commission of the European Communities. (7)

Nettenbreijer, A.H., Gemert, L.J. van.
Compilation of odour threshold values in air and water.
Voorburg, National Institute for Water Supply, 1977. (8)

Piet, G.J., Slingerland, P., Grunt, F.E., Heuvel, M.P.M. van der, Zoeteman, B.C.J.
Determination of very volatile halogenated organic compounds in water by means of direct head-space analysis.
Analytical Letters, A 11(5), 437-444 (1978). (9)

Onuska, F.L., Comba, M.E., Bistricki, T. and Wilkinson, R.J.
Journal of Chromatography, 142, 117, (1977). (10)

ADSORPTION OF INSECTICIDE RESIDUES - IMPORTANCE IN ENVIRONMENTAL SAMPLING AND ANALYSIS

J.R.W. Miles

Agriculture Canada, Research Institute

London, Ontario, Canada N6A 5B7

Insecticides can become incorporated in the water of streams and lakes in several ways, such as spray drift or spillage, industrial or municipal waste, or from wind and water erosion of contaminated soil particles with adsorbed chemical residues. It would be impossible to document the amounts involved in spray drift and spillage, but by measuring soil insecticide residues we can examine the potential for contamination of water systems by erosion of soils. In Figure 1 are shown the insecticide residues found in organic soil of the Holland Marsh, an important vegetable growing area north of Toronto, Ontario, during a 4 year study (Miles et al, 1978). These data are average ppm in the soil of 13 farms, and the levels are very significant, with DDT averaging about 29 ppm. Some individual farm soils had as high as 60 ppm DDT and 25 ppm ethion. There appeared to have been little degradation of the DDT in these organic soils. A few percent had converted to DDE, but p,p'-DDT comprised 76% of the total DDT and o,p'-DDT 15% - very similar to the analysis of technical DDT (Schecter et al, 1945). We initiated studies of water and sediment in the area of the vegetable marsh in an attempt to determine the contribution of these contaminated farm soils to environmental pollution.

This marsh is surrounded by dikes and the farm tiles drain into a ditch from which the water is pumped when rising water levels require it. We analysed the water at weekly intervals and combined these concentration data with amounts of water pumped to quantitate the amounts of insecticides being pumped from the marsh. Whole water samples, including suspended sediment, were analysed for the above quantitation. But parallel water samples were filtered and the insecticides adsorbed to the sediment were separately analysed.

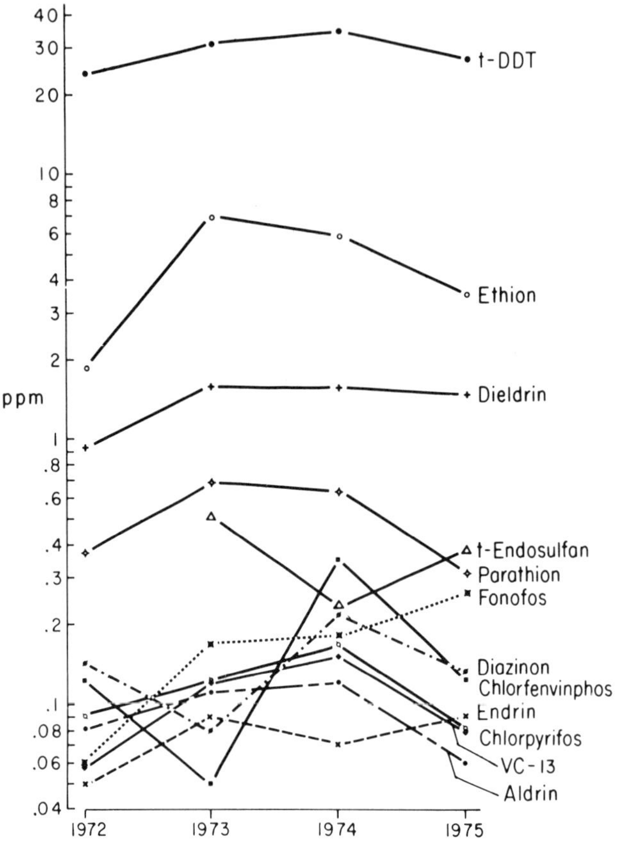

Figure 1. Insecticide levels in soils of Holland Marsh 1972-75.

In Table 1 are shown the percent of insecticides in the total water samples which was adsorbed on the suspended sediment. The Holland Marsh is the organic soil area described above. The Big Creek (Norfolk County, Ontario) drains mostly mineral soil of a tobacco and vegetable growing area, and flows into Lake Erie. DDT was adsorbed on the suspended sediment to a much higher degree than was dieldrin and the amounts adsorbed are in inverse relation to their water solubilities. In the Holland Marsh drainage water the organophosphorus insecticide diazinon was present at concentrations greater than dieldrin, but no diazinon was detected on the sediment. Since diazinon is soluble in water at about 40 ppm, the lack of diazinon adsorption agrees with the inverse water solubility ratio as noted above for DDT and dieldrin. The adsorption data for DDT and dieldrin on Big Creek sediment agree very closely with that of the Holland Marsh.

Table 1 Insecticides on Filtered Sediment, Reported as % of Total Water Analysis

		Total DDT	dieldrin	diazinon
Holland)	aver.	62%	12%	0%
Marsh	range	(31 - 93)	(3 - 42)	
Big)	aver.	59%	18%	-
Creek	range	(30 - 90)	(10 - 29)	
Solubility in water (ppm)		0.0012	0.186	40

Laboratory studies with dieldrin, ethion, lindane, diazinon and carbaryl generated the adsorption isotherms shown in Figure 2. The absorbent used in the experiments for Figure 2 data is sediment from the Big Creek described above. DDT was also included in this experiment but the data for DDT does not show on the graph because the solubility of DDT in water is only 0.001 μg/ml. However, DDT was adsorbed to the extent of 98 - 100% and would therefore generate a curve paralleling the ordinate. Comparing the curves for the other five insecticides the adsorption appears to be in inverse proportion to their water solubilities which agrees with the field data reported above. Dieldrin and ethion are represented by the same curve. They have similar water solubilities (0.2 and 0.6 ppm respectively) but dieldrin is an organochlorine, and ethion is an organophosphorus insecticide. Lindane, the most water soluble of the organochlorine insecticides, generates a curve midway between dieldrin and the insecticides diazinon and carbaryl, which are included to show the lesser adsorption of the more water soluble organophosphorus and carbamate insecticides. The emphasis of this symposium is on hydrocarbons and halogenated hydrocarbons, but when conducting environmental insecticide work we must compare the organochlorines (most now banned from agricultural usage) with the organophosphorus and carbamate insecticides which have replaced them.

Besides adsorption on the Big Creek sediment we examined adsorption on three soils from southwestern Ontario. In Figure 3 the adsorption of dieldrin is shown on 1 g each of a high organic muck, the mineral sediment, a sandy loam soil and Plainfield sand. The organic matter contents of these 4 adsorbents were 75, 2.8, 2.5 and 0.7% respectively, so the adsorption of this series is in direct relation to the % organic matter of the adsorbents. This is an over-

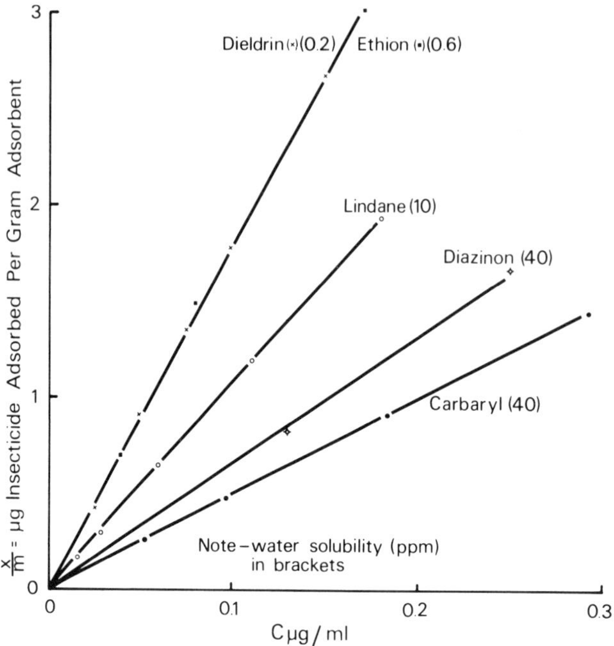

Figure 2. Insecticide adsorption isotherms (adsorbent = Big Creek Sediment).

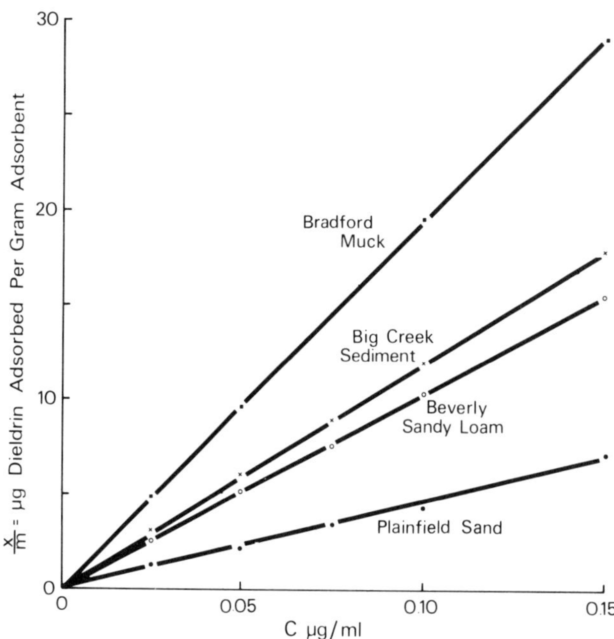

Figure 3. Dieldrin adsorption on four adsorbents.

Table 2 Ratios of Metabolites to p,p'-DDT in Bed Material, Holland Marsh 1973

	DDE/DDT	DDD/DDT
April	0.2	2.6
May	0.2	2.1
June	0.4	2.8
July	0.6	6.4
August	0.6	7.7
September	1.4	11.0
October	1.4	13.0

simplification because it is known that the mineral fraction of soils is also important in adsorption and certainly the adsorption by the muck is not in proportion to its 75% organic matter content.

In our environmental studies it appears that once the insecticides are adsorbed on sediment their condition does not remain static. We have noted significant conversion of DDT to its metabolites DDE and DDD (TDE) in bottom sediment measured through the seasons from spring through autumn (Table 2). It can be seen that the ratio DDE/DDT increased 7-fold while DDD/DDT increased 5-fold. We have also noted such increases in these ratios on bottom sediment (bed material) of a creek measured on the same day from source to the mouth of the stream. Evidently during the time the bottom sediment is moved downstream microorganisms are able to convert DDT to its metabolites. We have found an increase of 20 times in the bottom sediment of Big Creek from source to mouth!

Insecticides adsorbed on soil may be subjected to desorption action by irrigation water, rain, or when the soil is eroded into a stream. We examined the desorption of six insecticides by treating sediment with 2 ppm of each chemical and equilibrating with 200 ml of water. After centrifuging, 190 ml water was removed and analysed. One hundred and ninety ml of fresh distilled water was added and the above steps repeated three more times. The data are presented in Figure 4. It can be seen that very little DDT was removed from the sediment by the water washes. Significant amounts of ethion and dieldrin still remained adsorbed after the 4th rinse. Only 17% of lindane remained adsorbed after the 1st rinse and <10% of the more soluble OP and carbamate insecticides dizainon and carbaryl remained adsorbed after the 1st rinse.

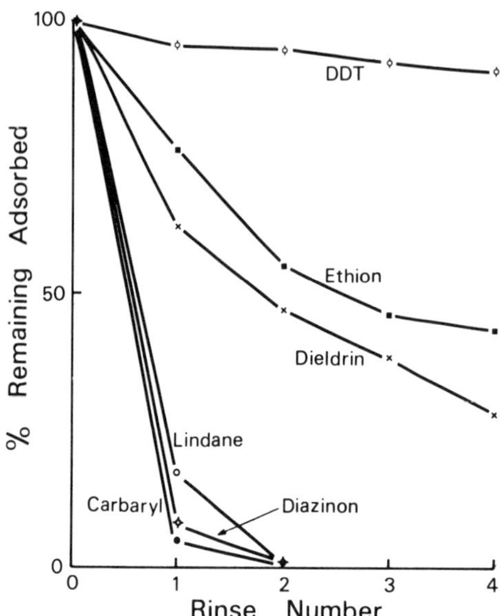

Figure 4. Successive desorption of six insecticides from 1 g Big Creek sediment.

Closely related to adsorption and desorption is the mobility of insecticides through soil. We examined the mobility of DDT, dieldrin, ethion, lindane, diazinon and carbaryl through Plainfield sand and a high organic muck treated with 2 and 4 ppm of the insecticides respectively. Ten g of the treated soil was placed in a 4.5 cm diameter glass cylinder inserted 1 cm into 300 g of wet sand contained in a Buchner funnel. Distilled water was allowed to flow into the cylinder to maintain 1 cm depth above the treated soil. Suction was applied to the receiving flask and 10 - 200 ml fractions of eluate were collected and analyzed. Table 3 shows the amounts of six insecticides leached from Plainfield sand.

A large proportion (52%) of the carbaryl was eluted in the first fraction with a total for ten fractions of 86%. Diazinon and lindane were almost quantitatively removed from the sand by the 2 L of water. Between the lindane (10 ppm water solubility) and ethion (0.6 ppm solubility) there is a significant drop in amount leached, - and note that no DDT was leached. Table 4 shows comparable data for leaching from organic muck. Much smaller percentages of each insecticide were leached from the muck than from the sand. The total carbaryl leached from the muck was 53%. essentially the same as was leached from sand by the first 200 ml fraction! Still note the marked difference between the amount of lindane leached, and those of ethion and dieldrin which are <1 ppm soluble

Table 3 Percentage of the Applied Insecticide Leached through Plainfield Sand with Successive 200 ml Fractions of Water

Fraction No.---		1	2	3	4	5	Total 10 Fractions
		percent leached					
Carbaryl	(40)	52	14	9	4	3	86
Diazinon	(40)	26	22	11	11	7	95
Lindane	(10)	20	18	11	9	7	93
Ethion	(0.6)	0	0.4	0.6	1.0	1.5	14
Dieldrin	(0.2)	0	0.1	0.4	0.8	0.9	10
p,p'-DDT	(0.001)	0	0	0	0	0	0

Note: Bracketed figures are water solubilities in ppm.

in water. Again, of course, no DDT was removed from the organic muck. Our field data indicates that some DDT is leached through organic soil. In Table 5 are the results of analyses of tile drain water collected from four farms in the Holland Marsh. The concentrations of DDT and dieldrin in the farm soils are listed, along with the concentrations of the insecticides we analyzed in the tile drain water. These concentrations in the tile drain water are significant even though the 101 ppt values for DDT in farm B is only 1/10th its solubility in distilled water. These water samples were crystal clear so none of the DDT or dieldrin was adsorbed on sediment and must have been in solution. The tile drain waters were very brown in colour, probably saturated with humates which reportedly enhance the solubility of insecticides like DDT in water,

Table 4. Percentage of the Applied Insecticide Leached through Bradford Muck with Successive 200 ml Fractions of Water

Fraction No.		1	2	3	4	5	Total 10 Fractions
		percent leached					
Carbaryl	(40)	10	10	8	6	5	53
Diazinon	(40)	3	4	11	9	7	50
Lindane	(10)	1	3	5	5	4	34
Ethion	(0.6)	0	0	.03	.09	.11	1.2
Dieldrin	(0.2)	0	0	.01	.07	.08	0.8
p,p'-DDT	(0.001)	0	0	0	0	0	0

Note: Bracketed figures are water solubilities in ppm.

Table 5 Insecticide Residues in Muck Soil and Tile Drain Water

Farm	Total DDT		Dieldrin	
	Soil	Tile water	Soil	Tile water
A	57 ppm	26 ppt (pp10^{12})	2.0 ppm	6 ppt (pp10^{12})
B	42	101	1.4	15
C	16	30	0.4	10
D	43	45	4.0	12

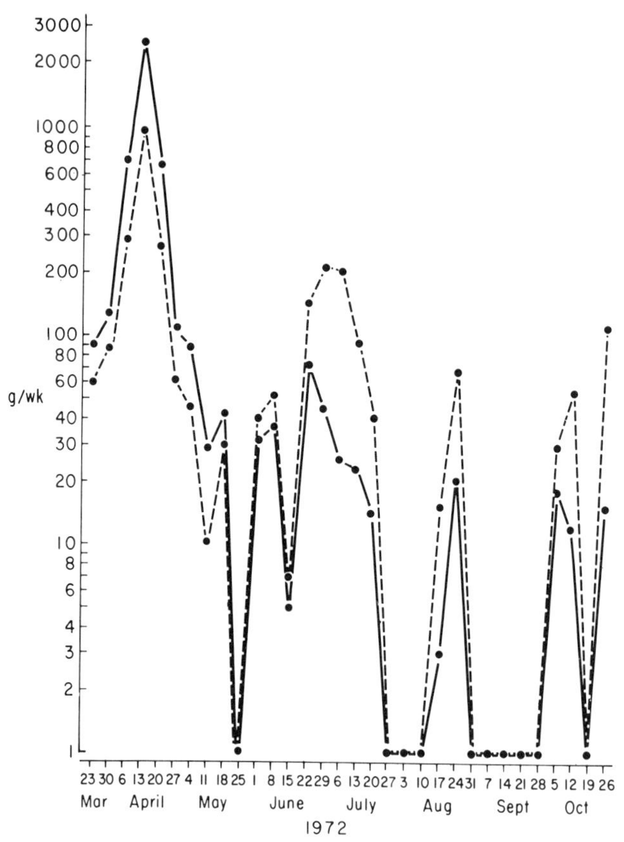

Figure 5. Insecticides pumped from Holland Marsh 1972.
——— Total organochlorine. – – –Total organophosphorus.

(Wershaw et al, 1969). This data contrasts with the laboratory studies above, where no DDT was leached from muck or sand.

What are the environmental considerations of these adsorption, desorption and mobility data? Figure 5 shows the amounts of organochlorine and organophosphorus insecticides pumped from the Holland Marsh in 1972. We have more recent data (Miles and Harris 1978) but 1972 data showed considerable contrast between the amounts of organochlorine and organophosphorus insecticides pumped. The greatest quantities of insecticides are pumped from the marsh in spring when melting snow and spring rains combine to erode considerable amounts of soil and sediment. In summer, pumping is only necessary when summer rains raise the level of water in the drainage ditches. Note the very significant amounts of organophosphorus insecticides which were pumped in the spring. This is before any of the current year's insecticides had been applied, so these organophosphorus insecticides, often reported as easily degradable, had persisted in the soil and water environment since the previous season, ready to be pumped out in the spring. The sediment load is highest in springtime and the OC's being associated with the sediment are pumped in greater amounts than are the OP's. However, from about the end of May through the summer, OP's are pumped in greater quantities that OC's. With less pumping the sediment settles and the more water-soluble OP's are present in water in greater concentration than the OC's. Summer sprays of diazinon and parathion probably also contribute to the higher concentrations of OP's. Similar data (not shown here) calculated for individual organophosphorus insecticides show that the less water soluble "ethion", is pumped in greater quantities in spring because of its association with sediment while the more soluble diazinon and parathion predominate during the summer months.

Our laboratory and field studies have shown that organochlorine insecticides being generally of lower solubility in water are adsorbed to sediments to a greater degree than the more soluble organophosphorus insecticides. But within the OC's the adsorption of individual insecticides is in inverse relation to their water solubilities, e.g. adsorption of DDT>dieldrin>lindane. And in the OP class, adsorption of ethion>diazinon or parathion.

Some OP insecticides degrade readily but we have shown (Miles et al, 1978) that ethion is fairly persistent, and not being very water soluble it will associate with sediment similar to the OC's. However, that portion of the organophosphorus insecticides and their metabolites which do not degrade are generally somewhat soluble in water and will remain in solution. Insecticides adsorbed on sediment and carried downstream settle into the bed material of bays when the streams enter the receiving lake. Most organophosphorus and carbamate insecticides will not behave in this manner but will remain in solution in the lake water - we may therefore have a different pattern of distribution of insecticides in the aqueous environment in the future.

REFERENCES

Miles, J.R.W., and C.R. Harris. 1978. J. Econ. Entomol. 71: 125.

Miles, J.R.W., C.R. Harris, and P. Moy. 1978. J. Econ. Entomol. 71: 97.

Schechter, M.S., S.B. Soloway, R.A. Hayes, and H.L. Haller. 1945. Ind. Eng. Chem. Anal. Ed. 17: 704.

Wershaw, R.L., P.J. Burcar, and M.C. Goldberg. 1969. Environ. Sci. and Technol. 3: 271.

METHODOLOGY FOR THE ISOLATION OF POLYNUCLEAR AROMATIC HYDROCARBONS FOR QUALITATIVE, QUANTITATIVE, AND BIOASSAY STUDIES

R.F. Severson, M.E. Snook, R.F. Arrendale, and O.T. Chortyk

Tobacco Laboratory, Science and Education Administration-Federal Research, United States Department of Agriculture, P.O. Box 5677, Athens, GA 30604, U.S.A.

Sources of polynuclear aromatic hydrocarbons (PAH) in the environment generally also contain heteroatom analogues of PAH. These O- and N-PAH also contribute to air pollution (Dong et al, 1977; Lao et al, 1975) and are found along with S-PAH, such as dibenzothiophene, in petroleum (McKay et al, 1976) and coal-derived fuels (Sharley, 1976). They often interfere in the analyses of PAH because of similar chromatographic characteristics. Therefore, in choosing a method for the analysis of PAH in an environmental sample, it is important to consider the source of the PAH and the possible presence of heteroatom aromatics.

The PAH fraction, generally obtained by the silicic acid Rosen (1955) method or by solvent partitioning methods, also contains heteroaromatics. As shown in Table 1, solvent partitioning of a PAH fraction from cigarette smoke between DMSO and cyclohexane did not result in the quantitative recovery of the PAH into the DMSO, mainly due to differences in partition coefficients of the various members. For cigarette smoke PAH, about 10% of the parent compounds and 20% of the methyl derivatives unexpectedly remained in the cyclohexane. Consequently, solvent partition separations of PAH should be avoided.

We have found gel permeation chromatography to be one of the most efficient means of isolating PAH. The Bio-Beads SX-12-benzene system that we use not only separates on a molecular weight basis but also absorbs aromatic compounds and quantitatively separates

Table 1 Distribution of Cigarette Smoke PAH after DMSO-Cyclohexane Partitioning

PAH	% Original PAH[a]	
	Cyclohexane	DMSO
Phenanthrene/Anthracene	6	94
Methylphenanthrenes/Methylanthracenes	18	82
Dimethylphenanthrenes/Dimethylanthracenes	16	84
Fluoranthene	7	93
Pyrene	6	94
Methylpyrenes	20	80
1,2-Benzanthracene/Chrysene/Triphenylene	7	93
Methyl-1,2-Benzanthracenes/Methyl-chrysenes/Methyltriphenylenes	17	83
Benzofluoranthenes	10	90
BaP/BeP	10(0.3)[b]	90(87)[b]

[a]Determined by GC PAH quantitation method.

[b]Determined by recovery of added ^{14}C-BaP.

them from non-aromatic material (Snook et al, 1975). These polystyrene gels also partially separate alkyl and parent PAH by the so-called methyl effect in which increasing numbers of methyl groups on a PAH moiety decrease its elution volume (Snook, 1976; Snook, 1978). We have found this behaviour to be very useful in the identification of components in complex mixtures.

The fractionation scheme shown in Figure 1 was developed to isolate the tumorigenic PAH of cigarette smoke condensate (CSC) (Severson, 1978). The Bio-Beads-benzene system was used to isolate a pure PAH fraction F-55B (Figure 1; gel fraction 36-60, Figure 2) with tumor-initiating activity equivalent to that of whole CSC (Severson, 1978). The gas chromatographic (GC) profile of the PAH of F-55B is given in Figure 3, and the corresponding identifications of major components are listed in Table 2. The gel chromatography system also isolated pure PAH fractions from the related neutral CSC fractions F-BPE, F-19, F-20, and F-54 (Severson, 1978). Based on the methodology used to isolate the tumorigenic PAH of CSC, we developed both qualitative and quantitative analytical methods for PAH and heteroatom PAH.

ISOLATION OF PAH AND HETEROATOM PAH

Figure 4 schematically details our gel chromatography system. As shown in Table 3, the Bio-Bead gels isolated the aromatics, containing both the PAH and the heteroatom PAH, as a class from earlier-eluting non-aromatics. We have also found that silicic acid (SA) chromatography can be used to separate the PAH from the N-aromatics.

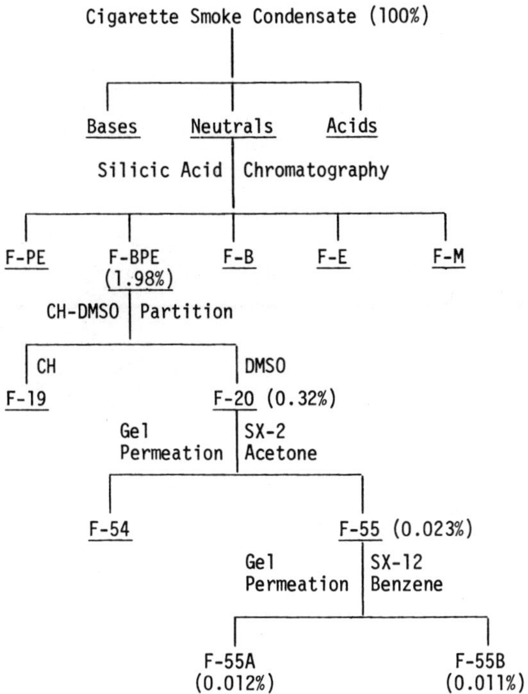

Figure 1. Cigarette smoke fractionation scheme for bioassay studies (F-fraction; PE-petroleum ether; BPE-25% benzene/petroleum ether, B-benzene, E-ethyl ether M-methanol).

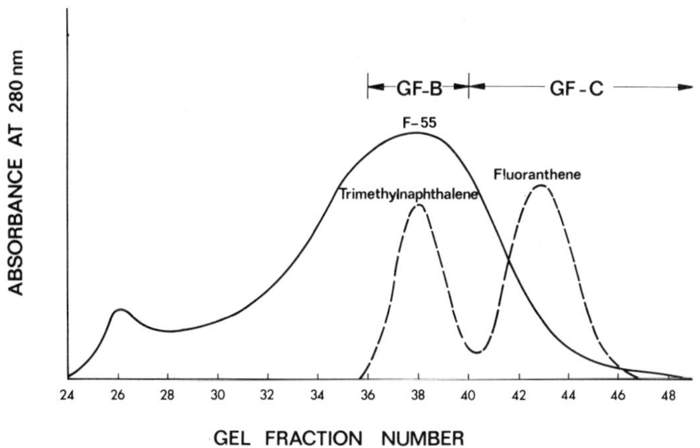

Figure 2. Separation of PAH fraction F-55 into multialkylated PAH (GF-B) and paretn (GF-C) fractions.

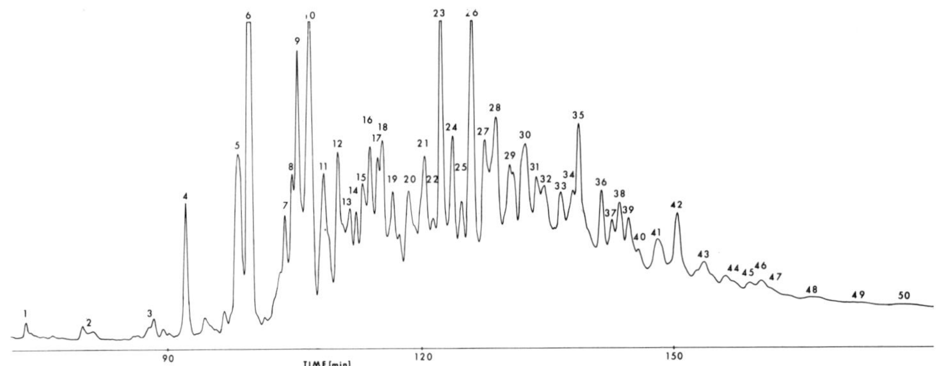

Figure 3. GC profile of PAH in F-55B (GF 36-60 on 15 ft Dexsil 300 GC column. See Table 2 for peak identifications.

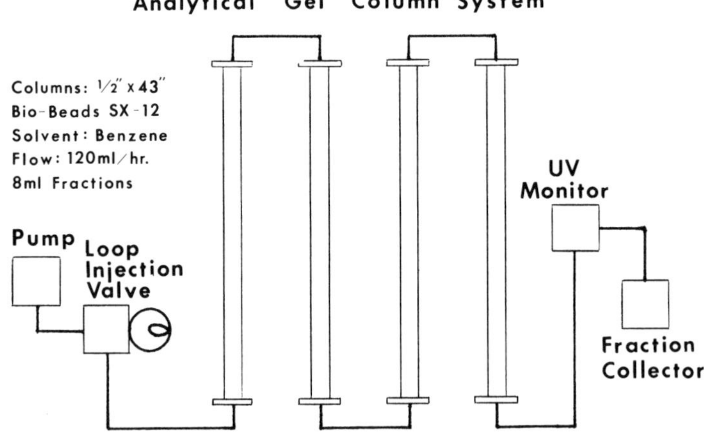

Figure 4. Schematic of Bio-Beads gel chromatographic system.

Table 2 Major PAH Components in Figures 3, 8, 9, and 10

Compound	Peak Number			
	Fig. 3	Fig. 8	Fig. 9	Fig. 10
Naphthalene		1	1	1
Methylnaphthalenes		4,5	2	2,3
Dimethylnaphthalenes		8-12	3	4
Vinylnaphthalenes		8,9		
Acenaphthylene		11		5
Acenaphthene		12	4	6
Dibenzofuran		13	5	
Methylacenaphthylenes		17-20		
Fluorene		18	6	7
Methylfluorenes		19,25,26		8
Dibenzothiophene				9
Phenanthrene	1	30	7	10
Anthracene		31	8	
Methyldibenzothiophenes				11
Methylphenanthrenes/ Anthracenes	2	32-34	9	12
Dimethyldibenzothiophenes				13
Dimethylphenanthrenes/ Anthracenes	3	35-38		14
Fluoranthene		39	10	15
Acephenanthrylene		40		
Pyrene	4	41	11	16
Methylfluoranthenes		44,45	12	17
Benzofluorenes		44,45		17,18
Methylpyrenes	5,6	46,47	13	18,19,20
Dimethylpyrenes	7-10	50-53	14	21
9-thia-1,2-benzofluorene				22
9-thia-3,4-benzofluorene				23
1,2-Benzanthracene	11	54	15	
Chrysene/Triphenylene	11	55	16	24
Methyl-9-thia-benzofluorenes				25
Methyl-1,2-benzanthracenes/ Chrysenes	15-18	57-59	17	26
Dimethyl-9-thia-benzofluorenes				27
Dimethylchrysenes				28
Benzo(b,j&k)fluoranthenes	23	63	18	
Benzo(a)fluoranthene	25	64	19	
Benzo(e)pyrene (BeP)	26	65	20	30
Benzo(a)pyrene (BaP)	25	65	21	31
Perylene	27	66	22	32
Methylbenzofluoranthenes	27,28	66,67		
Methylbenzopyrenes	29-32	68-70	23,24	33
Dibenz(a,c&a,h)anthracenes	34	73	25	
Indenopyrene	35	74	26	
Picene	35		27	
Benzo(g,h,i)perylene	36	75	28	
Anthanthrene/ Methylindenopyrene	37	76	29	
Dibenzofluoranthenes	43,44		32,33	
Dibenzopyrenes	45-48		34-36	
Coronene	47		34	

The silicic acid column readily separates the PAH from sp^3 N-aromatics, such as indoles and carbazoles. As shown in Table 4, the PAH fraction eluted with 25% benzene in petroleum ether (1:3, B/PE) also contained the oxygen and sulfur aromatics, dibenzofuran, benzonaphthofuran, and dibenzothiophene. The indoles and carbazoles were eluted with the slightly more polar solvent system, 50% benzene in petroleum ether. We thus recommend the chromatographic scheme shown in Figure 5 for the isolation of polyaromatics from complex environmental samples. Details on the preparation and operation of the silicic acid and Bio-Beads columns have been given (Severson et al, 1978; Severson et al, 1976).

The gel isolation of cigarette smoke PAH from the 25% benzene-petroleum ether SA fraction is shown in Figure 6. The PAH cut, contained in gel fractions 30 to 40, represented much less than one percent of the total weight of the starting fraction.

Table 3 Chromatographic Isolation of PAH and Heteroatom PAH on Bio-Beads SX-12 Gels

	% Distribution								
	Gel Fraction Number[a]								
Compound	41	42	43	44	45	46	47	48	49
Naphthalene		0.4	25.7	53.7	19.3	0.9			
Quinoline			9.4	47.4	37.8	5.3	0.1		
Anthracene		0.2	11.2	46.9	34.4	6.7	0.5		
Phenanthrene		0.4	1.1	22.2	45.0	27.3	4.7		
Acridine		1.5	26.4	48.5	19.2	4.1	0.3		
Phenanthridine			6.3	37.4	36.9	17.1	2.3		
5,6-Benzoquinoline			1.4	24.7	46.4	22.7	4.2		
Pyrene				0.1	4.3	26.9	42.1	21.7	4.9
1-Azapyrene				1.7	11.7	34.1	31.5	11.7	7.4
Indene	2.2	40.6	45.6	10.1	1.6				
Fluorene	0.8	24.7	52.0	20.0	2.3	0.1			
Indole					0.7	11.3	44.0	34.0	9.7
Carbazole				0.1	3.9	28.7	44.0	19.8	3.5
Dibenzofuran		6.9	46.2	38.5	7.8	0.5			
Dibenzothiophene				19.9	46.0	32.3	1.9		
3-Methylindole				0.8	24.6	50.2	23.1	1.3	
2,5-Dimethylindole		1.4	36.1	49.2	13.0	0.3			
2,3,5-Trimethylindole	0.9	28.9	56.0	13.1	1.1				

[a]Number of 8-ml fractions collected from point of injection on four-column gel system.

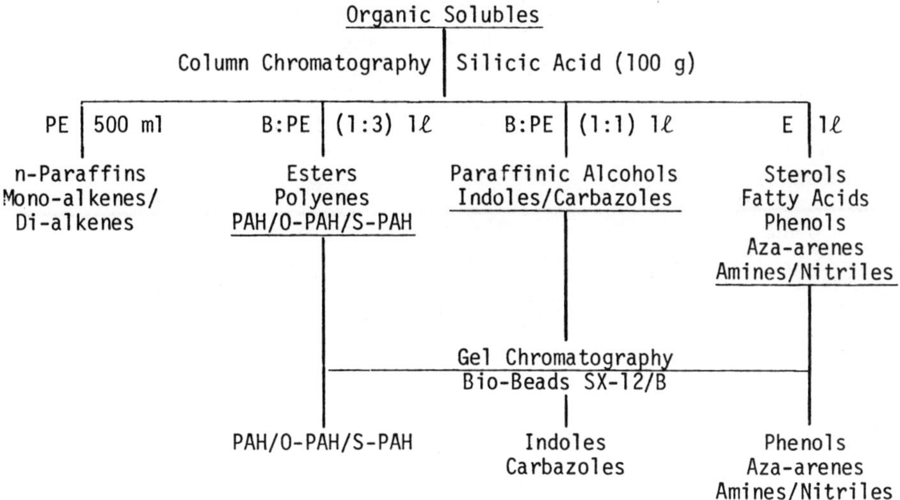

Figure 5. Chromatographic isolation of polyaromatics (PE-petroleum ether, B-benzene, E-ethyl ether).

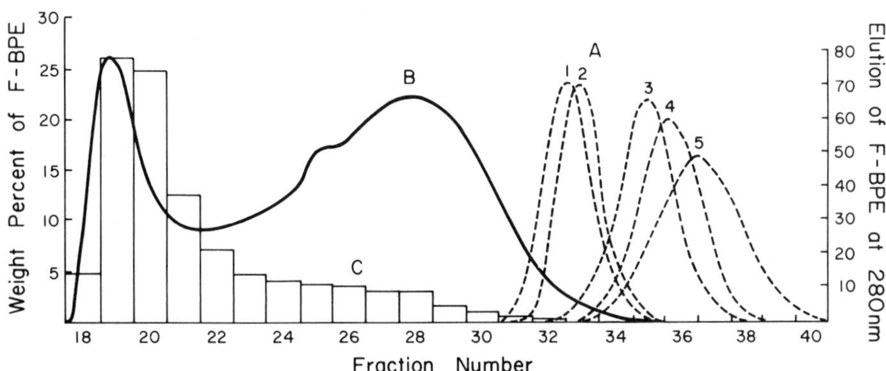

Figure 6. Gel chromatographic isolation (3-column Bio-Beads SX-12 system) of cigarette smoke PAH eluted from silicic acid with 25% benzene-petroleum ether. (A-elution curves of PAH standards; 1-naphthalene; 2-fluoranthene; 3-pyrene; 4-BeP; 5-dibenz(a,h)pyrene; B-UV absorbance curve; C-weight percent distribution).

Table 4 PAH and Heteroatom PAH Chromatography on Silicic Acid

	% Distribution				
	Eluting Solvent[b]				
Compound	B/PE (1:3)	B/PE (1:1)	B	B/E (1:1)	E
Naphthalene-^{14}C	100[c]	-	-	-	-
Anthracene-^{14}C	100[c]	-	-	-	-
Benzo(a)pyrene-^{14}C	100[c]	-	-	-	-
Dibenzofuran	100	-	-	-	-
Benzonaphthofuran	100	-	-	-	-
Dibenzothiophene	98	1.5	-	-	-
Indole	-	98	2	-	-
Indole-^{14}C	0.08[c]	68.9[c]	2.1[c]	15.3[c]	8.6[c]
3-Methylindole (Skatole)	-	100	-	-	-
5-Methylindole	-	100	-	-	-
5-Ethylindole	-	99	1	-	-
2,3-Dimethylindole	-	100	-	-	-
Carbazole-^{14}C	-	99[c]	1[c]	-	-
2-Methylcarbazole	-	100	-	-	-
2,3-Benzocarbazole	10	90	-	-	-
1-Methylindole	48	52	-	-	-
1,2-Dimethylindole	-	100	-	-	-
9-Ethylcarbazole	95	5	-	-	-
Dibenzo(c,g)carbazole	-	9	91	-	-
Quinoline	-	-	-	79	21
6-Methylquinoline	-	-	-	83	17
Acridine	-	-	-	100	-
Benz(c)acridine	-	-	69	31	-
Phenanthridine	-	-	-	100	-
1-Azapyrene	-	-	-	100	-

[a]Determined by GC.

[b]PE-petroleum ether, B-benzene, E-ethyl ether.

[c]Percent distribution of radioactivity.

The purification obtained by gel chromatography can be illustrated with the N-PAH fraction obtained from the SA separation. The GC profile of this crude aza-arene fraction from CSC is shown in the top portion of Figure 7. Quinoline and isoquinoline can be barely detected in the envelope of other components. However, after refinement of this fraction by gel chromatography, a clean aza-arene GC profile was obtained (bottom portion of Figure 7).

PAH QUANTITATION

The quantitative aspects of the PAH chromatographic isolation method are given in Table 5. For cigarette smoke condensate,

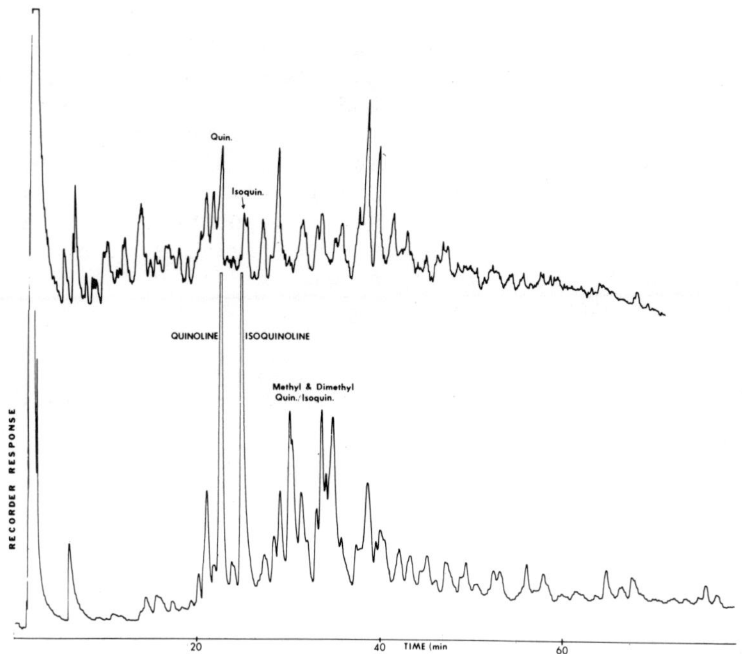

Figure 7. GC profiles of silicic acid aza-arene fraction (top) and of gel chromatography-purified isolate (bottom) of cigarette smoke (15 ft Dexsil 300 GC column).

naphthalene, anthracene, BaP, and carbazole were quantitatively recovered, however, some indole was lost in the SA step. When the procedure was applied to coal tar pitch and crude oil samples, we obtained similar recoveries of ^{14}C BaP.

For true quantitation, the recovery of the environmental aromatics from the trapping system must also be determined. Often standards are added to the trapping medium and the recoveries are used to quantitate the PAH in the environmental sample. We found that (Table 6) 500 ml of methylene chloride quantitatively recovered BaP from clean fiber filters of the type used to trap samples of air pollutants. However, recovery of the standards must also be determined in the presence of the trapped environmental sample. When a sample of ^{14}C-BaP was placed on a glass fiber filter previously used to collect a sample of polluted air, the recovery of the radioactive BaP was less than 65%. Thus, when extraction recovery is determined, standards should always be added to the environmental sample.

For quantitative analysis of PAH, the Bio-Bead gel system was standardized with trimethylnaphthalene, the earliest eluted commercially available PAH, and with fluoranthene, the earliest eluted major parent PAH (Severson et al, 1976). In order to quantitate the major parent compounds in complex PAH mixtures, it is necessary to separate them from the multialkylated PAH. This was accomplished by collecting gel cuts GF-B and GF-C (Figure 2). GF-C contained the major parent PAH and their monomethyl derivatives, while GF-B contained the complex multialkylated compounds (Figure 8). Obviously, an analysis of the total isolate, without prior separation into the two gel cuts, would not have given quantitative results for the parent PAH. Also the identification of all of the multialkylated components would have been impossible. (Hexatriacontaine, the C_{36} hydrocarbon, was used as an internal standard for GC quantitation.)

Table 5 Percent Recovery of ^{14}C-Labelled PAH and N-PAH

Fraction	% Recovery/Step				
	Naphthalene	Anthracene[a]	BaP[a]	Indole[a]	Carbazole[a]
Organic Solubles	100.0	100.0	100.0	100.0	100.0
Silicic Acid	100.0[b] (97.0)[a]	98.0	98.0	65.7	99.0
Gel Isolate	100.0[b] (95.0)[a]	100.0	97.0	98.3	99.0
	Total Recovery[c]				
	93.0	98.0	95.0	64.6	98.5

[a]Recovery after solvent reduction.

[b]Recovery before solvent reduction required to carry sample into next step.

[c]Based on ^{14}C added to organic solubles.

Table 6 Recovery of ^{14}C-BaP from Glass Fiber Filter[a]

CH_2Cl_2 (ml)	Filter Blank	Filter Plus Air Sample
100	52.2	
200	84.1	
300	94.6	58.0
400	97.8	
500	99.0	63.2

[a]Based on ^{14}C-BaP placed onto blank filter and filter containing trapped air sample pollutants.

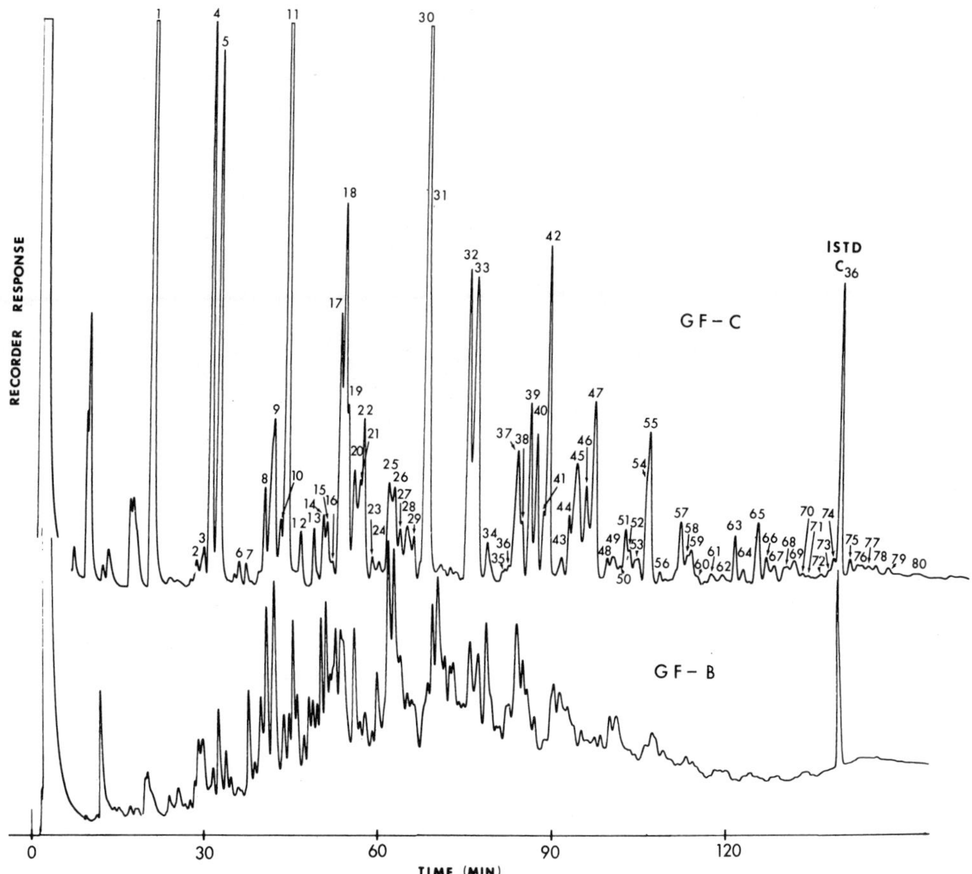

Figure 8. Gas chromatograms of PAH isolate GF-B and GF-C from 700°C tobacco pyrolyzate on 15 ft Dexsil 300 GC column (see Table 2 for peak identifications).

Our PAH isolation-analysis method was also applied to other samples. The gas chromatogram of the PAH isolated from a sample of coal-tar pitch is shown in Figure 9. The absence of interfering material and the lack of multialkyl PAH did not necessitate the gel chromatographic separation. However, the complex PAH isolates from crude oil samples required the gel refinement before GC analysis. The capillary column GC profiles of the parent and multialkylated PAH isolates from a Venezuelan asphalt base crude oil sample is shown in Figure 10. This example shows that the procedure also isolated S-PAH along with the PAH. Even though this sample contained high levels of the parent and alkylbenzothiophenes and 9-thiabenzofluorenes, they did not interfere with the analysis of the major parent PAH or their monomethyl derivatives.

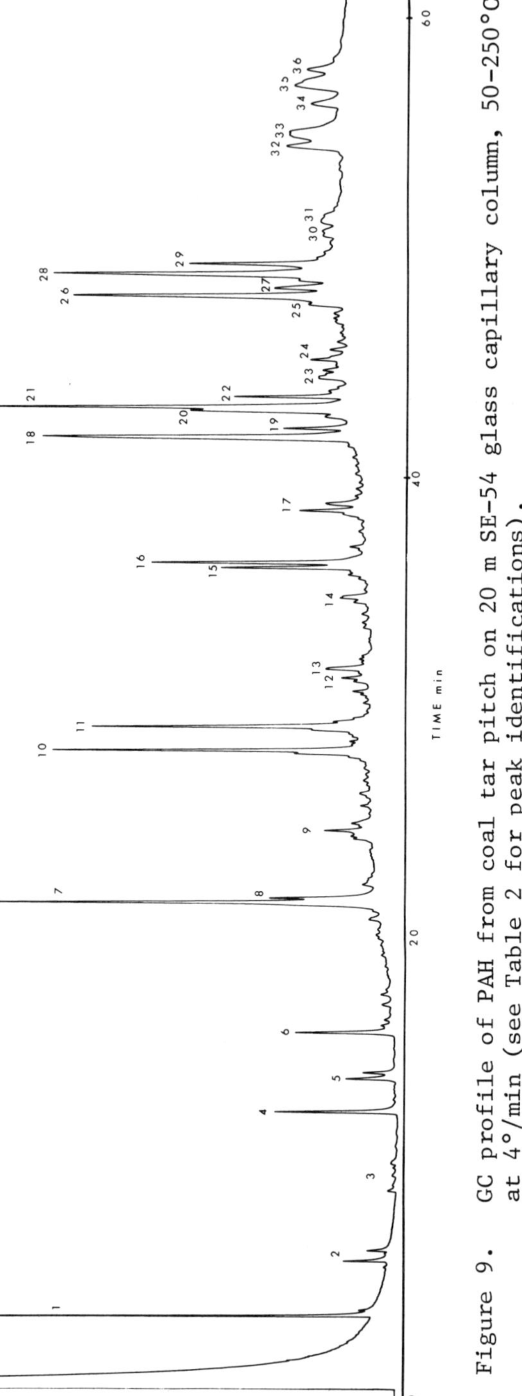

Figure 9. GC profile of PAH from coal tar pitch on 20 m SE-54 glass capillary column, 50-250°C at 4°/min (see Table 2 for peak identifications).

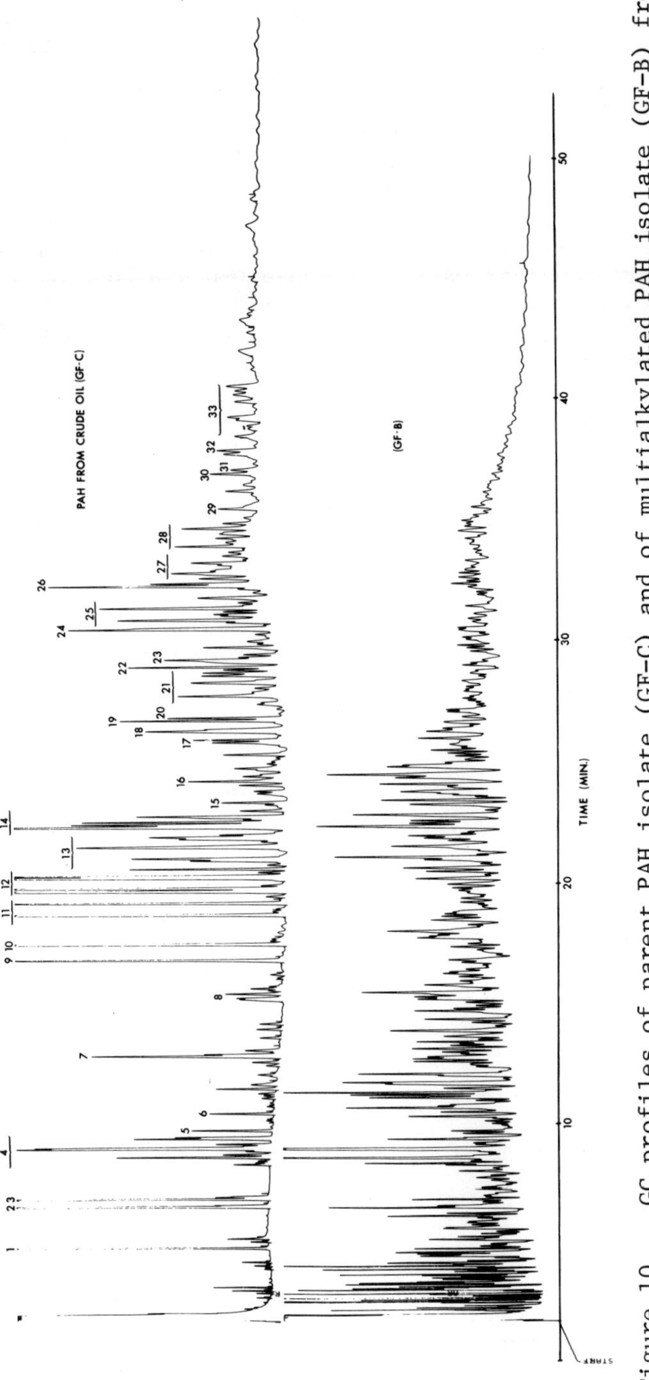

Figure 10. GC profiles of parent PAH isolate (GF-C) and of multialkylated PAH isolate (GF-B) from crude oil sample on 20 m SE-54 glass capillary column, 50-250°C at 4°/min (see Table 2 for peak identifications).

It is evident that in order to profile and quantitate the parent and monomethyl derivatives (upper chromatogram of Figure 10), the PAH in this crude oil sample must be separated by gel filtration into the two cuts.

IDENTIFICATION OF PAH

The complexity of PAH mixtures, like those isolated from petroleum, make identification of individual components difficult. The behaviour of polyaromatics in the Bio-Beads/benzene system (where increasing numbers of aromatic rings in a molecule increase its elution volume, and increasing numbers of alkyl substituents on the PAH moiety decrease their elution volume) is a great asset in separating and identifying components in complex aromatic mixtures. Thus, dimethyl PAH tend to elute before monomethyl compounds which in turn eluted before the parent PAH. This effect concentrated the differently alkylated PAH in different gel fractions and facilitated subsequent identification by GC, GC-MS, and UV.

The general elution order of PAH in various chromatographic systems is shown in Figure 11. In the Bio-Beads/benzene system, alkylated PAH are eluted before the parent PAH. Thus, the combination of these systems, with gel filtration followed by preparative GC and then high-pressure liquid chromatography (HPLC), generally allows the characterization of the complete family of PAH. For complex PAH mixtures, identifications can best be made by the analysis of each gel fraction (Snook et al, 1976). GC-MS is primarily used for the detection of molecular ions of PAH and fragmentations associated with their alkyl derivatives. Since many PAH can have the same molecular weight and essentially identical GC retention data, MS data alone do not unambiguously characterize a GC peak. We found preparative GC cuts can be further separated by HPLC to give collected fractions, corresponding to HPLC peaks, that can be analyzed by UV. Thus, UV and MS data combined with GC retention data, allow the unambiguous assignment of PAH structures. In this manner, we were able to identify over 900 PAH in cigarette smoke. (Snook et al, 1976; Snook et al, 1977; Snook et al, 1978).

This step-wise method of separation and analysis, involving silicic acid, gel, and gas chromatography, can be briefly illustrated by the identifications of the PAH of cigarette smoke. The gas chromatograms of gel fractions 45 and 47 of total cigarette smoke PAH are shown in Figure 12. The later elution of the larger ring compounds and the operation of the methyl effect of the gels is apparent on inspection. In GF-47, the monomethylpyrenes have almost completely disappeared. A decrease in the relative amounts of the alkylbenzopyrenes is also observed. In GF-47 the dibenzoanthracene, indenopyrene, benzoperylene, and ananthrene peaks were almost completely free of the multimethylbenzopyrenes.

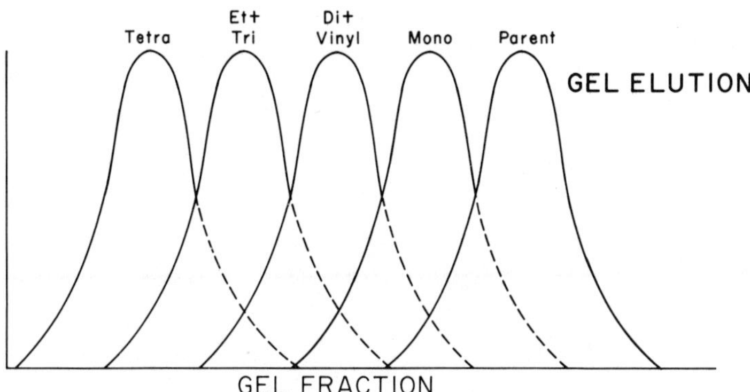

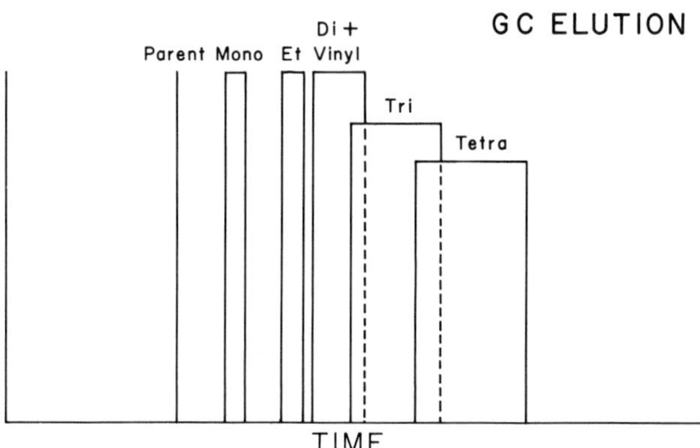

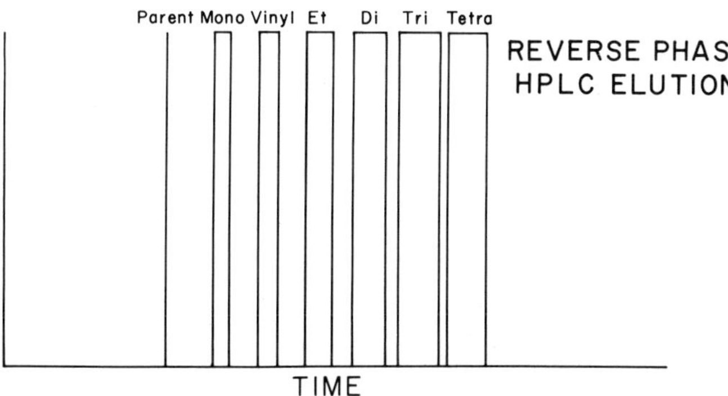

Figure 11. Elution order of PAH in various chromatographic systems.

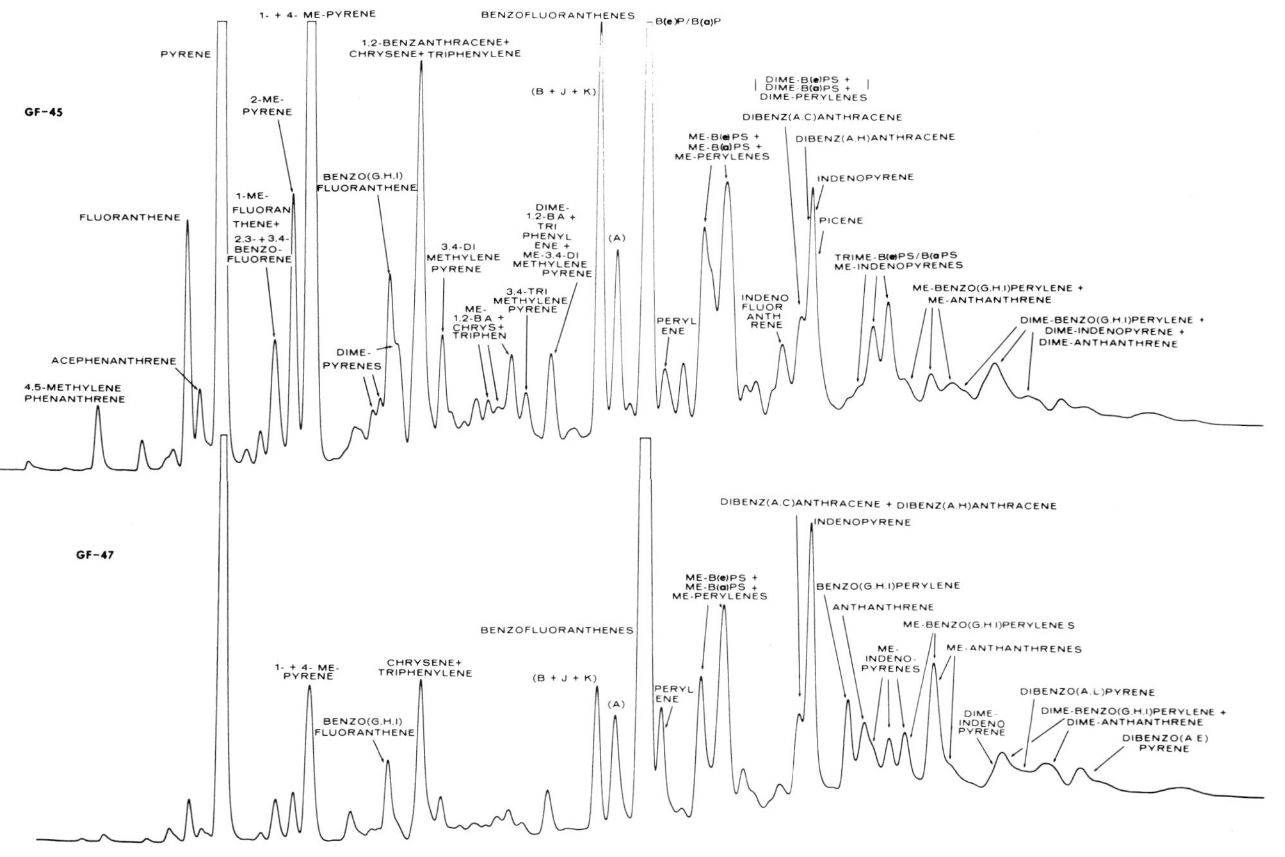

Figure 12. GC profiles of gel fractions 45 and 47 from cigarette smoke on a 15 ft Dexsil 300 GC column.

Sometimes, GC analysis of individual gel fractions was still not sufficient to identify the constituents. For those cases, HPLC was required. For example, the complete characterization of the indenopyrene peak was only possible by the use of HPLC on the preparative GC cut of the peak. HPLC separation results of the cut corresponding to the upslope portion of the indenopyrene GC peak in Tube 45 were in good agreement with GC-MS data and showed the presence of a number of components (Figure 13). UV and MS data on the various collected HPLC peaks allowed the identifications shown. The HPLC separation of the preparative GC cut corresponding to the backslope of the indenopyrene peak showed the presence of indenopyrene, primarily. (Snook et al, 1977).

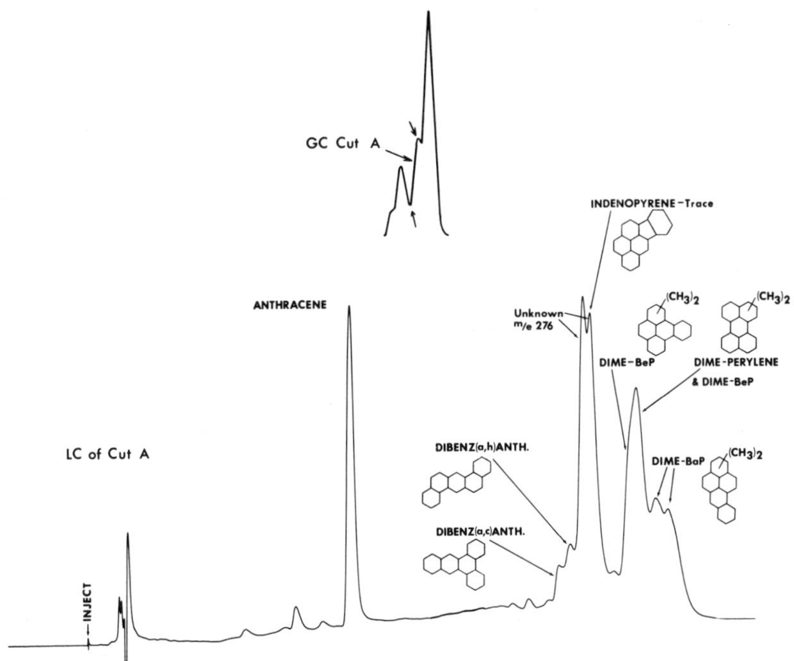

Figure 13. HPLC of the fraction corresponding to front slope of the indenopyrene peak in Gel Fraction 45 (25 cm ODS column, 65 to 85% MeOH-H_2O, anthracene added as retention standard).

The described methodology was developed to quantitatively isolate PAH and heteroatom PAH from complex mixtures for qualitative, quantitative, and bioassay studies. Most of the data presented were obtained in the analyses of cigarette smoke. However, since we have also successfully applied the methodology to coal tar pitch and crude oil samples, it should therefore be equally applicable to aquatic or other environmental samples.

REFERENCES

Dong, M., D.C. Locke, D. Hoffmann. 1977. J. Chromatogr. Sci. 15, 32.

Lao, R.C., R.S. Thomas, and L. Dubois. 1975. Anal. Chem., 45, 908.

McKay, J.R., J.H. Weber, and D.R. Latham. 1976. Anal. Chem., 48, 891.

Rosen, A.A., and F.M. Middleton. 1955. Anal. Chem., 27, 790.

Severson, R.F., M.E. Snook. F.J. Akin, and O.T. Chortyk. 1978. Carcinogenesis, Vol. 3, P.W. Jones and R.I. Freudenthal, Raven Press, New York, p.203.

Severson, R.F., M.E. Snook, R.F. Arrendale, and O.T. Chortyk. 1976. Anal. Chem., 48, 1866.

Sharley, A.F., Jr. 1976. Carcinogenesis, Vol. 1. R.I. Freudenthal and P.W. Jones, Raven Press, New York, p.341.

Snook, M.E., W.J. Chamberlain, R.F. Severson, and O.T. Chortyk. 1975. Anal. Chem., 47, 1155.

Snook, M.E. 1976. Anal. Chim. Acta, 81, 423.

Snook, M.E. 1978. Anal Chim. Acta, in press.

Snook, M.E., R.F. Arrendale, H.C. Higman, and O.T. Chortyk. 1978. Anal. Chem., 50, 88.

Snook, M.E., R.F. Severson, H.C. Higman, R.F. Arrendale, and O.T. Chortyk. 1976. Beitr. Tabakforsch., 8, 250.

Snook, M.E., R.F. Severson, R.F. Arrendale, H.C. Higman, and O.T. Chortyk. 1977. Beitr. Tabakforsch., 9, 79.

Snook, M.E., R.F. Severson, R.F. Arrendale, H.C. Higman, and O.T. Chortyk. 1978. Beitr. Tabakforsch., in press.

BENZO(A)PYRENE IN THE MARINE ENVIRONMENT: ANALYTICAL TECHNIQUES AND RESULTS

Bruce P. Dunn

Dept.of Medical Genetics, University of British Columbia

Cancer Research Centre, 601 W. 10th Ave., Vancouver, B.C.

INTRODUCTION

The class of compounds known as polycyclic aromatic hydrocarbons (PAH) includes a number of potent carcinogens. Knowledge of the origin, distribution, and fate of PAH in the aquatic environment requires sensitive and reliable procedures for their measurement. Unfortunately, measurement of PAH in aquatic samples is frequently more difficult than analysis of these compounds in other samples such as air pollution particulates, cigarette smoke condensate, or coal tar or petroleum products. Levels of individual compounds are often low, and they may be intimately mixed with a complex organic matrix, such as the tissues of marine organisms or in the organic material of sediments.

We have been investigating the occurrence and distribution of PAH in the marine environment, using a single carcinogen, benzo(a) pyrene (B(a)P), as an indicator compound for this class. (Dunn and Stich 1975, 1976a and 1976b, Dunn and Young 1976). This paper describes the methodology that we have used for analyzing over 600 samples of water, tissues of marine organisms, and sediments for benzo(a) pyrene. In addition, results of the analysis of sediments from different areas are presented, as well as results of biochemical determinations of the levels of aryl hydrocarbon hydroxylase in the livers of fish living in areas with different degrees of carcinogen contamination.

METHODS

Tissue and sediment samples were digested in alcoholic KOH followed by addition of water and partitioning of the hydrocarbons into iso-octane. Hydrocarbons were extracted from water samples directly into iso-octane. Interfering materials were removed by column chromatography on Florisil, followed by selective extraction of PAH from iso-octane into DMSO. PAH were recovered from DMSO by addition of water and back-extraction into iso-octane. B(a)P was separated from other PAH by thin-layer chromatography on cellulose acetate. The compound was eluted from the adsorbent and measured fluorimetrically in hexadecane using an artificial baseline technique. An aliquot of radioactively labelled B(a)P was added to each sample before extraction, and recoveries of the compound through the extraction and purification procedures were determined by measuring the amount of radioactive label remaining in the final fluorimetry sample. All measurements of B(a)P were corrected for losses during extraction and purification steps. The methods used are more fully described in Dunn (1976).

In contrast to pure DMSO which is practically odourless (Windholz 1976), commercial "spectral" grade DMSO from different suppliers generally had a foul odour. The impurity or impurities causing this odour was a foul smelling, non-volatile, non water-soluble oily liquid which was present in amounts of up to 50 μl per 10 ml of some batches of DMSO. This unidentified impurity was extracted from DMSO/water mixes by iso-octane during sample purification procedures, and thus contaminated extracts. The impurity however did not interfere with thin-layer chromatography and thus commercial grades of DMSO were used as received from the supplier. For purposes requiring uncontaminated PAH extracts, batches of DMSO were purified by adding an equal volume of water, and extracting impurities with two volumes of iso-octane. The DMSO-water mix was then rotary evaporated at aspirator vacuum and a bath temperature of 90° to remove the water, along with approximately 20% of the DMSO (which tends to co-distill). The amount of residual water in the DMSO can be checked by cooling a small sample and recording the freezing point. 5% water depresses the freezing point from 18.3° to approximately 10° (Mellan, 1977). DMSO containing 1 or 2% water is adequate for extraction of PAH from iso-octane. Alternate methods of purification of DMSO involving chemical pretreatments followed by fractional distillation under reduced pressure (Riddick and Bunger, 1970) have not been investigated, but may also be useful in the preparation of adequately pure DMSO.

When comparing levels of B(a)P found in sediments from the same general area, there was a trend for higher levels of B(a)P to occur in "mud" and "silt" samples, compared with "sand" samples. To facilitate comparison of B(a)P levels in different sediment types,

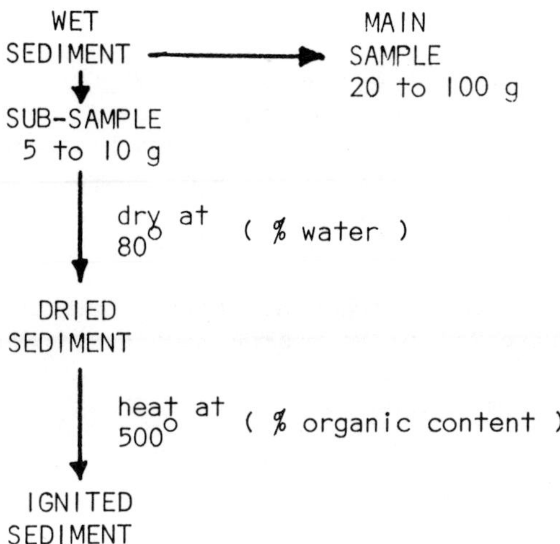

Figure 1. Treatment of sediment samples for determination of organic content by ashing at 500°.

the approximate organic content of samples was determined by heating them to 500° for 24 hours (Figure 1). The weight loss during this procedure gives a somewhat arbitrary but very easily measured figure indicating the amount of organic material present in the sediment. Sand samples typically lose less than 1% weight during ashing, while mud samples lose from 5 to 15% and samples containing decaying vegetation lose up to 60%. When a series of samples from a given area are examined, the levels of B(a)P expressed as μg/kg organic content were generally much less variable than levels expressed on the basis of wet or dry weight.

COMMENTS ON METHODS

Radioactive Internal Standard

^{3}H-Benzo(a)pyrene is used as an internal standard in preference to ^{14}C-B(a)P because of its higher specific activity. At the specific activity of commercial available ^{14}C-B(a)P, the labelled internal standard contributes substantially to the amount of B(a)P present in the sample (adding 1000 dpm of ^{14}C-B(a)P with a specific activity of 20 mCi/mmol adds approximately 5 ng of B(a)P to the sample, an amount which may be more than the pre-existing level of contamination of the sample). The tritium labelled material has a specific activity approximately 3 orders of magnitude higher, and contributes correspondingly less to the amount of carcinogen in the sample.

Recoveries

Recoveries through the procedure were generally of the order of 60 to 80%, with most losses occurring during Florisil chromatography and during thin-layer chromatography. Recoveries were sometimes lower with particular types of samples, especially those containing only small amounts of extractable organic material (e.g. sandy sediments). In such cases, losses were attributable to irreversible adsorption onto Florisil, and could be reduced by using smaller columns and more highly deactivated adsorbent. The variable recoveries encountered in processing samples of different origins emphasize the need for the use of internal standards when measruing PAH from environmental samples.

Alkaline Digestion and Florisil Chromatography

For marine organisms, alkaline digestion followed by partitioning of hydrocarbons into an aliphatic solvent is faster and simpler than Soxhlet extraction, and provides good extraction. The procedure is also adaptable to sediments (Dunn, 1976) allowing one type of extraction procedure to serve for all samples.

Florisil is used in preference to silica gel or alumina as it does not appear to catalyze the removal of tritium from the generally labelled ^{3}H B(a)P used as an internal standard. Careful preparation and deactivation are necessary to minimize chemisorption of PAH (Dunn, 1976), and batches of Florisil should be pretested before being used for samples. Prolonged heating at 250° during Florisil preparation accentuates chemisorption problems - we now routinely activate for 6 hours (vs. 18 hours in Dunn, 1976).

DMSO Extraction

This procedure has proved essential for adequate sample clean-up prior to thin-layer chromatography (Dunn, 1976). Recent investigations have been made in our laboratory on the use of high pressure liquid chromatography and UV detection for the separation and quantitation of PAH in extracts of aquatic samples. For this procedure Florisil chromatography and DMSO extraction were not sufficient to give usable PAH fractions - it was necessary to follow these clean-up steps with chromatography on Sephadex LH-20 (Giger and Blumer, 1974). Experience with seafood samples has suggested that DMSO extraction and chromatography on Sephadex LH-20 are complementary procedures - each removes non-PAH UV absorbing materials not removed by the other.

Thin Layer Chromatography

Chromatography on cellulose acetate has only a limited ability to resolve the complex mixture of PAH from aquatic samples, but is excellent as a specialized system capable of isolating benzo(a) pyrene from other PAH. The compound with the closest R_f value is benzo(b)fluoranthene, which barely separates from benzo(a)pyrene during normal chromatography (Figure 2, left side). Attempts to improve the separation between these two compounds by altering the solvent composition to give larger R_f values were unsuccessful. However, it was found that if the tic plate was placed in a modified chromatography tank so that the top 2 cm of the plate projected through an opening into the open air (Figure 2, right side), migration of the benzo(a)pyrene and benzo(b)fluoranthene bands would continue due to evaporation of the developing solvent at the exposed top of the plate. Under extended chromatography using the modified tank, benzo(a)pyrene and benzo(b)fluoranthene were separated in a 5 hour run.

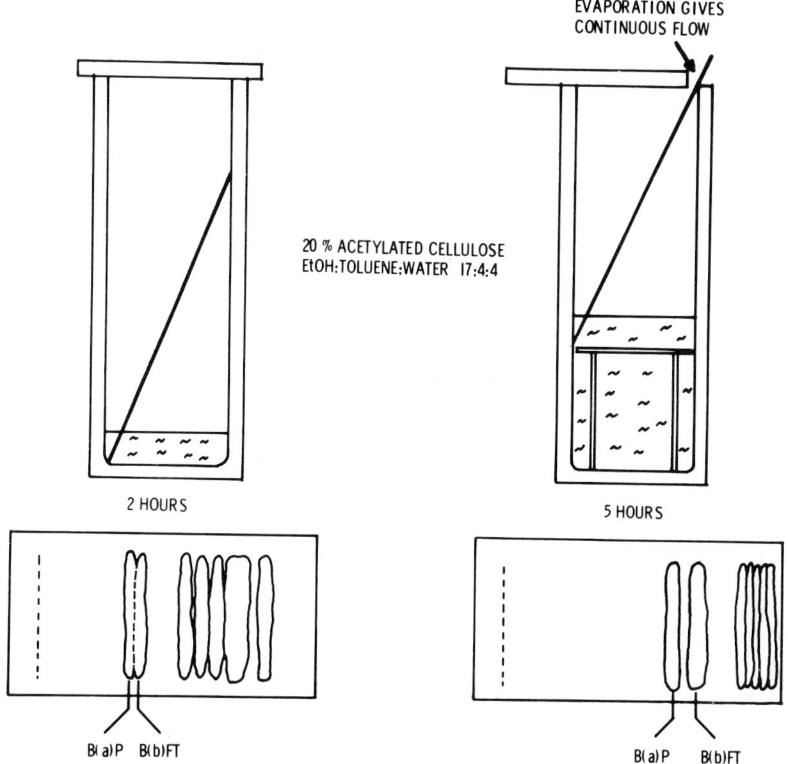

Figure 2. Chromatography of PAH on Cellulose Acetate Thin Layer Plates in Normal and Modified Chromatography Tanks

Fluorimetric Measurement of B(a)P in Hexadecane

Benzo(a)pyrene is eluted from the chromatography adsorbent with hot methanol - recoveries from the adsorbent are not always quantitative, although this does not affect the determination of B(a)P because of the use of an internal standard. Attempts to utilize less polar solvents for better recoveries have not been successful - solvents such as acetone make the cellulose acetate gummy, clogging the filter used to separate the adsorbent from the eluting solvent. Even less polar, non-water miscible solvents fail to properly wet the cellulose acetate, which retains some water from the chromatography solvent.

The methanol extract is evaporated, and the B(a)P transferred into hexadecane for fluorimetric and radioactivity determinations. This solvent is used as it is non-volatile (allowing samples to be stored without volume changes due to evaporation), and is not a strong quencher of the liquid scintillation process used to determine the amount of remaining radioactive label. A complete fluorescence spectrum of B(a)P is recorded, and the amount of B(a)P calculated using the height of one of the peaks in the B(a)P emission spectrum over an artificial baseline drawn between two minima in the spectrum (Dunn, 1976). This procedure minimizes the contribution of background fluorescence arising from samples and not attributable to B(a)P (since B(a)P is the slowest migrating compound on the cellulose acetate thin-layer plates, any streaking of faster migrating compounds leaves small amounts of other compounds in the B(a)P band area).

RESULTS

Benzo(a)pyrene in Marine Sediments

In order to gain some perspective on a world-wide basis as to levels of B(a)P in sediments from polluted and non-polluted areas, shallow water coastal sediments were obtained from a number of different locations. Samples came from 9 harbour and 9 non-harbour sites on the Pacific coast of North America (mainly in British Columbia and Washington state), from the delta of the Mackenzie river and the adjacent Beaufort Sea in northern Canada, from the north western Alaska coastline (Seward peninsula), and from Wilhelmshaven, an oil port in Northern Germany. Sites were classified into three somewhat arbitrary groups according to their potential for man-made pollution: harbour areas, non-harbour inhabited areas (i.e. areas of coastline with human habitation but no direct harbour activities), and non-harbour uninhabited areas.

Figure 3 shows B(a)P levels in individual samples taken from different areas - median values are indicated. It is clearly

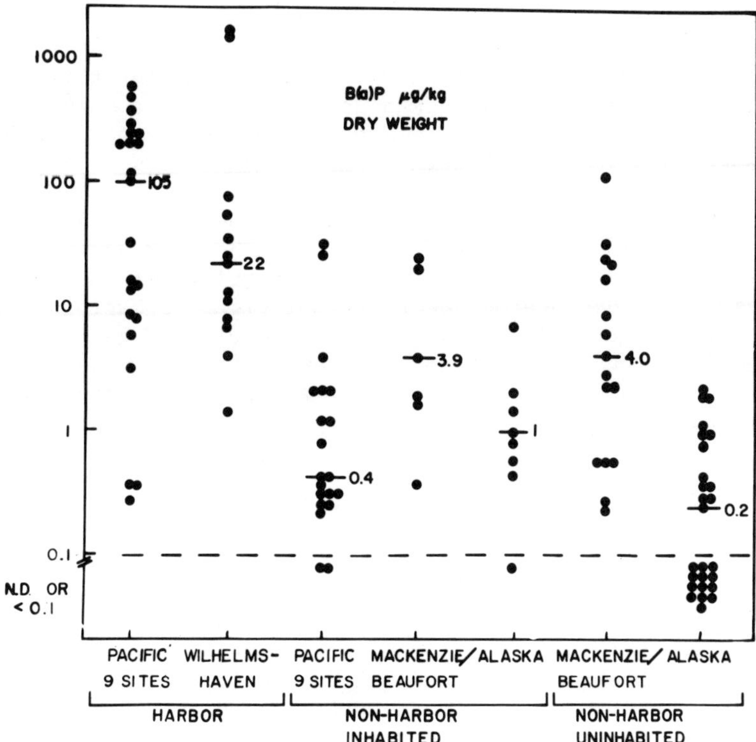

Figure 3. Levels of benzo(a)pyrene in sediments from different locations

evident that there is a general trend towards higher levels of B(a)P in harbour areas, compard with non-harbour or remote areas. For example, the median level of B(a)P in 22 sediment samples taken from 9 Pacific harbour locations was 105 µg/kg, while the median value for samples from 9 non-harbour locations on the Pacific coast was 0.4 µg/kg. The highest levels of B(a)P recorded were in two samples from a particularly contaminated part of Wilhelmshaven oil harbour (approximately 1500 µg/kg).

Samples taken from both inhabited and uninhabited sites in the Mackenzie Delta and Beaufort Sea coastline in northern Canada (Mackenzie/Beaufort in Figure 3) contained median levels of carcinogen approximately 10 times higher than levels found in non-harbour areas on the Pacific coast (3.9 and 4.0 µg/kg, vs. 0.4 µg/kg). The substantial carcinogen contamination of these sediments is not a universal feature of sediments from high latitudes - samples taken from non-inhabited areas in the Seward peninsula areas of Alaska contained levels of carcinogen more comparable to non-harbour areas on the Pacific coast (median value of 0.2 µg/kg for Alaskan samples compared with a median of 0.4 μg/kg for samples from 9 Pacific sites). Thirteen out of 26 samples contained either no detectable B(a)P, or an indicated level of less than 0.1 μg/kg. Against this low background, the contribution of human activity to the amount of B(a)P in the environment is evident - 7 out of 8 samples taken from inhabited Alaskan areas showed clearly measurable amounts of B(a)P.

The high levels of B(a)P in Mackenzie/Beaufort Sea samples compared with Alaska and Pacific Coast non-harbour areas, may reflect the role of the Mackenzie river in transporting oil and tar contaminated silt from hydrocarbon bearing geological formations such as the Athabaska tar sands. Overall, the results suggest that although human activity such as that found in harbours will substantially contaminate sediments, in more remote areas levels of naturally occurring hydrocarbons may dominate the contamination picture.

Aryl Hydrocarbon Hydroxylase in Lemon Sole

Since the data on benzo(a)pyrene levels suggests that there are substantial differences in the carcinogen contamination in different coastal areas, it seems reasonable to ask if these differences are reflected in any biological parameters measurable in organisms living in the areas. The inducible enzyme complex aryl hydrocarbon hydroxylase is involved in the detoxification and excretion of benzo (a)pyrene and related PAH compounds - levels of this enzyme rise on exposure of organisms to PAH and other inducing chemicals. An examination was therefore made of the levels of aryl hydrocarbon hydroxylase (AHH) in the livers of juvenile lemon sole (*Parophrys vetulus*), caught by bottom trawl in coastal Pacific waters. At the same locations, sediment samples were taken for the determination of benzo(a)pyrene. AHH was measured in liver homogenates by the radiometric method of DePierre (1975).

Levels of AHH in populations of fish taken from one location were quite variable. However, in general there was a pronounced trend for there to be higher average levels of AHH in fish taken from areas substantially contaminated by B(a)P (Figure 4). This suggests induction of the AHH enzyme complex by B(a)P and other PAH, or by compounds occurring in company with PAH. It is of interest to note that at the one location where low B(a)P levels in sediments are not correlated with low levels of AHH (data point at AHH = 28, B(a)P = 0.3), the area is contaminated by the effluent from a large pulp mill. Possibly there are non-PAH enzyme inducers in this effluent material.

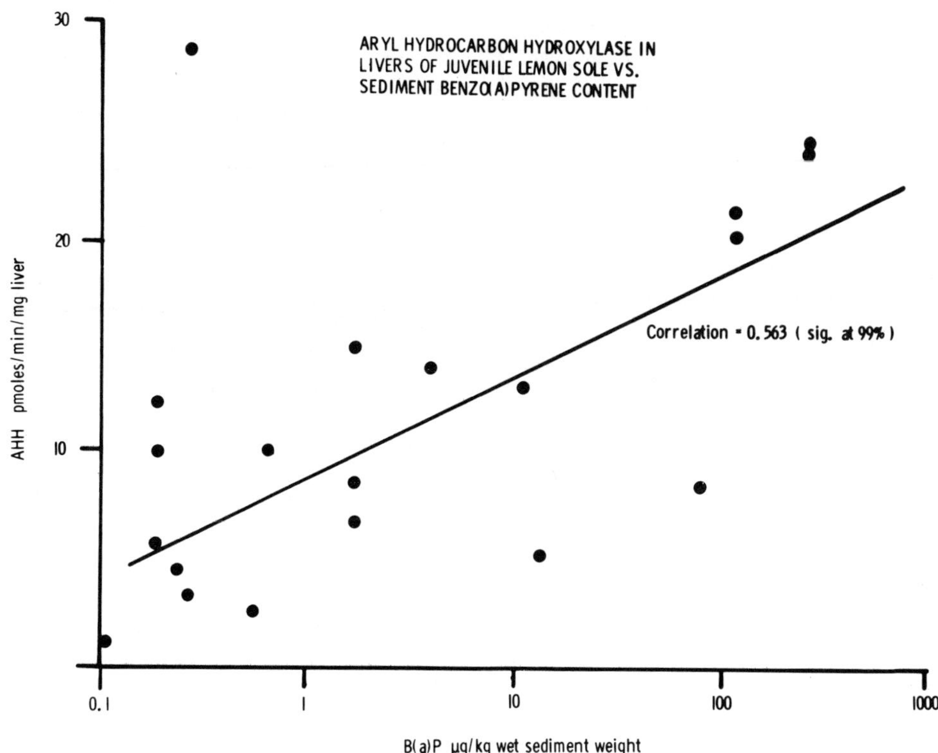

Figure 4. Aryl hydrocarbon hydroxylase in livers of juvenile lemon sole. Individual data points represent the mean of enzyme levels in 5 to 20 fish, plotted against the mean of benzo(a)pyrene levels measured in 2 or more sediment samples from each area.

CONCLUSIONS

Some of the relevant features of the techniques for analysis of benzo(a)pyrene are shown in Figure 5. The techniques are sensitive and reliable, and with the exception of a scintillation counter and fluorimeter (which are required only on a part-time basis and could be shared instruments), do not require expensive instrumentation. On the negative side, the chromatography procedures on cellulose acetate thin-layer plates are labour intensive, and are limited to the measurement of only one PAH compound (benzo (a)pyrene). Current investigations in this laboratory are focusing on the use of high pressure liquid chromatography coupled with the extraction and clean-up steps described as a procedure for quantitating a wider range of PAH carcinogens.

BENZO(A)PYRENE ANALYSIS BY ALKALINE DIGESTION, THIN-LAYER CHROMATOGRAPHY, AND FLUORIMETRY

OPERATOR TIME	: 3 to 4 hours including reagent preparation
REAGENT COSTS	: $ 10
SENSITIVITY	; about 0.1 μg/kg for 25 g samples
PRECISION	: ± 6% (S. D. of analysis of replicate samples)
SENSITIVITY TO VARIABLE RECOVERIES:	none (internal standard)
SPECIAL EQUIPMENT	: scanning grating fluorimeter, liquid scintillation counter (2 hours/week each)
DISADVANTAGES	: limited to one compound, t. l. c. procedure labour-intensive

Figure 5. Characteristics of the assay procedure

ACKNOWLEDGEMENTS

This work was supported in part under National Health Research and Development Project number 610-1138-40 of the Department of National Health and Welfare, Canada. The author also wishes to acknowledge the receipt of a Research Scholar award from Health and Welfare Canada.

REFERENCES

DePierre, J.W., M.S. Moron, K.A.M. Johannesen, and L. Ernster. 1975. A rapid, sensitive, and convenient radioactive assay for benzpyrene monoxygenase. Anal. Biochem. 63: 470.

Dunn, B.P. 1976. Techniques for determination of benzo(a)pyrene in marine organisms and sediments. Environ. Sci. Technol. 10: 1018.

Dunn, B.P., and H.F. Stich. 1975. Use of mussels in estimating benzo(a)pyrene contamination of the marine environment. Proc. Soc. Exptl. Biol. Med. 150: 49.

Dunn, B.P., and H.F. Stich. 1976a. Release of the carcinogen benzo(a)pyrene from environmentally contaminated mussels. Bull. Env. Contam. Toxicol. 15: 398.

Dunn, B.P., and H.F. Stich. 1976b. Monitoring procedures for chemical carcinogens in coastal waters. J. Fisheries Res. Board Canada 33: 2040.

Dunn, B.P., and D.R. Young. 1976. Baseline levels of benzo(a) pyrene in southern California mussels. Mar. Poll. Bull. 7: 231.

Giger, W., and M. Blumer. 1974. Polycyclic aromatic hydrocarbons in the environment; isolation and characterization by chromatography, visible, ultraviolet, and mass spectrometry. Anal. Chem. 46: 1663.

Mellan, I. Industrial Solvents Handbook, 2nd Edition. Noyes Data Corporation, Park Ridge N.J., p.169.

Riddick, J.A., and W.B. Bunger. 1970. Techniques of Chemistry Vol. 11. Organic Solvents. Physical properties and methods of purification, 3d Edition, Wiley-Interscience, New York, p.857.

Windholz, M. 1976. Marck Index, 9th Edition, p.443, entry #3249.

CHROMATOGRAPHIC AND SPECTRAL ANALYSIS OF POLYCYCLIC AROMATIC HYDROCARBONS IN THE AIR AND WATER ENVIRONMENTS

Morris Katz, Takeo Sakuma and Helle Tosine

Centre for Research on Environmental Quality

York University, 4700 Keele Street, Downsview,

Ontario M3J 2R3, Canada

INTRODUCTION

Polynuclear aromatic hydrocarbons (PAH) are widely distributed in the atmospheric and water environments. These compounds occur in diverse sources such as combustion products of fossil fuels, motor vehicle exhaust, fumes from coke ovens, incineration of refuse, liquid effluents from petroleum, petrochemical and many other industrial operations. Airborne particlate matter, which contains PAH, contributes to the pollution of water supplies by sedimentation, impaction and precipitation on land and water surfaces. The sources, reactivity, metabolism and carcinogenic properties of PAH have been reviewed in a monograph by the U.S. National Academy of Sciences (1972). Some PAH compounds have been shown to be potent carcinogens in biological assays but isomers of a specific group may differ greatly in potency. Thus, benzo(a)pyrene is a strong carcinogen, whereas the isomeric benzo(e)pyrene is non-carcinogenic; benzo(b)fluoranthene is carcinogenic but benzo(k) fluoranthene is negative. Similarly, benz(a)anthracene is positively carcinogenic, whereas the isomeric chrysene has been reported as negative or only weakly carcinogenic.

Pierce and Katz (1975a) have found that approximately 70 to 90% of the total PAH content of airborne particulates are associated with particles in the respirable size range of less than 5.0 μm in diameter. Such respirable particles may exert a significant influence on the inhalation health hazards of toxic airborne substances. It has been estimated that about 60 percent or more of all human cancers are due to environmental agents (Train, 1977).

Consequently, it is important to develop effective quantitative separation techniques for the analysis of PAH compounds in air and water samples, including the isolation of the individual components of isomeric groups that differ greatly in toxicity. Reported separation schemes employing thin-layer chromatography (Novotny et al, 1974; Klimisch, 1973), adsorption chromatography and charge-transfer complexation (Giger and Blumer, 1974), paper chromatography (Hluchán et al, 1974) and mass fragmentometry (Van Cauwenberghe and Cautreels, 1976) have met with little success in the separation of isomers.

EXPERIMENTAL

Accurate analytical methods have been developed in this laboratory, (Pierce and Katz, 1975a, 1975b, 1976) for the isolation and identification of PAH in particulate samples collected from the urban atmospheric environment of several cities in Southern Ontario by staff of the Ontario Ministry of the Environment. The methods involve gas (GC) and thin-layer chromatography (TLC) of organic liquid extracts, ultraviolet and fluorescence spectroscopy and mass spectrometry (MS).

Resolution of PAH by Gas Chromatography - Mass Spectrometry

The following five columns were examined for the analysis of PAH compounds by gas chromatography (GC), followed by mass spectrometry (MS) for spectral identification of the resolved compounds:

1. 5.4 m, 4 mm I.D. glass column packed with 1% OV-7 on Chromosorb W-HP (80-100 mesh) developed by Lane, Moe and Katz (1973).

2. 3.6 m, 2 mm I.D. glass column packed with "Aue" treated Carbowax 20 M on Chromosorb W-AW (80-100 mesh).

3. 4.6 m, 2 mm I.D. glass column packed with 2% Dexsil 410 on Chromosorb W-HP (80-100 mesh).

4. 15.2 m, 0.5 mm I.D. stainless steel capillary SCOT column coated with OV-1 (Perkin-Elmer Corp., Norwalk, Connecticut).

5. 3.8 m, 4 mm I.D. glass column packed with 0.7% OV-25 on Chromosorb W-HP (80-100 mesh).

Equipment for recording of mass spectra consisted of a DuPont mass spectrometer, Model 21-491-B, with a glass jet separator, interfaced to a Varian GC, Model 1700. The spectra were recorded on UV sensitive paper. The MS ion source and focusing unit were taken apart for clean-up purposes after every 5 - 10 runs, since viscous samples tended to deposit on the electrical plates and to cause a loss of sensitivity.

When a mixture in benzene solution of 21 PAH standards was injected into each of the above columns, after proper conditioning, only the OV-7 column (No. 1 above) indicated full resolution of the individual components, as shown in Table 1. After 5 years of ues, this column was still able to reproduce its original resolution.

Table 1 Separation of 21 PAH Standards by Glass Column of 1% OV-7 on Chromosorb W

Peak Number	Compound	RRT* (Found)	RRT** (Lane)
1	Fluorene	0.547	0.345
2	9-Fluorenone	0.716	0.578
3	Phenanthrene	0.738	0.612
4	Anthracene	0.749	0.625
5	Acridine	0.776	0.665
6	Carbazole	0.811	0.720
7	Fluoranthene	0.962	0.984
8	Pyrene	1.000	1.000
9	11-H-Benzo(a)fluorene	1.063	1.087
10	11-H-Benzo(b)fluorene	1.075	1.101
11	Benz(a)anthracene	1.218	1.274
12	Chrysene	1.225	1.284
13	Naphthacene	1.314	1.304
14	Impurity	-	-
15	Benzo(k)fluoranthene	1.470	1.551
16	Benzo(e)pyrene	1.540	1.638
17	Benzo(a)pyrene	1.564	1.650
18	Perylene	1.592	1.685
19	3-Methylcholanthrene	1.707	-
20	Dibenz(a,h)anthracene	2.065	2.222
21	Benzo(ghi)perylene	2.185	2.345
22	7-H-Dibenzo(c,g)carbazole	2.279	2.455

*RRT = Retention times relative to pyrene

**RRT = Retention times relative to pyrene, reported by Lane et al.

The column 5, OV-25, was developed for the separation of benzo(a) pyrene diones. This column separated effectively the 1,6-, 3,6- and 6,12-diones of benzo(a)pyrene.

Particulate samples obtained by filtration with a High Volume sampler on the roof of the Ontario Ministry of Environment research laboratory in Toronto were extracted with the organic solvents, hexane, benzene and ethanol. The extracts were placed in a liquid chromatographic column packed with silica gel. The eluents were collected, concentrated by evaporation at room temperature in a stream of nitrogen to a small volume, and injected into a gas chromatograph. Mass spectrometric identification of the products yielded the following compounds. The oxygenated compounds were identified by means of chemical ionization mass spectrometry with isobutane.

<u>In the hexane fraction</u>. Phenanthrene (178), methylphenanthrene (192), 9-cyanoanthracene (203), fluoranthene (202), pyrene (202), 11-H benzo(a)fluorene (216), 11-H benzo(b)fluorene (216), o,m,p-terphenyls (230), benzanthracene (228), chrysene (228), methylbenzanthracene (242), 3-methylchrysene (242), benzo(k)fluoranthene (252), benzo(b)fluoranthene (252), benzo(a)pyrene, benzo(e)pyrene, perylene (252), 3-methylbenz(j)aceanthrylene (266), dibenz(a,h) anthracene (278), benzo(ghi)perylene (276).

<u>In the benzene fraction</u>. Naphthonitrile (153), fluorenone (186), phenanthrene (178), acridine (179), carbazole (167), 3-methylbenzoquinone (193), 2-methylcarbazole (181), anthraquinone (208), fluoranthene (202), methylpyrene (216), o,m,p-terphenyls (230), benz(a)anthracene (228), chrysene (228), naphthacene (228), 7-methylbenz(a)anthracene (242), anthracene-benzene centre ring adduct (254), benzo(k)fluoranthene (252), biphenyl (154), dibenzfurane (168), methylfluorene (180), hexadecahydropyrene, methylbiphenyl, benzo(a)pyrene, benzo(e)pyrene, perylene (252), benzo (ghi)perylene (276).

<u>In the ethanol fraction</u>. 9-fluorenone (180), acridine (179), 7,8-benzquinidine (179), methylbenzquinidine (193), phenanthrene quinone (208).

Quantitative Resolution of PAH by Thin Layer Chromatography (TLC)

In the TLC method of PAH resolution, atmospheric samples of particulate matter were extracted for 8 hours with dichloromethane. The extracts were evaporated at room temperature under a stream of nitrogen and the weighed residue was dissolved in spectrograde toluene. Preliminary separation by TLC was carried out on neutral aluminum oxide plates (Brinkman), using pentane-ether (19:1 v/v) as the mobile phase. The developed chromatograms, consisting of four groups of compounds, were dried, the fluorescent areas removed and the PAH extracted with spectrograde dichloromethane. After evaporation of the solvent under a stream of nitrogen, the residue of each group was dissolved in a specific volume of toluene. Resolution of the individual PAH compounds in each group was accomplished by use of 40% acetylated cellulose plates with n-propanol-acetone-water (2:1:1 v/v/v) as the developing phase. Each relevant spot was removed from the dried plate, dissolved in anhydrous diethyl ether, the ether evaporated under a stream of nitrogen and the dried residue dissolved in spectrograde pentane for fluorescence analysis using excitation and emission spectra. This technique has been applied for the quantitative analysis of benzo(a)pyrene, benzo(e)pyrene, benzo(b)fluoranthene, benzo(k)fluoranthene, perylene, dibenzo(def,mno)chrysene, benzo(ghi)perylene, naphtho(1,2,3,4,def) chrysene, benzo(rst)-pentaphene and dibenzo(b,def)chrysene in samples of airborne particulates, collected in Toronto, Hamilton, Sarnia and Sudbury (Katz, Sakuma and Ho, 1978).

A number of polycyclic quinones derived from corresponding PAH's have also been determined by TLC, spectrophotometry and MS analysis (Pierce and Katz, 1976). Benzo(a)pyrene (BaP) in benzene solution was oxidized on a TLC aluminum plate with 1 ppm ozone by volume, under irradiation with a lamp that simulates the UV spectrum of sunlight (Lane and Katz, 1977). The oxidized material was removed from the plate, dissolved in chloroform, filtered and the liquid concentrated to a small volume at room temperature under a stream of nitrogen. The concentrated solution was spotted on a TLC plate of magnesium hydroxide and the chromatogram developed with chloroform as the mobile phase. The separated quinones were removed, dissolved in chloroform, filtered, concentrated and subjected to MS analysis, using an electron impact of 70 eV and a DS-60 automatic data processing system. This served to identify BaP-1,6-, BaP-3,6- and BaP-6,12 quinone.

Determination of PAH in Water Samples

For the determination of PAH in water, one-litre samples were allowed to percolate through purified XAD-2 resin (Rohm and Haas), contained in a column. The column was washed with several 20 ml portions of deionized water. The adsorbed PAH's were eluted from the resin with diethyl ether and the washings collected. Traces of water in the ether solution were removed by passage through pre-washed sodium sulphate. The ether eluate was then evaporated to 0.5 ml in a rotary evaporator at room temperature. Resolution of the PAH in the organic extract was accomplished by GC analysis, using a 2.5% OV-25 on chromosorb W (80-100 mesh) glass capillary column (0.25 mm x 10 m) mounted in a gas chromatograph equipped with a flame ionization detector and temperature programming.

Samples were also run on a glass packed column of 3% OV-17 on chromosorb W (80-100 mesh) interfaced with a computerized MS to confirm the identity of the various contaminants found in the samples.

RESULTS AND DISCUSSION

PAH and Derivatives in Environmental Samples from Southern Ontario

The combination of GC and MS represents a very powerful technique for the determination of PAH compounds in the environment. In addition to the five GC columns described above, a number of other columns were tested for resolution of PAH in standard mixtures. Most of these columns, with the exception of the OV-7 column, suffered from the defect that isomers of compounds of similar molecular weight could not be resolved, such as benzo(a)pyrene and benzo(e) pyrene; chrysene, benz(a)anthracene and naphthacene, as well as other isomers. However, a glass capillary column (0.3 mm I.D. x 27 m) treated with a 1% solution of benzyl-triphenylphosphonium chloride, then coated with a 5% solution of OV-17 in dichloromethane, showed the base-line separation of the 21 PAH standards listed in Table 1. This column was applied for the separation of contaminants in an extract of waste water from a pulp and paper industry and yielded excellent resolution of phenols and PAH of lower molecular weight.

A glass capillary column (0.25 mm I.D. x 35 m) etched with a 1% solution of KOH in ethanol, deactivated with Carbowax 20 M and coated with OV-101, proved to be an efficient column for the separation of polychlorinated biphenyls (PCB) in a sample of Aroclor 1254. Because of the ubiquitous nature of both PAH and PCB, there is a possibility of analytical interferences being manifested by members of these two classes, if the same GC column is employed.

A solution containing the 21 PAH standards and PCB Aroclor 1260 was prepared and introduced to the OV-7 column. A GC-MS analysis of the mixture indicated that the majority of the PCB, e.g. tetra-, penta- and hexa-chlorobiphenyls, appear in the range of pyrene and benzo(k)fluoranthene. Thus, 11-H-benzofluorenes, benz(a)anthracene, chrysene, naphthacene and similar compounds eluting in this range would be completely masked. This observation emphasizes the need to use combined GC-MS for the analysis of PAH, since misleading results may be produced by GC analysis alone.

A considerable number of samples collected from several urban areas of Southern Ontario were analyzed by GC-MS in order to compile a list of PAH detected in the environment. As shown in Table 2, the number of PAH compounds and derivatives is very large. The

Table 2 List of PAH and Derivatives Found by GC-MS in Various Environmental Samples Collected from Southern Ontario

1. Aceanthrylene
2. Acenaphthene
3. Acridine
4. Anthanthrene
5. Anthracene
6. Benz(a)anthracene
7. Benz(a)anthracenedione(s)
8. Benzindene
9. Benzo(a)carbazole
10. Benzo(b)chrysene
11. Benzo(b)fluoranthene
12. Benzo(ghi)fluoranthene
13. Benzo(j)fluoranthene
14. Benzo(k)fluoranthene
15. 11-H-Benzo(a)fluorene
16. 11-H-Benzo(b)fluorene
17. 11-H-Benzo(c)fluorene
18. Benzo(ghi)perylene
19. Benzo(c)phenanthrene
20. Benzo(a)pyrene
21. Benzo(e)pyrene
22. Benzoquinoline
23. Binaphthyl
24. Carbazole
25. Chrysene
26. Coronene
27. Dibenzacridine(s)
28. Dibenzanthracene(s)
29. Dibenzocarbazole(s)
30. Dibenzofluorene(s)
31. Dibenzofuran
32. Dibenzopyrene(s)
33. Dibenzothiophene
34. Dihydroanthracene (s)
35. Dihydrobenz(a)anthracene(s)
36. Dihydrobenzo(a)carbazole
37. Dihydrobenzofluoranthene(s)
38. Dihydrobenzofluorene(s)
39. Dihydrochrysene(s)
40. Dihydrofluoranthene(s)
41. Dihydrofluorene(s)
42. Dihydrophenanthrene(s)
43. Dihydropyrene(s)
44. Dimethylacenaphthene(s)
45. Dimethylanthracene(s)
46. Dimethylbenz(a)anthracene(s)
47. Dimethylbenzofluoranthene(s)
48. Dimethylbenzopyrene(s)
49. Dimethylchrysene(s)
50. Dimethyldibenzofuran(s)
51. Dimethylphenanthrene(s)
52. Diphenylene sulfide
53. Diphenylpyridine
54. Ethylanthracene(s)
55. Ethylbenz(a)anthracene(s)
56. Ethylphenanthrene
57. Fluoranthene
58. Fluorene
59. Hydroxyfluorene(s)
60. Indane
61. Indene
62. Isoquinoline
63. Methyoxyfluorene
64. Methylacenaphthene(s)
65. Methylanthracene(s)
66. Methylbenz(a)anthracene(s)
67. Methylbenzofluoranthene(a)
68. Methylbenzofluorene(s)
69. Methylbenzophenanthrene(s)
70. Methylbenzopyrene(s)
71. Methylcarbazole(s)
72. Methylfluorene(s)
73. Methylnaphthalene(s)
74. Methylphenanthrene(s)
75. Methylpyrene(s)
76. Naphthacene
77. Naphthalene
78. Naphthathionaphthalene
79. Perylene
80. Phenanthrene
81. Pyrene
82. Quinoline
83. Terphenyl
84. Tetrahydrophenanthrene
85. Tetramethylbiphenyl
86. Trimethylanthracene(s)
87. Trimethylfluoranthene(s)
88. Trimethylphenanthrene(s)
89. Trimethylpyrene(s)

identifications listed in the table are tentative in many cases because final identification cannot be achieved until the spectra of more authentic standards become available. Many of the compounds mentioned in the table are present in the data published by other investigators around the world, suggesting the ubiquitous nature of these environmental contaminants. A number of alkylated and hydrogenated derivatives of PAH are present but the exact positions of substituents on the aromatic rings cannot be determined because of the lack of appropriate standards.

TLC Methods of Analysis

Of the methods of PAH analysis presented here, the most efficient on a quantitative basis is the two-step TLC method involving preliminary separation of PAH in the organic matter by neutral aluminum oxide plates, followed by resolution of individual compounds on 40% acetylated cellulose plates and subsequent analysis by fluorescence excitation and emission spectra. Recoveries of specific PAH compounds have ranged from 81.1 to 94.2 percent in a mixture of 13 standard compounds (Katz, Sakuma, Ho, 1978). This TLC method has been successful in determining, quantitatively, isomers that are incompletely resolved by other methods (Klimish, 1973; Giger and Blumer, 1974; Hluchán et al, 1974).

ACKNOWLEDGEMENT

This work was partially supported by the Ontario Ministry of the Environment.

REFERENCES

Giger, W., M. Blumer, 1974, Anal. Chem., 46, 1663.

Hluchán, E., M. Jenik, E. Malý, 1974, J. Chromatogr., 91, 531.

Katz, M., T. Sakuma, A. Ho, 1978, Environ. Sci. Technol., 12, 909

Klimisch, H.J., 1973, J. Chromatogr., 83, 11.

Lane, D.A., H.K. Moe, M. Katz, 1973, Anal. Chem., 45, 1776.

Lane, D.A., M. Katz, 1977, In Vol. 9, "Advances in Environmental Science and Technology, Fate of Pollutants in the Air and Water Environments", I.H. Suffet, Ed., pp. 137-154, Wiley-Interscience, New York, N.Y.

National Academy of Sciences, 1972, "Particulate Polycyclic Organic Matter", Chap. 2.

Novotny, M., M.L. Lee, K.D. Bartle, 1974, J. Chromatogr. Sci. 12, 606.

Pierce, R.C., M. Katz, 1975a, Anal. Chem., 47, 1743.

Pierce, R.C., M. Katz, 1975b, Environ. Sci. Technol., 9, 347-353.

Pierce, R.C., M. Katz, 1976, Environ. Sci. Technol., 10, 45-51.

Train, Russell E., 1977, "Environmental Cancer", Science, 195, 443.

Van Cauwenberghe, K., and W. Cautreels, 1976, Advances in Mass Spec. Biochem. and Med., Vol. I, Chap. 42, p. 485, Spectrum Publ. Inc.

PRESENCE AND POTENTIAL SIGNIFICANCE OF o-o'-UNSUBSTITUTED PCB ISOMERS AND TRACE AROCLOR® 1248 AND 1254 IMPURITIES

David L. Stalling, James N. Huckine, and J.D. Petty

Columbia National Fishery Research Laboratory,

RR#1, Columbia, Missouri 65201

Determination of the potential impact of PCB contamination on aquatic ecosystems is facilitated by information on the toxicological significance of specific PCB constituents. In support of this approach, recent studies, (Goldstein et al 1977 and Poland and Glover 1977), indicate that physiological activity associated with the technical PCB mixtures may be largely due to the presence of trace levels of PCB isomers having no o-o'-chlorine substitution. Hepatic enzyme induction from exposure to the most active of these isomers may approach that of the chlorinated-dibenzofurans and -dioxins. Separation and isolation techniques for structurally similar components of Aroclor[1/] mixtures are prerequisites for structure-toxicity correlation studies.

Jensen and Sundström (1974) initially used an active carbon for the fractionation of PCBs according to their degree of o-o'-chlorine substitution. We separated many PCB components (Table 1) using columns of an experimental activated carbon, AMOCO PX-21, and an incremental gradient of toluene in cyclohexane (1.5 - 100 v/v%). PCB components eluted primarily as an inverse function of the number of o-o'-chlorines. Within each of the six possible o-o'-chlorine substitution groups, elution volumes generally increased with additonal chlorine substitution.

AMOCO PX-21 carbon was obtained from the AMOCO Research Corporation, P.O. Box 59, 10-A, Chicago, Illinois. Carbon was sieved into fractions >325 mesh and <325 mesh. Particles <325 mesh were used in the preparation of carbon/foam adsorbent and those >325 mesh were used without modification. A Varian 3700 gas chromatograph equipped with ^{63}Ni electron capture detector and 28 meter glass capillary WCOT Apiezon-L (Quandrex Corp., P.O. Box 3881

Table 1 Carbon (PX-21, >325 Mesh) Chromatography of Aroclor 1254[1]

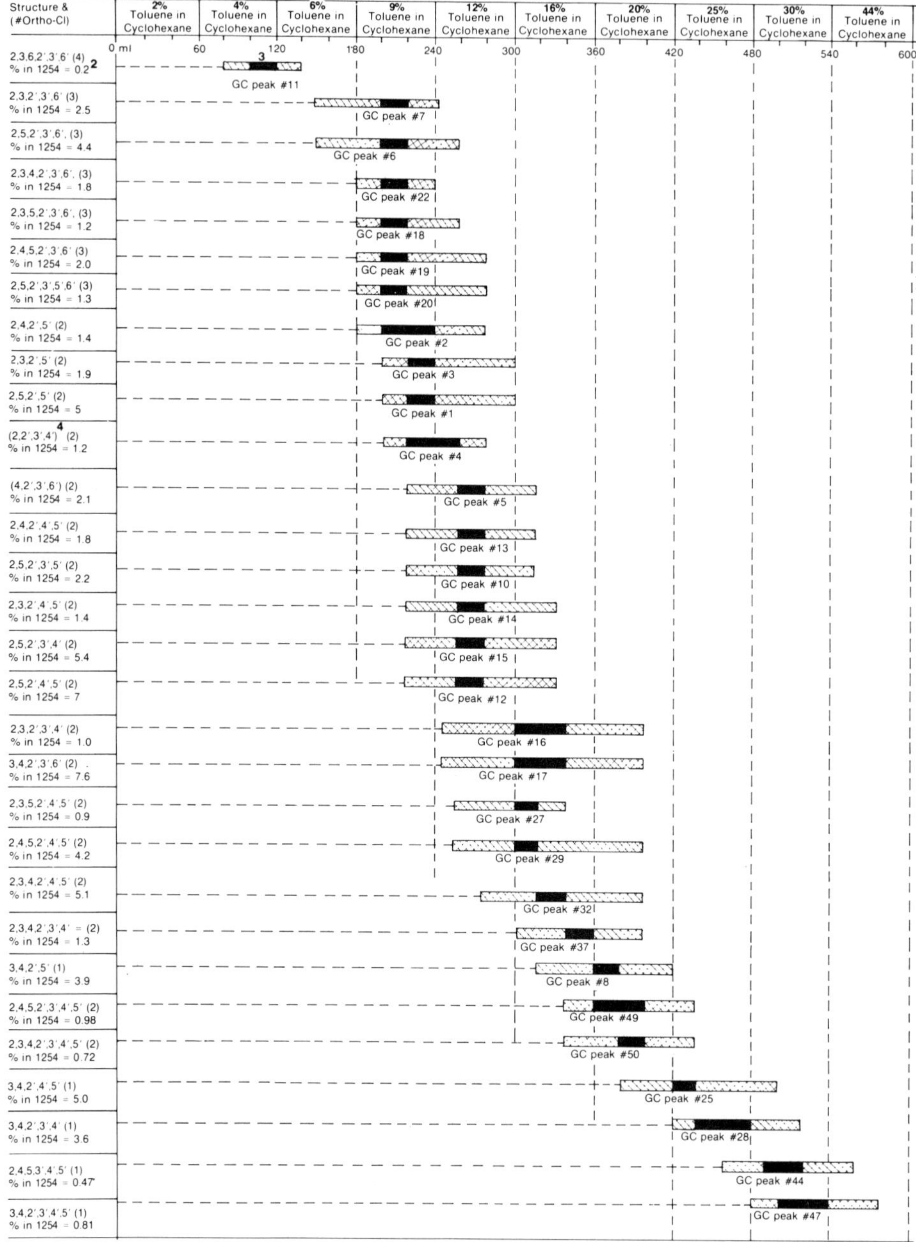

1Stepwise elution of 1 mg of Aroclor 1254 on 1 cm id x 10 cm PX-21 carbon (1.75g) column.
2% of total 1254 residue from Jensen and Sundström, ref. **3** in text.
3Dark portion of bar represents centroid or maximum concentration of PCB component, GC peak numbers from Jensen and Sundstrom, ref. **3** in text.
4Tentative structural assignments by Jensen and Sundström are given in parenthesis.

New Haven, Connecticut) was used for PCB component resolution. Injector split ratio was 1/20 and column temperature was maintained at 240°C. A Finnigan GC-MS 4000 equipped with an INCOS-Rev. 3.0 data system was used to characterize trace impurities in Aroclor 1248. The GC column was 6' x 2 mm id, 3% SE-30 on Chromosorb W HP.

In preparation for planned chronic toxicity and uptake kinetic studies with fish and other aquatic organisms, Aroclor mixtures (200 mg) were fractionated using 1.75 g PX-21 carbon columns (>325 mesh, 1 cm id x 10 cm). Because of the strong affinity of planar PCB constituents for the carbon surface, efficient removal (>95%) of PCBs lacking o-o'-Cl substitution from 1 g quantities of Aroclors was possible. Extraction efficiency of PCBs lacking o-o'-CL substitution was based on spikes of purified Aroclor mixtures.

Analysis of Aroclor 1248 and 1254 fractions for PCBs lacking o-o'-Cl substitution revealed the presence of substantial amounts of 3,4,4'-trichlorobiphenyl, 3,4,3',4'-tetrachlorobiphenyl, and a pentachlorobiphenyl 3,4,3',4',5' (Table 2). However, we did not detect 3,4,5,3',4',5'-hexachlorobiphenyl. Using carbon cleanup, GC analysis of Aroclor components lacking o-o'-Cl substitution was feasible at concentrations less than 1 μg/g. The presence of

Table 2 Concentration of PCB Isomers Lacking Ortho-chlorine Substitution in Two PCB Mixtures and a Fish Extract

Sample type (amount spiked)	Column[1] Chromatography	3,4,4' (μg/g PCB)	3,4,3',4' (μg/g PCB)	3,4,5,3',4',5' (μg/g PCB)
FDA Std. 1248 (2g)	10 x 2.5 cm id[2] PX-21 carbon (>325 mesh)	1180	2,200	N.D.[3]
FDA Std. 1254 (1.18g)	10 x 1 cm id PX-21 carbon (>325 mesh)	105	173	N.D.
Fish Extract[4] 1248 (26μg) 1254 (11.6μg) 1260 (1.8μg)	10 x 1 cm id PX-21 carbon (<325 mesh) /foam	N.D.	86.0 or 10.2 ng/g fish tissue	N.D.

[1]Column elution profiles and configurations used for isolation of no-ortho-Cl PCBs are given in Table II, see text for recovery values.

[2]1 L of toluene required for elution of no-ortho-Cl PCBs from 2.5 cm id column.

[3]None detected.

[4]Composite whole body sample consisting of 0.22g goldfish and 0.12g largemouth bass from the Hudson River, represents environmental PCBs.

chlorinated-naphthalenes and -methylnaphthalenes in Aroclors 1248 and 1254 was confirmed using GC-MS. Tables 3A and B summarize data from GC-MS analysis of trace Aroclor 1248 impurities. The reconstructed ion chromatogram is shown in Figure 1. Numbers above peaks relate to peak numbers in Tables 3A and B.

Stalling et al, U.S. Patent #733500 (1978) and Huckins et al (1978) developed a new cleanup procedure for chlorinated-dibenzofurans and -dibenzo-p-dioxins, using PX-21 carbon dispersed on the surface of shredded polyurethane foam. This dispersed carbon adsorbent is applicable to the analysis of non-ortho substituted PCBs in Aroclor mixtures and environmental extracts. The volume of aromatic solvent required for the recovery of planar constituents is greatly reduced (Figure 2). Carbon/foam chromatography is necessary because some ortho-substituted PCB isomers coelute even on high resolution glass capillary Apiezon-L, SE-30, and C_{87} columns (Figure 3). Using a combination of carbon/foam chromatography and high resolution capillary GC, we measured 3,4,3',4'-tetrachlorobiphenyl (Table 2) in an extract of a composite fish sample from the Hudson River.

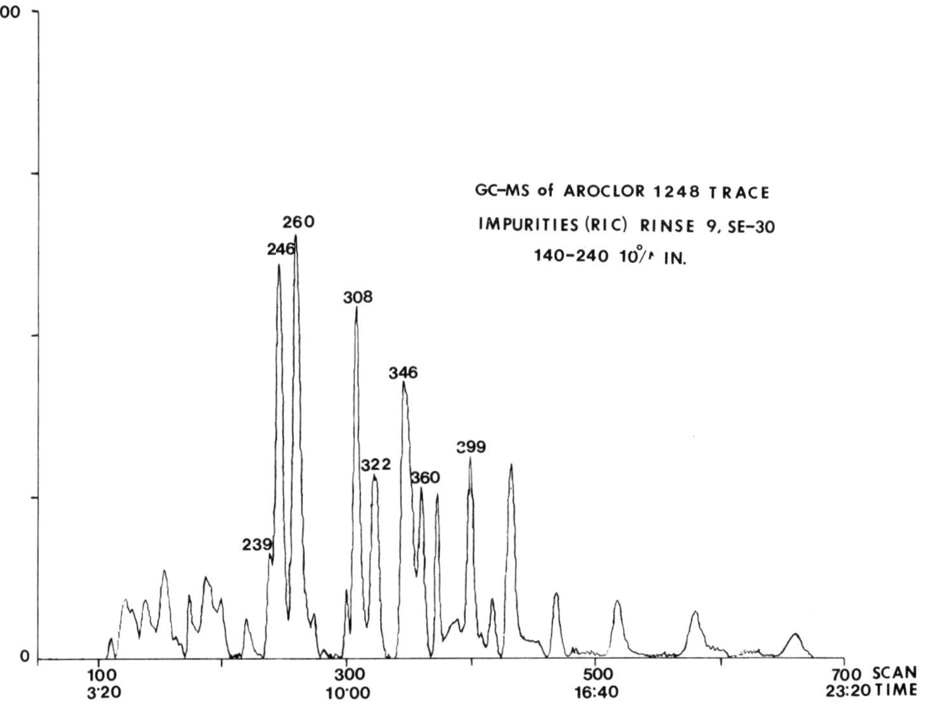

Figure 1. Reconstructed ion chromatogram of ninth 200 ml toluene wash of PX-21 carbon column (10 x 2.5 cm id) spiked with 2 g of Aroclor 1248. Numbered peaks are identified in Tables 3A and B. GC-MS conditions are given in text.

Table 3A Mass Spectral Analysis of Trace Aroclor 1248 Contaminants[1/]

RIC Peak No.	First Peak In Ion Cluster	Abundance %	No. of Cl	Compound[2/]
239	298 (GFW)	60	5	
	228	20	3	Pentachloro-naphthalene
246	298 (GFW)	65	5	
	263	10	4	Pentachloro-naphthalene
	228	40	3	
260	298 (GFW)	60	5	
	263	10	4	Pentachloro-naphthalene
	228	40	3	
308	332 (GFW)	50	6	
	262	40	4	Hexachloro-naphthalene
322	332 (GFW)	50	6	
	262	40	4	Hexachloro-naphthalene

[1/] Sample from the fractionation of 2 g of 1248 on a 2.2 cm id x 10 cm AMOCO >325 mesh carbon column, 9th 200 ml toluene wash.

[2/] Isomer composition unknown.

Table 3B Mass Spectral Analysis of Trace Aroclor 1248 Contaminants[1/]

RIC Peak No.	First Peak In Ion Cluster	Abundance %	No. of Cl	Compound[2/]
346	346 (GFW)	50	6	
	311	39	5	Hexachloro-methylnaphthalene
	275	30	4	
360	366 (GFW)	50	7	
	296	40	5	Heptachloro-naphthalene
	226	30	3	
399	380 (GFW)	10	7	
	345	5	6	Heptachloro-methylnaphthalene
	311	5	5	
417	400 (GFW)	30	8	
	330	30	6	Octachloro-naphthalene
	260	30	4	

[1/] Sample from the fractionation of 2 g of 1248 on a 2.2 cm id x 10 cm AMOCO >325 mesh carbon column, 9th 200 ml toluene wash.

[2/] Isomer composition unknown.

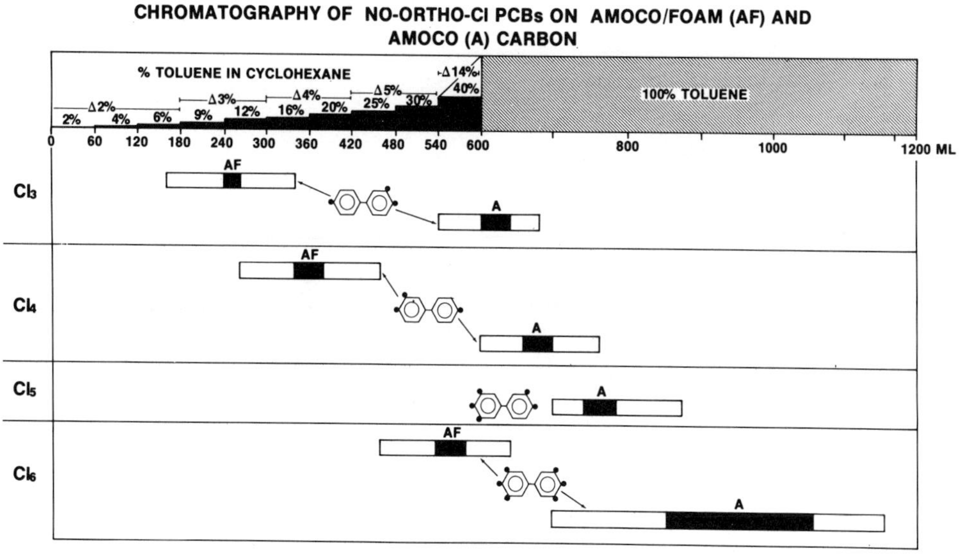

CHROMATOGRAPHY OF NO-ORTHO-Cl PCBs ON AMOCO/FOAM (AF) AND AMOCO (A) CARBON

Figure 2. Comparison of 10 x 1 cm id columns of PX-21 >325 mesh AMOCO carbon and <325 mesh PX-21 dispersed on shredded polyether foam (15/85,w/w) for the isolation of PCBs containing no ortho-Cl substitution. Dark areas on bars represent maximum concentration.

Until recently, the most complete resolution of PCB congeners has been provided by purified Apiezon-L (Jensen and Sundström, 1974) in packed columns, and support-coated open tubular (SCOT) Apiezon-L capillary columns (Sisson & Welti, 1971). Apiezon-L wall-coated open tubular (WCOT) glass columns provide superior PCB resolution (Figure 1). However, elution order for a pair of GC peaks reversed after extended use at 220 - 240°C, and some decrease in retention times was observed. Preliminary comparison of WCOT Apeizon-L and C_{87} liquid phase (Riedo et al, 1976) columns suggested that C_{87} columns provide comparable resolution without reversals of component elution order at higher temperatures.

Detailed qualitative and quantitative PCB component analyses are required to determine levels of non-ortho substituted PCBs in aquatic organisms. These residue data, when correlated with chronic toxicity data generated on structurally related PCB components, may improve the ability of investigators to interpret the significance of environmental PCB contamination.

1/ Reference to trade name does not imply government endorsement of commercial product.

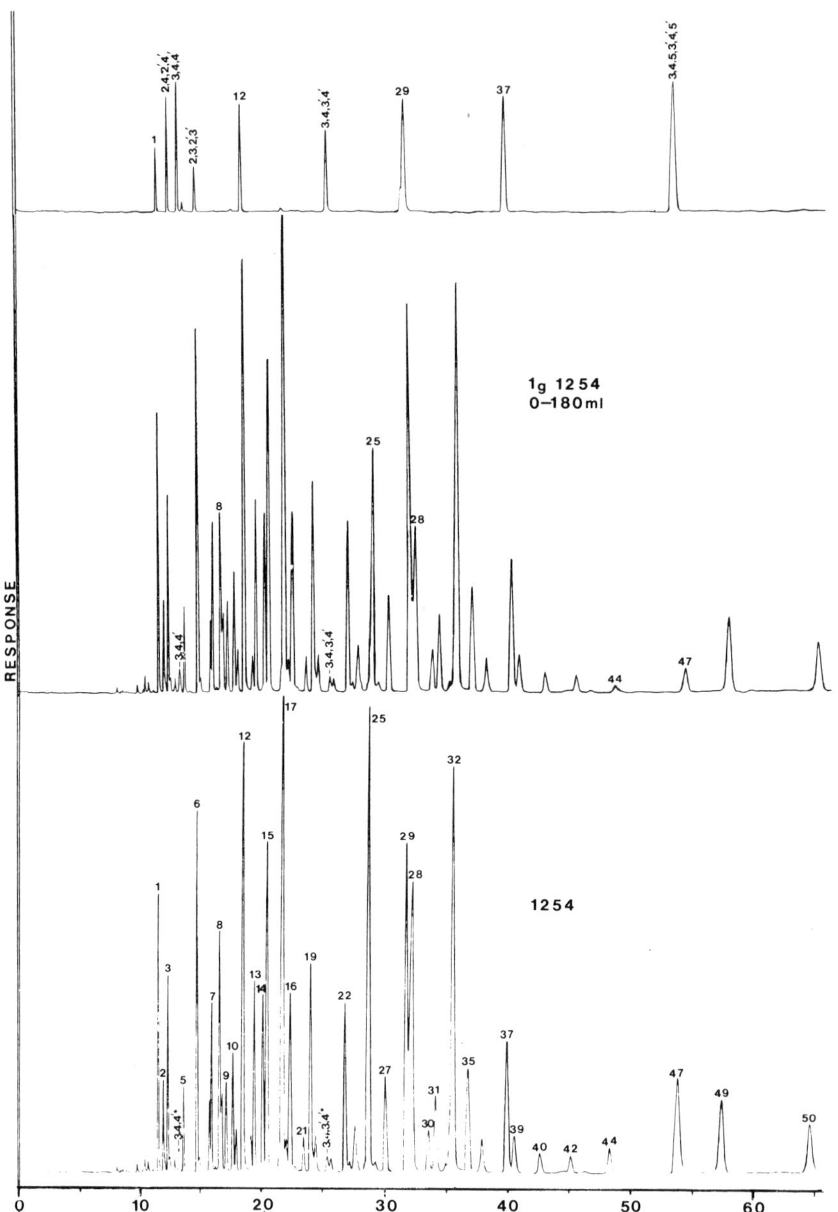

Figure 3. Bottom chromatogram; capillary Apiezon-L analysis of Aroclor 1254; see Table 1 for identity of numbered peaks. Middle chromatogram; Aroclor 1254 after carbon removal of PCB without ortho-Cl substitution. Similar 3,4,3',4' peak heights in the 1254 traces indicates that high resolution GC may not resolve 3,4,3',4' from other PCB components.

REFERENCES

Goldstein, J.A., P. Hickman, H. Bergman, J. McKinney, and M. Walker, (1977), Chem.-Biol. Interactions, 69-87, 17.

Huckins, J.N., D.L. Stalling, and W.A. Smith, (1978), JAOAC, 32-38 32-38, 61.

Jensen, S., and G. Sundström, (1974), AMBIO, 70-76, 3.

Poland, A., and E. Glover, (1977), Mol. Pharmacol., 924-938, 13.

Riedo, F., D. Fritz, G. Tarjan, and E. Kovats, (1976), J. Chromatogr., 63-83, 126.

Sissons, D., and D. Welti, (1971), J. Chromatogr., 15-32, 60.

Stalling, D.L., J.N. Huckins, and W.A. Smith, (1978), Patent Application Serial No. 733,500, USDI, Office of the Solicitor, Washington, D.C.

ANALYSIS OF MIREX IN LAKE SEDIMENTS: PROBLEMS AND SOLUTIONS

Alfred S.Y. Chau

Department of Environment Canada

P.O. Box 5050, Burlington, Ontario

Since the discovery of mirex in Bay of Quinte fish, (Kaiser, 1974), considerable effort has been devoted to the analysis of this and related compounds in environmental samples. In order to include mirex in the routine analysis of organochlorine pesticides and polychlorinated biphenyls (PCBs) in environmental samples, several laboratories have modified or adapted existing pesticide residue analytical methodologies (Task Force on Mirex, 1977). Due to the increasing demand on mirex analysis, methods that are more convenient or less subject to operational error for routine analysis have been developed.

At present, there are several key problems that must be addressed, namely:

1) separation of mirex from interfering PCBs;
2) direct confirmation procedure for mirex residue; and,
3) better recovery of mirex from turbid waters.

Although considerable efforts have been made to solve these problems, in this presentation discussion is confined to the separation of mirex from PCBs and confirmation of mirex identity with specific reference to sediment samples.

In sediment analysis for trace organics, one common and well known problem is the presence of sulfur or sulfur compounds which are co-extracted. It is sufficient to note that established procedures of using elemental mercury or activated copper power as described in an EPA analytical Manual (U.S. EPA, 1972) are available to eliminate or minimize these interferences, the copper powder being more efficient.

The column chromatography of mirex and PCBs using established Florisil, silica gel and coconut charcoal column procedures either do not give sufficient separation of mirex from PCBs or the variability of the adsorbent from batch to batch renders these procedures difficult to apply on a routine basis. These problems are discussed in more detail in the Mirex Task Group Report (1977) which also mentions the adoption of two well known chemical procedures dealing with the separation problems. They are: the perchlorination (Armour, 1973; Hutzinger et al, 1973; Hallett et al, 1976) and nitration (Jensen, 1969; Risebrough et al, 1969; Erro et al, 1967) procedures originally developed for PCBs in the late 60's. Because mirex is not reacted under these conditions and the PCBs are converted to compounds easily separable from it, mirex can be quantitated after the chemical reactions. These two approaches are cumbersome on a routine basis and to a certain degree sacrifice the quantitation of PCBs in favour of mirex. The use of HPLC as practiced in Water Quality Branch laboratories, offers a convenient approach. Another convenient approach without the need of sophisticated equipment and training of operators is the application of a charcoal-foam adsorbent, based on the idea of Stalling et al (1978). This version of the adsorbant (Reynolds, 1969), contains a 6:4 (W/W) ratio of Norit charcoal and polyurethane foam. Elution with cyclohexane recovers mirex and photomirex while subsequent elution with benzene recovers PCBs. This micro charcoal-foam column is used in conjunction with the florisil column as described by Reynolds (1969) for clean-up and fractionation of 16 o.c. and PCBs. The latter column is incorporated in the methods for water, fish, and sediment analysis (Chau and Wilkinson, 1974). Thus, sediment samples are extracted in the presence of activated copper powder by an ultrasonic apparatus using acetone/hexane and the extract in hexane is passed through a florisil column as usual. The first fraction containing PCBs, mirex, and photomirex is further treated by a micro charcoal-foam column to separate PCBs from these constituents.

The recoveries of all the deterimants from spiked lake sediments were at least 90%. Specifically, recoveries of mirex, photomirex and PCBs from sediment were respectively 100%, 95% and 100%.

The next problem is the confirmation of mirex identity. Perchlorination and nitration procedures mentioned above can only be regarded as an indirect technique, because mirex is not reacted. The use of chromous chloride is suggested by past success in reductive dechlorination of several cyclodiene pesticides (Chau, 1970; Chau et al, 1971; Chau and Cochrane, 1971; Chau and Wilkinson, 1972; Chau, 1972; Chau et al, 1974; Chau, 1970; Cochrane and Chau, 1971) which contains bridge-head chlorines and gem-dichloro moiety similar to those in the mirex molecules. Three procedures were developed (Chau et al, 1978), each giving a predominant peak at a different retention time on a 3% OV-101 column. These three procedures utilized three different chromous chloride and acetone ratios at room temperature and at 60°C, all carried out in the same manner as described for organochlorinated pesticides (Chau, 1970; Chau et al, 1971; Chau and Cochrane, 1971; Chau and Wilkinson, 1972; Chau, 1972). Based on peak height measurement, the sensitivities of the procedures are respectively 1.1, 0.7, 0.5 times the levels of mirex that could be analyzed by the particular analytical method in use.

In conclusion, the above discussion describes procedures suitable for routine analysis for the analysis of mirex in the presence of interfering PCBs and the confirmation of mirex residue identity. Based on these findings, a complete procedure for the analysis of 16 o.c., PCBs mirex and photomirex in lake sediments has been developed (Chau and Babjak, 1979).

REFERENCES

Armour, J.A. 1973. JAOAC, 56, 982-986.

Berg, O.W., P.L. Diosady, and G.A.V. Rees, 1972. Bull. Environ. Contam. Toxicol., 7, 338-347.

Chau, A.S.Y. 1972. JOAC, 55, 519-525.

Chau, A.S.Y. 1970. Proceeding to the Eastern Canada Seminar on Pesticide Residue Analysis, May 20-22, Ottawa.

Chau, A.S.Y., 1970. Bull. Environ. Contam. Toxicol., 5, 429-434; idem, ibid, 5, 435-439.

Chau, A.S.Y., and L. Babjak. 1979. JAOAC, in press.

Chau, A.S.Y., J.M. Carron, and H. Tse. 1978. JAOAC, 61, 1475-1480.

Chau, A.S.Y., and W.P. Cochrane. 1971. JAOAC, 54, 1124-1131.

Chau, A.S.Y., A. Demayo, J.W. ApSimon, J.A. Buccini, and A. Fouchier. 1974. JAOAC, 57, 205-216.

Chau, A.S.Y., and W.J. Wilkinson. 1974. "Procedure for Analysis of Organochlorinated Pesticides in Waters" and "Procedures for the Analysis of Organochlorinated Pesticides and PCBs in Fish and Sediments", in Analytical Methods Manual, Inland Waters Directorate, Water Quality Branch, Ottawa, Canada.

Chau, A.S.Y., and R.J. Wilkinson. 1972. Bull. Environ. Cont. Toxicol., 8, 105-108.

Chau, A.S.Y.,H. Won, and W.P. Cochrane. 1971. Bull. Environ. Contam. Toxicol. 6, 481-484.

Cochrane, W.P., and A.S.Y. Chau. 1971. Advan. Chem. Ser., 104 11-26.

Erro, F., A. Bevenue, and H. Beckman. 1967. Bull. Environ. Contam. Toxicol. 2, 372-379.

Fisheries and Environment Canada and Health and Welfare, Canada. 1977 1977. "Mirex in Canada", a report of the Task Force on Mirex, April 1.

Hallett, D.J., R.J. Norstrom, F.I. Onuska, M.E. Comba, and R. Sampson. 1976. J. Agric. Food Chem., 24, 1189-1193.

Hutzinger, O., W.J.J. Jamieson, S.S. Safe, and V.Z. Zitko. 1973. JAOAC, 56, 982-986.

Jensen, J. 1967. Private communication as quoted in Reynolds, L.M. 1969. J. Agric. Food Chem., 4, 128-143.

Kaiser, K.L.E. 1974. Science, 185, 523.

Reynolds, L. 1969. Bull. Environ. Contam. Toxic., 4, 128-143.

Risebrough, R.W., P. Reiche, and H.S. Olcott. 1969. Bull. Environ. Contam. Toxicol., 4, 192-201.

Stalling, D.L., A. Smith, and J.N. Huckins. 1978. JAOAC, 61, 32-38.

U.S. Environmental Protection Agency. 1972. "Analysis of Pesticide Residues in Human and Environmental Samples. Section 11-B.

CHLORO-ORGANIC COMPOUNDS IN THE LOWER FOX RIVER, WISCONSIN

P.H. Peterman, J.J. Delfino, D.J. Dube, T.A. Gibson
and F.J. Priznar
Laboratory of Hygiene, University of Wisconsin-Madison,
465 Henry Mall, Madison, WI 53706

ABSTRACT

The Lower Fox River, Wisconsin is one of the most densely developed industrial river basins in the world. During 1976-77 about 250 samples were analyzed by GC and GC/MS including biota, sediments, river water and wastewaters from 15 pulp and/or paper mills and 12 sewage treatment plants. A total of 105 compounds were identified in selected extracts by GC/MS with another 20 compounds characterized but not conclusively identified. Twenty of the 105 compounds are on the EPA Priority Pollutant List. Other compounds identified in pulp and paper mill wastewaters, including chloroguaiacols, chlorophenols, resin acids and chloro-resin acids have been reported toxic to fish by other investigators. Several compounds apparently not previously reported in wastewaters are chloro-syringaldehyde, chloroindole, trichlorodimethoxyphenol, and various 1-4 chlorinated isomers of bisphenol A. Concentrations of the various compounds, when present in final effluents, ranged from 0.5 to <u>ca</u>. 100 μg/L. An exception was dehydroabietic acid, a toxic resin acid not found on the EPA Priority Pollutant List. It was frequently found in pulp and paper mill effluents in concentrations ranging from 100 to 8500 μg/L. PCBs were found in all of the matrices sampled. Sixteen of the 35 fish exceeded the FDA limit of 5 mg/kg while 31 of the 35 exceeded the Canadian limit of 2 mg/kg. Concentrations of PCBs and other chloro-organics were related to point source discharges. There was a direct correlation of the concentrations of these compounds in wastewater with suspended solids values.

INTRODUCTION

Concern over the sources, distribution and fate of organic compounds in natural waters has increased considerably in recent years. With the development of GC/MS/DS instrumentation, thousands of compounds have been identified in industrial and municipal wastewaters, receiving waters and biota (Donaldson, 1977). Also, due to the extensive use of chlorination, numerous chlorinated organic compounds are being formed and these are now a matter of interest to researchers and regulatory officials (Jolley, 1976).

Because of this interest, the U.S. Environmental Protection Agency (Great Lakes Program Office, Region V) contracted with the State of Wisconsin (Department of Natural Resources and Laboratory of Hygiene) to assess the sources and distribution of organic compounds, particularly polychlorinated biphenyls (PCBs) and other chloro-organics, in the 64 km Lower Fox River in northeastern Wisconsin (Figure 1). This river drains into Green Bay-Lake Michigan and is one of the most densely developed industrial river basins in the world. Pulp and paper mills predominate; many of these use extensive amounts of chlorine. Five of the paper mills

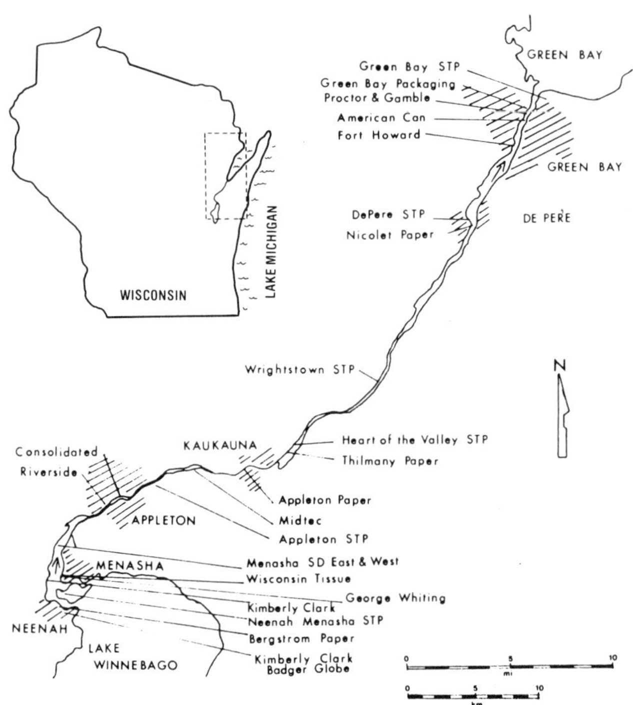

Figure 1. Effluent discharges to the Lower Fox River.

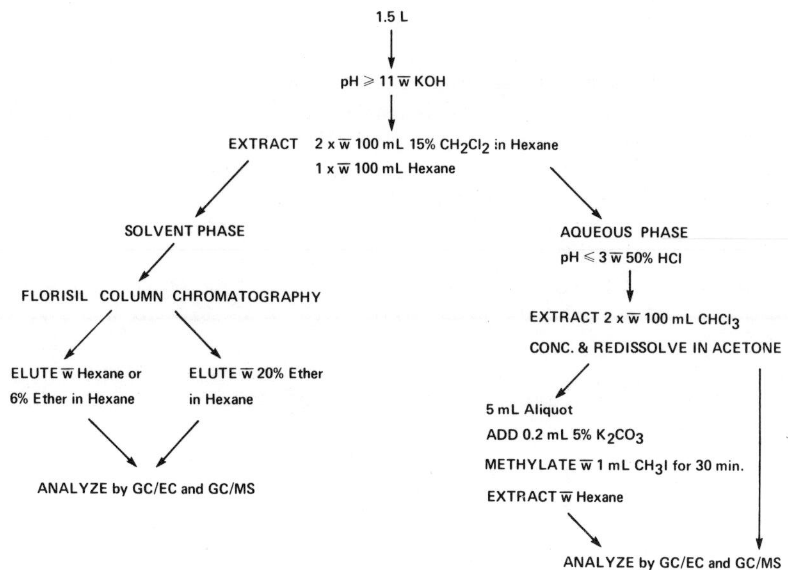

Figure 2. Water and wastewater extraction sequence.

de-ink and recycle paper to produce pulp, leading to the discharge of PCBs in their wastewaters.

EXPERIMENTAL

During 1976-77, ca. 250 samples were analyzed, including river and lake bottom sediments, snowmelt, biota (seston, clams and fish), river water and wastewaters from 15 pulp and/or paper mills and 12 municipal sewage treatment plants. Four of these municipal plants also treat pulp and/or paper mill wastewaters. Wastewaters comprised the majority of the samples received and were analyzed as described in Figure 2. Wastewater samples (1.5 L) were extracted at pH $\geq$ 11 with methylene chloride/hexane, fractionated on Florisil and screened by gas chromatography with electron capture detection (GC/EC) using procedures for chlorinated base-neutral compounds (USEPA, 1973). The remaining aqeuous phase of each sample was acidified to pH 3 and extracted with chloroform. The solvent was evaporated and the residue dissolved in acetone. This fraction was analyzed for chlorophenols, chloroguaiacols, and related chlorinated compounds. Selected extracts were derivatized with methyl iodide to facilitate analysis of acidic compounds and to confirm compounds identified by GC prior to methylation. Fractions whose GC/EC chromatograms exhibited significant unknown peaks were analyzed with a Finnigan 3100D Gas Chromatograph/Mass Spectrometer (GC/MS) and 6000 Data System.

Electron impact mass spectra ranging from m/z 35-500 were acquired every 3-4 sec. at an emission current of 0.35 ma, electron energy of 70 eV, amplification of 10^{-7} amp/V and electron miltiplier setting of 2.10 kV. Calibration with perfluorotertiarybutylamine was carried out according to the instrument manual and to specifications given by Carter (1976). PCBs and other base-neutral compounds were chromatographed on glass columns (1.8 m x 2 mm i.d.) packed with 3% SE-30 or 3% SP-2100 and temperature programmed from 100 to 220° C. Methylated and non-methylated acid fractions were analyzed on 1.8 or 3 m x 2 mm glass columns packed with Ultra-Bond 20M and programmed either from 90 to 210°C or 110 to 250°C, respectively.

Compounds were identified by (a) comparison of retention time and mass spectrum of a suspected constituent with those of a standard of that compound; (b) comparison of the full or partial mass spectrum (8 peaks) of a constituent with published spectra (e.g. Eight Peak Index, 1974; or (c) by interpretation of the mass spectral fragmentation pattern. Many of the compounds were available commercially, while some were provided by other researchers.

One compound that was unavailable from any source was chlorosyringaldehyde. Therefore, an experimental chlorination of syringaldehyde was performed. A commercial standard of syringaldehyde was added to a solution of 5.25% sodium hypochlorite (commercial bleach) in aqueous acetic acid. The reaction proceeded 16 hours, after which time the reaction product was extracted with methylene chloride, evapo-concentrated to dryness with a gentle stream of air, redissolved in acetone, and then injected into the GC/MS. A total ion chromatogram indicated both unreacted syringaldehyde and newly formed chlorosyringaldhyde. A very small amount of dichlorosyringaldehyde was also detected in the reaction product.

RESULTS AND DISCUSSIONS

Chlorosyringaldehyde was one of the 105 compounds identified by GC/MS (Table 1). Compounds in final effluents which were detected several times by GC/MS were quantitated by GC/MS, GC using flame ionization detection (FID) or GC/EC and listed in Table 2. Various effluents and extraction efficiencies were experienced, therefore only concentration ranges are given. The concentrations of these compounds generally corroborate earlier investigations of pulp and paper mill effluents (Rogers, 1973; Keith, 1976). Compounds detected and quantitated by GC/EC in fish, clams, river water, seston and sediments are also included in Table 2. A complete set of these data appears in a technical report (WI DNR, in press).

Table 1 Compounds Identified but not Quantified in Samples from the Lower Fox River System

	Compound
	Acetone, Tetrachloro-
c	Acetovanillone
	Aniline, Trichloro-
	Benzene, Dichloro-diethyl-
	Benzoate, Dimethyl-
	Benzoate, Methyl-methoxy-
	Benzoic Acid
	Benzoic Acid, Isopropyl-
	Benzophenanthrene, Methyl- or (Benzanthracene, Methyl-)
	Benzophenone
c	Benzyl Alcohol
	Biphenyl
	Biphenyl, Methyl-
	Bisphenol A
	Bisphenol A, Chloro-
	Bisphenol A, Dichloro- (2 isomers)
	Bisphenol A, Tetrachloro-
	Bisphenol A, Trichloro-
	Borneol, Iso-
c	Caffeine
	Camphor, Oxo-
	Carbazole
a	Chlordane
a	DDD
a	DDE
a	DDT
c	Dodecane
	FATTY ACIDS AND THEIR METHYL ESTERS
c	Heptadecanoic Acid
c	Lauric Acid
c	Myristic Acid
b	Oleic Acid
c	Palmitic Acid
c	Stearic Acid
c	Methyl Palmitate
c	Methyl Stearate
c	Guaiacol
b	Guaiacol, Dichloro- (3 isomers)
	Heptadecane
a	Hexachlorocyclopentadiene
c	Hexadecane
	Indole, Chloro-
	p-Menth-4-ene-3-one
	Naphthalene, Isopropyl-
	Naphthalene, Methyl-
	Nonadecane
	Octadecane
c	Pentadecane
	Phenanthrene, Methyl-
a	Phenol
	Phenol, p-Tertiary Amyl-
a	Phenol, Chloro-
	Phenol, p-(α-chloroethyl)-
	Phenol, Decyl-
c	Phenol, Ethyl-
	Phenol, Nonyl- (3 isomers)
	Phenol, Trichloro-dimethoxy-
	Phenol, Undecyl-
	Phenyl Decane
	Phenyl Dodecane
	Phenyl Undecane
	Phosphate, Tributyl-
	PHTHALATES
a	Dibutyl Phthalate
a	Diethyl Phthalate
a	Dioctyl Phthalate
c	Propan-2-one, 1-(4-hydroxy-3-methoxy phenyl) or guaiacyl acetone
	RESIN ACIDS
b	6,8,11,13-Abietatetraen-18-oic Acid
b	8,15-Isopimardien-18-oic Acid
b	Oxo-dehydroabietic Acid
b	Pimaric Acid
b	Sandaracopimaric Acid
	RESIN ACIDS, METHYL ESTERS
b	Methyl Dehydroabietate

(Continued next page)

Table 1 (Cont.)

	RESIN ACIDS, CHLORINATED
b	Chlorodehydroabietic Acid (2 isomers)
b	Dichlorodehydroabietic Acid
	RESIN ACID METHYL ESTERS, CHLORINATED
	Methyl Chlorodehydroabietate
	Methyl Dichlorodehydroabietate
c	Salicyclic Acid
c	Syringaldehyde
	Syringaldehyde, Chloro-
	Tetradecane
	Toluene, Dichloro-
	Toluene, Trichloro-
c	Vanillin
c	Vanillic Acid
c	Veratrole, Dichloro-
c	Veratrole, Trichloro-
	Xylene, Dichloro-
	Xylene, Trichloro-

a Compounds on EPA Consent Decree Priority Pollutant List
b Compounds in paper mill wastewaters reported toxic to fish
c Other compounds previously reported in paper mill wastewaters

To assess the significance of the compounds detected in this study, certain classifications were assigned. Twenty of the 105 compounds, including PCBs, appear in the EPA Consent Decree Priority Pollutant List (USEPA, 1977). Although the commercial use of Aroclar 1242 and other forms of PCBs in printing inks and carbonless copy paper apparently ended in 1972, PCBs are still being released into the Lower Fox River Basin. Deinking-recycling processes of five paper mills are some of the main sources. PCBs were detected in all of the various matrices sampled. All 35 fish fillet samples, consisting of both rough and sport fish, contained detectable levels of PCBs which were correlated to their fat content. Sixteen of the 35 fish exceeded the U.S. Food & Drug Administration tolerance limit of 5 mg/kg, while 31 of 35 exceeded the Canadian Food & Drug Directorate tolerance limit of 2 mg/kg.

PCBs are lipophilic and accumulate in fat tissue. Clams seeded in the Lower Fox River for 9 - 28 days showed that PCBs can rapidly bioaccumulate (WI DNR, in press). The mean uptake rates varied from 10 to 24 μg/day. The higher PCB uptake rate in the

Table 2 Compounds Identified and Quantified in Samples from the Lower Fox River System

	Compound	Environmental Matrix	Concentration Range	Units
	Anisole, Pentachloro-	Wastewaters	0.05 - 0.38	µg/L
		River water	0.002 - 0.02	µg/L
		*Seston	0.02 - 0.05	µg/L
		Fish	0.005 - 0.06	mg/kg
	Anisole, Tetrachloro-	Wastewaters	0.04 - 0.08	µg/L
	Benzothiazole	Wastewaters	10 - 30	µg/L
	Benzothiazole, Hydroxy-	Wastewaters	10 - 30	µg/L
c	Benzothiazole, Methylthio-	Wastewaters	10 - 40	µg/L
b	Dehydroabietic Acid	Wastewaters	100 - 8500	µg/L
		Sediment	2.7	mg/kg
a	Dieldrin	Fish	0.008 - 0.022	mg/kg
b	Guaiacol, Tetrachloro-	Wastewaters	10 - 50	µg/L
b	Guaiacol, Trichloro- (3 isomers)	Wastewaters	10 - 60	µg/L
a	Hexachlorocyclohexane (Lindane)	Wastewater	0.04	µg/L
a	Phenols, Dichloro- (2 isomers)	Wastewaters	15 - 40	µg/L
a	Phenol, Pentachloro-	Wastewaters	0.1 - 40	µg/L
		Sediments	0.22 - 0.28	mg/kg
	Phenol, Tetrachloro-	Wastewaters	2 - 20	µg/L
a	Phenols, Trichloro- (2 isomers)	Wastewaters	5 - 100	µg/L
a	Polychlorinated Biphenyls (Aroclor 1242, 1248 and 1254)	Raw wastewaters	0.2 - 8200	µg/L
		Final effluents	0.1 - 56	µg/L
		River water	0.05 - 0.85	µg/L
		*Seston	0.002 - 0.029	µg/L
		Sediments	0.05 - 61	mg/kg
		Clams seeded	0.26 - 0.74	mg/kg
		Fish	0.5 - 90	mg/kg
a	Polycyclic Aromatic Hydrocarbons (Acenaphthene, Anthracene, Chrysene, Fluoranthene, Pyrene)	Wastewaters	0.5 - 10	µg/L

* Entire aqueous sample filtered
a Compounds on EPA Consent Decree Priority Pollutant List
b Compounds in paper mill wastewaters reported toxic to fish
c Other compounds previously reported in paper mill wastewaters

clams occurred at locations having relatively high PCB concentrations in the river sediments. These locations were downstream from discharges containing PCBs.

Other compounds identified in pulp and paper mill wastewaters were those found to be toxic to fish by other investigators (Rogers and Keith, 1974; Leach and Thakore, 1977). These toxicants included chloroguaiacols, resin acids, chloro-resin acids and oleic acid. Our observations of chlorophenols corroborates work by Lindström and Nordin (1976). The source of the chlorophenols in the mill wastewaters investigated in this study has not yet been determined. Chlorophenols may have been used by paper mills for slime control or been present as wood preservatives. Phenolic compounds could also have been chlorinated in the bleaching or wastewater treatment stages. Chlorocatechols may have been present, but the analytical method employed did not appear to give any significant recovery of these compounds.

Previous investigations of toxicants in paper mill wastewater have always involved a pulp mill that derives its pulp from wood, thus releasing wood extractives and lignin-derived compounds such as resin acids, guaiacols and other phenolics, some of which become chlorinated in the bleach plant. In our survey, the highest levels of chloroguaiacols occurred at a sulfite pulp mill which produces mostly bleached pulp. These compounds eluted cleanly from an unmethylated acid extract on the Ultra-Bond 20M column which resolved 3 apparent trichloroguaiacol isomers. Chloroguaiacols were not detected in some Fox River paper mill wastewaters, presumably because these mills either do not bleach their wood-derived pulp, or else they bleach purchased pulp and/or deinked recycled paper. The last two should contain lesser amounts of the wood extractives and lignin-derived compounds.

The GC/MS analysis of a sample extract of a paper mill which bleached either purchased bleached pulp and/or deinked recycled paper is shown in Figure 3. The total ion chromatogram (TIC) of a methylated acid extract of the mill's final effluent shows 39 identified compounds, most of which are methylated derivatives. Thus, chloroanisoles were originally present as chlorophenols, chloroveratroles as chloroguaiacols and the dimethyl ether derivatives as various bisphenol A isomers. Likewise, the fatty and resin acid methyl esters were originally present as the corresponding acids in the acid extract. The scale of the TIC has been limited to 30% of full scale to better show the compounds at lower concentrations. For reference, 75 ng of aldrin (peak 15) were injected as an external standard. Peak 8, representing tetrachloroguaiacol, was quantitated by GC/EC at 14 μg/L, while peak 25, representing the resin acid dehydroabietic acid (DHA), was quantitated by GC/FID at 3200 μg/L. This concentration approached that of 8500 μg/L which was seen in the aforementioned sulfite pulp mill's

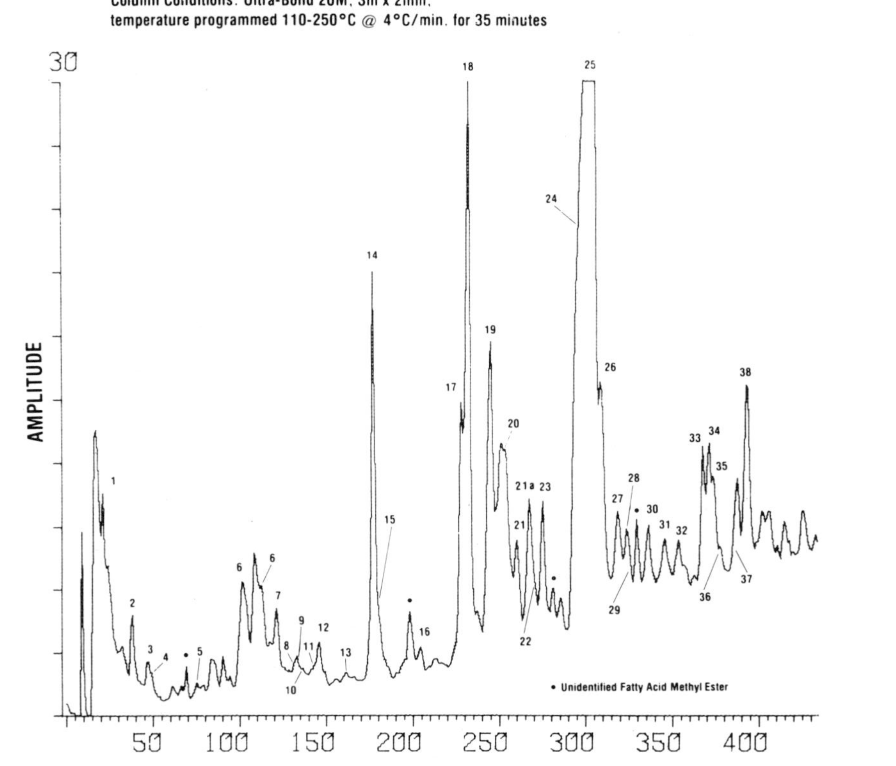

Figure 3. Total ion chromatogram of a methylated acid extract of a paper mill effluent

final effluent. Since the mill whose effluent is represented in Figure 3 lacks a wood pulping process, the relatively large amounts of fatty and resin acids present, especially DHA, could have come from its use of resin sizing (Merck Index, 1976) in the papermaking process. This water-intensive process could have diluted the available chlorine, thereby reducing the effectiveness of formation of chloro-resin acids (peaks 31, 32 and 37, Figure 3).

DHA appears to be the most stable of the resin acids (Brownlee and Strachan, 1977; Fox, 1977). The toxicity of resin acids to fish has been known since 1936. The 96 hour LC_{50} concentrations of DHA for young Sockeye Salmon are 2000 μg/L (Rogers, 1973) and 750 μg/L for Coho Salmon (Leach and Thakore, 1977). The latter investigators also reported even lower 96 hour LC_{50} concentrations for mono- and dichlorinated DHA.

Other compounds previously reported in paper mill wastewaters were also found in this study including acetovanillone, guaiacol, methyl thiobenzothiazole, syringaldehyde, vanillin and vanillic acid. Several compounds commonly used in industry that were identified including benzothiazole, an antioxidant; bisphenol A, a fungicide or an intermediate in the production of epoxy resins; and nonyl phenol, present in surfactants (Merck Index, 1976).

Several compounds which apparently have not been reported before in the environment are chlorosyringaldehyde, 5 separate chlorobisphenol A isomers, and trichloro-dimethoxyphenol, while chloroindole was found apparently for the first time in a sewage treatment plant effluent. Chlorosyringaldehyde was identified together with syringaldehyde in a semi-chemical pulp and paper mill untreated wastewater, as seen in another TIC (Figure 4). Syringaldehyde, a hardwood lignin degradation product would be anticipated to come from a pulp mill using hardwood. The apparent chlorination reaction within the plant compares with similar examples of the chloroguaiacols and chloro-resin acids in other plants. Chlorosyringaldehyde was identified in an acid fraction without derivatization when chromatographed in an Ultra-Bond 20M column. The mass spectrum of chlorosyringaldehyde has isotopic molecular ions of m/z 216 and 218 which are consistent for a compound with one chlorine atom (Figure 5). These two ions as well as the fragment ions m/z 215, 201, 173, 145, 130, 127 and others have been shifted 34 mass units higher than for syringaldehyde which is also consistent with the addition of a chlorine atom to the benzene ring. In addition, the laboratory chlorination of syringaldehyde described in the Experimental Section yielded a mono-chlorinated compound which matched not only the identical retention time but also the mass spectrum of the apparent chlorosyringaldehyde in the untreated wastewater (Figure 4).

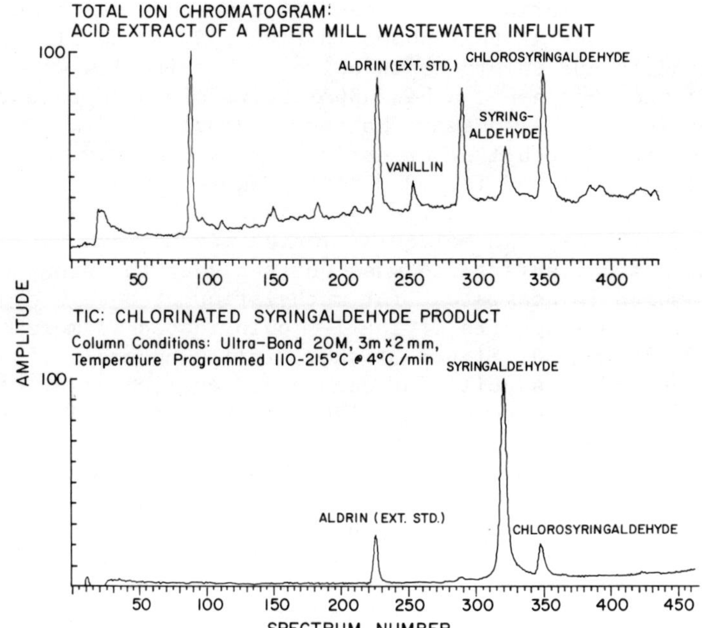

Figure 4. Total ion chromatograms: syringaldehyde and chlorosyringaldehyde

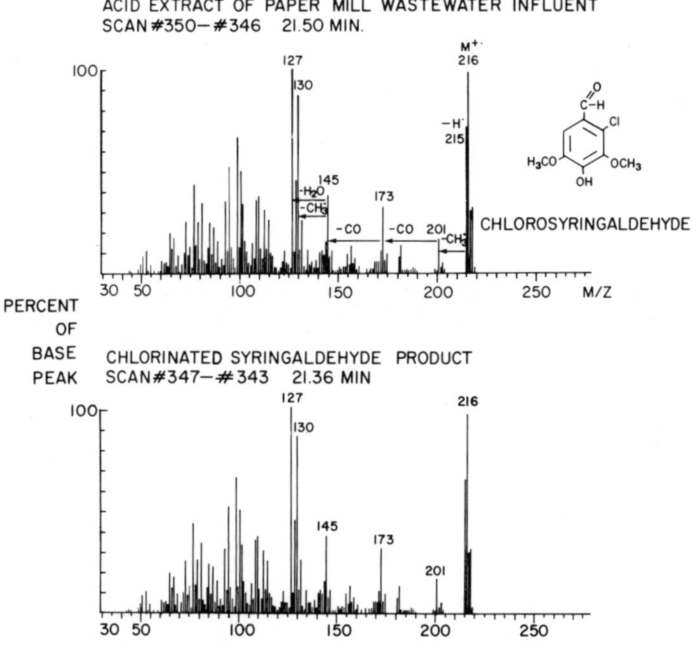

Figure 5. Mass spectra of chlorosyringaldehyde

A group of chlorinated bisphenol A compounds was identified in the same extract as that shown in Figure 3, which also contained chloroguaiacols and chloro-resin acids. Despite its widespread industrial use, bisphenol A has apparently only been recently identified in the environment (Matsumoto et al, 1977). The mass spectrum of its dimethyl ether derivation is compared to that of peak 21a (Figure 3) and included here (Figure 6) because it was apparently not available in the literature. The mass spectrum of its derivative shows the molecular ion at m/z 256 and the base peak $(-CH_3)^+$ at m/z 241 shifted 28 mass units higher than that of bisphenol A, which is consistent for methylation of both hydroxyl groups. The mass spectra of the 1-4 chlorinated isomers (peaks 28, 29, 33, 35 and 36, Figure 3) show similar upward shifts of the two main ions 34 mass units for each additional chlorine atom. The mass spectra of these compounds also show the respective isotopic clusters corresponding to the number of chlorine atoms present. Using ^{35}Cl, the molecular ion and the base peak of monochlorobisphenol A dimethyl ether were, respectively, 290 and 275; those of both dichlorobisphenol A dimethyl ether isomers were 324 and 309; those of trichlorobisphenol A dimethyl ether were 358 and 343; and those of tetrachlorobisphenol A dimethyl ether were 392 and 377. A pure standard of tetrachlorobisphenol A was commercially available, as it is apparently used as a flame retardant. The mass spectrum of its methylated derivative is compared with

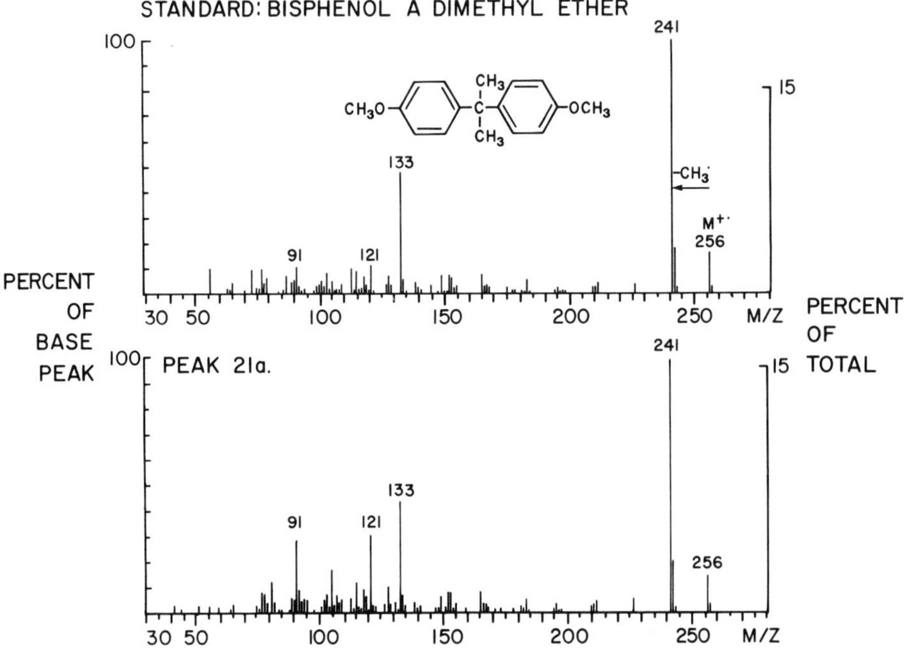

Figure 6. Mass spectra of bisphenol A dimethyl ether.

PEAK 33. TETRACHLORO-BISPHENOL A DIMETHYL ETHER

STANDARD: TETRACHLORO-BISPHENOL A DIMETHYL ETHER

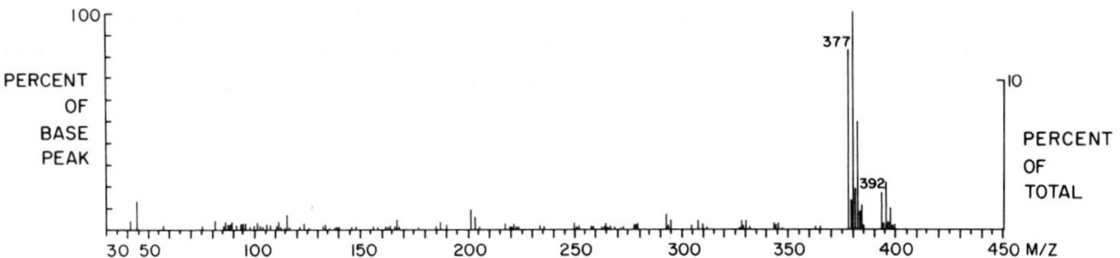

Figure 7. Mass spectra of tetrachlorobisphenol A dimethyl ether.

that of peak 33 (Figure 3) in Figure 7. In a laboratory chlorination of bisphenol A performed similarly to that of syringaldehyde, it was shown that 2,4,6-trichlorophenol was the main product formed, although various chlorinated bisphenol A isomers having 1-4 chlorines were also formed.

Trichloro-dimethoxyphenol was tentatively identified in the acid extract (Figure 3). Its methylated derivative was peak 11 which eluted just after a close congener, trichloroveratrole (peak 10). Trichloro-dimethoxyphenol was another compound which was detected in the acid extract chromatographed directly without derivatization on an Ultra-Bond 20M column. It eluted just after tetrachloroguaiacol. Its mass spectrum showed abundant molecular ions at m/z 256-260 consistent for 3 chlorine atoms, a similar cluster at ions m/z 241-245 $(M-15)^+$ and m/z 198-202, $(M-58)^+$. The mass spectrum of the methylated derivative, peak 11 (Figure 3) showed abundant isotopic molecular ions now at m/z 270-274 with subsequent fragments $(M-15)^+$, $(M-43)^+$ and $(M-58)^+$.

Chloroindole apparently has not been previously detected in wastewaters, but it has been isolated from a bacterium (*Pseudomonas pyrrocinia*) (Neidleman, 1975) and from an acorn worm (*Ptychodera flava laysanica*) from the Pacific Ocean (Higa and Scheuer, 1975). In our study of untreated wastewater from a municipal sewage treatment plant, this compound was detected in the 20% ether in hexane Florisil eluate at a concentration of *ca.* 30 μg/L. The mass spectrum showed abundant isotopic molecular ions m/z 151 and 153, base peak of m/z 89, and less abundant ions at m/z 124, 116 and 63. In comparing the compound from our sample with a commercial standard of 5-chloroindole, both mass spectra matched well but their retention times differed by several minutes.

In addition to the 105 compounds identified in this study, another 20 or so compounds were detected but not conclusively identified to date. A group of related compounds was consistently detected. The most prominent were two apparent isomers with a molecular weight of 196. In wastewater samples of the paper mill represented in Figure 3, these two isomers were followed by about nine chlorinated isomers with apparent molecular weights of 230, 264 and 298. The mass spectra of all of these are included in a technical report (WI DNR, in press). Mass spectra of the two non-chlorinated isomers are similar to diphenylacetaldehyde and trans stilbene oxide. Although the compounds with molecular weight of 196 have been detected in various extraction fractions, they and the chlorinated isomers primarily have been found in the first Florisil eluate (6% ether in hexane). Their concentrations have been sufficiently high to mask some of the PCB peaks detected with GC/EC. This mill's extensive deinking, recycling and bleaching processes could conceivably release the compounds with molecular weight 196 which ultimately become chlorinated.

This GC/MS study was aided by the use of a low-loaded, Ultra-Bond 20M column packing (ca. 0.3% Carbowax 20M) similar to that first discovered by Aue (1973). Elutions were characteristically sharp, with polar phenolic compounds eluting quite well. Baseline separation of pentachloroanisole from tetrachloroveratrole was achieved, contrary to the case for 3% SP-2100, 3% OV-17, or the mixed phase packing 4% SE-30/6% OV-210 designed for pesticide analyses. Very low bleed on temperature programmed analyses aided background substraction resulting in optimum mass spectra.

The fate and long-term health and ecological implications of many of the 105 compounds identified requires further research. For PCBs and some other chloro-organics, sampling data and laboratory experiments show a direct correlation of their concentrations in wastewaters with suspended solids concentrations. Suspended solids reduction in wastewater treatment plant also reduces the chloro-organic concentration in the final effluent (WI DNR, in press). For example, the untreated wastewater of a paper mill which deinks and recycles paper contained 25 μg/L PCBs and 2,020 mg/L suspended solids. Following primary clarification, concentrations were reduced to 2.2 μg/L PCBs and 72 mg/L suspended solids. After secondary treatment the final effluent contained only 1.4 μg/L PCBs and 10 mg/L suspended solids. Now the final disposal of the treatment plant sludge containing PCBs must be resolved.

ACKNOWLEDGEMENTS

This study was supported in part by a contract from the U.S. EPA Great Lakes National Program Office, Region V, (EPA Contract No. 68-01-4186). Additional support was provided by the Wisconsin Department of Natural Resources and the Wisconsin Laboratory of Hygiene.

REFERENCES

Aue, W.A., C. Hastings, and S. Kapila. 1973. J. Chromatog. 77, 299-307.

Brownlee, B., and W.M.J. Strachan. 1977. J. Fish Res. Bd. Can. 34, 838-843.

Carter, M.H. March, 1976. EPA Report No. 600.4-76-004.

Donaldson, W.T. 1977. Environ. Sci. & Technol, 11, 348.

Fox, M.E. 1977. J. Fish Res. Bd. Can. 34, 798-804.

Higa, T., and P.J. Scheuer. 1975. Naturwissenschaften, 62, 395-396.

Jolley, R.L. (Ed.) 1976. Proceedings of the 1975 Conference on the Environmental Impact of Water Chlorination, ORNL CONF 751096, Oak Ridge, Tennessee.

Keith, L.H. (Ed.) 1976. Identification & Analysis of Organic Pollutants in Water. Ann Arbor Science Publishers, Inc., Ann Arbor, MI, 718 p.

Leach, J.M., and A.N. Thakore. 1977. Prog. Water Tech., 9, 787-798.

Lindström, K. and J. Nordin. 1976. J. Chromatog.,128, 13-26.

Mass Spectrometry Data Centre. 1974. Eight Peak Index of Mass Spectra. 2nd ed., Mass Spectrometry Data Centre, Aldermaston, U.K., 2933 p.

Matsumoto, G., R. Ishiwatari, and T. Hanya. 1977. Water Res. 11, 693-698.

Merck and Co., Inc. 1976. The Merck Index, 9th ed., Merck and Co., Inc., Rahway, N.J.

Neidleman, S.L. 1975. CRC Critical Reviews in Microbiology 3 (4), 333-358.

Rogers, I.H. Sept. 1973. Pulp Paper Mag. of Can., 74 (9).

Rogers, I.H., and L.H. Keith. 1974. "Organochlorine Compounds in Kraft Bleaching Wastes. Identification of Two Chlorinated Guaiacols", Environment Canada, Fisheries & Marine Service, Res. & Dev., Technical Report No. 465.

USEPA. 1973. Fed. Reg. Part II, Vol. 38, Nov. 28.

USEPA. 1977. "Sampling and Analysis Procedures for Screening of Industrial Effluents for Priority Pollutants", EMSL, Cincinnati, Ohio, 69 p.

WI DNR, "Investigation of Chlorinated and Nonchlorinated Compounds in the Lower Fox River Watershed". EPA Contract No. 68-01-4186 (in press).

THE DETERMINATION AND LEVELS OF POLYCYCLIC AROMATIC HYDROCARBONS IN SOURCE AND TREATED WATERS

R.I. Crane, B. Crathorne and M. Fielding

Water Research Centre, Medmenham Laboratory, P.O. Box 16

Henley Road, Medmenham, Marlow, Bucks, SL7 2HD, England

INTRODUCTION

Polycyclic aromatic hydrocarbons (PAHs) form a class of organic compounds which has been shown to be ubiquitous in the environment. They are detectable in air, foodstuffs, soil, sediments, water and in many products and wastes from industrial activities. The concern over PAHs is that many have been shown to be carcinogenic to animals and substantial data exists which incriminates them as being carcinogenic to man (International Agency for Research on Cancer, 1973; Jones and Matthews, 1974). It has been estimated that at least 80% of cancers in man originate from environmental substances (Higginson, 1968). Thus, if contact with such substances can be reduced by reduction of their presence in the environment then a significant decrease in the incidence of cancer could follow.

Techniques for the analysis of PAHs in various environmental samples have been reviewed (Sawicki, 1964). The analysis of PAHs in water, as in most substrates, is difficult owing to the complex mixtures that are encountered, the similarity of the chemical and physical properties of the components of these mixtures and the low detection limits required. In air particulate samples, over 100 components have been identified (Lao et al, 1973), and in rivers it is likely that the PAH mixture is equally complex. The complexity of environmental PAH mixtures invariably necessitates chromatographic separation before quantification. However no chromatographic technique is capable of separating in one experiment all the PAHs encountered in a complex environmental mixture. Furthermore, each technique separates different PAHs preferentially.

A frequent approach to overcome this problem is to select for routine monitoring certain PAHs which are relatively easily measurable by a particular analytical technique. These compounds then serve as indicators of PAHs in general.

This approach forms the basis of the World Health Organization standard for PAHs in drinking water (WHO, 1971) which was established as a result of extensive studies on PAH levels in the aqueous environment by Borneff and co-workers (Borneff and Kunte, 1969; Borneff, 1967 and Borneff, 1969). The WHO standard recommends that the total of six specified PAHs (termed 'Total PAHs') in drinking water derived from surface water should not exceed 200 ng/l. The six PAHs monitored are fluoranthene, benzo(ghi)perylene, benzo(k)-fluoranthene, indeno(1,2,3-cd)pyrene, benzo(b)fluoranthene and benzo(a)pyrene. The structures of the six specified PAHs are shown in Figure 1. A level of more than 200 ng/l of 'Total PAHs' indicates unacceptable contamination resulting from insufficient treatment.

This standard was used as a guide to water quality in this work. Some problems became apparent in interpreting the results in relation to the standard which suggest that the standard may not be universally applicable. However, these problems are beyond the

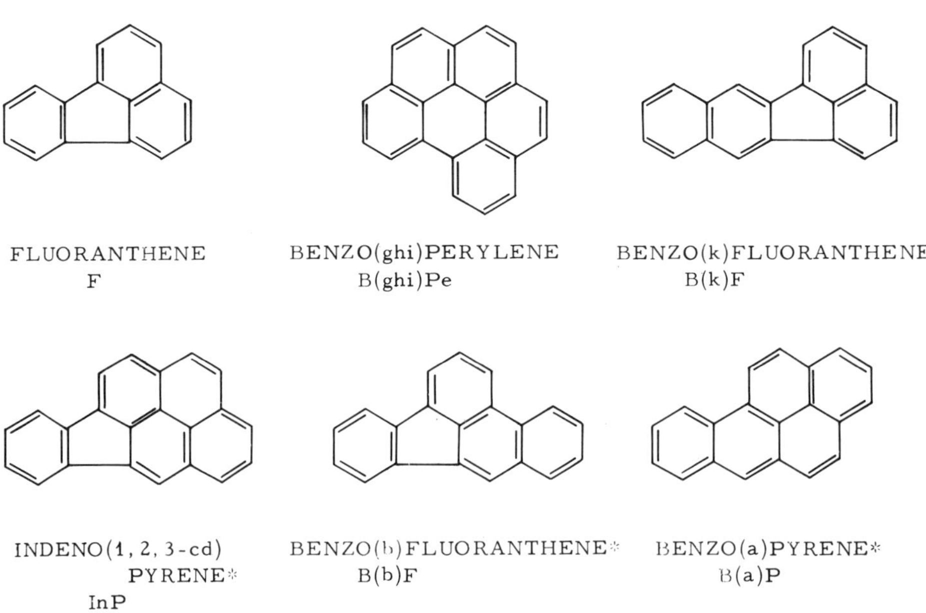

Figure 1. WHO specified PAH; structures, names and abbreviations

scope of this paper and have been reported elsewhere (Crathorne and Fielding, 1978).

Because little information existed on levels of PAHs in British waters, the Water Research Centre has carried out a study on this subject which was supported by the Department of the Environment.

This paper reports on certain aspects of this study. Information is given on analysis of PAHs including a new approach based on high-performance liquid chromatography (HPLC). Levels of PAHs in source and treated water and the effectiveness of water treatment are reported and the pick-up of PAHs during distribution is discussed.

ANALYTICAL TECHNIQUES

The technique employed at WRC for the survey of PAHs in various water samples was based on separation by two-dimensional thin-layer chromatography. The method (Crane, Crathorne and Fielding, 1978) was a modified version of that developed by Borneff and Kunte (1969) and recommended by the WHO (1971) in their standard for PAHs in potable water. This technique, however, has some disadvantages notably the semi-quantitative nature of the data resulting from the fact that quantification is based on a visual estimate of the intensity of illuminated TLC spots. The exact magnitude of the overall error is difficult to determine but is typically $<\pm 25\%$. For this reason an alternative analytical technique was developed based on separation of the PAHs by high-performance liquid chromatography (HPLC), using an absorption type column system and fluorimetric detection (Crane, Crathorne and Fielding, 1978).

Two-dimensional Thin-layer Chromatography

The sample of water (typically 2l) was extracted with cyclohexane (100 ml) and the extract concentrated to 50 μl. The whole of this extract was then spotted onto one corner of a TLC plate having an absorbent layer of alumina (7 parts) and acetylated cellulose (3 parts). The first development of this plate used hexane-toluene (92:8 v/v). The plate was then air-dried, and developed in methanol-ether-water (4:4:1 v/v/v) at 90° to the first development. The plate was dried once more and then illuminated with u.v. light (366 nm). The amounts of individual PAHs were estimated by visual comparison of the fluorescent intensity of each spot with the intensity of the spots on standard chromatograms.

Figure 2 shows a typical chromatogram obtained from the analysis of a standard solution of the six PAHs. A chromatogram obtained from the analysis of a typical polluted surface water sample is shown in Figure 3.

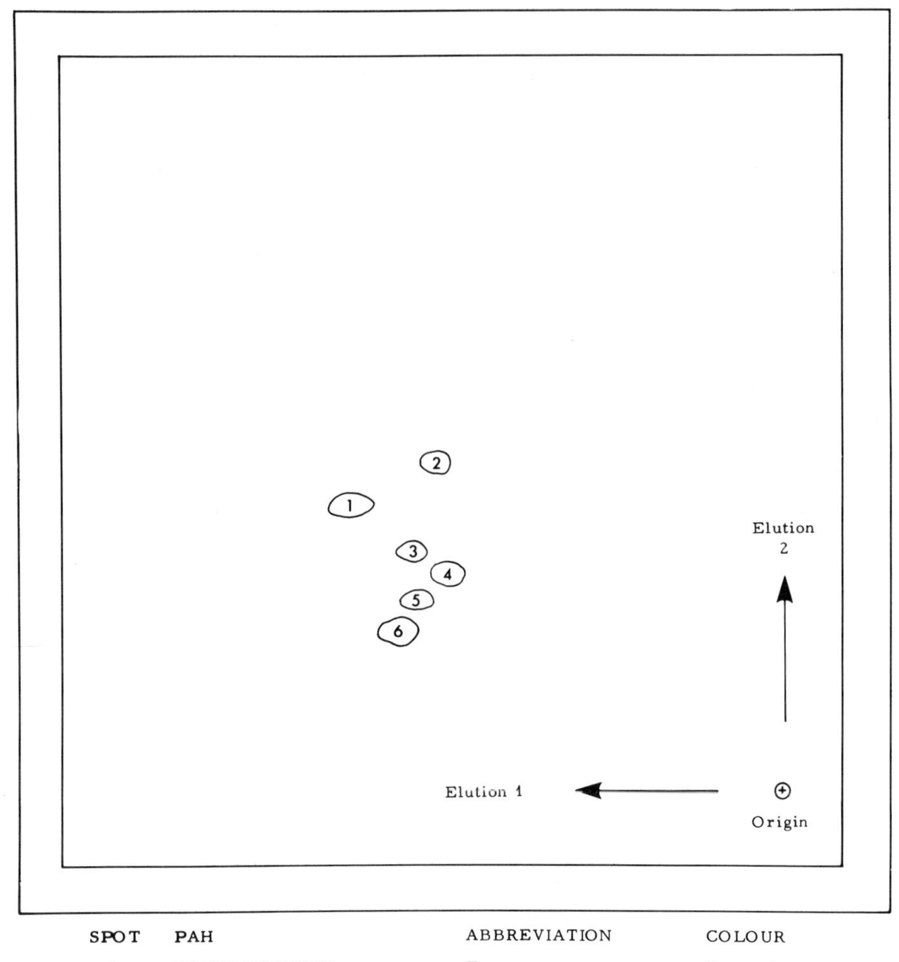

SPOT	PAH	ABBREVIATION	COLOUR
1	FLUORANTHENE	F	Turquoise
2	BENZO(ghi)PERYLENE	B(ghi)Pe	Violet
3	BENZO(k)FLUORANTHENE	B(k)F	Violet
4	*INDENO(1, 2, 3-cd)PYRENE	In P	Yellow
5	*BENZO(b)FLUORANTHENE	B(b)F	Turquoise
6	*BENZO(a)PYRENE	B(a)P	Violet

* **Carcinogenic**

Figure 2. Standard PAH chromatogram

High-performance liquid chromatography

The sample of water (5l) was extracted with cyclohexane (1 × 100 ml; 2 × 50 ml) and the organic extract dried over anhydrous magnesium sulphate. The extract was then concentrated to 0.5 ml and 10 µl of this injected onto a column of Partisil-5 (25 cm × 4.6 mm I.D.) and eluted with a mobile phase of 0.01% acetone in

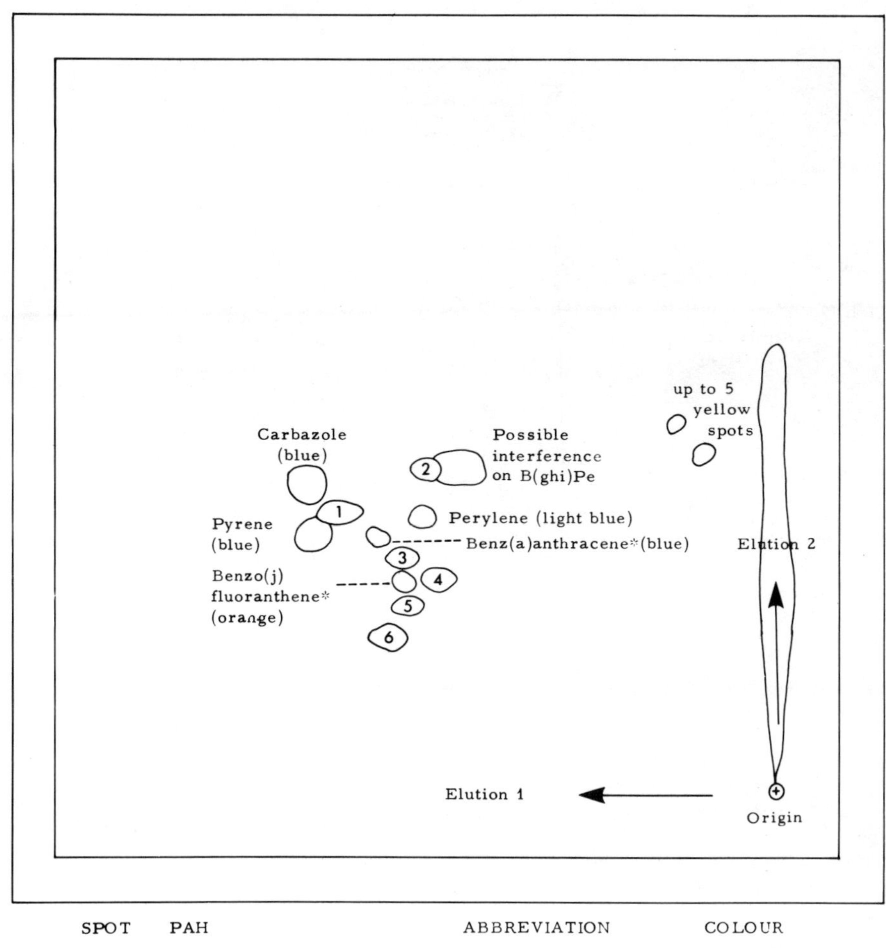

SPOT	PAH	ABBREVIATION	COLOUR
1	FLUORANTHENE	F	Turquoise
2	BENZO(ghi)PERYLENE	B(ghi)Pe	Violet
3	BENZO(k)FLUORANTHENE	B(k)F	Violet
4	*INDENO(1, 2, 3-cd)PYRENE	In P	Yellow
5	*BENZO(b)FLUORANTHENE	B(b)F	Turquoise
6	*BENZO(a)PYRENE	B(a)P	Violet

* Carcinogenic

Figure 3. Chromatogram of a typical polluted surface water sample

n-hexane at a flowrate of 2.0 ml/min.

The PAHs were detected in a fluorimeter and quantified by comparison with calibration curves constructed from the analysis of standard solutions. Figure 4 shows the chromatograms from the analysis of a standard solution of the six specified PAHs. Two injections are required in order to monitor each PAH as the optimum

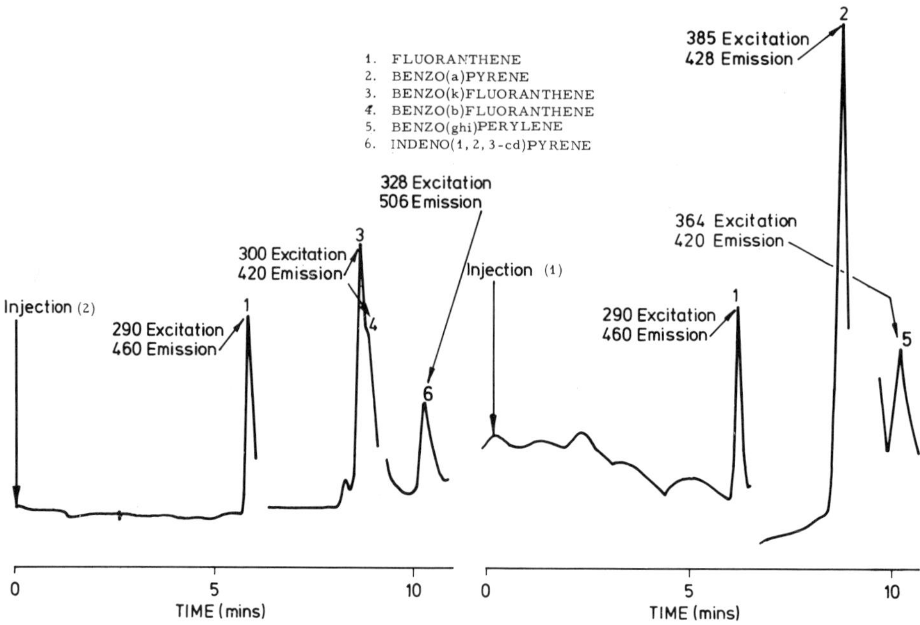

Figure 4. HPLC separation of a standard solution of PAHs using a Partisil-5 column with fluorescence detection.

excitation and emission wavelengths. Although it was possible to monitor all six PAHs with one set of excitation and emission conditions (and therefore carry out the analysis with one injection) this gave less sensitivity and specificity. Figure 5 shows the chromatograms from the analysis of a typical water sample. The breaks in the chromatograms occur when the excitation and emission wavelengths are altered for the monitoring of each PAH under its optimum conditions.

Two types of water (river water and de-ionised water) were spiked at levels of 10 ng/l and 100 ng/l of each of the six PAHs specified in the WHO standard.

Four replicate analyses were carried out at each level. The overall recovery was approximately 65% and the standard deviations varied from 0.5 - 6.7%.

RESULTS

Many source waters, treated waters and tapwaters were analyzed using the two-dimensional TLC technique.

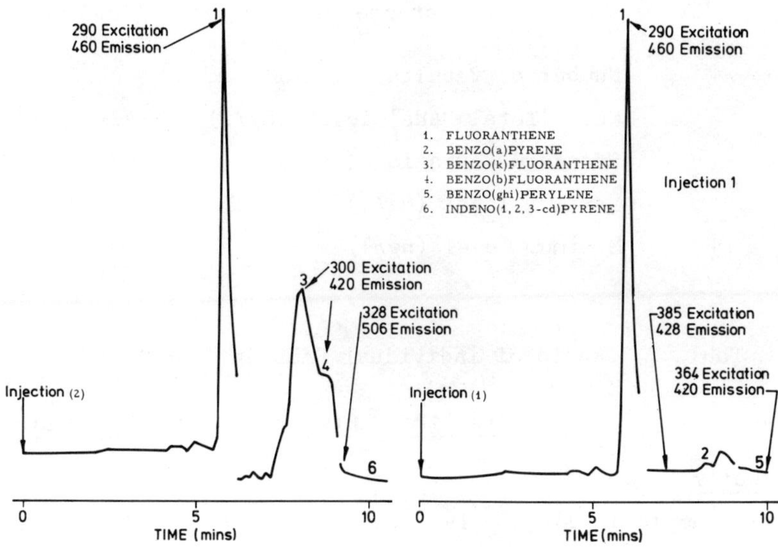

Figure 5. HPLC analysis of a water sample with fluorescence detection.

Source Water

Many source waters of different types were studied during the survey of PAHs in water. The results are summarized in Table 1.

Groundwater was found to contain very low levels of PAHs. Upland catchment reservoirs and feeder streams also contained low levels. Lowland rivers receiving sewage effluents typically contained much higher levels.

Table 1 'Total PAHs' in source waters

Type of water	'Total PAHs' (ng/1)
Groundwater	ND - 12
Upland catchment reservoirs and feeder streams	3 - 50
Lowland rivers	40 - 300
Lowland rivers in high flow conditions	50 - 3000

ND = not detected

Table 2 'Total PAHs' levels at one sampling site (lowland rivers)

Number of results	18
Mean 'Total PAHs' level (ng/l)	147
Standard deviation (ng/l)	63
Maximum level (ng/l)	282
Minimum level (ng/l)	50

Table 3 Ratio of individual PAHs in 'Total PAHs'

	F	B(ghi)Pe	B(k)F	InP	B(b)F	B(a)P
River Water						
mean (71 samples)	35	19	7	10	12	17
standard deviation	9	5	2	3	4	5

At one sampling site on a lowland river, repeated analyses over a period of time gave the results shown in Table 2. Although considerable variations in PAH levels was observed, no clear seasonal trend was found. The analysis of lowland rivers in high flow conditions indicated that when such conditions exist, relatively massive levels of PAHs occur (see Table 1). A rapid increase by a factor of up to 10 × appears typical. This increase appears to be related to the high level of suspended matter present. An additional factor may be the input from surface run-off. A very high level of PAH (60 μg/l) was found in a road surface drainage into a lowland river.

The ratios of PAHs encountered in lowland rivers are given in Table 3 and probably reflect the ratios of these PAHs in the major inputs into the aqueous environment. Quite different ratios exist in treated water.

Treated Surface Water

Several waterworks treating inland river water were studied. The water at the point of abstraction, after treatment stages and the finished water were examined. Generally, it was found that conventional treatment removed PAHs effectively and the finished water leaving all the works examined contained very low levels of PAHs. Table 4 gives a summary of the data for eight works. Treatment stages where particulate matter is removed, including

Table 4 Summary of effect of conventional treatment on 'Total PAHs' levels in river water

Treatment Works	A	B	C	D
Raw Water (ng/1)	174 ; 202	88 ; 105	90 ; 63	271
Finished Water (ng/1)	ND ; 2	4 ; 2	ND ; ND	10 ; 7
Treatment Works	E	F	G	H
Raw Water (ng/1)	122	986 ; 858*	290 ; 195	235 ; 280
Finished Water (ng/1)	11.5	ND ; ND	13	0.5 ; ND

* Very high flows.

reservoir storage, are effective in removing PAHs since much of the PAH in the source water is associated with such matter. Any PAHs detectable in the finished water consisted almost invariably of fluoranthene alone which reflects its greater water solubility. Figure 6 shows removal of PAHs in more detail.

Pilot plant reverse osmosis reduced PAHs to a level of 40 ng/1 (entirely fluoranthene) in the finished water from an intake level of 200 ng/1 'Total PAHs'.

Distribution Water

Since finished water is typically distributed over considerable distances through mains often lined with bituminous products, a study of the pick-up of PAHs during distribution was made. Samples were collected at domestic taps along several distribution systems.

Distributed surface water. Table 5 summarizes the results obtained for treated surface water taken at various points along distribution systems. Other systems gave almost identical results. Clearly no significant pick-up of PAHs in the distribution systems was encountered even though distances of more than 30 km from the treatment works were involved. If any PAHs were detected then this invariably consisted of fluoranthene alone. All distributed, treated surface water samples examined (over 50) indicated the same low pick-up.

Distributed groundwater. Table 6 summarizes the results obtained from the examination of samples of distribution groundwater. In the groundwater itself only very low levels were found (Table 1). Negligible pick-up of PAHs was found after distribution of several groundwaters from aquifers in different geological strata (samples

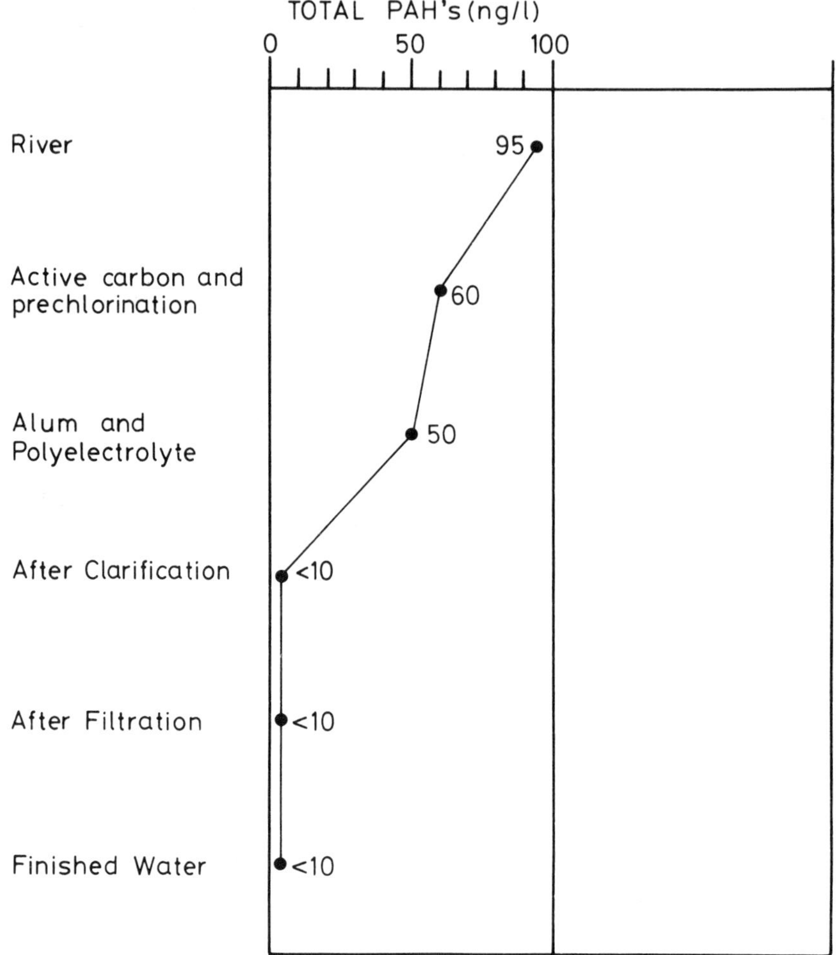

Figure 6. The effect of water treatment on PAH levels

E, F and G). However, a marked indication of pick-up from within a distribution system was found in water from several related aquifers (chalk) of which system H is typical. In Table 6 the mean level of 'Total PAHs' (consisting entirely of fluoranthese) found at a tap on system H is given. Samples from other points on the system gave similar findings.

The high level of fluoranthene pick-up seems almost certainly to involve the bituminous linings of distribution pipes. In follow-up research, laboratory experiments employing sections of pipe coated with such material produced a similar leaching effect when filled with groundwater from system H. This work is to be reported elsewhere.

Table 5 'Total PAHs' detected (ng/l) in distributed surface water

	System A	System B	System C	System D
Finished Water	ND ; 2	ND ; ND	8 ; 30	ND ; ND
Distributed Water				
mean	4(13 samples)	5(11 samples)	12(12 samples)	5(4 samples)
standard deviation	3	3	5	1

Table 6 'Total PAHs' detected (ng/l) in distributed ground water

	System E	System F	System G	System H*
Finished Water	4 ; 10	ND ; ND	3 ; 3	ND ; ND
Distributed Water				
mean	6(16 samples)	6(6 samples)	11(8 samples)	102(27 samples)
standard deviation	6	2	3	85

ND = Not detected
* Data for repeated analysis at one tap

CONCLUSIONS

Analytical procedures based upon TLC and HPLC have been developed which are suitable for determinations of the PAHs specified in the WHO standard for drinking water.

Of the various source waters examined only lowland rivers contained appreciable levels of PAHs. Urban run-off could be a significant source of PAHs in lowland rivers. However, conventional treatment was found to effectively remove such contamination.

Pick-up of PAHs during distribution was investigated but only in a few related groundwater systems was any appreciable effect found. This pick-up consisted almost entirely of fluoranthene (a non-carcinogen) and was probably due to leaching from bituminous pipe linings.

ACKNOWLEDGEMENTS

The work described was carried out under contract to the Department of the Environment and the authors wish to thank the Department and the Director of the Water Research Centre for permission to publish this paper.

REFERENCES

Borneff, J., 1967 GWF-Wasser Abwasser, 108, 1072.

Borneff, J., 1969, GWF-Wasser Abwasser, 110, 29.

Borneff, J., and H. Kunte, 1969, Arch. Hyg. Bakteriol, 220, 153.

Crane, R.I., B. Crathorne, and M. Fielding, 1978. Water Research Centre Technical Report, in press.

Crathorne, B. and M. Fielding, 1978, Proc. Anal Div. Chem. Soc., 15, 155.

Higginson, J., 1968, Canadian Cancer Conference Proceedings, Oxford, Pergamon, 40.

International Agency for Research on Cancer, 1973, Lyon, Volume 3.

Jones, D.W. and R.S. Matthews, 1974, Prog. Med. Chem., 10, 159.

Lao, R.C., R.S. Thomas, H. Oja, and L. Dubois. 1973 Analyt. Chem. 45, 908.

Sawicki, E., 1964, Chemist. Analyst, 53, part I 24, part II 56, part III 88.

MULTICOMPONENT POLYCYCLIC AROMATIC HYDROCARBON ANALYSIS OF INLAND WATER AND SEDIMENT

W.H. Griest

Oak Ridge National Laboratory

Oak Ridge, Tennessee, U.S.A., 37830

Many polycyclic aromatic hydrocarbons (PAHs) exhibit considerable carcinogenic, mutagenic, or toxic activity. Methods for reliable identification and quantification of PAHs in the aqueous environment are important for assessment of the potential environmental impact of processes releasing such compounds in aqueous discharges. This paper reports the development of multicomponent PAH analytical methods for inland water and sediment.

The method consists of separate steps for PAH extraction, isolation and concentration, and identification and quantification. Each step is considered separately below.

PAH EXTRACTION

A gross organic extract containing PAHs must first be obtained from the sample matrix. Ideally, the method should provide quantitative recoveries with a minimum of time and manpower. Traditionally, water samples are shaken (e.g., Strosher and Hodgson, 1975) with an immiscible organic solvent and sediments are soxhlet extracted (e.g., Giger and Schaffner, 1978) with an organic solvent. Radio-labeled PAH tracers were utilized to determine optimum conditions for extraction of PAHs from sediment and water. One litre samples of water were spiked in separate experiments with known activities of carbon-14 labeled naphthalene (^{14}C-Nap) and carbon-14 labeled benzo(a)pyrene, (^{14}C-BaP) and were extracted with 100 ml portions of water-equilibrated cyclohexane in 2 l separatory funnels equipped with Teflon stopcocks. Liquid scintillation counting of an aliquot of the pooled cyclohexane extract indicated quantitative recovery of both tracers in three extractions. Four extractions of water samples with 0.1 volumes of cyclohexane are carried out routinely to strive for complete recovery of all aqueous PAHs.

Sediments pose a greater extraction problem, but extended soxhlet-extraction with acetone is effective in recovering PAHs. One hundred gram aliquots of wet sediment were thoroughly mixed in separate experiments with ^{14}C-BaP or ^{14}C-Nap, packed into 43 x 123 ml cellulose thimbles, and were soxhlet-extracted with 250 ml of acetone. Periodically, aliquots of the acetone extract were withdrawn for tracer recovery measurements. Whereas aliquots of the water extracts could be directly analyzed by liquid scintillation spectrometry, the scintillation quenching from gross amounts of organic compounds in the raw sediment extracts necessitated a simple purification of an aliquot on a short layered Florisil/alumina column (3.5 cm x 1.1 cm OD each) with benzene prior to liquid scintillation tracer recovery measurements. Soxhlet extraction for 68 hours was sufficient to quantitatively recover both tracers. For convenience, such extractions are routinely extended to 72 hours.

Separate aliquots of sediment are dried at 110°C to constant weight to allow calculation of a dry/wet weight ratio for expression of PAH analytical results in terms of the dry weight of the sediment sample. Aliquots of sediment for PAH measurements are not dried; thus losses of volatile PAHs are reduced by this separate dry weight measurement.

PAH ISOLATION AND CONCENTRATION

The PAHs in the raw water and sediment extracts are next isolated by a two step adsorption column chromatography procedure. More complex isolation procedures (Kubota, Griest, and Guerin, 1975) incorporating liquid-liquid partitioning prior to adsorption column chromatography are not necessary to prepare a PAH isolate suitable for analysis by gas chromatography (GC). Cyclohexane extracts of water are concentrated to 10 ml and are applied to 10 g of Florisil in a 25 ml burette and are eluted with 150 ml of 6/1 hexane/benzene (volume/volume convention will be used throughout). Acetone extracts of sediment are reduced to 10 ml, diluted to 60 ml with water, and extracted four times with 20 ml of cyclohexane. These cyclohexane extracts are then treated identically to the water extracts described above.

The entire eluate from the Florisil column is then re-concentrated to 10 ml and passed through 20 g of neutral, deactivated (4 percent moisture) alumina, using a hexane/benzene step gradient of 6/1 (100 ml) and 2/1 (300 ml). The PAH isolate is obtained in two fractions defined by the tracers: a diaromatic fraction containing mainly alkyl naphthalens, acenaphthenes, and acenaphthalenes, and a polyaromatic fraction consisting of three to seven ring PAHs and their alkyl derivatives. Each fraction is separately reduced to a known volume by evaporative concentration with dry flowing nitrogen under reduced pressure and temperature.

PAH IDENTIFICATION AND QUANTIFICATION

Qualitative identification of PAHs in the purified PAH fractions is achieved by comparison of GC retention times and mass spectra with those of authentic PAH standards. Routine GC analyses are performed on a 6.6 m x 3 OD glass column packed with 3 percent Dexsil 400 on 100/120 mesh Supelcoport. Temperature programming from 100°C to 320°C at 1°C/min allows elution of PAHs ranging from two rings (e.g. naphthalene) through seven rings (e.g. coronene) in approximately four hours. Difficult to resolve PAH isomers which are separated with fairly good success include phenanthrene/anthracene, benz(a)anthracene/chrysene, and benzo(e)pyrene/benzo(a)pyrene.

PAH concentrations are calculated by comparing GC peak areas with those of an external standard. Recovery corrections for losses of PAHs in isolation and handling are estimated by liquid scintillation counting 0.1 volume aliquots of the fractions in 10 ml of scintillator solution prepared by dissolving 15 g of 2,5-diphenyloxazole and 190 mg of 1,4-bis-[5-phenyl-oxazole]-benzene in one gallon of reagent grade toluene.

EVALUATION

Initial evaluation of the method has focused upon PAHs in sediment because of the ability of stream and river sediments to concentrate PAHs from water by three to four orders of mganitude (Andelman and Suess, 1970). An experiment was conducted to determine the stability of sediment PAHs during storage of samples, and to define the accuracy and precision of the analytical method. For this experiment, eight 100 g replicate samples of stream sediment previously demonstrated to be free of detectable PAHs (limit of detection approximately 0.05 μg/g for this sample size) were spiked with multiple unlabelled PAHs at concentrations of about 3 μg/g each to simulate contaminated sediment. Each sample was spiked with ^{14}C-Nap and ^{14}C-BaP, and after addition of 10 ml acetone was stored at 4°C in the dark. Four samples were analyzed immediately; two others were analyzed after two weeks of storage and the final two after four weeks of storage. The results of the analyses expressed in terms of the percentage corrected recoveries of the polyaromatics are shown in Table 1. Data for picene and o-phenylene pyrene are not included because they were not recovered. The constant absolute recovery of the ^{14}C-BaP tracer and the corrected recoveries of the spiked PAHs indicate that there is no significant loss of PAHs (other than the two noted) from these sediment samples over a four week

Table 1 Corrected Recoveries of Polyaromatics and Tracer in Spiked Sediment as a Function of Storage Time

PAH	Corrected Percent Recovery at Storage Time			Cumulative Data	
	Initial	2 Weeks	4 Weeks	Avg. ± Std. Dev.	
Fluorene	56.8	41.1	67.8	55.6 ±	12.5
1-Methyl Fluorene	80.4	78.5	85.8	81.3	7.6
Phenanthrene	87.4	84.7	92.3	87.9	6.6
Anthracene	90.3	94.3	95.7	92.7	5.7
2-Methyl Anthracene	96.9	102.9	102.9	99.9	5.8
1-Methyl Phenanthrene	99.3	100.6	102.1	100.0	5.2
9-Methyl Anthracene	66.7	78.6	75.2	71.8	12.0
Fluoranthene	102.8	104.5	106.6	104.7	4.6
Pyrene	102.3	101.4	106.6	103.1	4.6
Benzo(a)fluorene	109.4	110.6	110.5	110.0	4.6
Benzo(b)fluorene	110.3	110.7	110.1	110.3	5.1
1-Methyl Pyrene	107.3	106.5	108.7	107.5	4.6
Benz(a)anthracene	111.7	111.1	110.0	111.2	4.9
Chrysene	111.8	108.0	107.7	109.8	5.9
Benzo(a)pyrene	111.2	112.0	109.8	111.2	4.9
Perylene	116.9	116.3	111.6	115.6	5.7
3-Methyl Cholanthrene	106.8	108.8	104.8	106.8	6.2
Benzo(ghi)perylene	101.4	102.2	102.5	101.9	4.4
Anthanthrene	86.4	105.4	85.5	90.9	12.8
^{14}C-BaP Absolute Recovery	96.2	98.6	97.6	97.1	3.0

period of storage under these storage conditions. Thus, we conclude that the combination of acetone, reduced temperature, and darkness suppress PAH degradation mechanisms for this particular sediment and that chemical analysis may be delayed at least one month with no appreciable effects on results. However, this conclusion may not necessarily hold true for sediments collected from other sites.

Analytical accuracy and precision are defined by the last two columns of data in Table 1. Recoveries of most polyaromatics are close to quantitative, for species ranging from alkylated three-ring PAHs through six-ring PAHs. These data indicate the ability of the multicomponent method to quantitate a large number of PAHs with reasonable accuracy. PAHs on both extremes of the polyaromatic fraction are recovered with a low bias, suggestive of evaporative losses for the more volatile PAHs (e.g. fluorenes), and incomplete sediment extraction for the very large PAHs (e.g. anthanthrene). Precision, as indicated by the relatively low standard deviations, is very good for such a wide range multicomponent method.

Table 2 shows the absolute recoveries of seven representative diaromatics. Absolute recoveries are reported because of the low and widely differing recoveries, which prevent any one species (such as ^{14}C-Nap) from accurately defining the recoveries of the others. The recoveries are inversely proportional to vapor pressures of the

Table 2 Absolute Recovery of Representative Diaromatics in Spiked Sediment

Diaromatic	Average $\pm$ Standard Deviation: Absolute Recovery (%)	Average $\pm$ Standard Deviation: Differential Recovery Factor
Napthalene	26.0 $\pm$ 11.1	1.0
2-Methyl Naphthalene	36.7 $\pm$ 11.7	1.30 $\pm$ 0.11
1-Methyl Naphthalene	44.0 $\pm$ 9.7	1.63 $\pm$ 0.35
Biphenyl	47.2 $\pm$ 12.6	1.71 $\pm$ 0.26
2,6-Dimethyl Naphthalene	50.0 $\pm$ 12.6	1.74 $\pm$ 0.27
1,5- + 2,3-Dimethyl Naphthalenes	56.8 $\pm$ 8.8	1.93 $\pm$ 0.34
Acenaphthene	61.5 $\pm$ 10.8	2.09 $\pm$ 0.48

diaromatics, suggesting evaporative losses during solvent concentration steps. Differential recovery factors relating the recovery of each diaromatic to a single species could be employed to correct the recoveries of the diaromatics, but the precision of these factors is poor. Further work on improving the analysis of diaromatics is in progress.

APPLICATIONS

The bulk of the multicomponent PAH analytical applications have been made with an older version of the present isolation procedure, which incorporated a serial solvent partitioning prepurification step prior to the adsorption column chromatography steps (Kubota, Griest, and Guerin, 1975). This prepurification step subsequently has been found to be unnecessary for sediment analysis, but results of the older procedure are equivalent to those obtained by the present procedure. Some of the former are described below.

Figures 1 and 2 show the polyaromatics in stream sediments collected in the vicinity of a coking plant, a steel rolling mill (and a nearby city), and a petroleum tank farm. A characteristic "PAH fingerprint" is obtained for each source. The coking plant releases mainly the parent, unsubstituted PAHs ranging up to at least six rings in size, characteristic of high temperature PAH formation processes (Blumer, 1976). Concentrations of PAHs in the sediment (Table 3) range from 0.1 to 31 μg/g (dry weight basis), and are very similar to those reported in an EPA study (Bass et al, 1974). Concentrations of PAHs in the effluent channel water are included in Table 3. Although they are three to four orders of magnitude lower than in sediment, (in agreement with data reviewed in Andelman and Suess, 1970) they do not parallel those of the

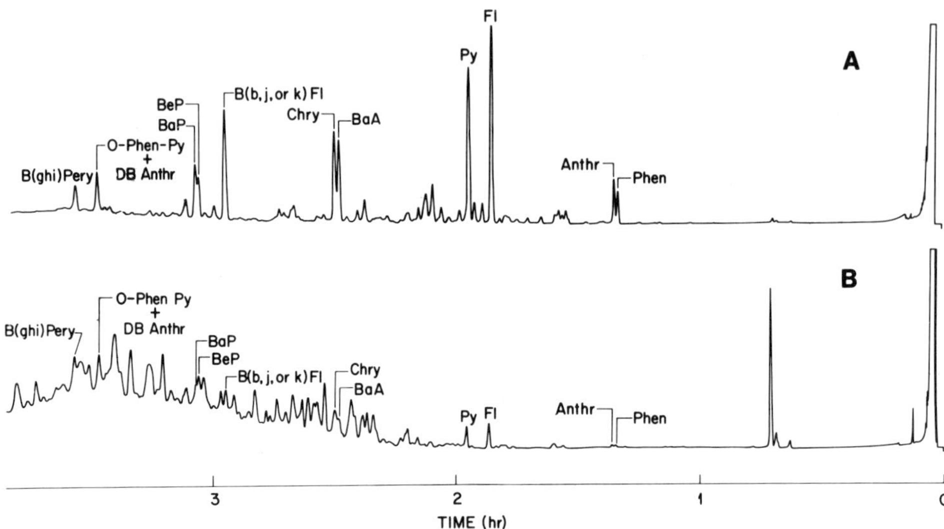

Figure 1. GC Separation of polyaromatic fraction from stream sediments collected near a (A) coking plant and (B) steel rolling mill.

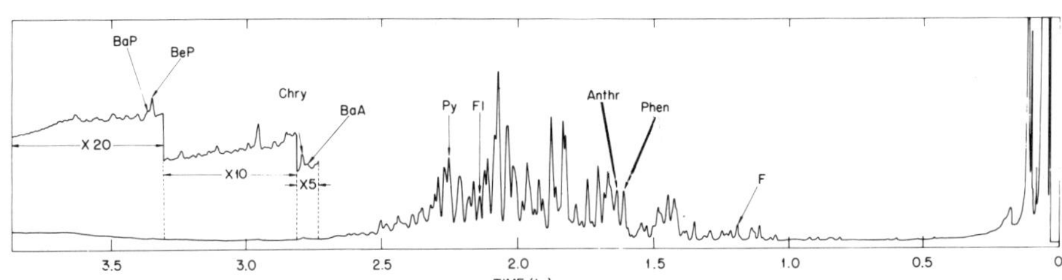

Figure 2. GC Separation of polyaromatic fraction of stream sediment collected near a petroleum tank farm.

Table 3 Concentrations of PAHs in Effluent Channel from a Coking Plant

	Concentration	
	Sediment	Water
	μg/g, dry sediment	μg/L
1-Methyl Fluorene	0.11	ND
Phenanthrene	3.6	0.40
Anthracene	6.7	0.31
2-Methyl Anthracene	1.4	2.4
1-Methyl Phananthrene	0.71	ND
Fluoranthene	31.	2.8
Pyrene	23.	4.0
Benzo(a)fluorene	7.2	0.80
Benzo(b)fluorene	3.2	0.81
Benzo(c)phenanthrene	2.1	ND
Benzo(ghi)fluoranthene	2.3	ND
Benz(a)anthracene	15.	2.0
Chrysene/Triphenylene	15.	1.7
Benzo(b,j, and/or k)fluoranthene	23.	ND
Benzo(a + e)pyrene[a]	19.	1.8
Perylene	3.8	ND
o-Phenylene pyrene + dibenz-(a,c and/or a,h)anthracene	8.6	0.95
Benzo(ghi)perylene	7.3	2.0
Anthranthrene	2.3	ND
TOTAL IDENTIFIED	175 μg/g	20 μg/L

[a]Incomplete resolution. BaP estimated at approximately 10 μg/g.

sediment. This result may arise from the fact that the samples were taken from a dynamic, flowing system in which sediment and water PAHs probably are not in equilibrium.

In marked contrast to the unsubstituted PAHs from the coking plant, the sediment PAHs near the steel rolling mills and the petroleum tank farm appear to be predominantly multialkylated derivatives as is commonly observed with PAHs formed at lower temperatures (Blumer, 1976). However, even between the latter two sources, significant differences in concentration as a function of PAH ring size are apparent. The multialkylated sediment PAHs sampled near the steel rolling mill are mainly four through six rings in size while those near the petroleum tank farm are limited to three and four rings. Concentrations of representative PAHs showing these different ring size distributions are presented in Table 4. More detailed analysis of other individual species is made very difficult

Table 4 Comparison of Representative PAHs from Three Sites

Comparison of Representative PAHs from Three Sites

Collection Site	Sediment Concentration, μg/g (dry weight basis)		
	Anthracene	Benz(a)anthracene	Benzo(a)pyrene
Coking Plant	6.7	15.	10.
Steel Rolling Mill	0.065	0.35	0.60
Petroleum Tank Farm	3.4	0.13	<0.049

by the chromatographic peak overlapping of multi-alkylated PAHs in the latter two samples, but this problem can be overcome by subjecting the polyaromatic isolate to gel filtration on Biobeads (Severson, Snook, Arrendale, and Chortyk, 1976) prior to GC analysis.

IMPROVEMENTS

Applications thus far have centered upon analysis of polyaromatics. Recoveries of diaromatics have been low and difficult to reproduce. One method of alleviating the diaromatic recovery problem is through the use of micro distillation for solvent removal. This method has been successfully employed in the analysis of diaromatics in tobacco smoke condensate (Schmeltz, Tosk, and Hoffmann, 1976), but does not seem suited for the routine analysis of large numbers of samples. An alternate to such procedures is a dual tracer technique in which two tracers (carbon-14 and tritium-labelled) bracketing the volatility range of the diaromatic fraction are added to the sediment and then are simultaneously measured by dual channel liquid scintillation counting after isolation of the fraction. The recovery of each diaromatic eluting between these two extreme tracers presumably could be determined by interpolation.

Studies examining the validity of the dual-tracer approach employed ^{14}C-BaP and tritium-labelled BaP (^{3}H-BaP) which are readily available commercially. These tracers differ only in the radiolabel, thus comparison of the recovery of ^{3}H-BaP with that of ^{14}C-BaP would indicate the utility of ^{3}H-labelled PAH tracers for such work. Known activities of ^{14}C-BaP and ^{3}H-BaP were supplied to a sediment. After extraction and isolation, the recoveries of the tracers were measured. The results expressed as the relative recovery of ^{3}H-BaP to ^{14}C-BaP are shown in Table 5. Included also are the results for column chromatography and gel filtration on commonly available packings. The results indicate that considerable losses of ^{3}H-BaP occur in the analytical procedure, particularly in alumina column chromatography. The data suggest losses of ^{3}H-BaP also during

Table 5 Relative Recovery of ^{3}H-BaP to ^{14}C-BaP

Experiment	Relative Recovery[a]
Sediment Extraction, Florisil Alumina Column Chromatography	0.71
Florisil Column Chromatography	∿ 1.0
Alumina Column Chromatography	0.88
Silica Gel Column Chromatography	0.85
Biobeads SX-12 Gel Filtration	0.97

[a] ^{3}H-BaP/^{14}C-BaP

soxhlet extraction. Thus the dual-tracer approach is not compatible with existing purification methods. It is felt that ^{3}H-^{1}H exchange between the ^{3}H-BaP and active surfaces on the adsorbent or sediment is responsible for the loss of ^{3}H activity. This hypothesis is supported by the observation that the more active adsorbents (alumina and silica gel) cause greater losses of ^{3}H activity than less active adsorbents (Florisil) or the "gentle" gels (Biobeads). Thus, a purification procedure featuring the latter two materials would allow a dual-tracer method if a means of suppressing ^{3}H-^{1}H exchange during sediment extraction could be devised. The Biobeads also offer the additional benefit of separating the complex multi-alkylated PAHs from the parent and simple alkylated PAHs, allowing interference-free analysis of the latter in complex PAH isolates (Severson, Snook, Arrendale, and Chortyk, 1976).

Another improvement to the method is the use of wall coated glass capillary columns. Figure 3 shows the superior resolution of a polyaromatic fraction by a glass capillary column than by a 6.6 m packed column, with both columns operated under the same column oven temperature programme (110°C to 320°C at 2°C/min). Normally, the packed column oven is temperature programmed at one-half the rate, and the separation requires twice the time of the capillary column separation. Thus, the capillary column can generate more detailed data in less time than the packed column.

CONCLUSIONS

The multicomponent PAH analytical methodology described in this paper generates a detailed characterization of the PAHs in sediment and water samples. Applications suggest that each source of PAH discharged into the aqueous environment is characterized by its own "PAH fingerprint" which can be differentiated from those of other sources.

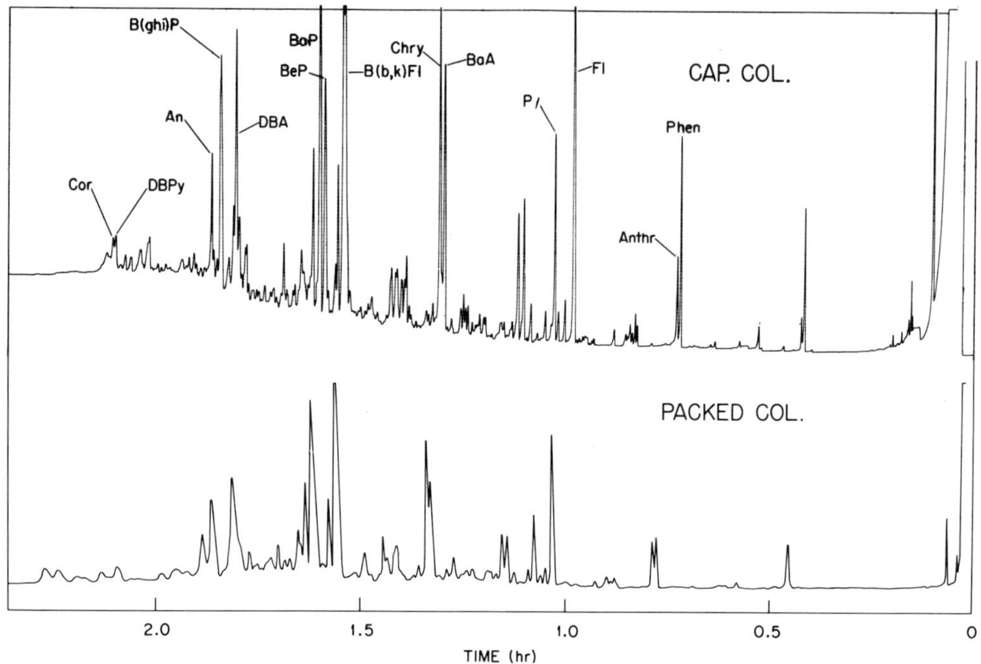

Figure 3. Resolutions of polyaromatic fraction from sediment on capillary and packed GC columns

ACKNOWLEDGEMENTS

The contributions of several colleagues and coworkers at Oak Ridge National Laboratory are gratefully acknowledged. Steve Herbes and George Southworth collected the samples analyzed in this report and Roberta Reagan provided invaluable technical assistance.

This research was sponsored by the Division of Biomedical and Environmental Research, U.S. Department of Energy, under contract W-7405-eng-26 with the Union Carbide Corporation.

REFERENCES

Andelman, J.B., and M.J. Suess (1970). Polynuclear Aromatic Hydrocarbons in the Water Environment, Bull. Wld. Hlth. Org., 43, 479-508.

Blumer, M. (1976). Polycyclic Aromatic Compounds in Nature, Sci. Amer., 234, 34-45.

Brass, H.J., W.C. Elbert, M.A. Feige, E.M. Glick, and A.W. Lington (1974). United States Steel, Lorain, Ohio Works Black River Survey: Analysis for Hexane Organic Extractables and Polynuclear Aromatic Hydrocarbons, Organic Chemistry Laboratories, National Field Investigations Center-Cincinnati, Office of Enforcement and General Council, U.S. Environmental Protection Agency, Cincinnati, Ohio.

Giger, W., and C. Schaffner (1978). Determination of Polycyclic Aromatic Hydrocarbons in the Environment by Glass Capillary Gas Chromatography, Anal. Chem., 50, 243-249.

Kubota, H., W.H. Griest, and M.R. Guerin (1975). Determination of Carcinogens in Tobacco Smoke and Coal-Derived Samples - Trace Polynuclear Aromatic Hydrocarbons, Trace Substances in Environmental Health, Vol. IX, D.D. Hemphill, Ed., University of Missouri, Columbia, p.281-289.

Schmeltz, I., J. Tosk, and D. Hoffmann (1976). Formation and Determination of Naphthalenes in Cigarette Smoke, Anal. Chem., 48, 645-650.

Severson, R.F., M.E. Snook, R.F. Arrendale, and O.T. Chortyk (1976). Gas Chromatographic Quantitation of Polynuclear Aromatic Hydrocarbons in Tobacco Smoke, Anal. Chem., 48, 1866-1872.

Strosher, M.T., and G.W. Hodgson (1975). Polycyclic Aromatic Hydrocarbons in Lake Waters and Associated Sediments: Analytical Determination by Gas Chromatography-Mass Spectroscopy, Water Quality Parameters, ASTM-STP-573, p.259-270.

AN EVALUATION OF PURGING METHODS FOR ORGANICS IN WATER

K.W. Lee, R.G. Oldham, G.L. Sellman, L.H. Keith,

L.P. Provost, P.H. Lin, and D.L. Lewis

Radian Corporation, PO Box 9948, Austin, Texas 78766

Our evaluation of the purge and trap method as employed for analysis of over 2,000 drinking water samples is outlined in detail in this paper. Necessary documentation of this evaluation includes detector linearity, reproducibility of standards and detection levels. Variations of the purge method include different types of instrumentation. Discussion of purging units, desorbing devices and a cryogenic loop are included. An evaluation would not be complete without a discussion of the problems one can encounter using the purge methods.

QUALITY ASSURANCE

Because the purge and trap method presents so many opportunities for errors, a stringent quality assurance program was found to be necessary for routine gas chromatographic analyses. As part of Radian's quality assurance program, Hall detector linearity, reproducibility in making and analyzing purgeable standards, blank sample interference, efficiency of purging, and differences in internal versus external standardization were determined.

Detector Linearity

We evaluated the U.S. Environmental Protection Agency's (EPA) "Purge and Trap" Method for volatile halogenated compounds.[1,2] The Hall electrolytic conductivity detector was chosen because of its stability during gas chromatographic oven programming, its response to only halogeanted compounds, and its linearity. The Hall detector is linear from 1 ppb to 200 ppb for most halogenated purgeable compounds from a 5.0 mℓ water sample. A least-squares regression analysis, assuming a linear model, was completed for several of the compounds using concentration as the dependent variable. The data used in developing these regression equations for chloroform is plotted in Figure 1. The 90 percent confidence interval for the regression intercept includes the origin. Therefore, this evidence indicates the Hall detector has a linear response in this range.

Analytical Reproducibility

The gas chromatographic conditions used in all studies are similar to those outlined in the EPA's April, 1977, Consent Decree Protocol.[3] The purge time for a 5.0 mℓ sample is 12 minutes with a

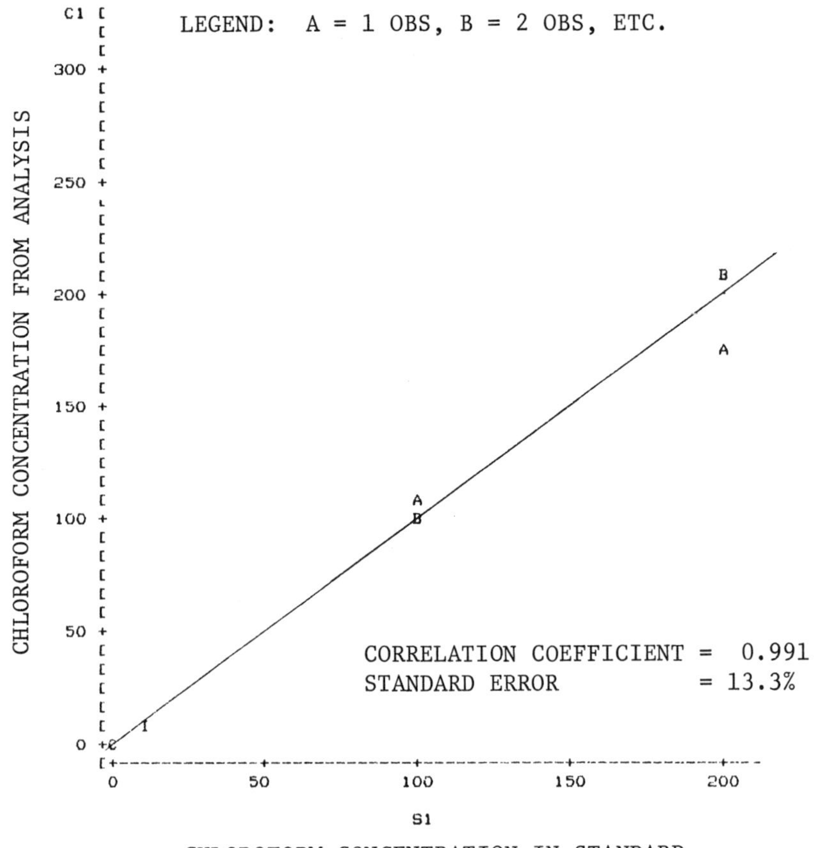

Figure 1. Plot of chloroform concentrations from 1 to 200 ppb.

40cc/minute gas flow as in the Protocol. During the shorter 3-minute desorption step, the gas chromatographic oven is held at room temperature instead of the recommended 60°C. The oven temperature is rapidly raised to 60°C with a 4-minute hold. The oven is then heated to 170°C at 8°/minute and held at this temperature for 10 minutes. A 3.5 m by 0.2 cm I.D. glass column was packed with 3.2 m of 0.2% Carbowax 1500 on Carbopack C preceded by 0.3 m of 2.0% Carbowax 1500 on Chromosorb W. Nitrogen was used for both the purging and the carrier gas. A typical chromatogram is illustrated in Figure 2 using the above conditions.

To test reproducibility, two standards were prepared at 10 μg/ℓ and 1 μg/ℓ levels (after final dilution in water). The standards contained 16 of the volatile halogenated compounds on EPA's Priority Pollutant list. The 10 μg/ℓ standard was run initially and entered into the memory of a Hewlett-Packard 3380A reporting integrator. The analyses were made in the order listed in Table 1. Table 1 also gives the reported concentrations in μg/ℓ from these runs. The data show that:

- most of the analyses of both the 10 μg/ℓ and the 1 μg/ℓ standard agree within 30 percent of the calculated values when corrected for the blank interferences, and

- only trichloroethylene, bromoform, and dibromochloromethane/1,1,2-trichloroethane/cis-1,3-dichloropropene had significant (greater than 5 percent) residues that could be obtained from a longer purge.

Purging Efficiency

Table 2 summarizes the repurging data for four haloform standards at different concentrations. As expected, an increase in recoveries from repurged samples occurs with increasing molecular weight of the purged compounds. Although the amounts of each compound repurged tend to increase with concentration, there is no statistical increase in the percent of each compound repurged over the concentrations studies.

Variability of Standards and Instrumentation

In this study, four standards were independently prepared with 10 μg/ℓ each (after final dilution in water) of the 16 compounds. Three standards (labeled Numbers 2, 3, and 4) were each analyzed after the first halogenated purgeable standard (about 2 months old, labeled Number 1) was run and entered into the memory of a Hewlett-Packard 3380A reporting integrator. This sequence of analyses was

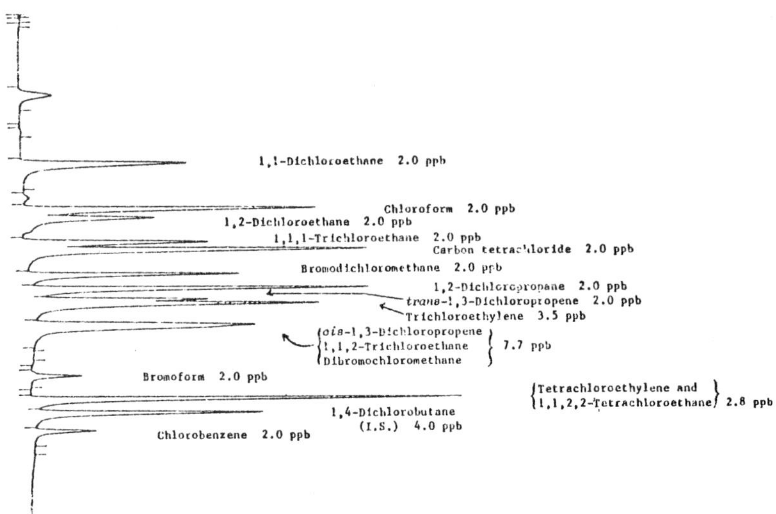

Figure 2. Typical gas chromatogram of purgeable halogenated priority pollutants using a Hall detector.

Table 1 Reproducibility of 10 ppb and 1 ppb Standards

Compound	Lower Detection Level (μg/ℓ)	Set-in Standard (μg/ℓ)	Second Analysis (μg/ℓ)	Third Analysis (μg/ℓ)	Repurge from Second Analysis (μg/ℓ)	Radian Blank Water (μg/ℓ)	First Analysis 1.0 ppb Standard (μg/ℓ)	Second Analysis 1.0 ppb Standard (μg/ℓ)	Third Analysis 1.0 ppb Standard (μg/ℓ)
1,1-Dichloroethane	0.1	10.0	10.4	11.0	N.D.	N.D.	0.6	0.7	0.6
Chloroform	0.1	10.0	10.3	10.7	0.2	0.4	1.1	1.1	1.0
1,2-Dichloroethane	0.1	10.0	10.4	10.8	0.1	N.D.	1.0	1.0	1.0
1,1,1-Trichloroethane	0.1	10.0	10.2	10.7	N.D.	0.1	0.7	0.7	0.7
Carbon tetrachloride	0.01	10.0	10.3	10.6	N.D.	N.D.	0.9	0.8	0.8
Bromodichloromethane	0.1	10.0	10.9	11.2	N.D.	N.D.	1.1	1.1	1.2
1,2-Dichloropropane	0.1	10.0	10.2	10.9	N.D.	N.D.	1.0	0.9	1.0
trans-1,3-Dichloropropene	0.1	10.0	10.4	10.8	N.D.	N.D.	1.0	1.0	1.0
Trichloroethylene	0.1	16.4	16.6	17.5	2.3	0.8	2.1	2.0	2.0
cis-1,3-Dichloropropene and 1,1,2-Trichloroethane and Dibromochloromethane	0.1	30.0	30.8	32.3	1.7	N.D.	3.4	3.1	3.2
Bromoform	0.1	10.0	10.3	10.5	1.0	N.D.	0.6	0.6	0.7
1,1,2,2-Tetrachloroethane and Tetrachloroethylene	0.1	8.0	8.2	8.7	0.2	0.2	0.8	0.7	0.7
Chlorobenzene	0.1	10.0	10.8	11.2	N.D.	N.D.	0.8	0.8	0.8

N.D. = Not Detected.

repeated after evaluating Standard Number 1 as an unknown. Table 3 gives the reported concentrations in μg/ℓ from these runs. Group 1 is the first four analyses and Group 2 is the second four analyses. The compounds are listed in order of peak elution as noted in Figure 2 with the omission of the internal standard, 1,4-dichlorobutane.

Table 2 Haloform Recoveries after Repurging

	Chloroform	Dichlorobromomethane	Dibromochloromethane	Bromoform
Standard A Concentrations:	1 μg/ℓ	1 μg/ℓ	1 μg/ℓ	1 μg/ℓ
Repurge Sample	0.09 μg/ℓ	<DL	<DL	<DL
Percent Repurged	9%	<DL	<DL	<DL
Standard B Concentrations:	10 μg/ℓ	10 μg/ℓ	10 μg/ℓ	10 μg/ℓ
Repurge	0.53 μg/ℓ	0.47 μg/ℓ	0.83 μg/ℓ	2.08 μg/ℓ
Duplicate Repurge	0.19 μg/ℓ	0.09 μg/ℓ	0.26 μg/ℓ	1.03 μg/ℓ
Average Repurge	0.36 μg/ℓ	0.28 μg/ℓ	0.54 μg/ℓ	1.56 μg/ℓ
Percent Repurged	3.6%	2.8%	5.4%	15.6%
Standard C Concentrations:	100 μg/ℓ	50 μg/ℓ	20 μg/ℓ	5 μg/ℓ
Repurge	0.27 μg/ℓ	0.05 μg/ℓ	<DL	<DL
Duplicate Repurge	2.18 μg/ℓ	0.87 μg/ℓ	0.63 μg/ℓ	0.51 μg/ℓ
Average Repurge	1.2 μg/ℓ	0.46 μg/ℓ	0.32 μg/ℓ	0.26 μg/ℓ
Percent Repurged	1.2%	0.9%	1.6%	5.1%
Standard D Concentrations:	200 μg/ℓ	50 μg/ℓ	20 μg/ℓ	5 μg/ℓ
Repurge Sample	3.98 μg/ℓ	0.77 μg/ℓ	0.74 μg/ℓ	0.63 μg/ℓ
Percent Repurged	2.0%	1.5%	3.7%	12.6%
Water Blank Concentrations:	0	0	0	0
Repurge Sample	0.07 μg/ℓ	<DL	<DL	<DL

An Analysis of Variance and Multiple Range Test were used to test the hypothesis that all four standards were equivalent for each of the 13 peak concentrations. Results of these analyses showed that there are significant differences between the old standard and the three new standards for a few of the measured concentrations. Table 4 summarizes the analytical variability for the 13 measureable concentrations for each of the four standards.

It was concluded that repeated GC analysis of the same purgeable standard should differ by less than 30 percent most of the time and that the variability introduced due to differences between standards evaluated is about the same magnitude as the instrument variability. However, most of this variability is due to differences between the old standard and the three new standards. Only chlorobenzene showed significant differences between the three standards independently prepared at the same time.

External Versus Internal Standardization

The Coefficient of Variation (CV) and repeatability were calculated from the analyses of four haloform standards. The CV is the percent standard deviation relative to the mean. The 95 percent repeatability statistic represents the maximum difference to be expected between two analyses of the same standard. These values were calculated for each of the four haloform standards after each had been analyzed at least in triplicate. The results are summarized in Table 5 with and without internal standardization. We conclude from the 95 percent repeatability averages that internal standardization gives better reproducibility by a factor of two than external standardization.

Detection Levels

Detection levels in this study are defined as integrable peaks. Since the purge and trap method needs to be as automated as possible

Table 3 Reproducibility among Equivalent Standards

Standard	Group	C2	C3	C4	C5	C6	C7	C8	C9	C10	C11	C12	C13	C14
1	1	10.0	10.0	10.0	10.0	10.0	10.0	10.0	10.0	16.4	30.0	10.0	18.0	10.0
1	2	9.1	8.9	10.5	8.2	10.8	11.7	11.1	11.1	12.6	32.7	10.5	16.4	4.2
2	1	10.4	7.3	11.4	11.3	11.2	13.7	12.5	11.9	16.4	33.5	10.0	19.4	7.6
2	2	11.7	8.8	11.0	10.5	11.8	12.1	11.4	11.7	16.3	31.1	9.7	19.2	15.5
3	1	12.4	7.8	12.5	12.9	13.9	15.1	12.9	13.1	16.2	31.8	10.6	20.4	9.2
3	2	12.0	8.6	11.1	11.6	12.1	13.3	11.7	12.1	18.1	34.2	10.8	21.1	18.2
4	1	10.1	6.9	10.6	11.0	12.4	13.2	12.1	13.0	12.6	32.7	9.6	19.4	6.8
4	2	10.1	7.7	9.4	10.7	10.9	11.0	10.8	10.9	15.1	28.8	9.2	19.3	13.9

Table 4 Variance due to Instrument/Analysis

Compound(s)	Variance Due to Instrument/ Analysis (μg/ℓ)	Percent of Total Variability	Standard Deviation as Percent of Mean	95% Repeatability[1]
C2	.33	22%	5.4%	15%
C3	.59	54%	9.3%	26%
C4	.48	51%	6.4%	18%
C5	.71	35%	7.8%	22%
C6	.81	53%	7.7%	22%
C7	1.69	60%	10.4%	29%
C8	.83	75%	7.2%	20%
C9	.83	72%	7.8%	22%
C10	3.04	78%	11.3%	32%
C11	4.25	100%	6.5%	18%
C12	.068	20%	2.6%	7%
C13	.38	17%	3.3%	9%
C14	28.43	100%	50.0%	140%
Median Value	.81	54%	7.7%	22%

[1]Repeated analyses of the same standard.

for large numbers of samples, it was necessary to define as measureable only those peaks which the Hewlett-Packard 3380A integrator would "see". The 16 component halogenated standard was simply diluted by factors of two to test the detection levels. The measured detection levels are shown in Table 1.

A problem we have seen sporadically is that of blank interferences. This may limit the reportable detection level in a sample, but due to its variability the blank must be measured with each set of samples. Further discussion and clean-up procedures of the blank are in the last section of this paper.

INSTRUMENTATION

The purge and trap technique consists of three steps: (1) purge and trap, (2) desorb, and (3) chromatograph and detect. These can be achieved through a variety of instrumentation. The purge step can be set apart from the desorption and chromatographic steps or all can be accomplished through connecting tubing. Both types of instrumentation have been evaluated at Radian Corporation.

Purge and Trap - Desorption Devices

Both commercially available and Radian-made devices have been evaluated for concentrating purgeable organic compounds from water samples. Tekmar LSC-1 concentrators used initially were later abandoned for GC analyses because of "memory effects" within the valves and transfer lines. Radian-made devices feature very short path lengths between the trap and the head of the gas chromatographic column and no solenoid valves.

In order to provide "instant" purged samples from field sites, Radian constructed a portable Field Purging Unit (Current Model FPU-2) which is shown in Figure 3. This device can purge two samples

Table 5 Coefficient of Variation and Repeatability of Haloform Analysis

	Internal Standardization		External Standardization	
Compound	Average CV	95% Repeatability	Average CV	95% Repeatabiity
Chloroform	11.8%	31 %	18.9%	52 %
Dichlorobromomethane	10.8%	30 %	22.6%	63 %
Dibromochloromethane	9.0%	25 %	19.3%	54 %
Bromoform	11.5%	32 %	22.4%	62 %
Average	10.6%	29.4%	20.8%	57.7%

at a time onto our Tenax-Chromosorb 102 traps. Using this device about 200 water samples from steam electric plants in Wyoming, West Virginia, and Florida were purged in the field. The traps were then transported to the Austin, Texas, laboratory (with no thiosulfate preservation or sample breakage) and analyzed at leisure.

Tests of storage and shipping were conducted on the traps used with the FPU-2. Table 6 contains data for traps stored for 10 days at ambient temperature and shipped 3,500 miles. No loss or decomposition within experimental error was observed. During these experiments several traps were left open to the air while others were capped closed with Swagelok fittings. Even those left open showed no loss of the haloforms. However, the traps which were not capped did absorb many contaminates from ambient air such as Freons from refrigeration units and solvents in the laboratory. Figure 4 shows chromatograms of samples stored in capped and uncapped traps. These traps must be capped during shipping for sample integrity to be maintained. The traps can also be used for ambient air monitoring of many compounds.

Ambient Temperature Traps

At the heart of these purging devices is the trap. We have abandoned the EPA recommended [1,2,3] trap packing of Tenax and silica gel in favor of Tenax and Chromosorb 102. The trap is made of 28 cm by 1/8 inch "OD", glass-lined, stainless steel tubing. This trap performed just as well for us as the EPA recommended trap and eliminated the problems associated with adsorption of water vapor by the silica gel. This adsorbed water can cause "steam cleaning" of the chromatographic columns with the disruptive results illustrated in Figure 5. The peaks in the "steam cleaned" chromatogram have had both their retention times and peak shapes altered.

Table 6 Purge Trap Storage Data

	PPB OF HALOFORMS (AVERAGED ± σ)*			
Description	$CHCl_3$	$CHBrCl_2$	$CHBr_2Cl$	$CHBr_2$
Haloform Standard (Immediate Analysis) (Start of Storage Test)	130.2 ± 5.3	61.6 ± 0.2	25.1 ± 0.2	5.6 ± 0.2
Haloform Standard (Immediate Analysis) (End of Storage Test)	128.2 ± 6.4	82.5 ± 1.1	27.2 ± 0.3	5.6 ± 0
Haloform Standard** (10-Day Storage - Shipped 3,500 Miles)	135.6 ± 3.5	79.3 ± 5.6	27.2 ± 0.7	5.3 ± 0.1

*All samples quantitated versus a 5 ppb standard. Averages based on three separate samples.

**One set of data was rejected based on Dixon Criteria for Testing Extreme Observations (20% level).

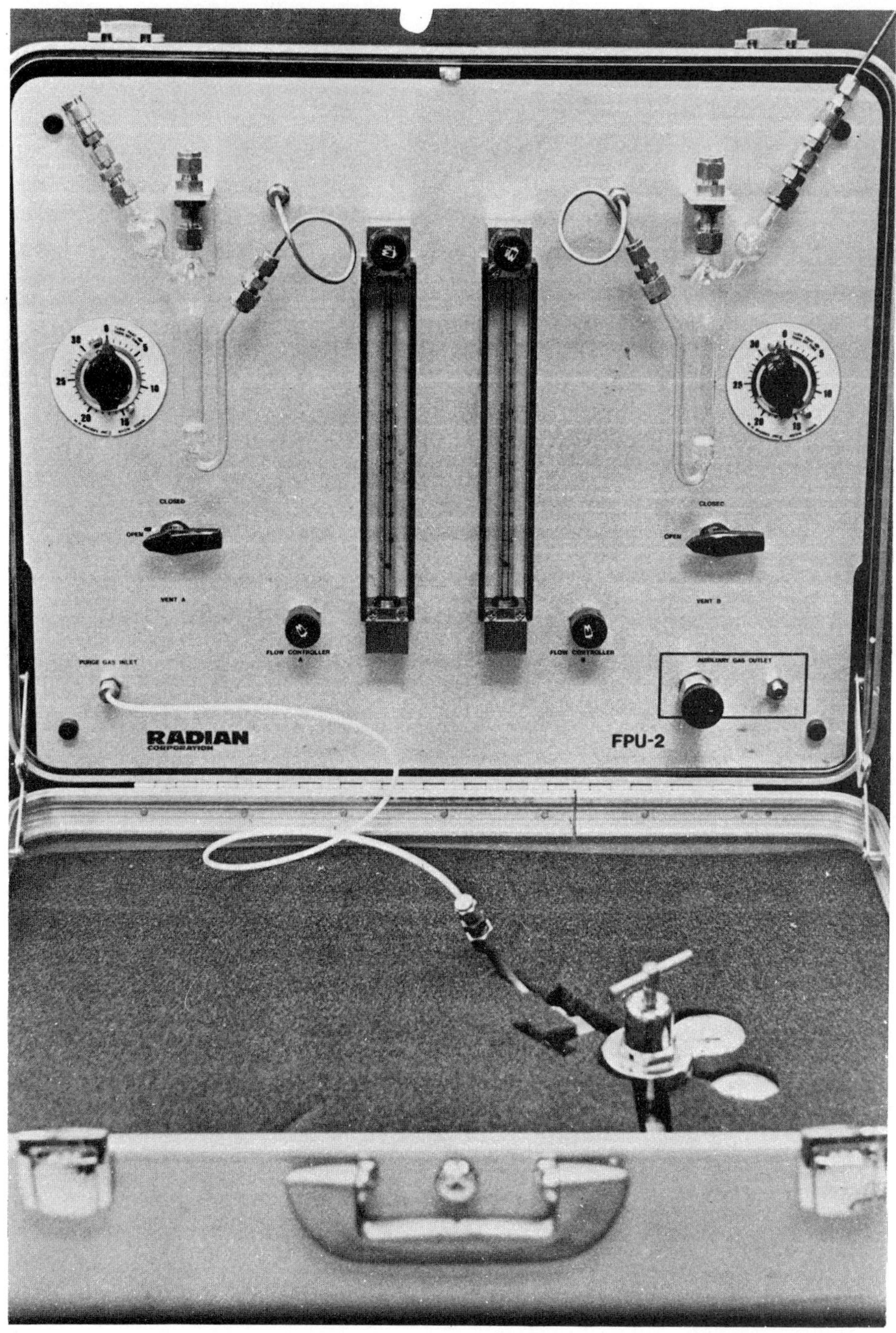

Figure 3. Field purging unit.

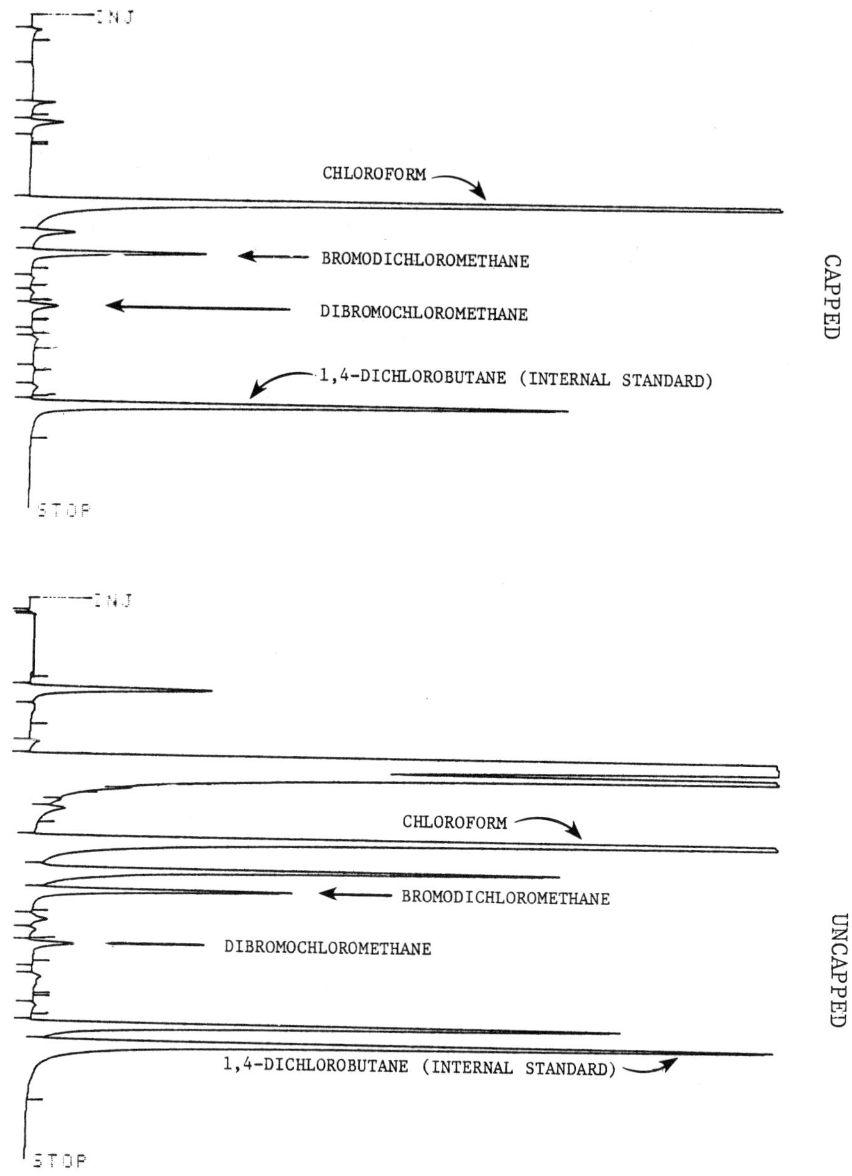

Figure 4. Chromatograms of a capped trap containing a sample of municipal water and an uncapped trap of the same water.

A

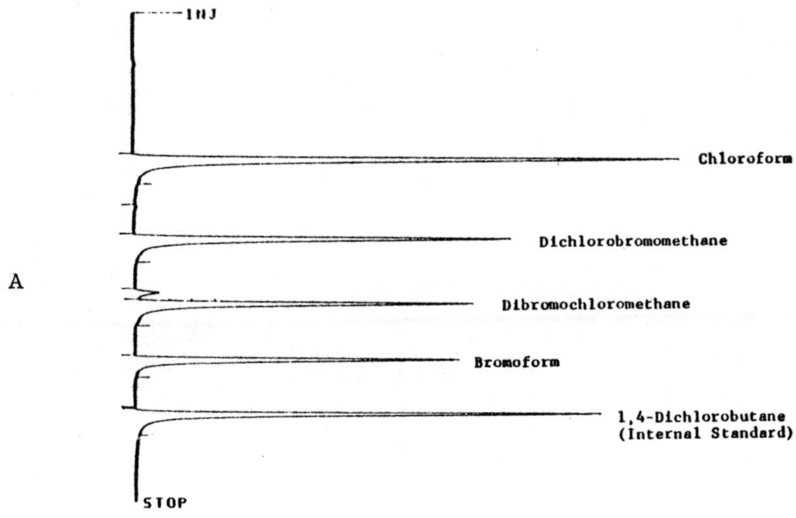

B

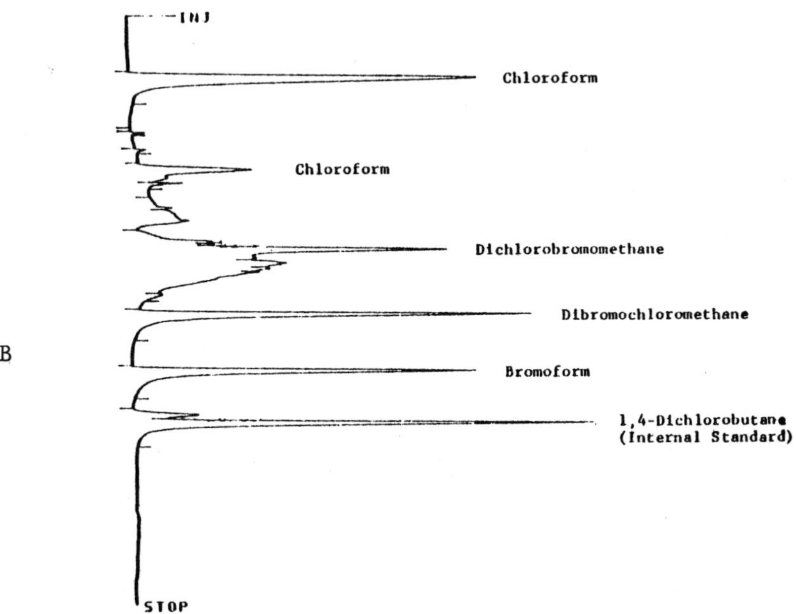

Figure 5. The effect of adsorbed water on a 10 μg/ℓ purgeable standard
A. Normal chromatogram using a new tenax/silica gel trap
B. Chromatogram using a degenerated tenax/silical gel trap.

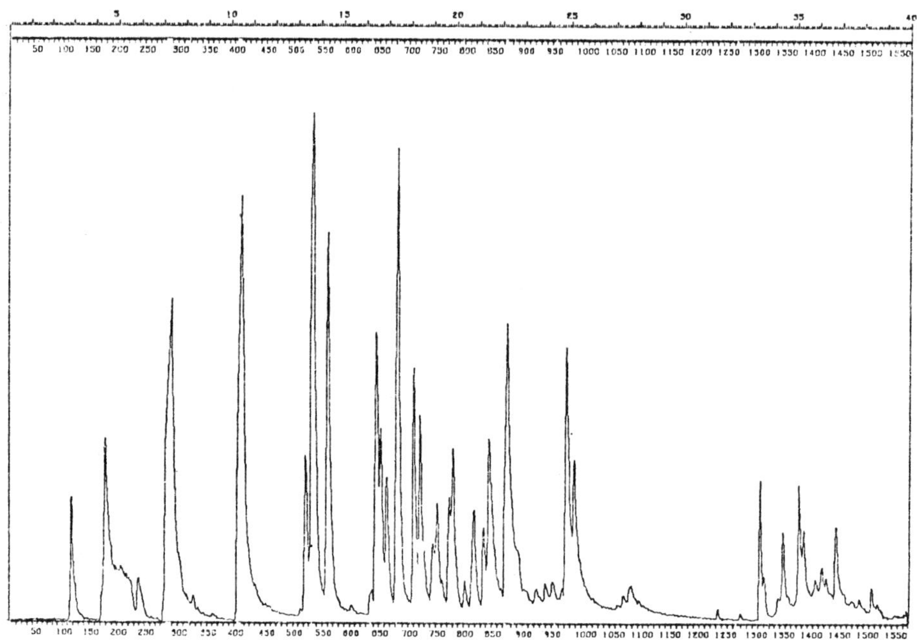

Figure 6. GC-MS total ion chromatogram obtained from a wastewater sample using a cryogenic trap and a Scot capillary column.

Cryogenic Traps

Radian has also built a cryogenic sample loop in order to provide high resolution gas chromatography of complex purged samples. The sample trap is inserted into an aluminum block containing a heating element and thermocouple connected to a digital thermometer. A 6-port stainless steel high temperature low dead volume valve allows thermal desorption of the organic materials from the resin trap with a helium flow of 20 cc/minute and a carrier gas flow through the capillary column of 3 cc/minute. The thermally desorbed compounds are trapped in a 1/16 inch OD stainless steel sample loop immersed in liquid nitrogen. Desorption usually takes 8 minutes at 220°C. When the sample loop is removed from the liquid nitrogen, the sample is flashed onto the head of a 55-meter SP2100 SCOT column with a heat gun or an oil bath. A typical chromatogram of a wastewater sample using this method is illustrated in Figure 6.

Desorption Units

When purging is complete, the trapped compounds are thermally desorbed onto the head of a gas chromatographic column held at ambient or subambient temperature. We have investigated several desorption devices including two Radian-designed units. One is a cylindrical oven that drops over the trap and the other is a hot air device. The cylindrical oven heats the trap to 180°C in less than 1.5 minutes and the hot air "gun" achieves the same temperature in less than 0.7 minutes.

BLANK PROBLEMS WITH THE PURGE AND TRAP TECHNIQUE

Many problems have been encountered using the purge and trap technique to analyze for μg/ℓ and sub μg/ℓ concentrations of EPA's Priority Pollutants. However, the single most persistent problem has been the elimination of contamination from the laboratory.

Possible Sources of Contamination

Laboratory air has numerous compounds which can easily interfere with sub μg/ℓ analyses of water samples. These compounds come from a variety of sources. The more obvious sources are solvents used in the laboratory itself and may come from extraction procedures, from vials of standards and samples and from organic rinses of glassware. These sources can sometimes be located many rooms apart from the analytical laboratory and still cause significant interferences.

Other sources of organics in the air of our laboratory facilities have included:

- Freons leaking from refrigerators,
- Freons used to dust electron micrographs,
- solvents for cleaning grease from metal parts in the machine shop, and
- building materials.

Other possible sources of contamination include the instrumentation, water for making standards, and analytical gases. Instruments with Teflon parts are sources of contamination if these parts are heated above 150°C. The instruments also may be indirect causes of blank problems. Once they have become contaminated from previous samples, the interfering compounds may bleed into the column and detector long after the sources of contamination have been eliminated.

Water is easily cleaned of organics but it is also easily contaminated again. Tap water usually has several compounds above the μg/ℓ level. These include chloroform, bromodichloromethane, dibromochloromethane and bromoform. A deionized water supply may also have these same compounds. We have found very high levels of chloroform compared to tap water in our deionized water system. A solution of Clorox® bleach had been used to clean the plastic lines and after

9 months, chloroform was still present in this system. Gases for purging and gas chromatography may also contribute to blank problems. Some gas manufacturers clean their tanks with Freons. Some instances have been observed where even zero grade gases will easily meet specifications yet will not be acceptable for sub μg/ℓ analyses.

Measurement of Sources of Contamination

Sources of contamination can be placed into three categories: 1) instrumentation including gases, 2) laboratory air and 3) water. Before measurement of the air and water contamination, the "instrument blank" must either be low and constant or negligible. The "instrument blank" is measured by going through the whole analytical scheme without a water sample.

After the "instrument blank" has been documented, the water sample for preparing standards and "blanks" containing an internal standard can be measured. In order to measure sources of contamination in air, a trap of Tenax and Chromosorb 102 as described earlier can be used. A large gas-tight syringe fitted with a short piece of Tygon or Teflon tubing is used to pull between 50cc and 400cc of air through the trap. The trap is then replaced into the instrument and the analysis scheme is started with the desorption step.

Elimination of Blank Problems

Elimination of interferences are discussed according to the categories used in the preceding discussion. Eliminating contamination from the instrumentation may be one of the most difficult problems. There may be more than one source of contamination and some parts of the system may be bleeding compounds into the system.

We have found that a systematic cleaning from the gases to the detector is the best approach. The gases are cleaned using large traps. In practice, a molecular sieve 5A trap or equivalent is placed after the gas regulators. Furthermore, a trap of 1/4 inch OD by 28 cm stainless steel packed with Tenax and Chromosorb 102 is placed in the gas lines just before they enter the instruments. These traps must be cleaned periodically by disconnecting them from the instrument and heating them to about 200°C with a hot air gun. The gas lines are also cleaned in a similar manner. Always start the cleaning at the gas sources and move along the route of gas flow.

Lines, fittings, and glass Bellar tubes may be cleaned by rinsing with high purity acetone. When very high concentrations of organics have contaminated the glass Bellar tube, we have resorted to placing 5 mℓ of acetone in the purging device and run it as a sample for a cleaning procedure.

As mentioned earlier, it is better not to have Teflon parts in the system. This is especially true around the trap area where temperatures reach about 200°C. Contamination can even occur from Teflon furrules used on the traps. Brass ferrules will eliminate this problem.

The water used for standards and blank checks is easy to prepare. The EPA procedure recommends activated carbon clean-up. Although this works, a simpler procedure is to take deionized water and purge it for 24 hours with high purity nitrogen or helium. A positive pressure is maintained in the large purging vessel so that the water remains clean after the 24-hour purge. We have observed that a beaker of prepurged water setting uncovered in the laboratory will become contaminated within 30 minutes from compounds in the air.

Contamination in the air cannot easily be eliminated. However, attention to technique and an enclosed system within the instrumentation will greatly decrease this problem. Rapid transfer of water samples from storage vials to a gas-tight syringe is one important step. All glassware such as the sample syringe, the syringes used for internal standards, and the glass Bellar tube must be rinsed each time after their use. The sample syringe should also be washed with acetone at the beginning of each day of use.

Instrumentation should be designed so that no parts which come into contact with the gases or water samples are exposed to the laboratory air. A Luer-Lok Valve is used on the Bellar tube to introduce the sample and then close out laboratory air in all Radian systems. Use of short transfer tubing from the desorption unit to the gas chromatographic column has eliminated most of the "memory" problems in the instruments.

Lastly, we have observed build-up of the blank contamination during long periods of shutdown. However, the instruments can be cleaned by the previously described procedures and several blank runs will usually decrease the contamination to acceptable levels. The "instrument blank" is reduced and eliminated with everyday use of the analytical instruments.

Summary

In summary, the quality assurance studies have shown that:

- the Hall detector is linear from at least 1 to 200 ppb;
- the variability between carefully prepared standards is small relative to instrument variability;
- repeated analyses of the same standards should agree within 30 percent most of the time when an internal standard is used;

- correlation between analytical results and standard values is consistently lower when an external standard is used and the results only agree within 60 percent most of the time;
- the median repurge percentage for the 1 to 200 ppb haloform standards was 1.7 percent and the average repurge percentage was 4.1 percent.

We have observed the purge and trap method to be fraught with pitfalls. Some of these include:

- incomplete purging will give incorrect values;
- breakthrough can occur if the purging time is too long;
- systems with solenoid valves and long lengths of transfer tubing may suffer from "memory" effects from previous samples;
- traps packed with Tenax and silica gel can degenerate and "steam clean" the chromatographic column when thermally desorbed;
- contamination of samples and blanks can occur from the instrumentation, the air, and the water.

However, with care and moderate expense, the purge and trap technique can be used successfully and relatively routinely to analyze organics in water at μg/ℓ to sub μg/ℓ levels.

References

1. T.A. Bellar and J.J. Lichtenberg, JAWWA 66 (12), 739, 1974.
2. T.A. Bellar, J.J. Lichtenberg and R.C. Kroner, JAWWA 66 (12), 703, 1974.
3. U.S. Environmental Protection Agency, "Sampling and Analysis Procedures for Screening of Industrial Effluents for Priority Pollutants," EMSL, Cincinnati, Ohio, March, 1977, Revised April, 1977.

DETERMINATION OF PURGEABLE ORGANICS IN SEDIMENT

David N. Speis, Chemist

U.S. Environmental Protection Agency

Region II, Edison, New Jersey 08817

The analysis of volatile organic compounds in sediment poses challenging problems. Previous methods have been insensitive to low concentrations of volatiles in sediments. The reasons behind this are twofold. Sediment analysis is usually limited to a 50 gm sample size for ease of sample handling. A liquid extraction of a sediment sample this large would be difficult. The large volumes of solvent required could not be concentrated by any solvent stripping technique without loss of the organics in question. Head space analysis is limited to small volume injections of low concentration resulting in a minimum detectable limit of 25 ppb "EMSL, Cincinnati, Ohio" (1977).

By modifying a Tekmar LSC-1 liquid sample concentrator to accommodate a 15 gram sediment sample and sample chamber heater, the analyst is able to thermally purge volatile organics from sediments. With this system, a minimum detectable limit of 0.1 ppb can be attained.

The purge and trap method is useful in the analysis of water samples for hydrocarbons and halogenated hydrocarbons. An inert gas (helium) is bubbled through the sample transferring those compounds favoring the vapor state from the aqueous phase to the gaseous phase. These gaseous compounds are then concentrated in a porous polymer trap at room temperature. (Figure 1) The trapped compounds are thermally desorbed into a gas chromatograph interfaced to an electron impact mass spectrometer, electron capture, or Hall electrolytic conductivity detectors "Bellar, Lichtenberg" (1974). (Figure 2).

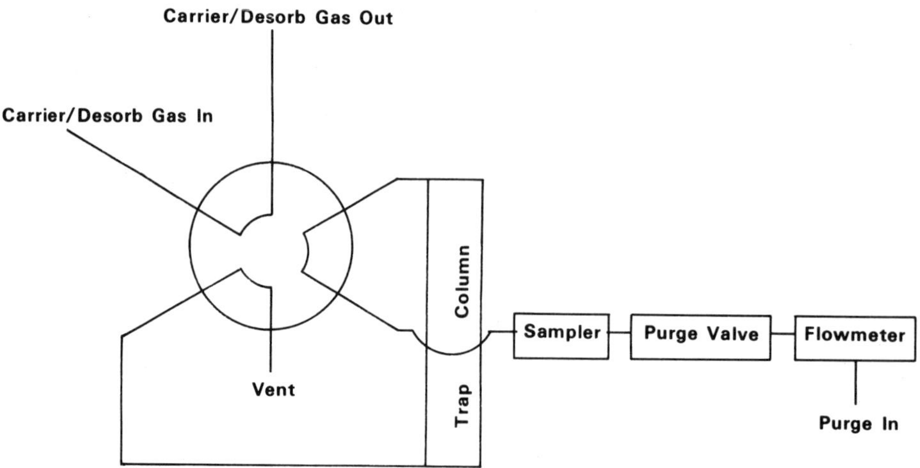

Figure 1. Purge Mode

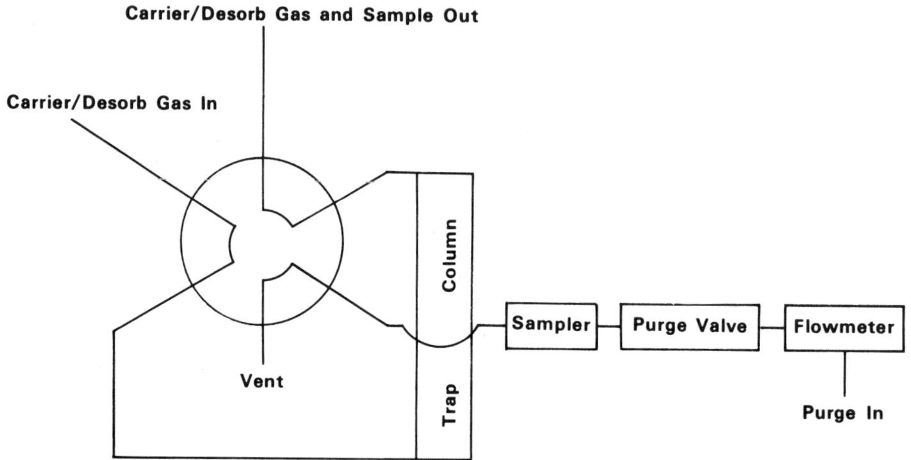

Figure 2. Desorb Mode

A Tekmar LSC-1 with modified sample container was interfaced to a Finnigan 3200 electron impact, gas chromatograph/mass spectrometer with a Systems Industries data system.

Sediment samples are collected in the field in pre-weighed Pierce (or equivalent) 20 ml hypovials with uncrimped aluminum seals and teflon backed septa. These vials have the capability

Table 1. Mean Recoveries

	Blank $\overline{X}$ NG	.100 μg Spike $\overline{X}$ Recovery	.500 μg Spike $\overline{X}$ Recovery	1.00 μg Spike $\overline{X}$ Recovery	2.00 μg Spike $\overline{X}$ Recovery	3.0 μg Spike $\overline{X}$ Recovery
Chloroform	**1.0 (1.7)**	**.42 (.06)**	**.34 (.10)**	**.41 (.08)**	**.46 (.11)**	**.39 (.12)**
1,1,1 Trichlorethane	**6.2 (2.7)**	**.52 (.05)**	**.44 (.10)**	**.50 (.04)**	**.50 (.09)**	**.46(.10)**
Toluene	**6.4 (0.17)**	**.32 (.07)**	**.43 (.05)**	**.44 (.13)**	**.34 (.07)**	**.37 (.09)**
Tetrachloroethylene	**1.6 (1.6)**	**.32 (.09)**	**.24 (.06)**	**.37 (.08)**	**.34 (.07)**	**.39 (.08)**
Chlorobenzene	**3.8 (3.3)**	**.32 (.08)**	**.25 (.03)**	**.44 (.11)**	**.28 (.09)**	**.43 (.15)**

N = 5 samples/compound concentration
( ) = standard deviation

of holding up to 15 grams of wet sediment. For best results, the vials should be filled to maximum capacity to reduce the amount of head space. The aluminum seals are crimped in the field after sample collection. All samples should be transported and stored at wet ice temperature and equilibrated to room temperature for weighing and analysis. Two holes are then drilled into the septum to allow the snug insertion of two 1/8" glass tubes to be used as a purge gas inlet and outlet. The purge gas inlet should be extended to the bottom of the septum vial. The purge gas outlet should extend ½" below the septum (Figure 3). The vial is wrapped in heating tape and the glass tubes are connected to the appropriate gas lines. The sample is then heated at 80°C for five minutes. At the conclusion of five minutes, the sample chamber is purged with helium for 4 minutes at a rate of 60 ml/min. This effectively traps volatilized organics on the polymer trap. The trapped organics are then desorbed onto the chromatographic column for analysis and data collection.

A pre-selected sediment whose consistency was that of a loose field soil, was muffled at 600°C to remove any volatile organics. This sediment served as a media for spiking. A solution of five volatile organics was prepared in methanol. This solution was injected directly into an empty septum vial to be used as a standard. This eliminates any matrix effects caused by the sediment so that an easy assessment of recovery can be made. Five dilutions of this standard were prepared in water. Ten mls of the diluted standard was pipetted into the septum vial containing 10-15 grams of pre-weighed sediment. The vial was sealed and allowed to stabilize for four hours. Five replicates of each concentration were analyzed as well as four unspiked sediments and four empty vials as blanks.

Minimum detectable levels obtained were 1.0 ng/15 grams of sediment (.07 ug/kg). Mean recoveries were calculated for each compound in each concentration group. Recoveries ranged from a high 52% to a low of 24%. The results are summarized in Table 1. The data was quite linear over the range of operation (Figure 4). The lowest correlation coefficient was .934 for chlorobenzene; however, three compounds had values greater than .99.

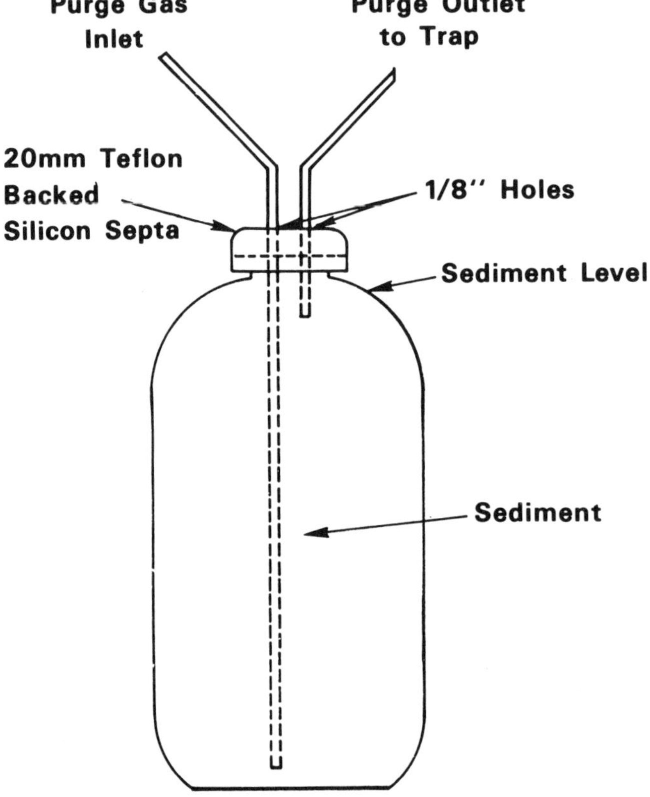

Figure 3. Diagram of Vial

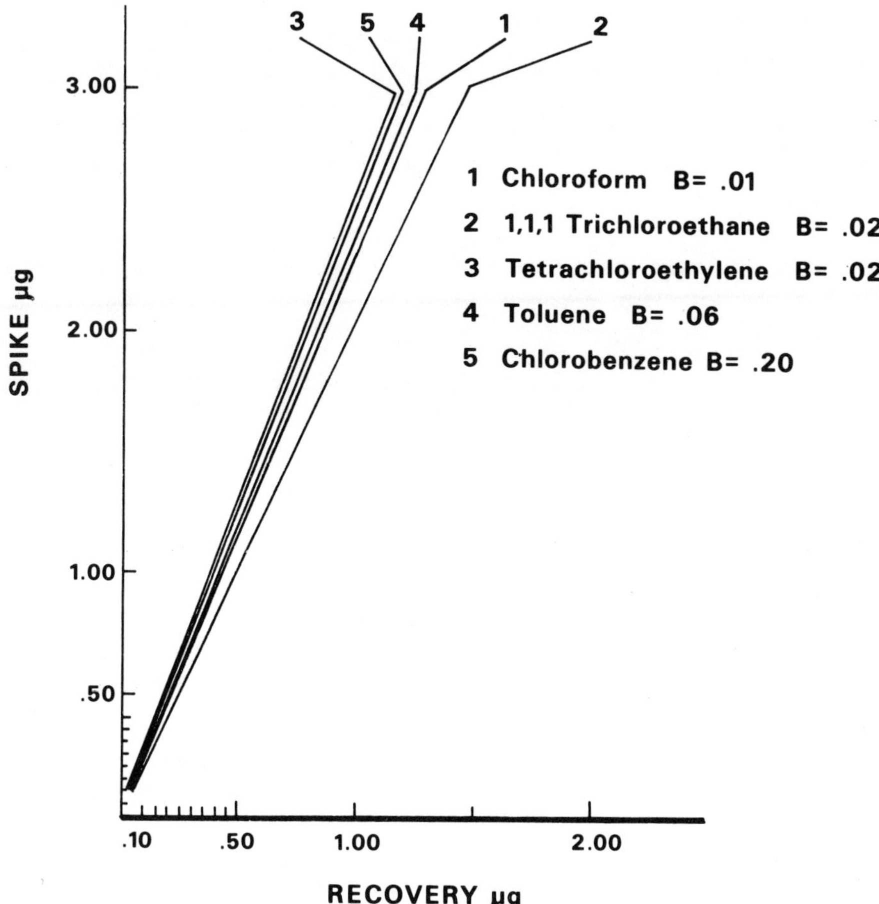

Figure 4. Recoveries

Minimum detectable levels obtained using this sediment purge and trap procedure represent a 350 fold improvement in detection limits over that reported for head space analysis techniques. Although recoveries are on the low side, they are linear and reproducible. The results indicate that this can be a reliable method for detecting and determining low levels of volatile organics in sediments. This soil matrix represented the type of sample that would be taken as the result of an organic spill in an uncontaminated area. The matrix used was very difficult to purge of organics. The muffling served as an activation of a sediment that, otherwise, might not have held onto a purgeable compound so tenaciously. Higher recoveries would be expected from a contaminated unmuffled sediment. Limited available data on spiked environmental samples gave recoveries of 80-100% using this method.

References

Bellar, T.A.; Lichtenberg, J.J., and Kroner, R.C., Determining Volatile Organics in Microgram per Liter levels by Gas Chromatography. Journal American Water Works Association, Volume 66, No. 12, December 1974.

Sediment Sampling for Volatile Organics, U.S. Environmental Protection Agency, Environmental Monitoring and Support Laboratory, Cincinnati, Ohio, August 2, 1977.

ANALYSIS OF HEADSPACE GASES FOR PARTS PER BILLION CONCENTRATIONS OF VOLATILE ORGANIC CONTAMINANTS IN WATER SAMPLES BY GAS CHROMATOGRAPHY

D.A.J. Murray

Department of Fisheries & the Environment

Freshwater Institute, 501 University Crescent

Winnipeg, Manitoba, Canada, R3T 2N6

INTRODUCTION

Trace organics in water samples have been encountered in distilled, deionized and drinking waters (Bevenue et al, 1971; Nowak et al, 1973 and Dowty et al, 1976). There is a need for rapid and accurate analytical methods because of the potential toxic and carcinogenic effects of these contaminants. These organics may be split into two classes, volatile and less volatile, as suggested by Nowak et al, (1973) each class requiring separate treatment to find the most suitable analytical techniques. Available methods of analyses for the volatile fraction may be subdivided under the following headings.

Stripping Methods

In this technique a pure gas is bubbled through the sample to sparge out the volatiles which are collected in a trap filled with an adsorbent. The contents of the trap are then analyzed by gas chromatography. Bellar & Lichtenberg (1974), Zlatkis et al (1973), Kopfler et al (1976) and Stevens & Symons (1977) used Tenax (2.6 diphenylphenylene oxide) as the adsorbent and back-flushed with heating to inject the trapped volatiles onto the gas chromatographic column. Dowty et al (1976) developed an automated system using Tenax and heat desorbed the volatiles onto a gas chromatographic column at -20°C. Nowak et al (1973) collected the volatiles in a cooled capillary sample loop which was heated rapidly to inject the organics onto the gas chromatographic column.

Solvent Extraction Methods

Henderson et al (1976) and Richard & Junk (1977) used 1 ml of pentane to extract halomethanes from 10 ml samples of water followed by gas chromatographic analysis using an electron capture (EC) detector. Mieure (1977) used 1 ml of methyl cyclohexane in a similar technique and Bevenue et al (1971) extracted 3.5 litre samples with 100 ml of hexane followed by concentration prior to analysis by gas chromatography.

Direct Injection Methods

Nicholson & Merez (1975) injected 10 $\mu\ell$ samples of water directly into a gas chromatograph (GC) with an EC detector to measure ppb concentrations of halogenated compounds with no deleterious effects on the detector. Fujii (1977) analyzed water samples by injecting 100 $\mu\ell$ into a GC with a mass spectrometer as a detector in the multiple specific ion mode to provide positive identification of the organohalides present.

Direct Adsorption Methods

In this technique water samples are passed directly through an adsorbent. The organics are subsequently eluted with a solvent, concentrated and analyzed by gas chromatography. Suffet et al (1976) used Amberlite XAD-2 macroreticular resin as an adsorbent and eluted with 200 ml of ether. Kissinger and Fritz (1976) used an acetylated XAD-2 resin and eluted with 1 ml of pyridine.

Headspace Methods

The water sample is equilibrated with a constant volume of clean inert gas by vigorously shaking and the headspace subsampled for gas chromatographic analysis. Kaiser & Oliver (1976) equilibrated at a reduced pressure and an elevated temperature followed by injection of a 5 $\mu\ell$ sample of the headspace into a GC with an EC detector to measure halogenated hydrocarbons in water samples. A stainless steel apparatus was developed by Mackay et al (1975) to equilibrate water samples with headspace gas with a gas sampling valve to inject an aliquot of the headspace. McAuliffe (1971) developed a headspace technique using glass syringes and a gas sampling valve. Multiple equilibrations of the sample with fresh gas were employed for quantitative analysis and also to show the preferential vaporization of hydrocarbons.

In general, the stripping techniques are cumbersome, time consuming and inadequate to purge all organics quantitatively (Kopfler et al, 1976). Heating of the sample can also cause an increase in chloroform concentration, (Kopfler et al, 1976). Solvent extractions introduce a solvent peak which can mask coeluting peaks and are most suited for EC detectable compounds. The direct injection method is less sensitive than other methods even for halogenated compounds and not all laboratories have access to a GC-MS. The direct adsorption techniques also introduce a solvent and, where the eluant requires concentration, losses of up to 80% of volatile materials can be encountered (Junk et al, 1974). The headspace methods seem to be the most promising for analysis of all the possible volatile contaminants in water samples.

The choice of detectors usually available in laboratories lies between electron capture and flame ionization, the former being particularly suitable for halogenated organic compounds. By splitting the column effluent and using both detectors in parallel, a more informative analysis of the volatile organic contamination can be achieved with a single injection.

Two different water purification procedures to produce clean water for toxicity studies on fish and benthic organisms were being studied. This paper describes the method developed to analyze water samples at the various stages of purification. The method is based on the syringe technique of McAuliffe (1971) and is rapid and sensitive.

EXPERIMENTAL

A 25 ml sample was placed in a 50 ml glass syringe and 25 ml of nitrogen were added. The mixture was equilibrated by vigorously shaking by hand the capped syringe for 2 minutes at room temperature. The gas phase was injected through a short 3 cm drying tube of magnesium perchlorate into the 5 ml sample loop of a gas sampling valve fitted prior to the injection port. The contents of the sample loop were then injected into the GC and temperature programmed from 60 - 150° C at 15°/minute using flame ionization and electron capture detectors in parallel. The concentrations were determined by integrating peak areas and by analysis of standard solutions.

The water to gas phase ratio was adjusted to 10 ml of water with 30 ml nitrogen for haloform analysis to avoid overloading the detector and isothermal GC conditions at 70°C were used.

Standard Solutions

Standard solutions were made by adding 10.0 mg of each halomethane to 2 litres of distilled water and shaking vigorously to give a solution of 5 ppm. Further dilutions in distilled water were made to give a range of lower concentrations. There were some losses due to the headspace of the volumetric flasks, but these were neglected. Since hexane is almost quantitatively transferred to the vapour phase (McAuliffe 1971), hexane standards were made by adding 1.5 μℓ (1.0 mg) to 4 ℓ of air in a glass bottle. After several hours equilibration further dilutions were made in syringes to give lower concentrations.

Gas Chromatographic Conditions

A Hewlett-Packard 5750 gas chromatograph was used with an Infotronics CRS 208 digital integrator for the quantitative analyses. A Carle No. 2018 micro sample loop valve was fitted to the helium line just prior to the injection block and a 3 m x 6 mm OD column containing 10% Dexsil 300 coated on Chromosorb W-AW was used.

Electron Capture	200 C	Helium	40 ml/minute
Pulse Interval	50 μ seconds	Hydrogen	30 ml/minute
Flame Ionization	150 C	Air	300 ml/minute
Injector	150 C	Argon 10% methane	30 ml/minute

RESULTS AND DISCUSSION

The chromatograms in Figure 1 illustrate the different contaminants in both distilled and tap waters using headspace analysis and the dual detector system. The limited temperature range and attenuation chosen permitted a satisfactory temperature programme with the EC detector. The sample of distilled water taken directly from the still contains only traces of EC detectable compounds, while the laboratory sample of distilled water which had passed through plastic delivery tubing contains hydrocarbon contaminants. The tap water chromatogram shows the presence of halogenated compounds but no hydrocarbon material. Haloforms in tap water appear to be the result of direct chlorination during the water treatment process (Bellar and Lichtenberg 1974) and are lost during distillation. Table 1 shows the compounds present in Figure 1 with some tentative identifications based on retention data. These identifications are supported by the results of Kissinger and Fritz (1976), Richard and Junk (1977) and Mieure (1977).

This procedure was used to analyze the water for haloforms at various stages of a pilot scale purification procedure which consisted of a diatomaceous earth filter, ozone treatment and an activated carbon filter. While the ozone treatment had little effect on the haloform concentrations, it is known to have an inhibiting effect on bacteria (Kinman 1972). Table 2 shows the drastic reduction of haloforms after the carbon filter.

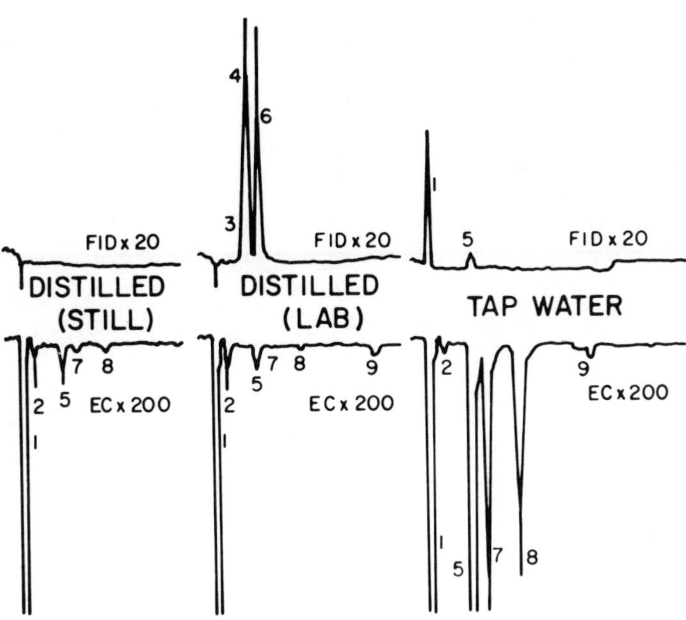

Figure 1. Chromatograms of headspace vapours of distilled and tap waters.

Table 1 Tentative Identification of Compounds Present in Headspace Analyses of Water Samples using F.I.D., E.C.D. and Retention Data

Peak Number	Compound	G.C. Response	
		F.I.D.	E.C.D.
1	Oxygen, methane	*	*
2	Unknown		*
3	Unknown	*	
4	n-Hexane	*	
5	Chloroform		*
6	Heptane isomer	*	
7	Carbon tetrachloride		*
8	Dichlorobromomethane		*
9	Dibromochloromethane		*

Table 2 Haloform Concentrations (μg/litre) Sampled from Pilot Scale Water Purification Process

	Inlet Water	After Ozone Treatment	After Carbon Filter
Chloroform	72	67	1.1
Carbon tetrachloride	2.0	2.4	0
Dichlorobromethane	4.2	3.0	0

A second water treatment process also was analyzed for residual chlorine and haloforms. This treatment consisted of a UV irradiation followed by a carbon filter. The results in Table 3 show the reduction in haloforms after the carbon filter and the reduction in chlorine after the UV treatment. A bed of 1.4 cu. ft. of granular activated carbon showed no loss of adsorptive power after 14 days continuous use at a daily rate of 900 U.S. gallons.

Table 3

	Days	$CHCl_3$	CCl_4	$CHCl_2Br$	Residual Chlorine
Inlet Water	1	24	0.1	1.5	353
	4	45	0.1	3.0	491
	8	38	0.1	1.8	425
	11	59	2.0	4.0	541
	14	41	0.1	2.0	392
After U.V.	1	21	0.1	0.5	0.2
	4	49	0.1	1.5	0.0
	8	37	0.1	0.5	0.0
	11	60	0.2	1.5	0.0
	14	41	0.1	0.5	0.0
After Carbon Filter	1	0.4	0.0	0.0	4.9
	4	0.9	0.0	0.0	2.9
	8	0.6	0.0	0.0	2.8
	11	1.0	0.0	0.0	1.6
	14	1.1	0.0	0.0	1.0

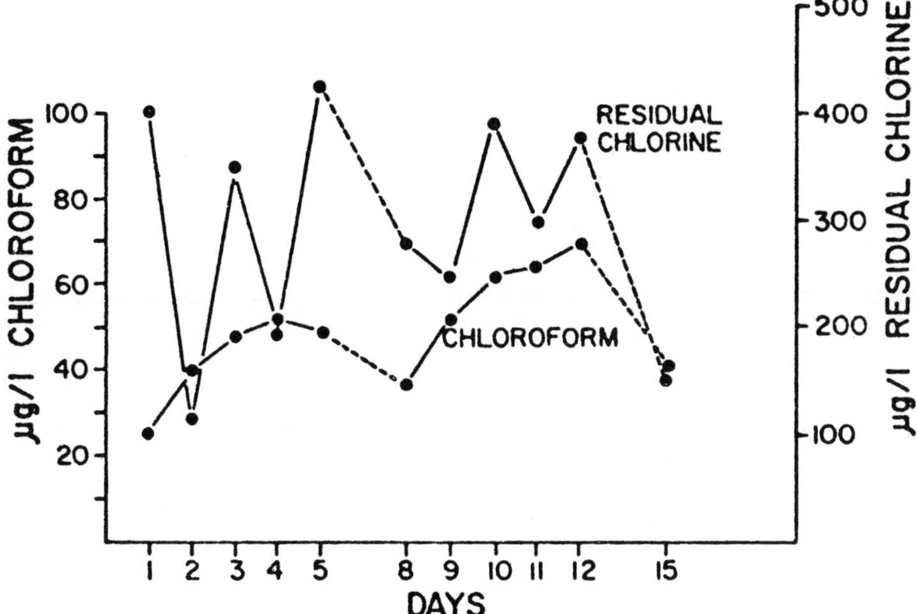

Figure 2. Chloroform and residual chlorine concentration on tap water.

The quality of tap water was followd by analyzing water samples daily over two weeks for residual chlorine and haloforms. The results in Figure 2 demonstrate the day-to-day changes in haloform contamination and show no correlation between residual chlorine and haloforms.

Distilled water was analyzed for hydrocarbons over a two week period and the results show concentrations for hexane and the heptane isomer in the range 3 - 20 μg/litre.

The calibration curves in Figure 3 show the variation of the response to the number of halogen atoms in each molecule as well as the non-linearity of the electron capture detector. The hexane calibration using the flame ionization detector was linear over the concentration range measured.

It was found that this technique would quantitatively measure volatile hydrocarbons and halogenated methanes in water samples down to 1 μg/litre with an accuracy of 5% at the 50 μg/litre level. Handling of the sample is reduced to a minimum and an analysis time of about 5 minutes gives a realistic value of 40 - 50 samples per man-day. The method is simple, accurate and convenient for the routine analysis of water samples.

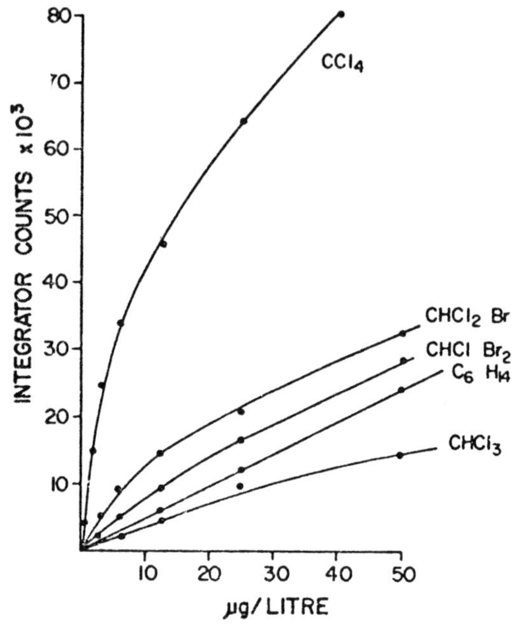

Figure 3. Calibration curves of haloforms and hexane using headspace analysis.

REFERENCES

Bellar, T.A., and J.J. Lichtenberg. 1974. Determining volatile organics at microgram-per-litre levels by gas chromatography. Jour. AWWA 66:12:739-744.

Bevenue, A., J.N. Ogata, Y. Kawano and J.W. Hylin. 1971. Potential problems with the use of distilled water in pesticide residue analysis. Jour. of Chromatogr. 60: 45-50.

Dowty, B., L. Green, and J.L. Laseter. 1976. Application of a computer-based chromatograph for automated water pollution analysis. Jour. of Chromatogr. Sci. 14: 187-190.

Fujii, T. 1977. Direct aqueous injection gas chromatography-mass spectrometry for analysis of organohalides in water at concentrations below the parts-per-billion level. Jour. of Chromatogr. 139: 297-302.

Henderson, J.E., G.R. Peyton, and W.H. Glaze. 1976. A convenient liquid-liquid extraction method for the determination of halomethanes in water at the parts-per-billion level. In: Lawrence H. Keith, ed. Identification and Analysis of Organic Pollutants in Water. Ann Arbor Science, Ann Arbor, Michigan.

Junk, G.A., J.J. Richard, M.D. Grieser, D. Witiak, J.L. Witiak, M.D. Arguello, R. Vick, H.J. Svec, J.S. Fritz, and G.C. Calder. 1974. Use of macro-reticular resins in the analysis of water for trace organic contaminants. Jour. of Chromatogr. 99: 745-762.

Kaiser, K.L.E., and B.G. Oliver. 1976. Determination of volatile halogenated hydrocarbons in water by gas chromatography. Anal. Chem. 48:14: 2207-2209.

Kinman, R.N. 1972. Ozone in water disinfection. In: Francis L. Evans III, ed. Ozone in Water and Wastewater Treatment. Ann Arbor Science, Ann Arbor, Michigan.

Kissinger, L.D., and S. Fritz. 1976. Analysis of drinking water for haloforms. Jour. AWWA 68:8:435-437.

Kopfler, F.C., R.G. Melton, R.D. Lingg, and W.E. Coleman. 1976. GC-MS determination of volatiles for the national organics reconnaissance survey (NORS) of drinking water. In: Lawrence H. Keith, ed. Identification and Analysis of Organic Pollutants in Water. Ann Arbor Science, Ann Arbor, Michigan.

Mackay, D., W.Y. Shiu, and A.W. Wolkoff. 1975. Gas chromatographic determination of low concentrations of hydrocarbons in water by vapor phase extraction. In: Water Quality Parameters, ASTM. STP 573.

McAuliffe, C. 1971. GC determination of solutes by multiple phase equilibration. Chem. Technol. 1:46-51.

Mieure, J.P. 1977. A rapid and sensitive method for determining volatile organohalides in water. Jour. AWWA 69:1:60-62.

Nicholson, A.A., and O. Meresz. 1975. Analysis of volatile halogenated organics in water by direct aqueous injection gas chromatography. Bull. Environ. Contam. Toxicol. 14: 453-456.

Nowak. J., J. Zluticky, V. Kubelka, and J. Mostecky. 1973. Analysis of organic constituents present in drinking water. Jour. of Chromatogr. 76: 45-50.

Richard, J.J., and G.A. Junk. 1977. Liquid extraction for the rapid determination of halomethanes in water. Jour. AWWA 69:1:62-64.

Stevens, A.A., and J.M. Symons. 1977. Measurement of trihalomethane and precursor concentration changes. Jour. AWWA 69:10:546-554.

Suffet, I.H., L. Brenner, and B. Silver. 1976. Identification of 1,1,1-trichloroacetone (1,1,1-trichloropropanone) in two drinking waters: A known precursor in haloform reaction. Environ. Sci. Technol. 10:13:1273-1275.

Zlatkis, A., H.A. Lichtenstein, and A. Tishbee. 1973. Concentration and analysis of trace volatile organics in gases and biological fluids with a new solid adsorbent. Chromatographia 6:2:67-70.

EVALUATION OF ANALYTICAL TECHNIQUES FOR ASSESSMENT OF TREATMENT PROCESSES FOR TRACE HALOGENATED HYDROCARBONS

Massoud Pirbazari, Mark Herbert and Walter J. Weber Jr.

Water Resources Program, University of Michigan

Ann Arbor, Michigan 48109

INTRODUCTION

Removal of trace amounts of halogenated hydrocarbons - materials suspect for potential toxic, carcinogenic, mutagenic and teratogenic characteristics - from public drinking waters is one of the most challenging problems facing the water supply industry today. Conventional treatment technologies are being re-evaluated in this regard, and new and/or modified processes, such as adsorption on activated carbon, are being examined for their effectiveness in accomplishing such treatment.

Precise quantitative determination of residual concentrations of organics from studies on treatment processes such as adsorption - information which is essential for evaluation of process effectiveness - requires application of analytical techniques which are both highly sensitive and compatible with experimental methodologies associated with process evaluation. There are general and specific characteristics inherent to particular analytical methods that must be considered in this regard. These characteristics include: efficiency, limits of detection, particular interferences, speed and ease of analysis, pretreatments required on samples prior to analysis, sample size, reproducibility, and cost and availability of materials.

Three principal factors generally affect the validity of the analysis: 1) type of detector system used; 2) method employed for sample collection and concentration; and 3) type of system to which sample collection and analysis is applied (i.e., configuration of the system, frequency of sampling, and type of analysis necessary for evaluation. Table 1 illustrates, by way of example, the

Table 1 Detection Limits for Trihalomethanes Using Different Detectors (μg/l) - Liquid-Liquid Extraction Technique*

Trihalomethanes \ Detector type	Electron Capture	Electrolytic Conductivity	Flame Ionization**
Chloroform	≤1	≤10	∿100
Bromodichloromethane	≤1	≤10	∿100
Dibromochloromethane	≤1	≤12	∿100
Bromoform	≤1	≤15	∿100

* extractant/sample = 1:10
** polar liquid phase gas chromatography

detection limits for trihalomethanes exploying different detectors and using a liquid-liquid extraction technique.

This work presents an evaluation of several analytical techniques for analysis of volatile halogenated hydrocarbons. These evaluations further relate to use of the techniques in conjunction with specific experimental methodologies required for measurement and assessment of adsorption equilibrium and kinetic data. Further, a rapid organic extraction technique compatible with adsorption equilibrium and rate methodologies has been examined for several halogenated hydrocarbons.

BACKGROUND

To develop a firm basis for understanding the process dynamics of treatments for removal of halogenated organic compounds by activated carbon at the extremely low levels found in water supplies, and for development of rational design criteria for such applications, it is imperative to assess the effectiveness of activated carbon for removing such compounds over the broad spectrum of conditions likely to be confronted in water treatment applications. As a first step, it is essential to establish appropriate adsorption equilibrium and rate data for the system(s) of interest, and to identify and quantify major process variables.

Adsorption equilibrium relationships are usually represented in terms of adsorption isotherms, which are useful both for representing the ultimate capacity of a carbon for adsorption of an organic compound and for providing a description of the functional dependence of capacity on the concentration of the compound. Rate, or kinetic, data are significant for evaluation of the time-dependent approach to equilibrium capacity, specification of contact time, and for design of continuous-flow adsorption systems.

ANALYTICAL TECHNIQUES FOR HALOMETHANES

Dynamic Headspace

This procedure, developed by Bellar and Lichtenberg (1974), was employed extensively in the USEPA National Organic Reconnaissance Survey. The method involves quantitative removal of volatile organics from water by purging with an inert gas and subsequently trapping the organics on a cold porous polymer trap. The organics are then desorbed into a cold gas chromatographic column, which is programmed up to operating conditions. An electron capture detector (^{63}Ni), an electrolytic conductivity, or a microcoulometric detector can be used for this analysis. Detection of the trihalomethanes in the limiting requirement, $<0.1\ \mu g/l$, is possible and consistently attained.

Static Headspace

This method of analysis has been suggested as a sensitive technique for halomethane detection (Morris and Johnson, 1976). The principles of the procedure consist of the equilibration of the compound of interest in a water sample and its head space in a closed container. After equilibration at constant temperature, a sample of the head space gas is removed and analyzed using ^{63}Ni electron capture gas chromatography.

A modification of this method was used by Weber et al (1977) in conjunction with activated carbon adsorption studies. These investigators employed scandium tritide electron capture gas chromatography in their investigations.

Liquid-Liquid Extraction

Liquid-liquid extraction techniques have received considerable attention since the development of specific modifications for the analysis of haloforms (e.g. Henderson et al, 1976; Mieure, 1977). In general, the techniques involve extraction into an organic solvent followed by measurement by electron capture gas chromatography. A detection limit of less than 1 μg/l is obtained by this method.

Aqueous Injection

In an attempt to develop a simple method for analysis of volatile halogenated organics, Nicholson and Meresz (1976) developed a technique whereby the aqueous sample is directly injected into the gas chromatograph using a Sc^3H electron capture detector. A detection limit of 1 μg/l was reported for the target haloforms. A similar procedure was conducted by Hammarstrand (1976) for chloroform using ^{63}Ni electron capture gas chromatography and a detection limit of 1 μg/l was specified.

Table 2 demonstrates a typical evaluation of several analytical methods for chloroform using different detectors.

Table 2 Evaluation of Analytical Methods for Chloroform

Analytic Technique	Detection Limit (μg/l)		Reproducibility (Coef. of Variance) %		Extraction Efficiency %	
	E.C.D.[d]	El. Cond. D.[e]	E.C.D.	El. Cond. D.	E.C.D.	El. Cond. D.
Static headspace[a]	≤0.3	≤2	3-8	2-6	N.A.	N.A.
Dynamic headspace[b]	≤0.1	≤1	3-8	5-10	95-100	95-100
Liquid-liquid extraction[c]	≤1	≤10	6-12	5-10	82-86	95-98

Note: Coefficient of Variance and Extraction Efficiency are calculated for a chloroform concentration range of 1 to 25 μg/l

a: 0.1 - 0.5 ml vapor injection
b: stripping 5 ml of sample
c: extractant: sample = 1:5
d: ^{63}Ni electron capture detector
e: electrolytic conductivity detector
N.A.: not applicable

ADSORPTION METHODOLOGIES FOR HALOMETHANES

Equilibrium Studies

Adsorption equilibrium experiments were conducted in 150 ml air-tight serum vials using carefully weighed amounts of activated carbon. Depending on the choice of analytical technique, the vials were filled with aliquots of solution either completely, permitting no headspace, or partially, leaving 50 ml headspace. The vials were then sealed, using teflon coated septums and crimped on aluminum caps, and agitated at constant room temperature. After achieving equilibrium, the samples were collected, concentrated, and analyzed using appropriate techniques.

For mathematical description and quantification of the adsorption equilibrium data, several theoretical and empirical equations were investigated. The Freundlich isotherm was found to provide the best description of the experimental data. The Freundlich equation has the form (Weber, 1972)

$$Q_e = K_f Ce^{1/n} \qquad (1)$$

where:

Q_e = the amount adsorbed per unit weight of adsorbent;

Ce = the amount of solute remaining in solution at equilibrium; and

K_f and $1/n$ = characteristic constants relating to adsorption capacity and intensity, respectively.

To quantify adsorption isotherm parameters, data which accord with Equation 1 are normalized by plotting the logarithm of Q_e vs. the logarithm of Ce. The resulting straight line yields a slope of 1/n and intercept log K_f.

It is important to note that the type of data required strongly indicates the proper choice of sample collection, concentration and analytical method employed. For example, to adequately describe the functional dependence of capacity on the concentration of the compound over a broad range of concentration, it is essential to generate many data points such as those presented in Figure 1. This figure demonstrates the experimental data and Freundlich isotherms for carbon tetrachloride, bromodichloromethane and chloroform using static headspace electron capture gas chromatographic analysis.

Further, in some instances, compound characteristics command pursuit of a specific method for collection, concentration and

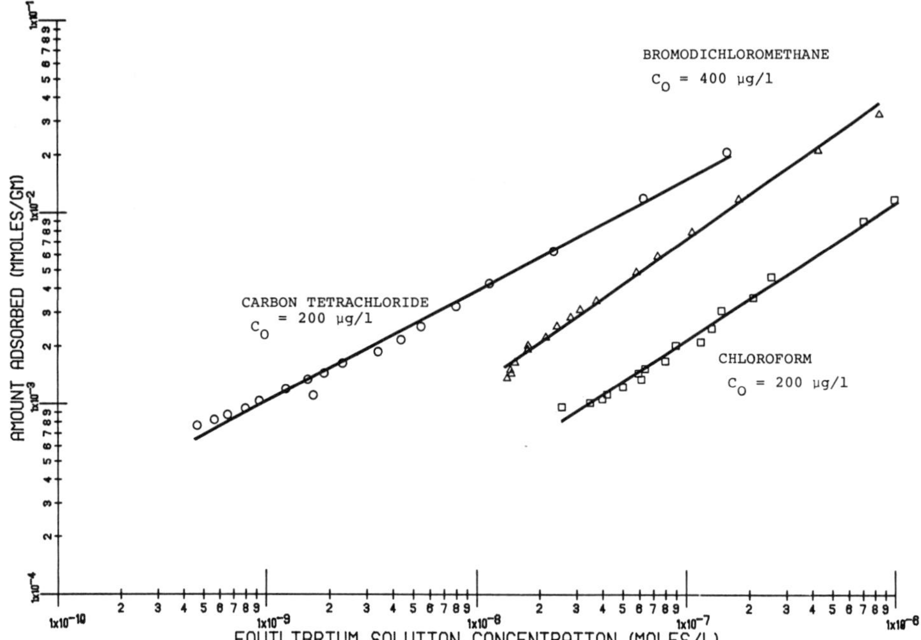

Figure 1. Experimental data and Freundlich isotherms for adsorption of selected volatile halogenated compounds on powdered activated carbon at pH 7; static headspace technique, electron capture gas chromatographic analysis.

analysis. Figure 2 represents the comparison of capacities and intensities of adsorption for benzene for different types of commercial activated carbons using dynamic headspace flame ionization gas chromatographic analysis.

Rate Studies

Completely mixed batch (CMB) reactor rate experiments for the volatile halomethanes were performed in carefully sealed 2.6 litre glass reactors. A weighed quantity of granular activated carbon was added to the experimental solution in the vapor-phase-free reactor. The carbon was dispersed by a motor-driven glass stirrer and 5 ml samples were withdrawn at fixed time intervals. A positive displacement plunger eliminated introduction of any headspace by sample volume displacement. The experimental set up for the reactor is presented elsewhere (Weber et al, 1978).

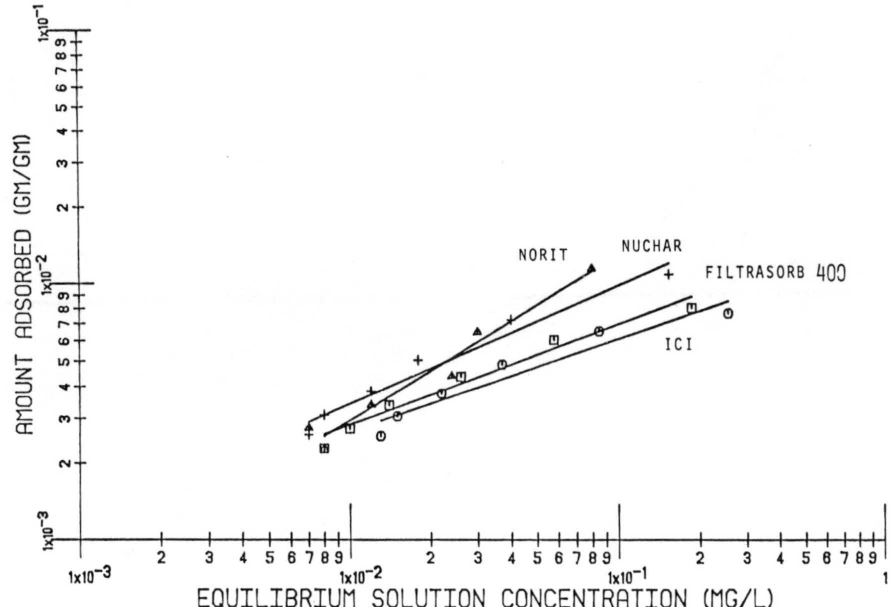

Figure 2. Experimental data and Freundlich isotherms for benzene using different commercial activated carbons at pH 7, C_o = 500 ppb; dynamic headspace technique, flame ionization gas chromatographic analysis.

The dynamic headspace and the liquid-liquid extraction techniques were found to be compatible with the CMB methodology for the analysis of volatile halogenated hydrocarbons. Nonetheless, the liquid-liquid extraction procedure suggested by Mieure (1977) was pursued due to its rapidity. The samples together with 2 ml of high-purity hexane, and 1 g of sodium chloride, were placed in 10 ml vials fitted with teflon-coated screw-on caps. The samples, as well as appropriate standards, were shaken vigorously for one minute, the hexane allowed to separate and solute analysis performed by injecting a few microlitres of the extract into a gas chromatograph equipped with Sc^3H electron capture detector. A calibration curve of the concentration of the standard solution versus detector response was used for the sample concentration determination. Figure 3 presents typical rate data for adsorption of several volatile halomethanes.

EVALUATION OF ANALYTICAL TECHNIQUES FOR HALOMETHANES IN CONJUNCTION WITH THE ADSORPTION METHODOLOGIES

The stripping and preadsorption technique developed by Bellar and Lichtenberg (1974) for volatile halogenated compounds was found to be highly efficient and reproducible. Nonetheless, this method

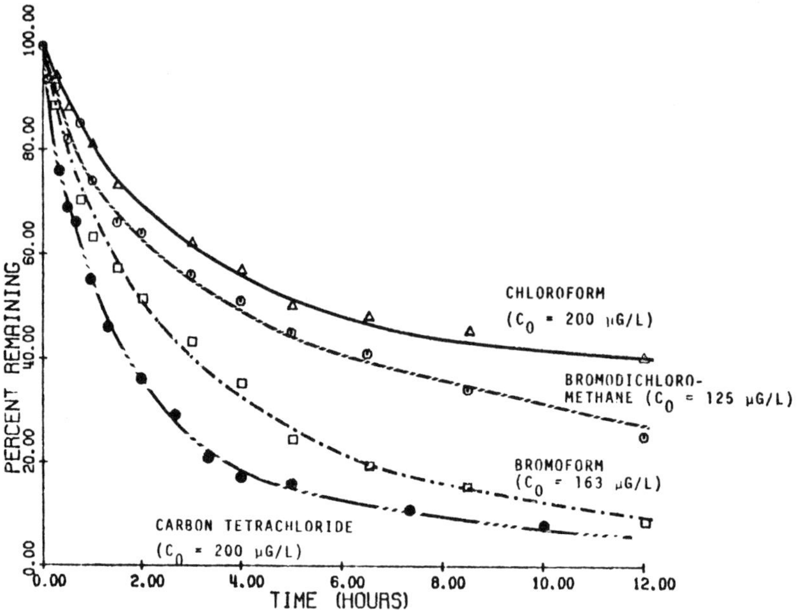

Figure 3. Rates of adsorption of typical halogenated compounds in CMB reactors on granular carbon at pH 7; liquid-liquid vial extraction technique, electron capture gas chromatographic analysis.

requires special equipment for the stripping, adsorption, and desorption procedures. Further, the technique demands considerable amounts of time for the large number of samples involved in the adsorption quantitation studies.

The static headspace procedure discussed by Weber et al (1977) was considered to be efficient and reproducible. However, the detection limit depends on a compound's vapor phase partial pressure. Further, in adsorption equilibrium experiments the presence of solid, liquid, and vapor phases requires a total mass balance estimation necessitating the preparation of two sets of calibration curves.

In the direct aqueous injection technique described by Nicholson and Meresz (1975 and 1976), haloform precursors decompose at high injector temperatures to produce haloforms. The concentration of halogenated compounds obtained by this method is therefore the sum of the free haloform content of the sample plus the quantity of haloforms produced in the actual analysis - referred to as "total potential haloforms". In cases where aqueous injection techniques are pursued, the choice of ^{63}Ni electron capture gas chromatography is recommended; aqueous injection is detrimental to Sc^3H electron capture detectors.

The liquid-liquid extraction technique described by Mieure (1977) was found to be rapid and efficient. However, this method is susceptible to interference by other extractable compounds in aqueous solution. Further, when powdered adsorbents are being evaluated, the isotherm hypo-vials must be centrifuged for solids separation prior to analysis.

The liquid-liquid extraction technique followed by electron capture gas chromatography was found to be the most attractive alternative, especially with regard to compatibility with the adsorption equilibrium and rate experiments reported herein.

ADSORPTION METHODOLOGIES FOR CHLORINATED HYDROCARBONS

Equilibrium Studies

Equilibrium studies for a PCB mixture (Aroclor 1016), dieldrin and p-dichlorobenzene were conducted in a 1 ℓ glass bottles. Carefully weighed amounts of carbon together with 1 ℓ of high quality water (deionized and glass-distilled water passed through a Milli-Q water purifier) were placed in the bottles. The bottles were then spiked with the compound of interest to achieve a desired aqueous cncentration. The bottles were sealed with screw-on caps, using aluminum foil for liner, and agitated to achieve equilibrium. It was found necessary for PCB equilibrium experiments to employ a spike-step procedure, whereby bottles with no detectable residual PCB were spiked periodically and allowed to reach equilibrium.

To investigate the equilibrium concentration a simple liquid-liquid extraction technique was pursued. In these experiments a 10 ml or 20 ml sample was withdrawn from each bottle and transferred to a 30 ml vial. The sample was extracted with 2 ml pesticide-grade hexane and the extract subjected to gas chromatography analysis using a 150 cm x 2 mm ID glass column packed with 3% SE 30 on 80/100 Gas Chrom-Q. Table 3 demonstrates an evaluation for the extraction procedure conducted for typical chlorinated hydrocarbons.

Again the Freundlich isotherm equation was found to provide the best fit to the experimental data. Figure 4 represents the capacity and intensity of adsorption for PCB conducted for several commercial activated carbons. Experimental data and Freundlich isotherms for dieldrin and p-dichlorobenzene are shown in Figures 5 and 6, respectively.

Table 3 Evaluation of Liquid-Liquid Extraction Technique for Typical Chlorinated Hydrocarbons

Compound		Lower Detection Limit (μg/l)	% Recovery*	Coefficient of Variation (%)*
PCB (Aroclor 1016)	10:2**	5	85	12
	20:2	3	74	12
Dieldrin	10:2	0.05	99	2
	20:2	0.03	99	2
1, 4-dichloro-benzene	10:2	10	98	3
	20:2	5	98	4

* estimated for an initial concentration of: Aroclor 1016: 30 μg/l; Dieldrin: 1 μg/l; PDB: 100 μg/l

**Sample- Extractant Ratio

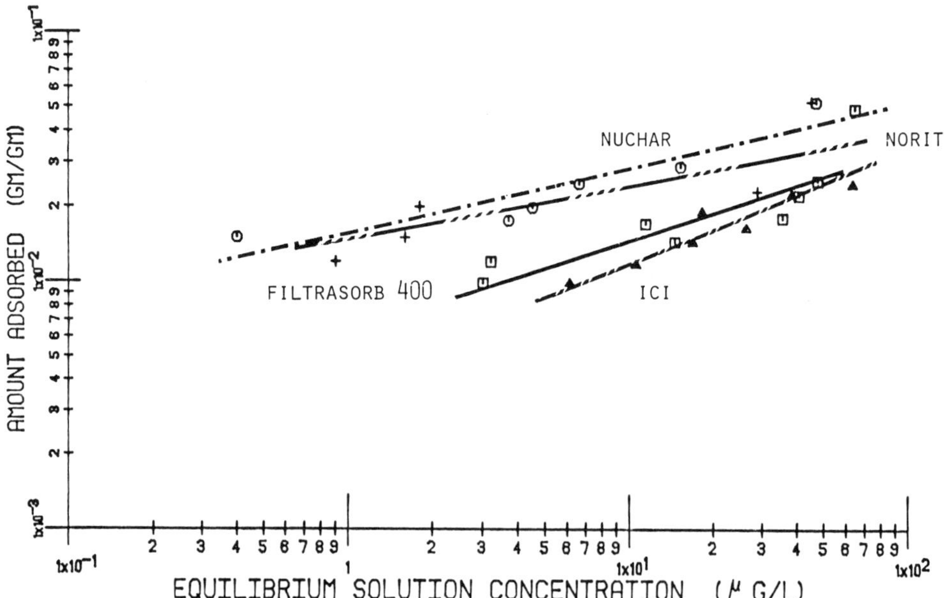

Figure 4. Experimental data and Freundlich isotherms for PCB (aroclor 1016) using different commercial activated carbons at pH 7, C_o = 80 μg/l; liquid-liquid vial extraction technique, electron capture gas chromatographic analysis.

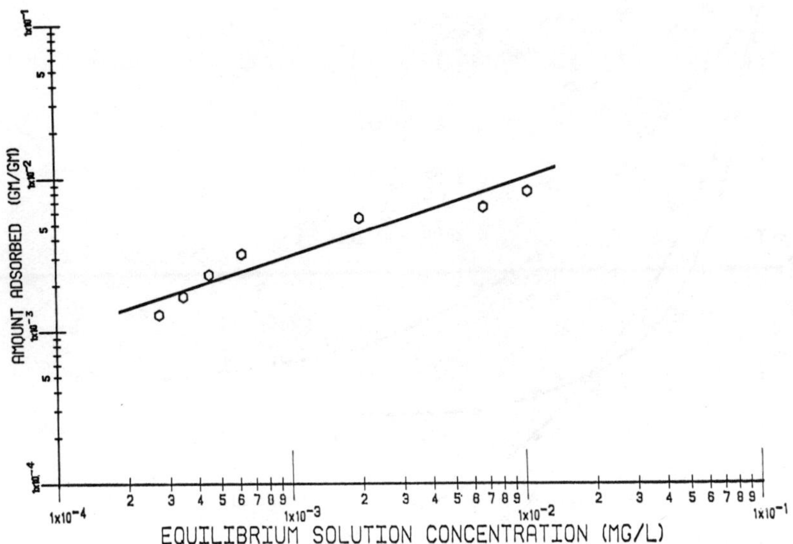

Figure 5. Experimental data and Freundlich isotherm for adsorption of dieldrin on granular carbon at pH 7; liquid-liquid vial extraction technique, electron capture gas chromatographic analysis.

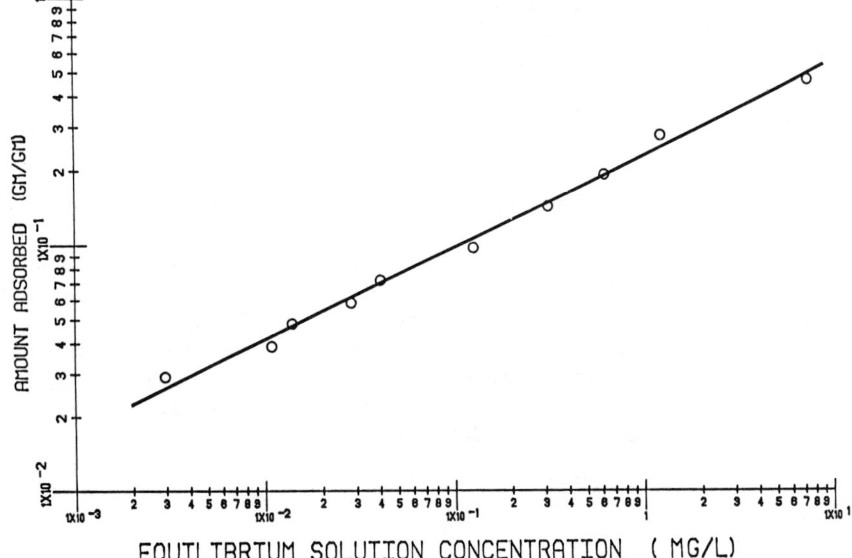

Figure 6. Experimental data and Freundlich isotherm for P-dichlorobenzene on granular carbon at pH 7; liquid-liquid vial extraction technique, electron capture gas chromatographic analysis.

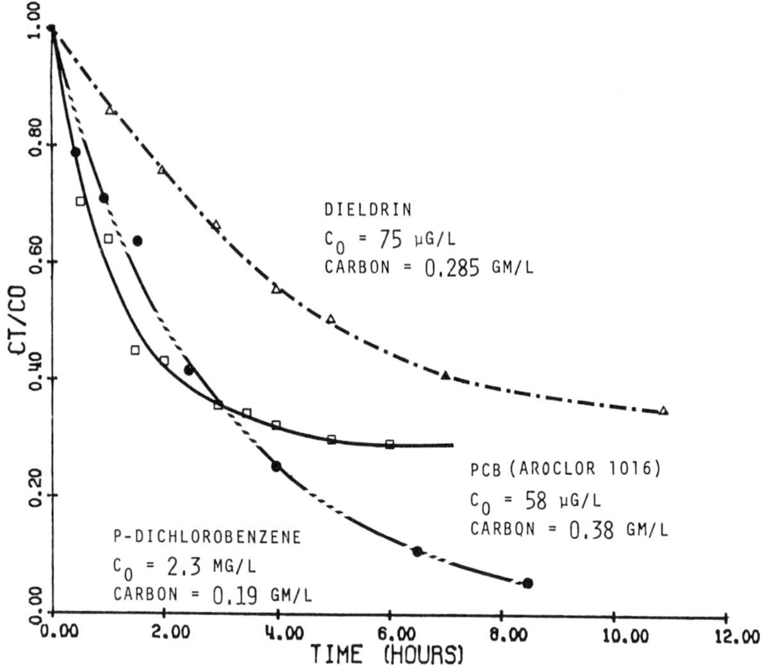

Figure 7. Rates of adsorption of selected chlorinated hydrocarbons on granular carbon in CMB reactors at pH 7; liquid-liquid vial extraction technique, electron capture gas chromatographic analysis.

Rate Studies

CMB rate studies for PCB, dieldrin, and *p*-dichlorobenzene were performed in air-tight glass reactors similar to the one described earlier for halomethanes; 10 ml samples were withdrawn periodically and placed in 20 ml vials containing 2 ml of pesticide-grade hexane. The vials were then sealed with screw-on caps lined with teflon septums, and agitated vigorously. A few microlitres of the extract was injected into a gas chromatograph equipped with Sc^3H electron capture detector. Figure 7 represents the concentration decay curve for the selected chlorinated hydrocarbons.

SUMMARY

Methodologies for adsorption equilibrium and kinetic experiments for volatile halogenated compounds are described, and evaluations and selection of appropriate analytical techniques based on the specific nature and requirements of the experiments arepresented.

A rapid and sensitive extraction technique which can satisfactorily replace cumbersome multiple extraction procedures for chlorinated hydrocarbons in adsorption experiments is discussed.

ACKNOWLEDGEMENTS

This work has been supported in part by Reserach Grant R 80436901 from the Division of Water Supply Research, U.S. Environmental Protection Agency. The authors wish to thank James Long and Alan Moore for their valuable assistance in the laboratory work reported herein.

REFERENCES

Bellar, T.A., and J.J. Lichtenberg. "The Determination of Volatile Organic Compounds at the μg/ℓ Level in Water by Gas Chromatography", USEPA, Cincinnati, Ohio (November 1974).

Hammarstrand, K. "Chloroform in Drinking Water", Varian Instrument Applications, Vol. 10, No. 2, 1976.

Henderson, J.E., G.R. Peyton, and W.H. Glaze. "A Convenient Liquid-Liquid Extraction Method for the Determination of Halomethanes in Water at the Parts-Per-Billion Level", Chapt. 7, Identification and Analysis of Organic Pollutants in Water, Ann Arbor Science, Ann Arbor, Michigan, 1976

Mieure, J.P. "A Rapid and Sensitive Method for Determining Volatile Organohalides in Water", J. Am. Water Works Assoc., (January 1977).

Morris, R.L., and L.G. Johnson. "Agricultural Runoff as a Source of Halomethanes in Drinking Water", J. Am. Water Works Assoc., (September 1976).

Nicholson, A.A., and O. Meresz. "Organics in Ontario Drinking Water: Part I. The Occurrence and Determination of Free and Total Potential Haloforms", Ontario Ministry of the Envir. Lab. Service Branch, presented at Pittsburgh Conf. of Anal. Chem. and Applied Spectroscopy (1976).

Richard, J.J. and G.A. Junk. "Liquid Extraction for the Rapid Determination of Halomethanes in Water", J. Am. Water Works Assoc., (January 1974).

Weber, W.J., Jr., Physiocochemical Processes for Water Quality Control, New York: Wiley Interscience, 1972.

Weber, W.J., Jr., M. Pirbazari, M. Herbert, and R. Thompson. "Effectiveness of Activated Carbon for Removal of Volatile Halogenated Hydrocarbons from Drinking Water", Chapt. 8, in Viruses and Trace Contaminants in Water and Wastewater, Ann Arbor Science, Ann Arbor, Michigan, 1977.

Weber, W.J., Jr., M. Pirbazari, and M.D. Herbert. "Removal of Halogenated Organic and THM Precursor Compounds from Water by Activated Carbon", presented at 98th Am. Water Works Assoc. Conference, Atlantic City, N.J., June 1978.

THE EXAMINATION AND ESTIMATION OF THE PERFORMANCE CHARACTERISTICS OF A STANDARD METHOD FOR ORGANO-CHLORINE INSECTICIDES AND PCB

I.W. Devenish and L. Harling-Bowen

Southern Water Authority

Guildbourne House, Worthing, Sussex, England

BACKGROUND

As early as 1880, following a report of the Government Water Examiner, the Council of the Society of Public Analysts invited members to send results of water analysis on a monthly basis. To make the results comparable, instructions on the methods of analysis were sent to all the analysts involved and were also given in the Analyst in 1881. This was particularly important since at the time no small measure of controversy existed, particularly over the determination of the organic purity of water supplies. It is reported that the Water Committee of the Society advised that results be expressed in grains per gallon since "the reports in many cases pass into the hands of those to whom a statement in parts per 100,000 is more or less unintelligible" (Chirnside and Hamence, 1974). One cannot but ponder on the reaction to results in parts per thousand million (μg/l) or indeed parts per million million (ng/l) which form the basis of this paper. These methods of water analysis were not supported by any collaborative studies but were the subject of much discussion by the members.

In 1923 the Society was asked to standardize methods of analysis. In view of the large amount of work that would be involved, the Council decided against such an idea but recommended that those concerned should get together and submit papers for publication in the Analyst. A committee was formed, known as the Standing Committee on the Uniformity of Analytical Methods.

In 1929 the publication in Chemical Analysis as Applied to Sewage and Sewage Effluents came out, based on the work of the Royal Commission on Sewage Disposal, which was published in 1904.

In 1935 the Committee on the Uniformity of Methods was reconstituted and renamed the Analytical Methods Committee. The main work was carried out by sub-committees, and the methods recommended were based on collaborative studies. In 1954 an appeal was launched for funds to give secretarial support to the sub-committees. This appeal was very successful and its first fruits were the Recommended Methods for Trade Effluents published in 1958 by a joint committee for the Society for Analytical Chemistry and the Association of British Chemical Manufacturers. Two years earlier the methods for sewage and sewage effluents were revised by a committee set up by the Ministry of Housing and Local Government and published in 1956.

In 1949 Approved Methods for Physical and Chemical Examination of Waters was published jointly in the journal by the Institute of Water Engineers, the Royal Institute of Chemistry and the Society of Public Analysts and other Analytical Chemists from which emerged the Society for Analytical Chemistry. This publication was revised in 1953 and 1960.

In 1964 the Ministry of Housing and Local Government set up committees to revise these 1956 and 1958 publications and to issue them in one book. Its membership was drawn from the Association of British Chemical Manufacturers, later called the Chemical Industries Association, the Society for Analytical Chemistry, the River Boards Association, the Institute of Sewage Purification, now the Institute of Water Pollution Control, the Water Pollution Research Laboratory, now the Water Research Centre, and the Laboratory of the Government Chemist. Also in 1964 the Society for Water Treatment and Examination suggested that the Ministry Committee should be enlarged to include potable water interests. This was agreed as it was recognised that a single book dealing with the analysis of water and effluents would be of benefit and make more realistic the connection between all the facets of the hydrological cycle. The results of the Committee's deliberations was the publication of what has become affectionately known as the Green Book. (H.M.S.O. 1972). In the Committee's report it was recommended that a Standing Committee of Analysts be established to consider new methods and modifications of existing ones and to arrange for them to be fully tested. Publication of these new methods or modifications in the technical press would, it was thought, encourage the widest possible use of them and facilitate production of a revised edition at an appropriate time when they would be included and recommended as official tests.

By 1975 several changes had taken place. The Ministry of Housing and Local Government had been replaced by the Department of the Environment; the whole water cycle in England, excluding private water supply undertakings, had become the responsibility of nine Regional Water Authorities and a National Water Council has been established. Also the "Green Book" was then out of print and unavailable.

In 1973 a Standing Committee of Analysts to Review Standard Methods for Quality Control of the Water Cycle was set up by the Department of the Environment. With the creation of the Regional Water Authorities in England, the Welsh National Water Development Council and the National Water Council, this committee became one of the Joint Technical Committees of the Department of the Environment and the National Water Council in 1974. Its first report covering the period from May 1973 to January 1977 was published towards the end of last year. (Department of the Environment, 1977). (Figure 1).

The membership of the Standing Committee of Analysts, to use its abbreviated title, includes the English and Welsh Water Authorities, the Water Research Centre, the Laboratory of the Government Chemist, Consultant Analysts, Industry, the Ministry of Agriculture Fisheries and Food, and other competent persons and interested bodies, and is listed on page 2 of the Report. Its terms of reference are: to review continuously the recommended methods for the analysis of water, sewage, effluents and their associated sludges, sediments, etc.; to update and provide new methods as necessary; to advise annually on needs and priorities for research in relation to problems experienced in practice; to encourage the application of research results and report at least once every two years.

Nine Working Groups have been set up to cover all aspects of the work. These groups have the responsibility of selecting and evaluating methods and approving drafts. They are specialist groups but in most cases are assisted by Panels of other analysts. The composition of these panels changes as the determinands under consideration change, but their Chairmen are always members of their relevant Working Group. Similarly the Chairmen of the Working Groups are members of the main Committee (Table 1).

Working Group 6.0 under the Chairmanship of Dr. B.T. Croll of the Anglian Water Authority has to consider all organic impurities except simple organo nitrogen compounds which come within the ambit of Working Group 5. Panel 6.3 is under the Chairmanship of Mr. D. Meek of the Water Research Centre and was previously under Dr. Croll. The panel's first task was to prepare and evaluate the method for the examination of organo-chlorine insecticides and PCB's in water. The final publication of the method has yet to take place but by the spring of 1977 it was in a form sufficiently well defined to allow tests to be made to evaluate its performance characteristics and to discover any refinements or changes that could usefully be incorporated in it. When this work was completed

Table 1
Standing Committee of Analysts, Working Groups and Panels

S.C.A. Main Committee
0.1 Editorial Board   0.2 Special Panels

Working Group 1.0 General Principles of Sampling and Accuracy of Results
- Panel 1.1 Sampling
- Panel 1.2 Assessment of Accuracy of Results

Working Group 2.0 Instrumentation and On-Line Analysis
- Panel 2.1 Continuous Flow Methods

Working Group 3.0 Empirical and Physical Methods
- Panel 3.1 Physical Parameters
- Panel 3.2 Oxygen Demand
- Panel 3.3 Total Organic Carbon

Working Group 4.0 Metals and Metalloids
- Panel 4.0a Mercury, Cadmium, Lead
- Panel 4.1 Arsenic, Selenium, Antimony, Tellurium
- Panel 4.2 Manganese, Chromium, Iron, Cobalt, Nickel, Vanadium
- Panel 4.3 Calcium, Magnesium, Barium, Aluminium, Sodium, Potassium
- Panel 4.4 Silver, Zinc, Copper, Tin and Thallium

Working Group 5.0 General Non-Metallic Substances
- Panel 5.1 Nitrogen Compounds
- Panel 5.2 Sulphur Compounds
- Panel 5.3 Phosphorus, Silicon and Boron
- Panel 5.4 Alkalinity, Carbon Dioxide and simple organic acids
- Panel 5.5 Disinfecting Agents
- Panel 5.6 Halides

Working Group 6.0 Organic Impurities
- Panel 6.1 Phenols
- Panel 6.2 Oils, Fats and Waxes
- Panel 6.3 Pesticides and PCB
- Panel 6.4 Polyaromatic Hydrocarbons
- Panel 6.5 Detergents

Working Group 7.0 Biological Methods
- Panel 7.1 Macroinvertebrates and Other Aquatic Animals
- Panel 7.2 Algae and Higher Plants
- Panel 7.3 Microbiological Methods
  - A Affecting Man
  - B Other Organisms and Microbiological Aspects
- Panel 7.4 Toxicity Tests
  - A Fish Toxicity
  - B Effects on Sewage Treatment Works
- Panel 7.5 Biodegradability Tests

Working Group 8.0 Sludge and Other Solids Analysis
- Panel 8.1 Inorganic Constituents
- Panel 8.2 Organic Constituents
- Panel 8.3 Physical Properties
- Panel 8.4 Sludge Biology

Working Group 9.0 Radiochemical Methods

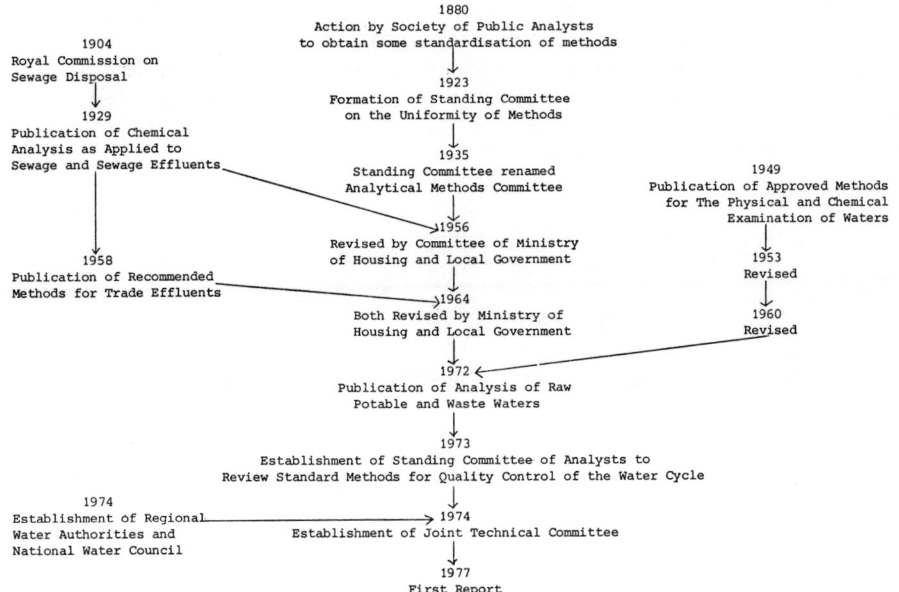

Figure 1

a final draft of the method would be submitted to the Main Committee and its editorial panel for its publication as a Recommended Method or a Tentatively Recommended Method. For the former, performance characteristics would have been obtained from at least five separate laboratories and an example is the method for lead in potable supplies. (S.C.A. 77). In the latter case the performance characteristics have not been so defined and further information is required. An example of this is the method for cadmium in potable supplies. (S.C.A. 77).

The performance characteristics required are given in Appendix I which constitutes the first page and a half of the method.

The author, with the consent of his employers, the Southern Water Authority, and with financial support from the Standing Committee of Analysts, undertook the task of obtaining some performance characteristics and reporting on the method. The work was done in the Resource Planning Laboratory in Brighton by its staff under the direction of the author. To allow the functions of the laboratory to continue largely unimpeded, a temporary staff member was recruited for a six month period. This was the time that it was thought the work would take and was unfortunately underestimated. This has left several lines of interest yet to be pursued.

Wilson (1972) recommends that at least five laboratories should obtain estimates of precision for levels of the determinands sought, each with at least ten degrees of freedom with the analysis spread over at least five days. It is further recommended that samples and standards be included in the scheme of analysis (Wilson, 1970), and such a scheme would be:-

Blank
Standard 10% of working range
Standard 100% of working range
Real Sample
Real Sample ("Spiked")

each solution being examined in duplicate on ten occasions.

In connection with insecticides and PCB several determinands could, it was thought, be examined simultaneously. It was also thought desirable to spike real samples at two levels. This would have required 12 solutions to be examined in one batch ten times. Further, as was stated above, it would be necessary to execute this work in at least five laboratories. The method (S.C.A. 1977) describes two means of extraction, shaking technique and rolling technique. To be all-embracing, the scheme could have involved the examination of 24 solutions, half by each technique, ten times. This was obviously far too great a task to be countenanced by a small laboratory which has other work to do, even with the addition of one to the staff. The Panel elected, therefore, to put in hand sufficient work to enable the method to be tentatively recommended.

TERMS OF REFERENCE AND METHOD

Objectives

To test and comment on the suitability and performance characteristics of the draft method provided by S.C.A. Panel 6.3 and described in document S.C.A. 71.

Programme

Using the shaking technique only of the Draft Method of Panel 6.3:

1. To examine the following solutions on 11 occasions:

    i. Blank (in duplicate)
    ii. Standard Solution at 10% of working range
    iii. Standard Solution at 90% of working range
    iv. Sample Solution

v. Sample Solution spiked at 10% working range
vi. Sample Solution spiked at 75% working range

for the following materials over the working ranges indicated:

γHCH	up to 5 ng/1
Aldrin	up to 5 ng/1
Dieldrin	up to 7.5 ng/1
pp'DDE	up to 7.5 ng/1
pp'TDE	up to 15 ng/1
pp'DDT	up to 30 ng/1
Aroclor 1260	up to 400 ng/1

2. To measure the recovery of added insecticide with time.

3. To comment upon the effectiveness of the Draft Method and to supply analytical data to support statements with regard to limits of detection, recovery and precision.

Method

The method is fully described in reference (8) and will, it is hoped, shortly be published by H.M.S.O.

Briefly, 2 1 of sample are shaken with 50 ml of hexane for five minutes. The layers are separated and the hexane concentrated to 1 ml. Most of the extraneous coextractions are removed by passing the concentrated extract successively through columns of alumina and alumina impregnated with silver nitrate. More hexane is used to elute the desired material from this column and the eluate is again concentrated to 1 ml. At this stage the gas chromatographic examination of an aliquot of the concentrated cleaned extract may be made to determine if partition is necessary to separate some insecticides from the PCB. Partition is achieved with silica gel. The 1 ml of concentrated extract is added to a column of this material and eluted with hexane. Previously the volume of eluate needed to separate pp'DDE and pp'DDT has been determined and this volume is collected, Fraction 1. The column is then further eluted with a mixture of diethyl ether and hexane, Fraction 2. The disposition between the two fractions is as follows:

Fraction 1	Fraction 2
Aldrin	γHCH
pp'DDE	Dieldrin
PCB	pp'TDE
	pp'DDT

Each of the two fractions is concentrated to 1 ml and the resultant concentrate examined by gas/liquid chromatography. By comparing the gas chromatographic traces of the extract and standards, the concentrations present in the extract are determined. Further calculation is used to relate the concentration in the extract to that in the original sample.

Equipment

For the gas/liquid chromatography a Pye G.C.V. fitted with a Ni_{63} electron capture detector was used. This was coupled to a Hewlett Packard Reporting Integrator Model 3380A. The G.C. conditions were as follows:

Column	- Glass 1.5 m x 4 mm id. packed with 80-100 mesh AW-DMCS chromosorb supporting 2% OV-1 plus 3% QF-1 by weight.
Ovens	- Injection 250°C
	Column 190°- 225°C
	Detector 250°C
Gas	- Nitrogen 40 ml/min
Injection	- 5 μl of final extract or standard solution.

The higher column temperature was used to speed the elution of PCB isomers. The reporting integrator was not available for the preliminary work but was used for each of the runs of the scheme.

PRELIMINARY WORK

Spiking of Test Solutions

It is generally accepted that the immediate extraction of added materials is not a justifiable means of determining extraction efficiency with regard to insecticides or indeed many other determinands. Enhanced recoveries will generally ensue and it is, or should be, common practice to allow the determinand to remain in contact with the material for a period before any attempt at recovery is made. Before the scheme of analysis in the programme could begin it was necessary to try to define this period and subsequently to employ the same period for all batches. Twenty litres of river water containing 20 mg/l of suspended solids were shaken for ten minutes. Two one-litre portions were withdrawn and extracted, each with ten millilitres of hexane to determine any background levels. To the remaining volume a composite mixture of insecticides in acetone was added and mixed by a further ten minutes shaking. Immediately two one-litre portions were withdrawn and similarly extracted. Subsequently, portions were withdrawn after mixing on successive days until the bulk was exhausted. The results are shown in Table 2.

Table 2 Recovery with Time

	γHCH ng/l	Aldrin ng/l	pp'TDE ng/l	pp'DDT ng/l
Spiking Level	100	100	300	600
Day				
0	77	60	147	254
1	73	51	90	110
2	88	64	114	149
3	74	59	94	115
4	68	38	106	137
5	68	44	102	132
6	64	35	58	81
7	92	56	112	124
8	81	40	122	141
mean	76	50	105	138
S_t	9.4	10.7	24.4	47.9

It was hoped that an immediate high recovery would be shown on day 0 followed by a reduction on subsequent days, but possibly giving a plateau after x days. It had tentatively been decided that three days' contact time should elapse before extraction after spiking the test solutions, and this time period was, in fact, used for the scheme of analysis. This broadly approximated to the location of the mean result but the results can be interpreted as showing that day one would have been equally appropriate given the variation of the results.

The spiked standard solutions and spiked river samples were prepared in the following manner. Composite standard solutions were first prepared by diluting the strong standards of each material to give a strength in proportion to 10%, 75% and 90% of the working range. The same volume (100 micro litres) from each of these composite solutions was then added to the relevant two-litre volumes of insecticide-free distilled water or river water. The solvent used for adding the materials was analar grade acetone. Each bottle was then shaken for five minutes before storage in the refrigerator. Although more complicated than adjusting the volume of composite standard added to each volume of water, this technique standardized the volume of solvent present in each solution, except the blanks, put up for extraction.

Linearity of Detector Response

The linearity of calibration was initially ascertained by injecting increasing volumes of standard insecticides solutions and relating the peak area expressed as a percentage to the quantity of insecticide. The results are shown in Table 3. This linear calibration allowed standards at 10% and 100% of the working range only to be required when examining test solutions. Three standards were used at the low end of the calibration range to correspond with the range of application of the scheme and a single strong standard used for the upper point. It is possible that with the equipment used the range may be larger, but this was not explored.

Following extraction of a two-litre sample and the concentration to 1 ml, the equivalent range in terms of ng/l are shown in parentheses. For concentrations in excess of this range the sample extract would normally be diluted to bring the concentration to within the normal working range.

Extraction Efficiency

Two 2 l pairs of cleaned glass bottles containing insecticide-free water were spiked with the six insecticides in acetone. One pair was extracted immediately and the second pair after 70 hours refrigerated storage. In some measure this repeated previous work but greater accuracy in estimation was achieved by using an integrator. The results are shown in Table 4.

A greater disparity in results between aliquots treated in the same manner was noted, but overall the results were of the same order. These results showed that adsorption by glass was not a particularly significant factor but that with a single extraction recoveries of the order of 50 - 75% only could be expected.

Table 3 Linear Range of Calibration

γHCH	250 pg	(25 ng/l)
Aldrin	375 pg	(37.5 ng/l)
Dieldrin	375 pg	(37.5 ng/l)
pp'DDE	375 pg	(37.5 ng/l)
pp'TDE	750 pg	(75 ng/l)
pp'DDT	1,500 pg	(150 ng/l)
Aroclor 1260	10,000 pg	(1000 ng/l)

Concentration

The possibility of losses occurring when concentrating the initial extracts was examined. Four rates were investigated, 2, 16, 25 and 32 ml/min when 50 ml of hexane spiked with each of the six insecticides from a single strong composite standard mixture were evaporated to 1 ml. The resultant extracts were then compared with a mixed standard and the results plotted as percentage recovery against distillation rate. It was demonstrated that at the lowest distillation rate approximately 20% of γHCH and Aldrin were lost in the distillate, increasing to about 40% at the highest rate. Recovery of pp'DDE appeared to improve as the distillation rate increased, and the other insecticides were largely unaffected. Concerning a reasonable throughput of samples, at the slowest distillation rate the time taken to reduce the 50 ml volume to 1 ml was more than 1½ hours, and at the fastest rate 2½ minutes. The temperature range for these distillation rates was 20°C. The use of the slowest rate in most working laboratories would be untenable, while above this rate there is little to be gained by not using the fastest.

A different concentration technique such as the use of a thin film evaporator may possibly improve the recovery of the more volatile insecticides. Neither equipment nor time were available to test this.

Table 4

Insecticide	Spiking level ng/ml 1 ml extract		Immediate recovery %	70 hr recovery %
γHCH	50		51 - 53	47 - 65
Aldrin	50		50 - 51	52 - 70
pp'DDE	75		50 - 56	54 - 67
Dieldrin	75		57 - 65	57 - 73
pp'TDE	150		60 - 69	61 - 75
pp'DDT	300		59 - 72	63 - 80
		mean	58 ± 7.9	63 ± 10.5
		S_t	7.5	10

Volume of Extract

The centrifuge tubes used were calibrated by the manufacturers over the range 0.1 to 10 ml in 0.1 ml steps. Some disquiet had been expressed by the panel members of the validity of these calibrations. Also the accuracy with which one can judge the position of the meniscus at the 1.0 ml calibration mark cannot be greater than 0.05 ml. To test this a batch of ten tubes were selected at random. The individual tubes were weighed before and after the collection of extract and evaporation to 1 ml of 50 ml portions of hexane. The weight per unit volume of the solvent was determined at the same time, 0.655 g/ml. The results showed that a range of 0.595 g to 0.665 g had been collected with a mean of 0.635 g and a standard deviation of 0.02 g. These weights correspond to a range of volumes from 0.908 ml to 1.015 ml, a spread of more than 10% with a mean of 0.969 ml $\pm$ 0.02 ml.

Repeatability of Injections

Five micro litres of a solution containing 10 ng/ml of Aldrin and 5 ng/ml of Dieldrin were injected onto the OV/QF1 column six times and the peak areas recorded. From these areas the standard deviation was calculated and was expressed as a percentage of the area:-

Aldrin $S_t = 4.5\%$
Dieldrin $S_t = 4.2\%$

These figures are probably about average for someone whose working life is not entirely devoted to G.C. work.

Silica Gel Partition

The object of the partition by silica gel is to separate most of the insecticides which give discrete gas chromatographic peaks from the isomers of the PCB compound. The draft method describes the procedure to be used to check the activity of each batch of silica gel. Of necessity this was undertaken at an early stage of the work and showed that most of the pp'DDE was eluted in the first 5 - 6 ml (with some tailing), pp'DDT in the next 4 ml and Dieldrin was only eluted by the ether/hexane mixture. A cut-off after 5 ml was, therefore, chosen for the scheme. However, when it was applied to samples, it was observed that considerable interference occurred with the insecticide gas chromatographic traces. Detailed examination of the gas chromatography records indicated that this contamination was related to the PCB. Further work on this important aspect of the draft method was indicated.

First the activity check was repeated using a mixed insecticide standard and PCB standard separately at the 5 ml cut; this showed that some Aldrin and pp'DDE came through in fraction 2 and not all the PCB was eluted in fraction 1, Table 5. Secondly the PCB and insecticides were added to the silica gel, mixed and again a 5 ml cut taken, Table 6.

In the ideal situation, there should be no PCB in fraction 2 and no pp'DDT in fraction 1. Separately a PCB standard at 670 ng/1 and pp'DDT standard at 200 ng/l were put through the silica gel columns and 1 ml portions of eluate collected and examined with the results shown in Table 7.

The results shown in Tables 5 and 6 have been calculated to 100% Aldrin basis to aid comparison.

For all these tests, great care was taken to add the materials to the silica gel in as tight a band as possible. To this end the test solutions were added dropwise by pipette and washed into the column with 1 ml hexane added dropwise also. Normally the 1 ml extract obtained after alumina/alumina silver nitrate clean-up would be poured from the graduated tube and the 1 ml rinse measured in the same tube and similarly poured onto the column.

From these results it is apparent that the draft method will not adequately separate PCB and pp'DDE if both are present in the sample under examination. Also some isomers of the standard Aroclor 1260 were transiently retained by the silica gel and only eluted by the ether hexane mixture. In the context of the work, this breakthrough seriously affected the calculations at the 10% spiking level. Because of this interference the result for pp'DDE had to be rejected. This breakthrough into fraction 2 of PCB isomers may not be obtained with natural water samples. The gas chromatographic records show that it was probably the lesser chlorinated isomers that were eluted late from the silica gel column.

Further work requires to be put in hand to examine the partition on silica gel particularly with regard to the degreee of dehydration that is required to separate the pp'DDE and the pp'DDT and also to reduce the tailing of the PCBs and their subsequent elution in fraction 2.

Table 5 5 ml Cut Separate Treatment of Insecticides and PCB

Fraction 1		Fraction 2	
Aldrin	88%	γ HCH	77%
pp'DDE	72%	Aldrin	11%
PCB	73%	DDE	26%
		Dieldrin	78%
		pp'TDE	69%
		pp'DDT	98%
		PCB	4%

Table 6 5 ml Cut on Mixed Insecticides and PCB

Fraction 1		Fraction 2	
Aldrin	92%	γHCH	85%
pp'DDE	83%	Aldrin	8%
PCB	90%	pp'DDE	17%
		Dieldrin	77%
		pp'TDE	79%
		pp'DDT	126%
		PCB	4%

Table 7

Fraction 1

mls collected	% collected	
	DDT	PCB
3	0	Not estimated
4	0	Not estimated
5	3	4
6	30	1
7	50	0.2
8	Not estimated	0.2
9	Not estimated	0
Fraction 2		
Total concentrated to 1 ml	0	1.5%

RESULTS

Blanks and Limits of Detection

Although the eleven batches were begun, only seven produced a complete set of pairs of results for all the determinands excluding pp'DDE. The results are summarised in Table 8.

Table 8 Blanks and Limits of Detection

Insecticide	Mean	S_w	S_b	S_t	Limit of Detection** ng/l	Criterion of Detection* ng/l
γ HCH	0.21	0.320 (7)	NS (6)	0.320 (13)	12	0.9
Aldrin	0.04	0.032 (7)	0.060 (6)	0.063 (6)	3	0.1
Dieldrin	0.28	0.205 (7)	0.430 (6)	0.480 (7)	14	0.6
pp'TDE	0.42	0.570 (7)	NS (6)	0.70 (11)	8	1.5
pp'DDT	0.03	0.125 (7)	NS (6)	0.125 (13)	15	0.3
PCB	2.8	1.8 (7)	3.2 (6)	3.7 (8)	106	4.8

These results were obtained by examining pairs of blanks on seven separate occasions.

(n) figures in brackets indicate numbers of degrees of freedom.
S_w = Within-batch standard deviation.
S_b = Between-batch standard deviation.
S_t = Total standard deviation.

* Criterion of Detection: that concentration which is unlikely to be exceeded unless the sample contains the determinand (95% confidence level).

** Limit of Detection: that concentration for which there is a desirably small probability that the result will be less than the criterion of detection.

The limits of detection were calculated from

$$100\,[\sqrt{2}.t.S_w + t'.\,S'] \div p \quad \text{(Wilson 1977)}$$

t is the single sided $t_{0.05}$ with 7 degrees of freedom
S_w is the within batch standard deviation of the blanks
S' is the total standard deviation for blank corrected results for river water spiked at the 10% level
t' is the single sided $t_{0.05}$ with degrees of freedom
and p is the percentage of recovery at the 75% spiking level of river water.

The criterion of detection, that is the level below which the determinand can be claimed to have been not detected, was calculated from

$\sqrt{2}\ t_{0.05} S_w$ (Wilson, 1970)

where the symbols have the same meaning as above.

These tests showed that the samples were subject to additional sources of random error not present in the blank tests. That is, the S_t of the river results at 10% spiking level was much greater than the S_t for the blanks. Hence it would have been wrong to give double the criterion of detection as the limit of detection. Also the calculated limits include an element reflecting less than quantitative recovery. This requires that the same element be applied to results of samples which is unusual in the U.K. in the context of insecticide determinations. Current work in the U.K. on discharges to coastal waters requires limits of detection, for aldrin 100 ng/l, and for dieldrin 50 ng/l. The 14th edition of the A.P.H.A. handbook gives 10 ng/l for γHCH and 20 - 25 ng/l for pp'DDT as limits of detection for clean samples. Thus the method produced limits of detection which were adequate.

These limits of detection, when compared with the results obtained with the samples spiked with the determinands shown in Table 9, demonstrate that the working ranges given in the objectives were pitched too low.

Results on Samples

Again a complete set of results for all the determinands for all batches was not obtained; the hazards of a technique involving several stages and manipulations repeated many times. It was never anticipated that all the programmed batches would be completed but that sufficient information would be obtained to define the required characteristics. The results are summarised in Table 9.

Table 9
Estimates of Total Standard Deviation (S_t) (Results expressed in ng/l)

	γHCH			Aldrin			Dieldrin		
	Spike	Mean	S_t (n)	Spike	Mean	S_t (n)	Spike	Mean	S_t (n)
Distilled water plus 10% spike	0.50	0.50	0.29 (4)	0.50	0.50	0.20 (4)	0.75	0.4	0.33 (4)
Distilled water plus 90% spike	4.5	2.7	0.75 (9)	4.5	2.6	0.85 (9)	6.75	3.8	2.05 (9)
River water	-	-	-	-	0.05	0.07 (7)	-	1.30	1.13 (7)
River water plus 10% spike	0.5	4.7	2.84 (7)	0.5	0.85	0.7 (7)	0.75	1.15	1.05 (7)
River water plus 75% spike	3.75	6.1	4.19 (7)	3.75	1.8	0.74 (7)	5.625	2.3	2.30 (7)

	pp'TDE			pp'DDT			PCB		
	Spike	Mean	S_t (n)	Spike	Mean	S_t (n)	Spike	Mean	S_t (n)
Distilled water plus 10% spike	1.5	1.85	0.5 (4)	3.0	2.2	1.47 (4)	40	42	22 (5)
Distilled water plus 90% spike	13.5	11.6	2.25 (9)	27	15.2	4.3 (9)	360	228	62 (5)
River water	-	0.35	0.40 (7)	-	0.15	0.20 (7)	-	6.0	7.0 (8)
River water plus 10% spike	1.5	3.0	2.20 (7)	3.0	2.75	3.4 (7)	40	25	21 (5)
River water plus 75% spike	11.25	8.2	1.38 (7)	22.5	10.4	5.25 (7)	300	133	49 (5)

Results expressed in ng/l.

S_t = estimate of total standard deviation with (n) degrees of freedom.

The results for γHCH in the unspiked river water were incomprehensible in that in each case they were considerably higher than in the spiked samples. Its widespread use and occurrence in the environment is acknowledged but the high levels found cannot solely be attributed to contamination. The only difference between spiked and unspiked samples was the presence of 100 μl of acetone in the 2 l of spiked samples (0.005%). With hindsight one can say that the examination of samples spiked with pure acetone would have been advantageous. It does, however, underline the dangers of creating an artificial situation, but there was no other way of adding very small amounts of the determinands to the river water in a repeatable standardised manner.

The results show that a higher degree of variability obtained with river water than with distilled, and recovery was worse. The percentage recovery at the higher spiking levels was less variable and it may be surmised that if the working ranges had been chosen such that the 10% spiking level was at the limits of detection shown in Table 9, a better set of results may have been obtained.

DISCUSSION

The loss of some of the materials sought would appear to be characteristic of the method. These losses can be attributed to:-

(a) Less than quantitative extraction from the samples. The extraction efficiency may be improved by re-extracting the aqueous phase. Logically there is no reason why this should not be so, but the degree of improvement would need to be measured before a second or even third extraction step could be recommended.

(b) Less than quantitative elution from the alumina clean-up column. Examination of Table A1 of the Appendix of the method and a return to the original source of the information provided to Panel 6.3 showed that the volume of eluent recommended was critical particularly with regard to dieldrin. The tolerance in volume of eluent could have led to variation of concentration eluting from one column to another and the non-elution of some or all of the dieldrin. The application of an excess of eluent could overcome this problem, but again further work is required to define the minimum volume that should be used.

(c) Volatilisation of the determinand during concentration. In order to achieve the required concentration ratio of 2000 : 1, the reduction of the volume of the extractant and eluants is imperative. The methods employed should be those which keep volatilisation losses to a minimum but this may be incompatible with the time allowable for this work, and some compromise may be required.

These three factors may also account for some of the variation in the results. To these must be added the other source of error previously identified.

Generalising, it may be stated that in the hands of a reasonably competent operator and at levels above the limit of detection, the method will:-

(a) give recoveries of approximately 50% ± 20% except for dieldrin, which has been referred to above, and

(b) give a relative standard deviation of approximately 50% for river water and 30% for distilled water.

Thus for levels above the limit of detection and correcting for recovery, results should be obtained no worse than within an order of magnitude. For low levels in natural and potable waters, this is probably sufficient. Problems arise, however, when one is working to an imposed standard, especially if that standard is approaching the limit of detection. In these cases, the examination of a single sample portion is out of the question. At least a pair of samples must be examined to reduce the confidence limits by $1/\sqrt{2}$. A spiked sample and/or standard examined at the same time as the unspiked sample would give some measure of the recovery obtainable, but again one has the problem of the artificial situation. These would all involve a considerable extra workload and the use of these alternatives would have to be a matter of policy determined in part by the degree of importance to be attached to the results. The situation would be eased somewhat if only one or two specific determinands were being sought.

RECOMMENDATIONS

Further work is required to be undertaken on those aspects of the method which are most likely to have adversely affected its performance. These are:-

(a) Extraction stage. Would a second extraction of the aqueous phase give a significantly improved efficiency?

(b) Elution from clean-up column. What is the minimum volume of eluent required to give quantitative recovery of the determinands sought commensurate with the retention of undesirable coextractives by the column?

(c) Silica gel partition. What is the degree of dehydration and rehydration of the silica gel required to effectively separate pp'DDE and pp'DDT and reduce the tailing of PCB?

When these questions have been satisfactorily answered and the method modified then the performance characteristics will need to be redetermined.

CONCLUSIONS

The method will determine levels of insecticide and PCB concentrations with a suitably low level of detection. Recoveries of about 50% $\pm$ 20% can be expected for most insecticides with relative standard deviations of about 50%. Recoveries could possibly be improved by making a double extraction of the sample and by increasing the volume of eluent applied to the clean-up column. Results should be obtained within an order of magnitude of the actual concentration taking into account of the percentage recovery. Results should be recorded in terms of concentration in the sample taking account of the percentage recovery or the fact noted that a recovery factor has not been used. Results more closely approximating to the concentration in the sample could be obtained by the use of standards and replicate analyses. 'In-house' recovery performance tests would also be of benefit. Further work is needed to improve the silica gel partition.

ACKNOWLEDGEMENTS

I am indebted to the staff of the Sussex Area Resource Planning Offices and the Water Research Centre for their considerable assistance with the analytical work, with the treatment of the raw data and with the production of this paper, and to the Southern Water Authority for allowing the work to be undertaken and this paper to be presented. The work was carried out for the Standing Committee of Analysts and financed by the Department of the Environment under Contract DGR 480/297.

REFERENCES

Chirnside, J.C., and Hamence, J.H. 1974. The Practising Chemist. Society of Analytical Chemistry, London. p.136.

Department of the Environment and National Water Council. D.o.E. London, 1977. Standing Committee of Analysts to Review Standard Methods for Quality Control of the Water Cycle, First Report 1973-1977.

Her Majesty's Stationery Office, London, 1972. Analysis of Raw, Potable and Waste Waters.

Her Majesty's Stationery Office, London, 1977. S.C.A. Lead in Potable Waters by Atomic Absorption Spectrophotometry. Methods for the Examination of Waters and Associated Materials.

Her Majesty's Stationery Office, London, 1977. S.C.A. Cadmium in Potable Waters by Atomic Absorption Spectrophotometry.

S.C.A. London, 1977. The Determination of Organochlorine Insecticides and Polychlorinated Biphenyls in Waters. Document S.C.A. 71.

Wilson, A.L. 1972. Criteria to be satisfied by Analytical Methods before their consideration as Standard Methods. Technical Memorandum 73. Water Research Centre, Medmenham, Buckinghamshire, p.5.

Wilson, A.L. 1972. Minimising and Estimating Analytical Errors, part 2: Estimation of Precision and Bias. Technical Memorandum 56. Water Research Centre, Medmenham, Buckinghamshire, p.5.

Wilson, A.L. 1977. Private Communication.

Wilson, A.L. 1970. Minimising and Estimating Analytical Errors, part 1: Simple Statistical Techniques for Analysis. Technical Memorandum 55. Water Research Centre, Medmenham, Buckinghamshire, p.23.

APPENDIX 1

The Determination of Organochlorine Insecticides and Polychlorobiphenyls in Waters

Performance Characteristics of the Method

1.	Substances determined	Organochlorine insecticides and polychlorobiphenyls. (PCB).
2.	Type of sample	Natural waters, potable supplies and sewage effluents.
3.	Basis of method	Extraction into hexane and removal of extraneous materials using a column of Alumina $AgNO_3$. Separation of most chlorinated insecticides from PCB by column chromatography on silica gel, followed by gas liquid chromatography using an electron capture detector.
4.	Range of application	Typically up to 250 ng/1.
5.	Calibration Curve	Range of linearity depends on the detector in use. The instrument used in the performance tests gave a linear response over the following ranges:

γHCH	0 - 250 pg
Aldrin	0 - 375 pg
Dieldrin	0 - 375 pg
pp'DDE	0 - 375 pg
pp'TDE	0 - 750 pg
pp'DDT	0 - 1500 pg
Aroclor 1260	0.1 - 10 ng

6.	Standard Deviation	See Tables 2 and 3.
7.	Limit of Detection	See Table 2.
8.	Sensitivity	Dependent on determinand and instrument in use.

9.	Bias	The recoveries of insecticides are variable and seldom quantitative, typically about 50% depending upon extraction efficiency which may vary with sample and determinand, see Table 3.
10.	Interference	Any electron capturing material which passes through the procedure and has similar gas chromatographic characteristics to the determinand.
11.	Time required for analysis.	a) Assuming all reagents prepared and the instrument already calibrated, extraction and clean-up 2 hours, gas chromatography up to a further 2 hours depending on determinand and instrument in use. b) Total analysis including preparation of reagents, apparatus, etc., and confirmation of identity of determinands: approximately 6 samples per man-week.

A SYSTEM FOR THE PREPARATION OF ORGANOHALOGEN REFERENCE MATERIALS WITH MARINE LIPID MATRICES

M.D. MacKinnon, W.R. Hardstaff and W.D. Jamieson

Atlantic Regional Laboratory, National Research Council of Canada, 1411 Oxford Street, Halifax, N.S., B3H 3Z1

Well characterized reference materials which closely resemble real biota, water, or sediment samples are useful in establishing the accuracy of analyses for trace levels of analytes in such natural samples. Natural samples, sometimes enriched by addition of specific analytes, have been used in studies to compare the accuracy of analyses done in different laboratories. Such intercalibration studies have been reported recently for organochlorine compounds (Harvey et al, 1974; Holden, 1975; Hom and Pavlou, 1976) in various matrices (biological materials, waters, sediments). Similar studies are now being conducted by the International Atomic Energy Agency, Monaco, and by the National Water Research Institute, Burlington, Canada. The samples used in such studies have often not been generally available, particularly on a continuing basis after the completion of an intercalibration project.

If a reference material were available on a continuing basis, it would be useful not only for intercalibration studies but also for the evaluation and development of new analytical procedures. Readily available reference materials also allow analysts to monitor the accuracy of results obtained with analytical procedures used routinely. We have examined the use of hydrogenated herring oil as a matrix for a system of such reference materials useful with analytical procedures using electron-capture detection gas chromatography. The hydrogenated herring oil was processed by vacuum stripping to yield material free from interfering substances detectable by an electron-capture detector. Known amounts of specific analytes were added to hexane solutions of this cleaned lipid to yield reference samples containing defined quantities of the analytes at levels below, at, and above the levels usually found in natural samples. For a set of such samples containing organochlorine analytes, we have studied effects of storage and the variability of replicate samples.

METHODS

Natural herring oil was cleaned of organochlorine compounds by vacuum (5-10 torr) stripping (220°C) after hydrogenation (Addison and Ackman, 1974). The cleaned lipid was dissolved in redistilled n-hexane (50 g/100 ml solution). Concentrated hexane solutions of highly pure (> 99% by GC, GC/MS) commercially obtained analytes [lindane, β-BHC, aldrin, heptachlor epoxide, DDE (o,p and p,p'), DDD (o,p and p,p') and DDT (o,p and p,p')] were prepared by weighing each analyte [usually, 35 (± 0.05) mg in 50 ml]. Portions of these solutions [pipette, ±2% accuracy] were combined to make stock lipid-hexane solutions which had the relative concentrations required. Three such stock solutions were prepared containing analyte concentration levels ranging between 0.10 and 6.0 ppm of lipid. A fourth stock hexane solution which contained only the cleaned lipid was used to prepare blank reference samples.

The solution containing the lipid plus analytes dissolved in hexane was placed in a glass reservoir (Figure 1; b). Aliquots of this solution were transferred to 10 ml glass ampoules (pre-cleaned by heating in air at 450° for about 10 hours) by the stainless steel filling-apparatus shown in Figure 1. In step A, the sample (0.62 ml) loop (i) and hexane (2.0 ml) flushing loop (h) were filled. In step B, the sample loop was flushed with nitrogen (j) into the sample ampoule (m). After 2.0 ml of n-hexane had been flushed through the sample loop (step C), the ampoule was flame-sealed.

A concentrated solution of the same analytes in hexane was prepared for use as a "check" solution. This solution was used to normalize the analytical results. Aliquots (0.62 ml) of this concentrated solution were prepared using the apparatus described above and were sealed in glass ampoules. When the contents of such an ampoule were diluted with n-hexane to 25 ml, a standard stock solution containing about 600-700 pg/μℓ of each analyte was obtained. Dilutions of this solution were used for instrument calibration.

When samples of the reference material were analyzed, the entire contents of each ampoule (containing about 300 mg lipid) were transferred to the head of a Florisil column used to separate the analytes from the lipid. The eluate volume was adjusted to 10 ml. One μℓ aliquots of this eluate (analyte concentrations of 4-200 pg/μℓ) were injected into the gas chromatograph.

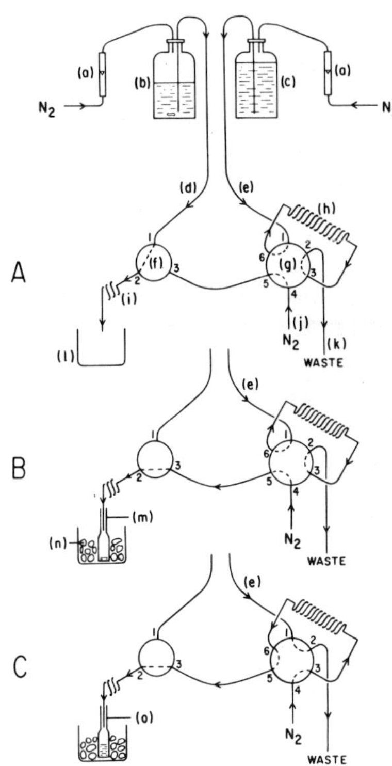

Figure 1. Apparatus used to prepare replicate aliquots of the reference material.
A. Fill sample (0.62 ml) and flush (2 ml) loops
B. Empty sample loop into glass ampoule
C. Flush sample loop with hexane.

a) Flowmeter; b) Sample in hexane; c) Redistilled n-hexane; d) Sample delivery line; e) hexane delivery line; f) Valco 3-way valve; g) Valco 6-way valve; h) Flush loop (2.0 ml); i) Sample loop (0.62 ml); j) N_2 line; k) Waste; l) Waste; m) Glass ampoule (10 ml) for sample; n) Solid CO_2; o) Filled ampoule (0.62 ml sample + 2.0 ml hexane)

The analytical procedure used was based on that of Addison et al (1972) with some modifications: solvents, redistilled; cleanup column, 25 g Florisil (60-100 mesh) activated at 130°; elution, 200 ml 6% diethyl ether/hexane (v/v) at 3-5 ml/min; concentration, rotary evaporation to about 10 ml (not to dryness); GC conditions, Hewlett-Packard 5750 with ^{63}Ni pulsed electron-capture detector, electronic integration; glass column, 2 m x 6 mm 3% OV1 on Gas Chrom Q (100/120); temperatures, 210° (column), 235° (detector), 220° (injector); carrier gas, 5% methane/argon at 50 ml/min.

Samples of the reference solutions were stored under varying conditions (-20°, dark; 25°, dark; 25°, fluorescent "Cool White" light [75 μE m^{-2} sec^{-1}; about 10% "full daylight"]) to study the stability during storage.

RESULTS

Samples prepared by the described procedure were analyzed by gas chromatography using electron-capture detection. Artifacts which would affect the analytical results were not detected. There was no observable alteration of the analytes which had been added (Figure 2).

The variance of the measured analyte levels in randomly chosen samples was not significantly different from that for replicate analyses of a single, large sample (Table 1).

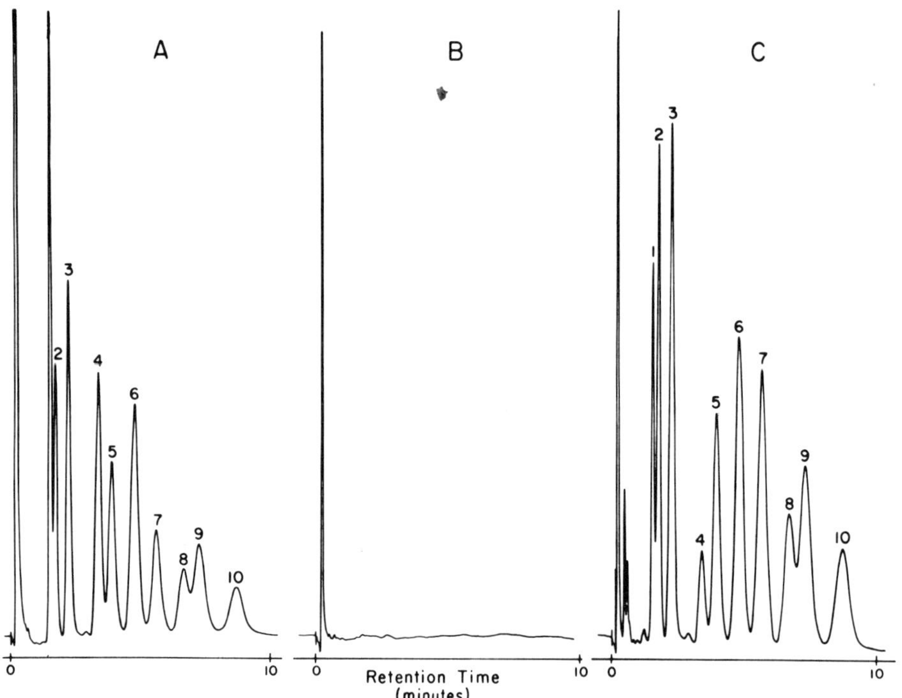

Figure 2. Gas chromatographic analyses of analytes in reference material
1 - Lindane; 2 β BHC; 3 - Aldrin; 4 - Heptachlor epoxide; 5 - o,p DDE; 6 - p,p' DDE; 7 - o,p DDD; 8 - o,p DDT; 9 - p,p' DDD; 10 - p,p' DDT.
A - "Check solution of analytes in hexane.
B - Blank of cleaned herring oil matrix material prepared and stored by the described procedure.
C - Reference material prepared by the described procedure.

Table 1 Precision of Analyses of Replicate Reference Samples

COMPOUND	PREPARED CONCENTRATION (PPM OF LIPID)	N	MEASURED CONCENTRATION (PPM OF LIPID)	COEFFICIENT OF VARIATION (%)
LINDANE	1.40	5	1.23	5.9
β BHC	3.21	5	3.17	6.6
ALDRIN	3.00	5	2.79	4.5
HEPTACHLOR EPOXIDE	1.28	5	1.01	6.4
o,p DDE	2.85	5	2.80	6.1
p,p' DDE	3.12	5	3.17	6.3
o,p DDD	5.74	5	5.83	6.6
o,p DDT	4.15	5	4.37	6.5
p,p' DDD	3.76	5	3.46	6.0
p,p' DDT	3.84	5	3.81	7.1

AVERAGE % RECOVERY = 96 ± 8%
AVERAGE COEFFICIENT OF VARIATION = 6.2 ± 0.7%

There was no significant difference in the analytical results obtained for samples of the organochlorine-containing reference material stored under dark conditions (at 25° or -20) for up to 150 days (Figure 3). However, for samples stored under the "Cool White" fluorescent light (75 $\mu E\ m^{-2}\ sec^{-1}$), a 50% loss of p,p' DDE after 3 months was observed; other analytes were not significantly affected.

The results of analyses of samples stored in the dark at -20° are shown in Table 2. The precision of the results is acceptable (Holden, 1975). Percentage recoveries of added analytes were high for each of the concentration levels of the individual analytes. Similar results were obtained for samples which had been stored under the other conditions, except for the loss of p,p' DDE in samples stored in the light.

DISCUSSION

It would be useful if samples of a reference material could always be available. The preparation of a sufficiently large sample which can be stored indefinitely without changing significantly is useful only if highly stable analytes are of interest, if future demand can be accurately assessed, and if sufficient resources are available. A reference material can also be continuously available if acceptably reproducible successive

Table 2 Results of Analyses of the Prepared Organochlorine Reference Material Stored in the Dark at -20°C for 150 days

CONCENTRATION OF ANALYTE (ppm of Lipid)

minimum - mean - maximum (Coefficient of Variation)

Compound	SAMPLE A PREPARED	SAMPLE A n	SAMPLE A Measured	SAMPLE B Prepared	SAMPLE B n	SAMPLE B Measured	SAMPLE C Prepared	SAMPLE C n	SAMPLE C Measured
Lindane	0.28	14	0.25-0.30-0.38 (11.8%)	0.56	14	0.50-0.58-0.73 (11.1%)	1.40	12	1.26-1.42-1.55 (6.8%)
β BHC	0.13	11	0.10-0.14-0.18 (17.9%)	0.64	14	0.47-0.61-0.84 (19.9%)	3.21	12	2.76-3.27-4.01 (12.7%)
Aldrin	0.58	14	0.57-0.64-0.71 (7%)	1.02	14	0.85-0.97-1.09 (8.2%)	3.00	12	2.82-3.29-3.6 (6.2%)
Heptachlor Epoxide	0.26	13	0.20-0.26-0.32 (14.4%)	0.64	11	0.47-0.53-0.61 (8.6%)	1.28	11	0.92-1.14-1.31 (10.0%)
o,p DDE	0.57	14	0.48-0.56-0.62 (6.3%)	1.43	14	1.13-1.24-1.42 (7.6%)	2.85	12	2.74-3.05-3.53 (7.6%)
p,p' DDE	0.31	14	0.25-0.33-0.40 (12%)	1.07	14	0.83-1.04-1.26 (11.4%)	3.13	12	3.10-3.47-3.88 (6.5%)
o,p DDD	1.44	14	1.04-1.20-1.40 (8.4%)	2.87	14	2.04-2.47-3.07 (14.2%)	5.74	12	4.59-5.57-7.00 (13.6%)
o,p DDT	0.55	14	0.39-0.49-0.57 (10.6%)	1.66	13	1.21-1.42-1.64 (9.9%)	4.15	12	3.70-4.47-5.27 (8.9%)
p,p' DDD	0.50	14	0.41-0.48-0.61 (12.7%)	1.50	12	1.07-1.30-1.56 (12.2%)	3.76	12	3.08-3.91-4.85 (12.3%)
p,p' DDT	0.64	13	0.48-0.58-0.76 (14.2%)	1.28	11	0.95-1.12-1.40 (12.6%)	3.84	12	3.53-4.20-4.99 (9.0%)
Average % Recovery =			98.9 ± 9.1%			90.7 ± 6.7%			103.8 ± 6.8%
Average Coefficient of Variation =			11.5 ± 3.5%			11.5 ± 3.6%			9.4 ± 2.7%

batches can be prepared as required. It must be possible to store a reference material and to distribute it to other laboratories without the material having been changed in ways which would affect analytical results.

In our procedure, analytes are mixed in a hexane solution of cleaned lipid. The reference material which results can be reproducibly sub-sampled to yield replicate samples. The reproducibility of this sampling procedure is indicated by the high precision of the results of our analyses of random samples. Since the material is prepared, handled, and stored as a hexane solution, problems resulting from biological degradation of the analytes are eliminated. Contamination from the container, evaporative loss of analytes, or introduction of contaminants are minimized by storage in sealed glass ampoules. Using this sample container, reference materials can be shipped with confidence as was indicated by the results of the storage experiments (Figure 3) where no significant loss of analytes was noted during about 5 months of storage under various conditions. However, a significant loss (about 50%) of p,p' DDE was noted with samples stored in the light (75 $\mu E\ m^{-2}\ sec^{-1}$): conditions of storage more harsh than would be expected under normal storage or shipping conditions. The relatively high recoveries of added analytes from samples stored in the dark show these samples can be analyzed accurately.

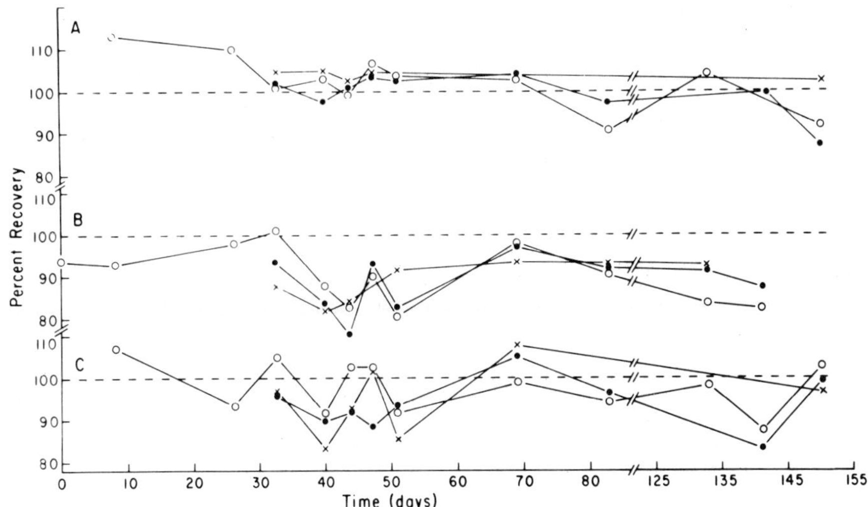

Figure 3. Averaged recoveries of the added analytes in the prepared reference material during storage under various conditions: O - dark, -20°C; X - dark, 25°C; ● - light, 25°C.
A - High concentration reference material (1.28-5.74 ppm of added analytes)
B - Medium concentration reference material (0.56-2.87 ppm of added analytes)
C - Low concentration reference material (0.13-1.44 ppm of added analytes)

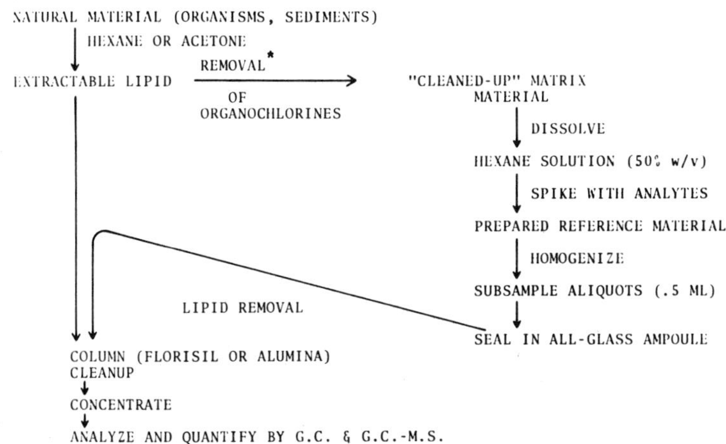

Figure 4. Role of a reference material with a cleaned lipid matrix in scheme for analysis of natural samples

The purpose of a reference material is to provide the analyst with a material containing known concentrations of chosen analytes prepared in a matrix and at concentrations similar to those of natural samples. During the analysis, the reference material is processed through all stages of the analytical procedures as would be required for the analysis of natural samples following their extraction (Figure 4). The amount of lipid (250-300 mg) in each glass ampoule is about the quantity recommended for the analysis of natural lipid extracts (Addison et al, 1972). To eliminate problems of sub-sampling, the analyst transfers the entire contents of each ampoule to the clean-up column used for the separation of analytes from lipid.

In this way, the reference material can be used to evaluate the column clean-up, concentration, and chromatographic procedures for the analysis of analytes in lipids (Figure 4). By normalizing the results with the "check" solution, the analyst can evaluate the precision and accuracy of his methods and compare his results with those of other workers.

For a reference material to simulate a natural sample, the matrix into which the analytes are added must be similar to that of real samples. If successive batches are to contain accurately known levels of analytes, the matrix material must be cleaned of interfering compounds before the analytes are added. Any clean-up step will alter the matrix material. In this study, a natural herring oil was cleaned of organochlorines by hydrogenation and vacuum stripping. Obviously, the natural herring oil was altered by this process. However, the lipid material which results is processed in the same way during the analytical procedure as would be done with natural extracts. While a method which resulted in no structural changes would be advantageous (e.g. gel permeation), the material described in this study seems to satisfy present requirements. The lipid matrix in our reference material is not identical to the natural lipid, but since it has been cleaned of organochlorines, the amount of analytes added to it can be accurately determined and the material can be sub-sampled with precision and stored with confidence until analysis.

CONCLUSION

A procedure has been described by which stable, characterized reference materials containing known trace levels of analytes in a lipid matrix can be prepared. These materials provide the analyst with a way to assess the precision and accuracy of all analytical methodology subsequent to extraction of the lipid from the whole natural sample (Figure 4).

In future studies, reference materials containing more complex mixtures of analytes will be prepared. The complexity of the resulting reference materials can be controlled to be similar to the required range of natural samples. Additional analytes may include metabolites, polychlorinated biphenyls, alkylated aromatic and polynuclear aromatic hydrocarbons. The reference material can be prepared to suit the requirement.

REFERENCES

Addison, R.F. and R.G. Ackman, J. Amer. Oil. Chem. Soc., 51, 192-194 (1974).

Addison, R.F. and M.E. Zinck, J. Fish. Res. Bd. Canada, 29, 349-355 (1972).

Harvey, G.R., H.P. Milas, V.T. Bowen and W.G. Steinhauer. J. of Mar. Res., 32, 103-118 (1974).

Holden, A.V. Proceedings of IUPAC 3rd International Congress of Pesticide Chemistry, Helsinki, 1974. In "Environmental Quality and Safety". Suppl. Vol. III, "Pesticides". pp.40-46. Published by Georg Thieme, Stuttgart (1975).

Pavlou, S.P. and W. Hom. Marine Chem., 4, 155-163 (1976).

CHEMICAL IONIZATION MASS SPECTROMETRY OF HYDROCARBONS AND HALO-HYDROCARBONS

Alex G. Harrison

Department of Chemistry, University of Toronto

Toronto, Canada, M5S 1A1

Since the technique of chemical ionization was first introduced by Munson and Field (1966), chemical ionization mass spectrometry (CIMS) has developed into a powerful tool for the identification and quantitaton of organic molecules and, thus, has found extensive application, particularly in the biomedical and environmental fields. In CIMS ionization of the sample of interest is effected by ion-molecule reactions rather than by electron impact, photon impact, or field ionization/desorption and often provides information complementary to these techniques rather than supplementary. Although much of the earlier work utilized positive ion-molecule reactions, in recent years there has been an increased interest in negative ion chemical ionization. In the present review we shall first consider positive ion chemical ionization, for which a much greater body of published data exists, and then briefly review the smaller amount of data concerned with negative ion chemical ionization. Throughout we will concentrate on studies involving hydrocarbons and halohydrocarbons, the species of major interest in the present symposium.

POSITIVE ION CHEMICAL IONIZATION

The essential reactions in positive ion chemical ionization are given in general form in (1) to (3). Ionization of the reagent gas (present in large excess), usually followed by ion-molecule reactions involving the primary ions and the reagent gas, produces the chemical ionization reagent ion (reaction (1)). The reagent ion, R_1^+, reacts with the additive A (present at low concentration levels) to produce the addition ion A_1^+ which may fragment by one or more pathways. The final array of ions A_1 to A_i constitutes the chemical ionization mass spectrum of the additive A.

(1) $e + R \xrightarrow{R} R_1^+$

(2) $R_1^+ + A \rightarrow A_1^+ + N$

(3) $A_1^+ \rightarrow A_2^+ + N_2$

$\rightarrow A_3^+ + N_3$

$\vdots \quad \vdots$

$\rightarrow A_i^+ + N_i$

The majority of reagent ions used have been those which react either by H^- abstraction (reaction (4)) or, more frequently, as Brønsted acids by proton transfer (reaction (5)).

(4) $R_1^+ + M \rightarrow [M-H]^+ + R_1H$

(5) $RH^+ + M \rightarrow MH^+ + R$

Reaction (5), resulting in initial formation of the protonated molecule MH^+, is by far the most common CI reaction. The utility of the CI procedure arises, in part, from the fact that by controlling the exothermicity of reaction (5) the extent of fragmentation of MH^+ can, in effect, be controlled. Furthermore, the fragmentation of the compound of interest begins from the even-electron MH^+, rather than the odd-electron $M^{+\cdot}$ of electron impact ionization, with the result that the fragmentation reactions are different, frequently involving elimination of stable neutral molecules characteristic of functional groups present in the molecule. Table 1 provides a partial list of proton affinities (taken mainly from Kebarle (1977)). The additive molecules normally will have proton affinities >180 kcal mole^{-1} and it can be seen that proton transfer from H_3^+ will be highly exothermic, leading to extensive fragmentation, proton transfer from CH_5^+/ $C_2H_5^+$ will be moderately exothermic, while protonation by $C_4H_9^+$ or NH_4^+ will be only mildly exothermic (or in some cases endothermic) resulting in little fragmentation of MH^+. (Note however that if endothermic, the proton transfer reaction may be extremely slow, resulting in low sensitivity of the CI process). Methane has been the most widely used reagent gas since, in general, it appears to give adequate ion signals in the molecular weight region for molecular weight determination, as well as the fragmentation necessary to provide structural information.

An alternative mode of ionization which as seen some use is that of charge exchange (reaction (6)).

(6) $R^{+\cdot} + M \quad M^{+\cdot} + R$

Table 1 Proton Affinities of Gaseous Molecules

M	MH^+	PA(M) (kcal $mole^{-1}$)
H_2	H_3^+	101
CH_4	CH_5^+	128
C_2H_4	$C_2H_5^+$	159
C_3H_6	$C_3H_7^+$	182
i-C_4H_8	$C_4H_9^+$	195
NH_3	NH_4^+	201
C_6H_5Cl	$C_6H_6Cl^+$	179
C_6H_6	$C_6H_7^+$	180
ROH	ROH_2^+	182 - 194
RCO_2R'	$RCO_2R'.H^+$	184 - 200
R_3N	R_3NH^+	210 - 220
RR'CO	$RR'COH^+$	195 - 210

In this mode the odd-electron molecular ion, characteristic of electron impact ionization, is the initial product ion. Consequently the fragmentation reactions observed are similar to those observed in EI mass spectrometry but with the difference that, whereas in EI the $M^{+\cdot}$ ions produced have a distribution of internal energies, in charge exchange ionization the internal energy of $M^{+\cdot}$ is determined primarily by the exothermicity of reaction (6). This is defined by the difference in recombination energy of the reactant ion and the ionization energy of the additive. Table 2 provides a partial list of charge exchange reagents with their approximate recombination energies. The potential uses of charge exchange ionization would appear to be two-fold. In mixtures with proton transfer reagents, which through formation of MH^+ provide molecular weight information, charge exchange can be utilized to produce fragment ions to assist in structure elucidation. Alternatively, low energy charge exchange can be used to produce predominantly $M^{+\cdot}$, without the reduction in sensitivity inherent in the use of low energy electron impact, or, in suitable cases, to selectivity ionize components of low ionization energy.

Table 2 Recombination Energies of Gaseous Ions

Ion	Recombination Energy (eV)
Ar^+	15.8, 15.9
Kr^+	14.0, 14.7
N_2^+	∿15.3
CO_2^+	13.8
Xe^+	12.1, 13.4
CS_2^+	∿9.5 or 10.0
NO^+	8.5 - 9.5
$C_6H_6^+$	9.3

A question of considerable importance is the sensitivity (ion current of additive per unit additive pressure) of the CI method compared to the sensitivity of electron impact ionization. This question has been addressed by Field (1972) and the following is derived largely from his work. The total ion current derived by electron impact, I_{EI}, is given by (7).

$$(7) \quad I_{EI} = I_e Q \ell N$$

where I_e is the ionizing electron current, Q is the ionization cross section, ℓ is the ionizing path length, and N is the concentration of molecules per cm³. For chemical ionization the additive ion current I_{CI} is given by the total number of reagent ions produced times the fraction which react with the additive. In CI the ionizing electron beam is completely attenuated with the result that the total number of reagent ions produced is given by the ionizing electron current, I_e, multiplied by the number, a, of ion pairs produced per electron. (Typically the energy required to produce an ion pair is ∿30 eV, so that a ≃ 10 for 300 volt electrons.) Thus

$$(8) \quad \frac{I_{CI}}{I_{EI}} = \frac{aI_e(1 - e^{-kNt})}{I_e Q\ell N} \simeq \frac{aI_e kNt}{I_e Q\ell N}$$

which, for identical values of I_e and N, gives

$$(9) \quad \frac{I_{CI}}{I_{EI}} = \frac{akt}{Q\ell} .$$

Substituting typical values a = 10, k = 2 x 10^{-9} cm^3 $molecule^{-1}$ sec^{-1}, t = 1 x 10^{-5} sec, ℓ = 1 cm, Q = 2 x 10^{15} cm^2, we obtain

$\frac{I_{CI}}{I_{EI}} \simeq 100.$ However, this represents the ratio of ion currents produced in the ion source. Assuming the extraction efficiency from the source is proportional to the relative areas of the ion exit slits, the ratio of collected ion currents can be reduced by a factor of ten due to the use of a smaller ion exit slit in the CI source. In addition, the ionizing electron current frequently is reduced in CI (through use of a smaller electron beam entrance slit) further reducing I_{CI} relative to I_{EI}. Nevertheless the rough calculations do show that the sensitivity of CI is comparable to or better than EI. Note, however, that I_{CI} is critically dependent on the rate coefficient for the ionizing ion-molecule reaction. The results to date show that exothermic proton transfer to polar molecules typically proceeds with the rate coefficients of the order of 2 x 10^{-9} cm^3 $molecule^{-1}$ sec^{-1} (Harrison, Lin, and Tsang (1976)) however, endothermic proton transfer reactions will be much slower. A further potential advantage of CI is that by suitable choice of the reagent ion(s) the ionization can be confined to a few product ions, whereas in EI it may be distributed over many ions. For quantitative work, involving single ion monitoring, the sensitivity of CI is even further enhanced in this case.

The major aspects of positive ion chemical ionization mass spectrometry are illustrated by the results for hydrocarbons and halohydrocarbons discussed in more detail in the following.

CI of Alkanes

The CH_4 CI of an extensive series of linear and branched alkanes have been reported by Field, Munson, and Becker (1966). The spectra of the n-alkanes are characterized by abundant $[M-H]^+$ ions (32% of total ionization) originating by reactions (10) and (11) and lower weight alkyl ions originating both by fragmentation of the $[M-H]^+$ ion (reaction (12)) and by alkyl ion displacement (reaction (13)).

(10) $CH_5^+ + n\text{-}C_6H_{14} \rightarrow C_6H_{13} + H_2 + CH_4$

(11) $C_2H_5^+ + n\text{-}C_6H_{14} \rightarrow C_6H_{13}^+ + C_2H_6$

(12) $C_6H_{13}^+ \rightarrow C_4H_9^+ + C_2H_4$

$\rightarrow C_3H_7^+ + C_3H_6$

$$(13)\ CH_5^+ + n\text{-}C_6H_{14} \rightarrow C_5H_{11}^+ + CH_4 + CH_4$$
$$\rightarrow C_4H_9^+ + C_2H_6 + CH_4$$
$$\rightarrow C_3H_7^+ + C_3H_8 + CH_4$$

Detailed mechanistic studies (Houriet, Parisod, and Gaumann, 1977) have shown that $C_2H_5^+$ reacts entirely by (11) while CH_5^+ reacts by both reactions (10) and (13). Although for most alkanes the $[M-H]^+$ ion intensity is much larger than ions in the molecular weight region in EI, branching of the alkane increases the probability of reaction (12) with a resulting decrease in intensity of the characteristic $[M-H]^+$ ion.

In an attempt to reduce this fragmentation, Hunt and Harvey (1975a) have studied the nitric oxide CI mass spectra of alkanes. In pure NO or N_2/NO mixtures NO^+ is the major reactant ion. The major product ion observed for most alkanes was $[M-H]^+$ formed by H^- abstraction, however, (usually) low intensity ions were observed at $[M-3H]^+$ and $[M-H_2 + NO]^+$. These create a problem in analytical applications since the NO CI of alkenes (vide infra) show abundant $[M-H]^+$ and $[M+NO]^+$ ion signals which are isobaric with these products derived from alkanes.

CI of Alkenes and Cycloalkanes

The CH_4 CI CI mass spectra of a number of alkenes have been reported by Field (1968). The spectra of mono-olefins are dominated by two series of ions, the $C_nH_{2n-1}^+$ alkyl ions consisting of the MH^+ ion and fragments derived therefrom and the $C_nH_{2n-1}^+$ alkenyl ions consisting of the $[M-H]^+$ ion and lower mass fragments. Allylic hydrogens apparently are most readily abstracted since the $C_nH_{2n-1}^+$ ions are more prominent in compounds with a large number of allylic hydrogens. Although for lower alkenes the MH^+ and $[M-H]^+$ ion intensities are sufficiently intense for molecular weight identification, their intensities become very low for larger olefins. Multiple double bonds or a cyclo-olefin structure enhances the MH^+ ion abundances. The CH_4 CI of cycloalkanes (Field and Munson, 1967) resemble the spectra of alkenes in showing the $C_nH_{2n+1}^+$ and $C_nH_{2n-1}^+$ series of ions.

From an analytical viewpoint, the MH^+ and $[M-H]^+$ ion intensities become too small for larger alkenes for molecular weight identification; further, the $C_nH_{2n+1}^+$ ions are isobaric with the alkyl ions which dominate the CI of alkanes. Hunt and Harvey (1975b) have shown that in the NO CI of internal olefins the $[M+NO]^+$, $M^{+\cdot}$ and $[M-H]^+$ ions dominate the spectrum (70 - 90% of total ionization). The spectra of the terminal olefins are much more complex, showing a series of fragment ions $C_nH_{2n}NO^+$ ($n \geq 3$) resulting from electrophilic addition of NO^+ followed by rearrangement and fragmentation. Because of this complication and the overlap of products in the NO CI of alkanes, the analytical utility of this system appears limited.

A problem unresolved by conventional proton transfer CI experiments concerns the location of double bonds in olefins. A novel approach to this problem has been reported by Ferrer-Correia, Jennings, and Sen Sharma (1976, 1978) who have shown that the vinyl methyl ether molecular ion forms a cyclic intermediate with olefins which subsequently fragments to form species characteristic of the double bond location (Scheme 1). From the results reported this approach appears to have several limitations and derivitization of the double bond prior to chemical ionization (see, for example, Blum and Richter, 1973) may prove equally useful. The charge exchange CI of a number of olefins has been studied (Li and Harrison, 1978). Figure 1 shows the spectra for 4-decane using a variety of reagent gases of different recombination energies. Charge exchange with Ar^+ leads to extensive fragmentation, however, with decreasing recombination energy of the reactant ion, the extent of fragmentation decreases, with the result that in the N_2/10% CS_2 CI spectrum (where CS_2^+ is the major reactant ion) the olefinic molecular ions constitute the major ions in the mass spectra. No differences were observed dependent on double bond location, however, isomeric linear and branched alkenes show substantial differences, particularly in the low energy N_2/CS_2 charge exchange spectra.

CI of Alkylbenzenes

In their initial studies of CI mass spectrometry, Munson and Field (1967) determined the CH_4 CI mass spectra of 21 alkylbenzenes.

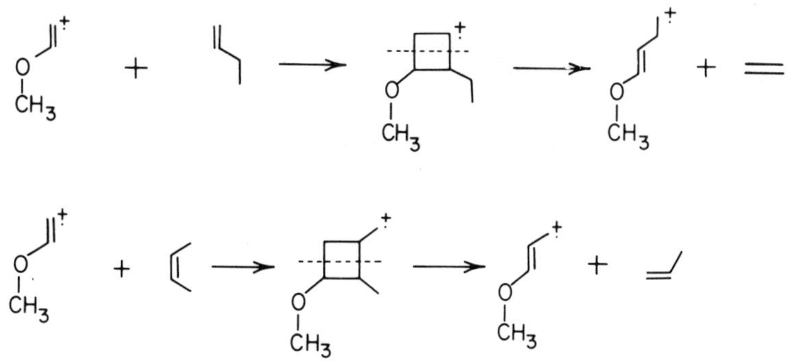

Scheme 1

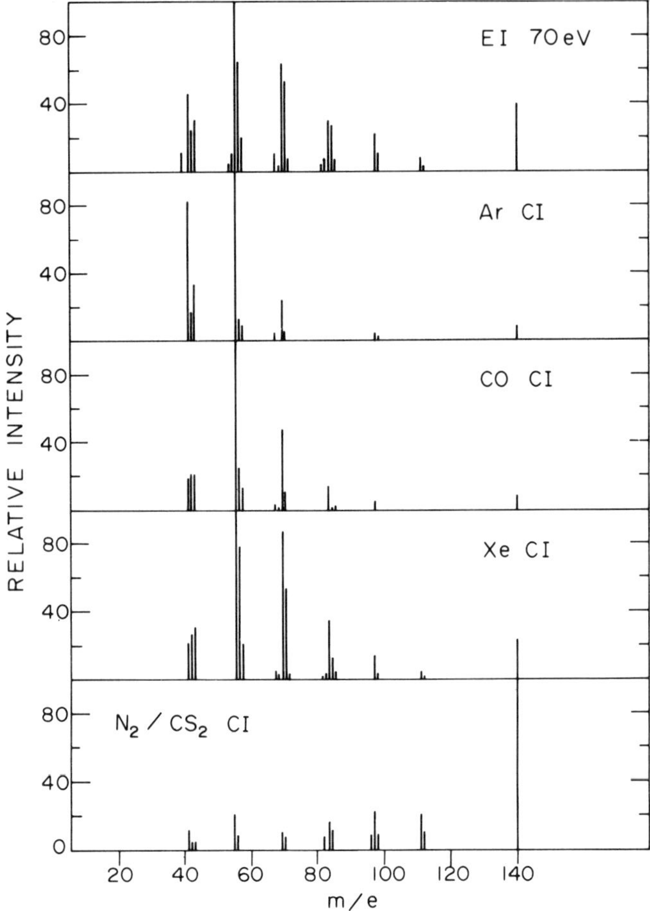

Figure 1. Charge exchange chemical ionization mass spectra of 4-decane

The major reaction modes of CH_5^+ and $C_2H_5^+$ with alkylbenzenes are illustrated in Schemes 2 and 3 respectively using s-butyl-benzene as an example. The major reaction of both CH_5 and $C_2H_5^+$ involves proton transfer to the basic π system to give MH; however, with increasing size of the alkyl group, fragmentation of MH^+ increases in importance, the major fragmentation modes being olefin elimination to give a protonated benzene or neutral benzene elimination to give the alkyl ion of the substituent. Thus, for the butyl benzenes $C_4H_9^+$ is the base peak in the CH_4 CI mass spectra. Hydride abstraction appears to involve only the benzylic hydrogens and increases in importance with increasing number of such hydrogens, $[M-H]^+$ being ∿24% of the total ionization in pentamethyl-benzene, where olefin elimination and alkyl ion formation do not occur.
A detailed study of the H_2 and CH_4 CI of the C_7 to C_{10} alkyl benzenes

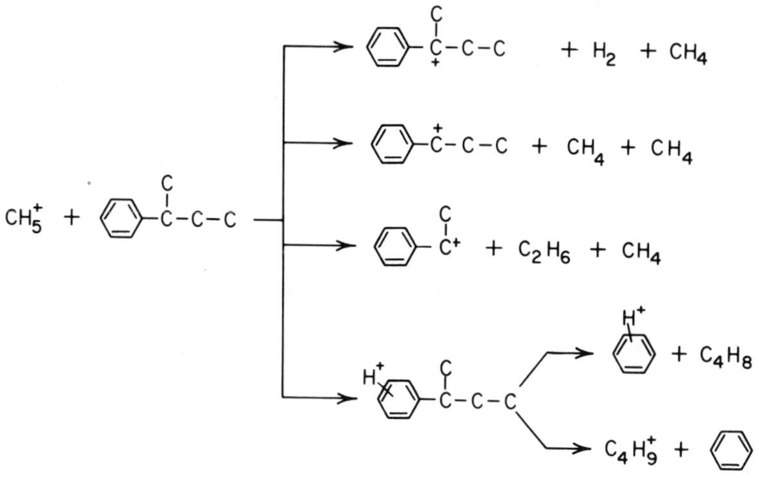

Scheme 2

has been reported (Harrison, Lin, and Leung, 1978). As anticipated, the extent of fragmentation is greatly increased in the H_2 CI where the initial protonation reaction is much more exothermic.

Charge exchange of CI of alkyl benzenes produces spectra similar to electron impact mass spectra, with the extent of fragmentation depending on the recombination energy of the reaction ion. With reactant ions of sufficiently low recombination energy exclusive formation of the molecular ion is achieved. The use of benzene ($C_6H_6^{+\cdot}$) as a charge exchange reactant has been proposed for the selective ionization of aromatics, such as alkylbenzenes, with ionization energies lower than that of benzene itself (Subba Rao and Fenselau, 1978; Sunner and Szabo, 1978).

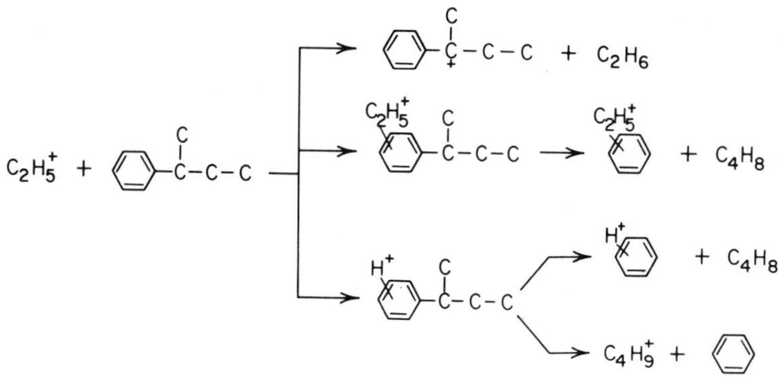

Scheme 3

CI of Polycyclic Aromatic Hydrocarbons (PAHs)

The electron impact mass spectra of isomeric polycyclic aromatic hydrocarbons are almost identical, making the structural identification of PAHs a difficult problem, particularly when they are encountered in complex mixtures. Similarly, because of the high stability of these aromatic compounds, conventional proton transfer chemical ionization also is unlikely to distinguish between isomers since, in general, only MH^+ will be formed.

Some success has been reported by Lee and Hites (1977) using a mixed charge exchange-proton transfer system. Using an Ar-CH_4 reagent mixture they observed that the $MH^+/M^{+\cdot}$ ratio increased with increasing ionization energy of the PAH for an isomeric series of compounds. In this mixed reagent Ar^+ serves as a charge exchange reagent to produce $M^{+\cdot}$ while the methane-derived $CH_5{}^+$ and $C_2H_5{}^+$ ions transfer protons to yield $MH^{+\cdot}$.

It appears possible that charge exchange chemical ionization using reagent ions of high recombination energy may be useful in the identification of PAHs. In the only report to date (Ryan, 1977) charge exchange using Ar has been utilized to effectively remove interferences by dissociative ionization in single ion monitoring of PAHs in complex mistures. Preliminary reports (Hunt et al, 1977) suggest that combined positive and negative ion CI using O_2 as reagent gas may be useful in identifying PAHs.

CI of Simple Halohydrocarbons

There have been no systematic studies of the chemical ionization of simple haloalkanes, however, facile loss of hydrogen halide from the initially formed MH^+ would be expected. As Field (1972) has pointed out, the ease of loss of X, as HX, from the protonated species, $RX.H^+$, should correlate inversely with the proton affinity of HX. This has been confirmed experimentally (Jardine and Fenselau, 1976; Harrison and Onuska, 1978). The hydrogen halides have low proton affinities and should be readily lost as neutral fragments from protonated species. As expected, then, the CH_4 CI mass spectra of the cyclohexyl halides show only $[MH^+-HX]$ (X = Cl,Br,I) as significant peaks (Jardine and Fenselau, 1976). This would be anticipated for all simple halohydrocarbons and, indeed, if the initial protonation is sufficiently exothermic further fragmentation of the $[MH^+-HX]$ species would be expected. Similarly, the CH_4 and H_2 CI mass spectra of the benzyl halides show $[MH^+-HX]^4$ as the only significant peak (Harrison and Lin, 1975; Leung and Harrison, 1976).

For the halobenzenes and halotoluenes with a halogen attached to the phenyl ring loss of HX from MH^+ is much less pronounced and

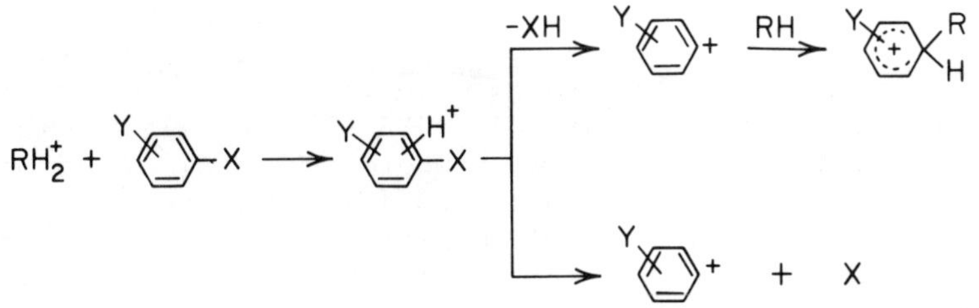

Scheme 4

MH^+ is observed in the CH_4 CI mass spectra, except for the iodo compounds (Harrison and Lin, 1975; Leung and Harrison, 1976). The major features of the chemical ionization process for these compounds are summarized in Scheme 4 where RH_2^+ represents CH_5^+, or H_3^+. CH_5^+ reacts with fluoroaromatics to yield abundant MH^+ ions but with significant fragmentation to $YC_6H_4^+$ which reacts by addition to CH_4 to give the appropriate protonated toluene or xylene. Some of this dehalogenated product arises by the alternative route of direct reaction of CH_5^+ with the halo aromatic (Leung, Ichikawa, Li, and Harrison, 1978). The importance of the fragmentation reaction leading to $[MH^+-HX]$ and $[MH^+-HX+CH_4]$ decreases in the order F > Cl > Br and these products are not observed for the iodo compound. Instead, for the iodo compounds loss of I from MH^+ to give the odd-electron $YC_6H_5^{+\cdot}$ is the dominant fragmentation reaction. $C_2H_5^+$ reacts with the fluoro and chloro compounds primarily by proton transfer or complex formation, however, for the bromo and iodo compounds significant loss of Br or I occurs to form the odd-electron $C_2H_5C_6H_4Y^{+\cdot}$ species. These are the most striking examples that fragmentation of MH^+ in chemical ionization does not always involve formation of even-electron species.

The H_2 CI mass spectra are similar to the CH_4 CI mass spectra but the extent of fragment ion formation is greater. An intriguing feature is that polyhalo aromatics, e.g. dichlorobenzene, trichlorobenzene, chlorofluorobenzene, undergo a series of hydrogen halide loss and H_2 addition reactions with eventually dehalogenate the haloaromatic to form $C_6H_7^+$ (protonated benzene) as a major product of the H_2 CI (Leung and Harrison, 1978).

CI of Polyhalohydrocarbons

There have been numerous investigations of the positive ion chemical ionization mass spectra of polychlorinated compounds of the DDT, polycyclodiene, and biphenyl types, largely as a result of their role as significant environmental pollutants.

Biros, Dougherty, and Dalton (1972) have reported the CH_4 CI mass spectra of ten polycyclic chlorinated insecticides or metabolites. Not surprisingly, the MH^+ ion intensities were extremely small, the base peak normally being $[MH^+-HCl]$ except for those compounds containing a hydroxyl group where $[MH^+-H_2O]$ constituted the base peak. Further loss of HCl from $[MH^+-HCl]$ was observed in several cases as was a retro-Diels-Alder fragmentation. It was suggested that the relative importance of this latter fragmentation might be useful in distinguishing between isomers, viz, aldrin vs isodrin, α-chlordane vs γ-chlordane. The main features of these spectra have been confirmed in work largely concerned with the oxygen-containing derivatives (McKinney, Oswald, Palaszek, and Corbett, 1974). This work also showed that less fragmentation occurred in the i-butane CI with more abundant ions (MH^+M $[M-H]^+$) in the molecular weight region. The CH_4 and i-butane CI mass spectra of mirex have been reported (Oswald, Albro, and McKinney, 1974). No MH^+ was observed in either spectrum and only a weak $M^{\cdot+}$ in the isobutane CI spectrum. The $C_{10}Cl_{11}^+$ ($[MH^+-HCl]$) ion intensity was reported to be weak in both spectra, the base peak in the i-butane CI being $C_5Cl_6^+$ and in the CH_4 CI $C_5Cl_5^+$. (However, note that a more recent report (Harless and Oswald, 1977) shows $C_{10}Cl_{11}^+$ ($[MH^+-HCl]$) as the base peak in the CH_4 CI mass spectrum).

Oswald, Albro, and McKinney (1974) also have reported on the CH_4 and i-butane CI mass spectra of the isomeric hexachlorocyclo hexanes. As expected, no MH^+ ions were observed, the major peaks in the CH_4 CI spectra corresponding to $[MH^+-HCl]$ and $[MH^+-2HCl]$, the latter being the base peak. The major fragment in the i-butane CI spectra corresponded to $[M^{\cdot+}-CHl]$ rather than $[MH^+-HCl]$. Proton transfer from $C_4H_9^+$ to $C_6H_6Cl_6$ undoubtedly is endothermic and the ionization observed may arise by charge exchange from other less abundant ions present in the high-pressure isobutane plasma. The authors noted that larger samples were necessary to obtain i-butane CI spectra, consistent with an ionization process of low probability. (This lower ionization probability may be true for the i-butane CI of all polychloro compounds not containing a strongly basic function).

Despite the presence of the basic aromatic system the CH_4 CI mass spectra of the DDT series of compounds (ClC_6H_4) CH-R (R= CCl_3, $CHCl_2$, CH_2Cl, CH_3) (McKinney et al, 1974) show no MH^+ ion signals, but rather fragmentation of MH^+ to $[MH^+-HCl]$, $[MH^+-RH]$, and $[MH^+-ClC_6H_5]$ is complete, the latter product usually comprising the base peak. The i-butane CI mass spectra of a number of these compounds also have been reported (Dougherty, Roberts, and Biros, 1975). The most significant feature is the high abundance of the $[MH^+-HCl]$ fragment. The base peak in the case of p,p'-DDT was reported to be $[M^{\cdot+}-HCl]$ rather than $[MH^+-HCl]$ which is characteristic of the other compounds including o,p-DDT. Obviously, this striking difference needs to be confirmed.

The CH_4 CI mass spectra of the unsaturated compounds $(ClC_6H_4)_2$ C=R (R = CCl_2, CHCl, CH_2) show more pronounced MH^+ and $M^{\dot{+}}$ ion signals, as well as [MH^+-HCl] and [MH^+-ClC_6H_5] ion signals, the latter being the base peak for DDMU (R = CHCl) and DDNU (R = CH_2). By contrast, the i-butane CI mass spectra of the DDE type compounds were dominated by $M^{\dot{+}}$ and MH^+ ions.

Brief reports of the CH_4 CI mass spectra of isomeric polychlorobiphenyls have appeared (Oswald, Levy, and McKinney, 1974; Oswald, Levy, Corbett, and Walker, 1974). The major features of the mass spectra included abundant MH^+, $M.C_2H_5^+$, and $M,C_3H_5^+$ ions, as might be expected from the aromatic nature of the molecules. The authors noted that the CI spectra offered no advantage over the EI spectra for identification purposes.

This brief summary illustrates that conventional positive ion CI using proton transfer reagent gases (CH_4 or i-butane) provides little advantage over EI in the identification of polyhalohydrocarbons. Further systematic studies are called for, particularly using different reagent gases. For example, a systematic study of the charge exchange CI of polychlorobiphenyls might prove rewarding.

NEGATIVE ION CHEMICAL IONIZATION

In contrast to positive ion chemical ionization, negative ion chemical ionization is a relatively recent development. This arose in part because negative ion electron impact mass spectrometry largely was neglected until recently because of the (usually) low sensitivity of negative ion production and the dependence of negative ion mass spectra on electron beam energy and, in part, because many commercial instruments were not equipped to operate in the negative ion mode.

The major route for negative ion formation by electron impact involves electron capture, either dissociative or non-dissociative. These are resonance processes usually involving electrons of thermal or near-thermal energy; hence at low sample pressures the yield of negative ions can be low and critically dependent on the electron beam energy. However, under the high pressure conditions of chemical ionization the incident high energy electrons rapidly are thermalized and additional low energy electrons are produced during the formation of the positive ions. These abundant quasi-thermal electrons may be used as the reagent species in negative ion CI or they may be utilized to form reactant ions from suitable additives or, indeed, from the compound of interest itself. Alternatively, electrons and/or negative ions may be produced in electrical discharges or by the ionizing radiation from radioactive materials as has been used in atmospheric pressure ionization (Horning, Horning, Carroll, Dzidic, and Stillwell, 1973).

The methodology of negative ion CI is not as well developed as positive in CI. However, it is known that the use of Ar, N_2, CH_4, or i-butane as reagent gases results in large yields of thermal energy electrons (Hunt, Stafford, Crow, and Russell, 1976), while the addition of small amounts of H_2O, HCl or CH_3ONO will lead, respectively, to OH^-, Cl^-, and CH_3O^- as major reactant ions. Ionization in an N_2/N_2O mixture yields $O^{\cdot-}$ as the major reactant ion (Jennings, 1977; Bruins, Ferrer-Correia, Harrison, Jennings, and Mitchum, 1978) while ionization in N_2O/H_2 or N_2O/CH_4 mixtures yields OH^- as the major reaction ion (Smit and Field, 1977). The use of O_2 in a discharge source or a radiation source produces both thermal energy electrons and O_2^- as reagent species (Hunt et al, 1976; Dzidic, Carroll, Stillwell, and Horning, 1975). Tannenbaum, Roberts, and Dougherty, 1975 have used dissociative electron capture in high pressures of CH_2Cl_2 as a source of Cl^-.

A major reaction of simple anions of the type X^- is the proton transfer reaction

(14) X^- HY $\rightarrow$ HX + Y^-

Extensive studies of the rates and equilibria of such reactions have been made (Kebarle, 1977) to establish a scale of negative ion proton affinities (or, equivalently, bond dissociation energies plus electron affinities, D(H-C) + EA(X)). A partial listing of proton affinities is given in Table 3 from which it can be seen that species such as H^-, OH^-, and CH_3O^-, with relatively high proton affinities, should react with most hydrogen-containing organic molecules of any complexity by proton abstraction. This process appears to occur with little excitation of the produce $[M-H]^-$, with the result that this species should be the major additive ion, a useful feature for establishing molecular weights.

Because of the relatively recent addition of negative ion CI to our repertoire of ionization techniques, few systematic studies of classes of compounds have been carried out. In the following review we will concentrate on those studies which have a particular relevance to the negative ion CI of hydrocarbons and halohydrocarbons.

Smit and Field (1977) have examined the OH^- chemical ionization of a variety of compounds including some aromatic hydrocarbons and some alkenes. The OH^- was produced either by electron impact on or electrical discharge in CH_4/N_2O mixtures. The spectra of the aromatic compounds showed $[M-H]^-$, $[M+43]^-$, and $[M+25]^-$. The latter products were attributed to reaction of the $[M-H]^-$ ion with N_2O to give $[M+43]^-$, which on H_2O elimination gives $[M+25]^-$. The olefins examined (C_8 to C_{11}) showed no $[M-H]^-$ although the dominant $[M+43]^-$ and $[M+25]^-$ ions presumably originate by reaction of the $[M-H]^-$ species with N_2O. It is apparent that the initial reaction

Table 3 Proton Affinities of Anions

$$X^-_{(g)} + H^+_{(g)} \rightarrow HX \quad PA(X^-) = -\Delta H$$

Anion	$PA(X^-)$ (kcal $mole^{-1}$)
H^-	394
OH^-	384
CH_3O^-	379
$NCCH_2^-$	367
$CH_3COCH_2^-$	364
$C_3H_5^-$	354
$C_6H_5CH_2^-$	351
$C_6H_5O^-$	347
Cl^-	333
CF_3COO^-	321

of OH^- with the compounds studied does lead almost exclusively to formation of $[M-H]^-$ and the failure to observe this product is due to its reactivity with N_2O. An alternative source of OH^- (possibly moist N_2) should simplify the spectra considerably.

Hunt et al (1976, 1977) have reported briefly on the O_2^- CI mass spectra of a variety of aromatic hydrocarbons. The product ions observed were $[M-H]^-$, M^-, $[M+30]^-$, $[M+31]^-$, and $[M+32]^-$. The relative intensities of these products were reported to be highly dependent on molecular structures.

In an atmospheric pressure ionization study of the negative ion CI of chlorinated aromatic compounds using air or N_2 (containing 0.5 ppm O_2) as the carrier gas, Dzidic, Carroll, Stillwell, and Horning (1975) observed phenoxide anions ($[M-Cl+O]^-$) fromed by either or both of the reactions

(15) $M^{-\cdot} + O_2 \rightarrow [M-Cl+O]^- + OCl$

(16) $O_2^{-\cdot} + M \rightarrow [M-Cl+O]^- + OCl$

Formation of the phenoxide anion (rather than Cl^-) in a series of chlorinated benzenes was important only for the tetrachloro and more highly chlorinated benzenes. The authors demonstrated a subpicogram detection level for 2,3,4,5,6,-pentachlorobiphenyl based on monitoring of the $[M-Cl+O]^-$ ion. An impressive demonstration of the sensitivity of negative ion CI and the potential of specific ionization reactions is the use of O_2 as reagent gas to analyze 2,3,7,8-tetrachlorodibenzo-p-dioxin (TCDD) at picogram levels through monitoring of a product ion of m/e 176 arising from reaction of the molecular anion of TCDD with O_2 (Hunt, Harvey and Russell, 1975).

The $O^{\overline{\cdot}}$ chemical ionization mass spectra of a variety of compounds have been determined (Jennings, 1977; Bruins et al, 1978). The $O^{\overline{\cdot}}$ was produced by electron impact on a N_2-10% N_2O mixture. The $O^{\overline{\cdot}}$ ion undergoes a variety of reactions, the relative importances being strongly dependent on the organic substrate. The major reactions with aliphatic and aromatic hydrocarbons are:

a) H atom abstraction to give OH^-, which reacts further by proton abstraction to give $[M-H]^-$.
b) Proton abstraction to give $OH^\cdot$ and $[M-H]^-$, which frequently is the base peak in the CI spectrum.
c) H_2^+ abstraction to yield $[M-H_2]^-$. This reaction is an important reaction channel with many 1-alkenes, diolefins, and aromatic compounds. Interestingly, it is observed as a major reaction with m-xylene but not with the ortho or para isomer.
d) H atom displacement leading to $[M-H+O]^-$ ($[M+15]^-$). This is an important reaction channel for all aromatic compounds, in several cases leading to the base peak in the CI mass spectrum. It also is more important with conjugated dienes than with unconjugated dienes, permitting, for example, a ready distinction between 1,3-cyclohexadiene and 1,4-cyclohexadiene.

Dougherty and colleagues (1972, 1975) have determined the negative ion CI mass spectra of a number of polycyclic chlorinated and aromatic chlorinated pesticides. Using CH_4 or isobutane as reagent gases, the major feature in all spectra was the group of peaks due to chloride attachment ($[M+Cl]^-$), together with low intensity $M^{\overline{\cdot}}$ ions for the polycyclic compounds. The Cl^- reactant ions presumably arise by dissociative electron attachment to the substrate molecules. Traces of water or other oxygen-containing impurities in the reagent gas gave rise to $O^{\overline{\cdot}}$ and the characteristic $[M-Cl+O]^-$ products. Tannenbaum, Roberts, and Dougherty (1975) also have examined the Cl^- attachment mass spectra of a variety of compounds using CH_2Cl_2 as reagent gas. For the polychlorinated compounds this reagent gas provided no advantages over using CH_4 or i-butane as reagent gas. The work with CH_2Cl_2 showed that olefinic and aromatic hydrocarbons exhibited no chloride attachment products, although it is not reported whether $[M-H]^-$ ions were observed.

The results obtained to date show that negative ion chemical ionization spectrometry is a useful addition to our collection of ionization techniques. It appears that the sensitivity is at least as great as positive ion CI and, in favourable cases, may be substantially larger. The ability to generate intense molecular or pseudo-molecular anions will be particularly useful in quantitation at low concentration levels. In addition, there appear to be distinct possibilities of developing ionization reactions specific to certain molecules or classes of molecules.

REFERENCES

Biros, F.J., R.C. Dougherty, and J. Dalton. 1972. Org. Mass Spectrom., 6, 1161.

Blum, W., and W.J. Richter. 1973. Tetrahed. Letts., 11, 835.

Bruins, A.P., A.J. Ferrer-Correia, A.G. Harrison, K.R. Jennings, and R.K. Mitchum. 1978. Adv. Mass Spectrom., 7, 355.

Dougherty, R.C., J. Dalton, and F.J. Biros. 1972. Org. Mass Spectrom., 6, 1171.

Dougherty, R.C., J.D. Roberts, and F.J. Biros. 1975. Anal. Chem., 47, 54.

Dzidic, I., D.I. Carroll. R.N. Stillwell, and E.C. Horning. 1975. Anal. Chem., 47, 1308.

Ferrer-Correia, A.J., K.R. Jennings, and D.K. Sen Sharma. 1976. Org. Mass Spectrom., 11, 867.

Ferrer-Correia, A.J., K.R. Jennings, and D.K. Sen Sharma. 1978. Adv. Mass Spectrom. 7, 287.

Field, F.H. 1968. J. Am. Chem. Soc., 90, 5649.

Field, F.H. 1972. In Mass Spectrometry, MTP Review of Science. Vol. 5, Series 1, A. Maccoll, Ed., Butterworths, London.

Field, F.H. and M.S.B. Munson. 1967. J. Am. Chem. Soc., 89, 4272.

Field, F.H., M.S.B. Munson, and D.A. Becker. 1966. Adv. Chem. Series, 58, 167.

Harless, R.L., E.O. Oswald. 1977. Paper presented at 25th Annual Conference on Mass Spectrometry, Washington, D.C.

Harrison, A.G., and P.-H. Lin. 1975. Can. J. Chem. 53, 1314.

Harrison, A.G. and F.I. Onuska. 1978. Org. Mass Spectrom, 13, 35.

Harrison, A.G., P-H., Lin, and C.W. Tsang. 1976. Int. J. Mass Spectrom., Ion Phys., 19, 23.

Harrison, A.G., P-H, Lin, and H.-W. Leung. 1978. Adv. Mass Spectrom., 7, 1394.

Horning, E.C., M.G. Horning, D.I. Carroll, I. Dzidic, and R.N. Stillwell. 1973. Anal. Chem., 45, 936.

Houriet, R., G. Parisod, and T. Gaumann. 1977. J. Am. Chem. Soc. 99, 3599.

Hunt, D.F., and T.M. Harvey. 1975a. Anal. Chem., 47, 1965.

Hunt, D.F., and T.M. Harvey. 1975b. Anal. Chem., 47, 2136.

Hunt, D.F., T.M. Harvey, and J.W. Russell. 1975. J. Chem. Soc. Chem. Comm., 151.

Hunt, D.F., W.C. Brumley, G.C. Stafford, and F.K. Butz. 1977. Paper presented at 25th Annual Conference on Mass Spectrometry, Washington, D.C.

Hunt, D.F., G.C. Stafford, F.W. Crow, and J.W. Russell. 1976. Anal. Chem., 48, 2098.

Jardine, I., and C. Fenselau. 1976. J. Am. Chem. Soc., 98, 5086.

Jennings, K.R. 1977. In Mass Spectrometry Vol. 4, Specialist Periodical Report, Chemical Society, London.

Kebarle, P. 1977. Ann. Rev. Phys. Chem., 28, 445.

Lee, M.L., and R.A. Hites. 1977. J. Am. Chem. Soc. 99, 2008.

Leung, H.-W., and A.G. Harrison. 1976. Can. J. Chem., 54, 3439.

Leung, H.-W., and A.G. Harrison. 1978. unpublished results.

Leung, H.-W., H. Ichikawa., Y.-H., Li., and A.G. Harrison. 1978. J. Am. Chem. Soc., 100, 2479.

Li, Y.-H., and A.G. Harrison. 1978. Unpublished results.

Lindholm, E. 1972. In Ion Molecule Reactions Vol. 2, J.L. Franklin, Ed., Plenum Press, N.Y.

McKinney, J.D., E.O. Oswald, S.M. Palaszek, and B.J. Corbett. 1974. In F. Biros and R. Haque, Eds., Mass Spectrometry and NMR Spectroscopy in Pesticide Chemistry, Plenum Publ., N.Y. pp 5-32.

Munson, M.S.B., and F.H. Field. 1966. J. Am. Chem. Soc., 88, 2621.

Munson, M.S.B., And F.H. Field. 1967. J. Am. Chem. Soc., 89, 1047.

Oswald, E.O., P.W. Albro, and J.D. McKinney. 1974. J. Chromatog. 98, 363.

Oswald, E.O., L. Levy, B.J. Corbett, and M.P. Walker. 1974. J. Chromatog. 93, 63.

Ryan, P.W. 1977. Paper presented at 25th Annual Conference on Mass Spectrometry, Washington, D.C.

Subba Rao, S.C., and C. Fenselau. 1978. Anal. Chem., 50, 511.

Sunner, J. and I. Szabo. 1978. Adv. Mass Spectrom., 7, 1383.

Tannenbaum, H.P., J.D. Roberts, and R.C. Dougherty. 1975. Anal. Chem., 47, 49.

IDENTIFICATION AND QUANTITATIVE ANALYSIS OF POLYCHLORINATED BIPHENYLS ON WCOT GLASS CAPILLARY COLUMNS

Francis I. Onuska and Michael Comba

National Water Research Institute

Burlington, Ontario, Canada

INTRODUCTION

During the past decade, polychlorinated biphenyls (PCBs) have become a matter of public concern, since they are one of the most persistent and widespread environmental pollutants in our ecosystem. As stated in the W.H.O. Environmental Health Criteria Report (1978), the cumulative world production of PCBs since 1930 is in the order of one million tonnes and have been found to bioaccumulate in low level organisms, higher level species, and human beings.

Their technology was introduced into the United States by Monsanto Company, after the Second World War, as a result of the War reparations from Germany. As published by Hutzinger et al (1974), PCBs have been manufactured in many countries and may be supplemented to include some of those listed in Table 1.

In general, PCBs are produced by direct chlorination of biphenyl resulting in a very large number of possible isomers. It is known (Onuska, unpublished data) that technical biphenyl used in the chlorination process may contain from 8 - 15 percent terphenyls and higher polyphenyls, which, during chlorination may undergo the same reaction as biphenyl-producing chlorinated polyphenyls, traces of dibenzofuranes and various aromatic organochlorine compounds. In view of different technological procedures, the composition of PCBs can vary from manufacturer to manufacturer and from batch to batch (Mieure et al, 1976).

Qualitative data for all the possible isomers from the different manufacturers has yet to be compared and the quantitative determination of the technical products has yet to be achieved. This

Table 1 Major Known Manufacturers of Polychlorinated Biphenyls and Terphenyls

Country	Manufacturer	Trade Name
Czechoslovakia	Chemko, n.p., Strážske	Delor and Delorene
East Germany	Deutchen Solvay Werken, A.G.	Orophene
West Germany	IG-Farben and Bayer, A.G.	Clophen and Clophenharz
Italy	Caffaro, s.p.a.	Fenclor
Great Britain	Dow Chemical Company	
France	Prodelac	Pyralene
Japan	Kanegafuchi	Kaneclor
U.S.A.	Mosanto Company	Aroclor Series
U.S.S.R.	--	Sovol and Sovtol

task would require the synthesis of all possible PCB isomers for their standardization and the use of wall coated open tubular column (WCOT) gas chromatography for their separation and characterization.

Gas chromatographic patterns of PCB isomers in environmental samples have somewhat resembled Aroclor 1254 and Aroclor 1260, although it is evident that Aroclor mixtures may be derived from the lowest Aroclor 1221 to the highest Aroclor 1268 or even contain polychlorinated terphenyls corresponding to the Aroclor 5400 series (Jensen et al, 1974; Jensen et al, 1972; Jensen et al, 1976; and Webb et al, 1972).

The separation and identification of these PCB components have been attempted by means of conventional packed column gas chromatography (Reynolds, 1969) and by WCOT column gas chromatography (Jensen et al, 1974; Jensen et al, 1972; Sissons et al, 1971; Zell et al, 1977; and Krupcik et al, 1977). Mass spectrometric characterization of some isomers was performed by Rote et al, 1973, Greichus et al, 1974, Eichelberger et al, 1974, and Ahnoff et al, 1973. Excellent review articles on the PCBs are also available (Krull, 1977; Fishbein, 1972).

Many of the quantitative procedures used for determining PCBs were surveyed by Chau and Sampson (1975), who found that their accuracy, with regard to environmental residues, remains a virtual unknown. The concentrations of PCBs in air have been reported in the order of 50 ng/m^3 to 1000 ng/m^3 while concentrations in water sources may vary between 0.5 ng/l to 500 ng/l. Concentrations in living organisms is dependent upon the extent of local pollution, fat content in the tissues and trophic level of the biota in food chains, although levels up to 4000 mg/kg in Herring Gulls have been detected (W.H.O., 1978). Human adipose tissues may contain concentration levels of PCBs in the range of 1 mg/kg although levels up to 1000 mg/kg have been found in persons exposed to PCBs in manufacturing, as reported by Stendell (1976).

The data given here should provide qualitative and quantitative information with reasonable analytical precision and accuracy for the identification and determination of individual isomers contained in Aroclor mixtures. Most of the PCB components have been separated by capillary gas chromatography at our laboratory and characterized by means of available standards and GC-MS. Corresponding retention times, retention indices and retention temperatures obtained by high performance capillary column gas chromatography were compared with sample data in order to identify and determine individual isomers in the residue.

EXPERIMENTAL

Materials. Individual isomers of chlorinated biphenyls were obtained from RFR Corporation, Hope, Rhode Island, U.S.A. and from Professor S. Safe, Unviersity of Guelph, Guelph, Ontario. Aroclor standards were obtained from RFR Corporation, Hope, Rhode Island, U.S.A. Standard solutions were prepared at concentrations of 200 ng/ml in n-hexane.

Gas Chromatography. Capillary columns of Pyrex glass with I.D. 0.26 mm were prepared according to the procedure developed by Onuska et al (1976). Two capillaries were used for this study; the first one, a 22 m column coated with SP-2100, having a separation number of 45 for C_{12}-C_{13} hydrocarbons and employing Ar-Ch_4 as a carrier gas, the second a 20 m long column, coated with OV-17, having a separation number 32 for C_{12}-C_{13} hydrocarbons employing helium as a carrier gas.

The measurements of the retention data were carried out by means of an Autolab system 1 integrator in tandem with a Carlo Erba Model 2100 gas chromatograph equipped with both flame ionization (FID) and electron capture (EC) detectors (Brechbueler, A.G., Urdorf, Switzerland) operating in the pulsed mode.

Response factors were calculated for both detectors relative to p,p'-DDE and Kovats retention indices were measured at three or

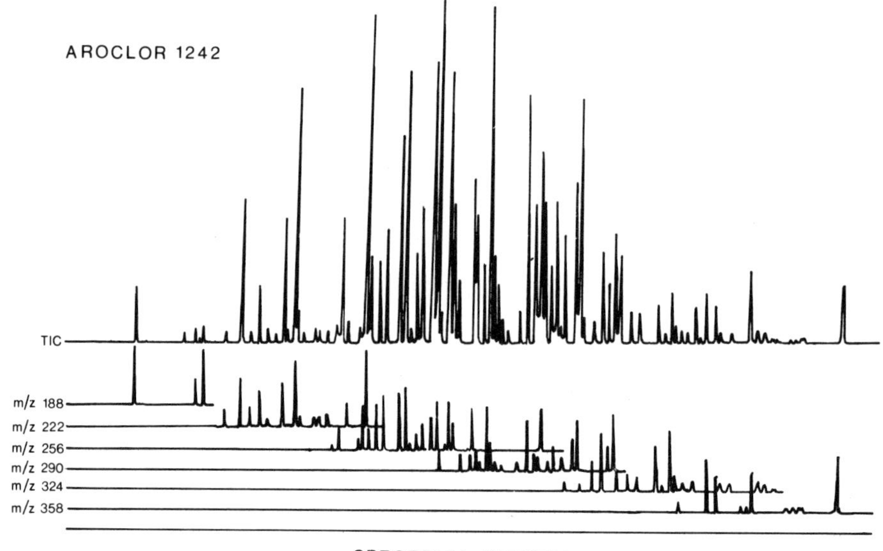

Figure 1. Mass chromatogram of Aroclor 1242

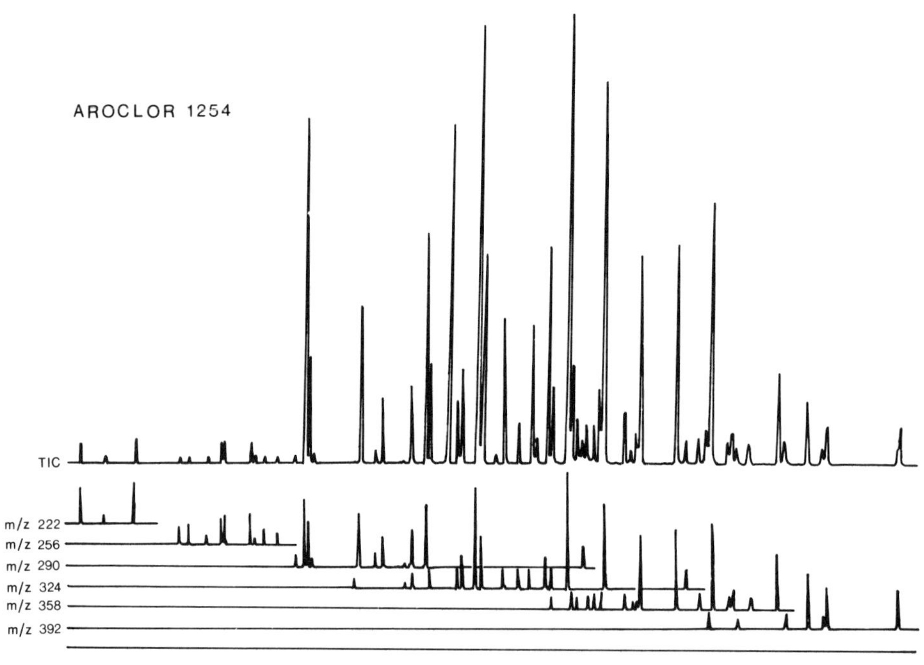

Figure 2. Mass Chromatogram of Aroclor 1254

four different temperatures. Obtained values were calculated in accordance with a program developed in this laboratory (Onuska and Comba, 1976).

DISCUSSION AND RESULTS

Gas chromatography-mass spectrometry. A Varian MAT 311A connected to the Varian 2700 GC via an open split interface (Holmes and Morrell, 1957; Henneberg and Annal, 1961) was employed. Total ion current traces for a 20 m OV-17 capillary column were obtained on Aroclor mixtures using the following conditions:

Electron Energy	65 eV
Electron Multiplier	2 eV
Transfer Line Temperature	240°C
Ion Source Temperature	240°C
Scan Speed	5 s/dec.

The stored spectra were then used to obtain dedicated mass plots for the monochloro to heptachloro biphenyl derivatives. The data were manually evaluated and assigned to the peaks in total ion current trace. As shown in Figures 1 and 2, it was possible to assign and subdivide the chromatograms into corresponding groups as recommended by Eichelberger et al (1974).

Retention indices, relative retention times and response factors relative to p,p'-DDE were measured and data calculated at three different temperatures. Results are summarized in Table 2.

Afterwards, seven different mixtures were prepared containing hydrocarbons and individual PCB isomers. They were run separately and as spikes with both Aroclor mixtures employing the same chromatographic conditions and temperature programming rate. This enabled us to match individual isomers to corresponding peaks and to calculate a retention index using temperature programming. Determined indices were then employed to calculate and predict the position of the remaining components in the chromatograms according to the method proposed by Sissons and Welti (1971).

It must be emphasized that the relative contribution values for 1/2 RI, for ring substitution patterns are applicable only to the capillary column, for which the calculations were made, although one may establish the elution order for those isomers, which are not commercially available, for that phase.

Chromatograms of Aroclars 1242 and 1254 on two different WCOT columns are shown in Figures 3 to 6.

Identification of individual positional isomers and their retention indices and retention temperatures are given in Tables 3 and 4.

Comparison of the PCB patterns to a chromatogram obtained on a well-prepared 10 ft., 3% OV-101 column at optimum conditions is shown in Figure 7. Only 16 distinct peaks are observed, demonstrating the necessity for using capillary column techniques.

Quantitative Analysis. Because of inadequate separation power of standard gas chromatographic techniques, quantitative methods for determining total PCB concentration have not considered the complete residue. Risebrough (1969) assumed that each PCB isomer produced the same peak height with an EC-detection in proportion to the amount by weight of p,p'-DDE. After running the total contribution of individual peaks, the sum was multiplied by a factor obtained from calibration with standard solutions. However, Zitko et al (1971) and Gregory et al (1961) showed that the EC-detector response was not the same for all isomers, but dependent on the degree of chlorination. Koeman (1969) measured PCBs by employing one peak in a commercial mixture as a standard. Reynolds

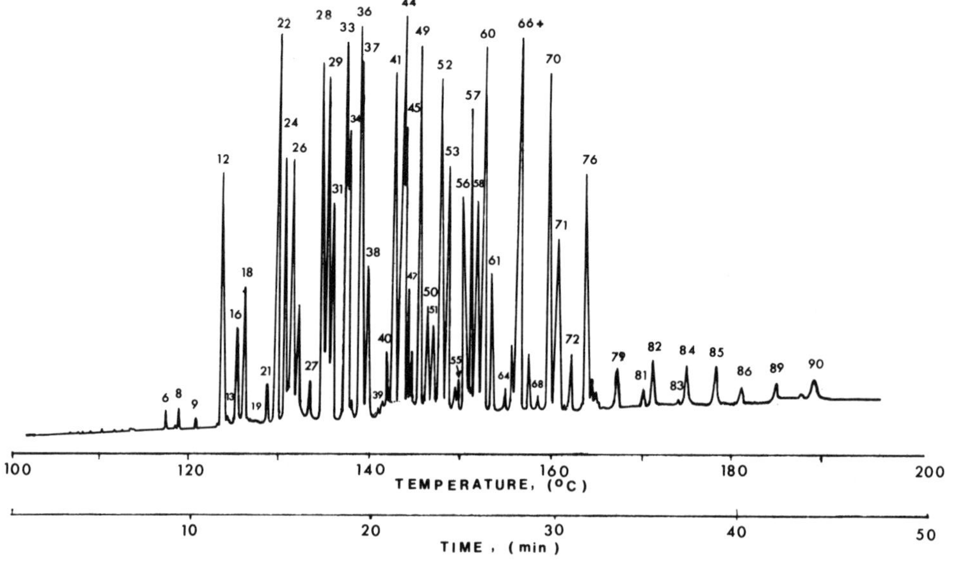

Figure 3. Separation of Aroclor 1242 on a 20 m SP-2100 WCOT column from 100°C - 200°C at 2°C/min. and 3 min. hold, EC-detector. Peaks are identified in Table 3.

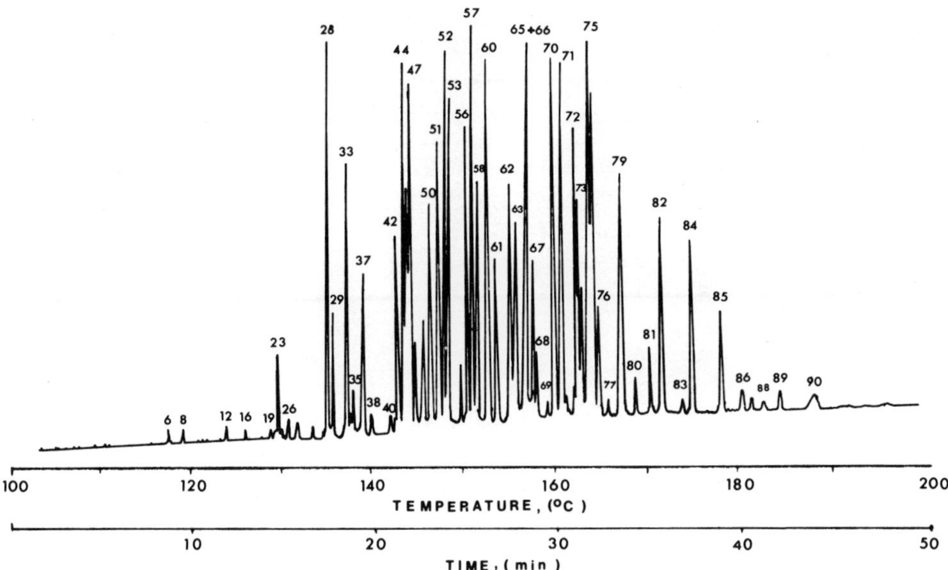

Figure 4. Separation of Aroclor 1254 on a 20 m SP-2100 WCOT column from 100°- 200°C at 2°C/min. and 3 min. hold, EC-detector. Peaks are identified in Table 3.

(1969) used two or more peaks from a standard mixture. Rote et al and Webb et al (1973) introduced p,p'-DDE as an internal standard to adjust detector sensitivity for variations during a run.

Perchlorination of PCB residues to form decachlorobiphenyl as described by Berg et al (1972) is probably the most accurate means for quantitating total PCBs. However, this procedure is quite tedious for routine analysis and does not provide any information on the composition of the mixture. We believe a more accurate quantitation can be achieved when employing WCOT columns. Each peak may be quantitated separately while still providing individual component identification. By employing p,p'-DDE as an internal standard, one may calculate with a high degree of accuracy and precision, either the individual components in the mixture of PCBs, or determine which components have undergone metabolic transformation or degradation in comparison to Aroclor residue mixtures.

Table 5 gives the quantitative data on the three methods employing FID-detection and packed and glass capillary chromatography.

Table 2. Relative Retention Times and Responses to p,p'-DDE on OV-17 WCOT Column

Structure	Range	RRT to DDE	RRF to DDE	I	∂I/∂T
2-mono	115-130	0.056±0.013	1.600	1665	9
3-	115-130	0.075±0.019	1.300	1730	6
4-	100-130	0.076±0.014	1.664	1738	9
2,2'-di	115-130	0.103±0.014	1.341	1794	12
3,3'-	115-130	0.199±0.025	0.930	1935	9
4,4'-	115-130	0.227±0.029	0.912	1954	12
2,4-	100-120	0.121±0.008	1.000	1855	8
3,4-	120-140	0.201±0.070	0.969	1945	22
2,5-	140-155	0.121±0.021	1.0605	1854	3
2,4,6-tri-	120-140	0.170±0.039	0.819	1912	23
2,5,2'	150-160	0.198±0.009	1.019	2080	59
2,5,3'	150-160	0.249±0.012	1.165	2114	53
2,4,5	120-140	0.264±0.099	1.162	2005	4
2,5,4'	150-160	0.264±0.011	1.165	2126	58
2,3,4	120-140	0.354±0.020	1.173	2061	6
2,4,6,2'-tetra	150-170	0.287±0.037	1.097	2062	31
2,6,2',5'	150-170	0.334±0.031	1.147	2095	31
2,5,2',5'	150-170	0.375±0.027	0.894	2177	50
2,4,2',4'	150-170	0.383±0.038	0.998	2192	90
2,4,6,4'	150-170	0.390±0.031	1.092	2144	31
2,4,2',5'	150-170	0.398±0.035	1.021	2148	31
2,3,5,6	150-170	0.403±0.028	1.025	2149	29
3,5,2',5'	150-170	0.456±0.029	0.709	2184	28
2,4,6,2',6' penta	150-170	0.438±0.032	0.883	2174	30
2,5,2',3'	150-170	0.473±0.032	1.006	2187	20
2,4,6,2',5'	150-170	0.516±0.023	0.796	2219	36
2,3,4,5	150-170	0.568±0.028	0.850	2273	36
2,3,6,2',5'	160-190	0.542±0.044	1.125	2228	43
2,3',4',5'	150-170	0.612±0.022	0.739	2256	24
2,4,6,2',3'	160-190	0.674±0.048	0.980	2273	30
2,4,6, 2',4',6'	170-200	0.739±0.099	1.072	2284	65
2,4,5,2',5'	150-170	0.745±0.010	0.840	2295	23
2,4,6,3',4'	150-170	0.809±0.017	0.854	2315	24
2,3,4,2',5'	150-160	0.985±0.007	0.850*	2357	13
2,4,5,2',4',6'	160-190	0.998±0.019		2385	27
3,4,3',4'	150-170	1.074±0.002	0.958	2402	38
2,3,5,6,2',5'	160-190	1.131±0.006	0.870	2421	36
2,4,5,2',4',5'	170-200	1.490±0.033	1.080	2495	41
2,3,4,5,2',3'	160-190	2.05 ±0.14	0.488	2572	32
2,3,4,5,6,2',5'	180-210	2.205±0.021	0.727	2658	106
2,3,4,5,3',4'	160-190	3.19 ±0.15	0.631	2670	40
2,3,4,5,2',3',4',5'	180-210	5.62 ±0.97	0.776	2905	44

* Average RRT and Deviations for Three Different Temperatures.

Table 3. Retention Indices and Retention Temperatures of Individual Isomers of Aroclor 1242 and Aroclor 1254 on SP-2100 WCOT Column

Identification	Retention Temperature (°C)	KI	Assigned Peak 1242	Assigned Peak 1254
2,5 and 2,6-	117	1642	6	6
2,3'-	118	1660	7	7
2,4'-	118.5	1663	8	8
2,6,2'-	121	1701	-	-
3,3'-	123	1729	11	-
2,5,2'-	124	1744	12	12
4,4'-	124.5	1748	13	-
2,3,2'-	125.5	1773	16	16
2,6,4'-	126	1776	18	-
2,6,2',6'-	127	1792	19	-
3,5,2'-	128	1803	21	19
2,5,4'-	129	1822	22	23
2,4,4'-	129.2	1823	23	24
2,3,3'-	130.5	1843	24	26
2,3',4'-	130.8	1845	25	-
2,4,2',6'-	131.5	1852	26	-
3,5,3'-	134	1873	27	27
2,5,2',5'-	135	1888	28	28
2,4,2',5'-	135.7	1896	29	29
2,4,6,2'-	136	1899	30	30
2,4,2',4'-	136.3	1901	31	-
2,3,2',5'-	137	1920	33	33
3,4,4'-	137.3	1924	34	34
2,3,2',4'-	138	1927	35	35
2,3,4,2'-	138.7	1940	36	-
2,3,6,4'-	139	1942	37	37
2,3,2',3'-	139.6	1952	38	38
2,5,3',5'-	141	1959	39	-
2,4,3',5'-	142	1963	40	40
2,4,5,3'-	143	1972	41	42
2,4,5,4'-	143.5	1974	41	42
2,3,3',5'-	143.7	1978	-	-
2,5,3',4'-	144	1986	44	44
2,3,6,2',5'-	144.3	1990	45	45
2,4,3',4'-	144.4	1991	45	45
2,3,6,2',4'-	144.6	1999	47	47
2,4,5,2',6'-	145	2005	48	48
2,3,4,4'-	146	2021	49	49
2,4,6,2',4',6'-	146.5	2030	50	50

Table 3 (Continued)

Identification	Retention Temperature (°C)	KI	Assigned Peak 1242	Assigned Peak 1254
2,3,5,2',4'-	147	2038	51	51
2,4,5,2',5'-	148	2047	52	52
2,4,5,2',4'-	148.5	2052	53	53
2,3,5,2',3'-	149	2063	54	-
2,3,5,6,2',6'-	149.5	2069	55	55
2,4,5,2',3'-	150.5	2079	56	56
2,3,4,2',5'-	151	2086	57	57
2,3,4,2',4'-	151.5	2092	58	58
2,3,6,2',3',6'-	151.7	2093	58	58
3,4,3',4'-	152.5	2102	60	60
2,3,6,3',4'-	153	2115	61	61
2,3,5,6,2',5'-	153.6	2120	64	62
2,3,5,2',3',6'-	153.8	2129	-	63
2,3,5,6,2',4'-	154	2135	64	-
2,3,4,6,2',5'-	155	2149	65	65
2,3,6,2',4',5'-	155.5	2150	66	66
2,4,5,3',4'-	157	2160	67	67
2,3,5,2',3',5'-	158	2172	68	68
2,3,4,6,2',3'-	159	2184	-	69
2,3,4,2',3',6'-	159.5	2195	70	70
2,4,5,2',4',5'-	160.5	2206	71	71
2,3,4,5,2',5'-	162	2223	72	72
2,3,4,6,2',3',6'-	162.5	2225	-	73
2,3,5,2',3',4'-	163	2230	-	74
2,3,4,2',4',5'-	164	2245	-	75
2,3,4,5,2',3'-	164.5	2252	76	76
2,3,5,6,2',3',5'-	166	2271	77	77
2,3,4,2',3',4'-	167.5	2287	79	79
2,3,5,6,2',4',5'-	168.5	2292	-	80
2,3,4,5,6,2',5'- 2,3,4,6,2',4',5'- }	170.5	2309	81	81
2,3,4,5,3',4'-	172	2339	82	82
2,3,5,2',3',4',5'-	174	2369	83	83
2,3,4,5,2',4',5'-	175.5	2381	84	84
2,3,4,5,2',3',4'-	178.5	2422	85	85
2,3,4,5,6,2',3',5'-	181.5	2451	86	86
2,3,5,6,2',3',4',5'-	182	2467	-	87
2,3,4,6,2',3',4',5'-	183	2486	-	88
2,3,4,5,6,2',3',4'-	186	2507	89	89
2,3,4,5,3',4',5'-	189	2556	90	90

Table 4 Retention Indices and Temperatures for Polychlorinated Biphenyls on OV-17 WCOT Columns

Identification	Retention Temperature (°C)	RRT[a]	KI	Assigned Peak 1242	#Cl[b]	1254	#Cl[b]
Biphenyl	140.0		1562	-	-	1	0
2-chloro	158.0		1701	1	1	2	1
2,6	168.0		1830	7	2	-	-
2,2'	173.5		1844	8	2	3	2
N.I.			1859	9	-	-	-
2,5			1872	10	2	-	-
2,4			1879	11	2	-	-
2,3'	182.0		1906	12	2	4	2
2,4'	182.5		1919	13	2	5	2
2,3	183.5		1928	14	2	6	2
N.I.			1964	15	-	-	-
2,6,2' & 3,3'	188.5		1985	(16-17)	2&3	7	3
3,4'	190.5		1990	18	3	-	-
2,5,2' & 2,4,2' & 4,4'	194.0	0.647	2016	19	3	8	3
2,6,3'			2037	20	3	-	-
N.I.	197.2		2057	21	3	9	3
2,3,2'	198.0		2067	22	3	10	3
2,5,3'	200.0		2083	23	3	11	3
2,4,4' & 2,5,4'	201.6	0.715	2103	24	3	12	3
	202.0	0.719	2108	25	3	13	3
2,3,4	204.0		2127	26	3	15	3
2,3,3'	204.5		2133	27	3	-	-
2,5,2',6'	206.0		2140	28	4	16	4
			2146	29	4	-	-
2,3,4' & 2,4,6,3'	207.2		2153	30	3&4	17	3&4
2,4,6,4'	209.0		2172	31	4	18	4
2,5,2',5'	209.5	0.784	2180	32	4	19	4
2,4,2',4'	210.3	0.788	2185	33	4	20	4
2,5,4',6'	211.5		2188	-	4	21	4
2,3,2',6' & 2,4,5,2'	212.0		2192	34	4	-	-
2,3,5,6			2200	35	4	-	-
2,5,3',5'			2204	36	N.I.	-	-
2,4,3',5'	214.8		2210	37	4	22	4
2,3,2',4' & 2,3,4,6	216.4	0.829	2239	38	4	23	4
2,3,6,4'			2251		-		
2,4,6,2',5'	218.0		2258	39	4	24	4&5

a RRT to p',p' DDE
b #Cl by mass spectrometry
N.I. Not Identified

Table 4 (Continued)

Identification	Retention Temperature (°C)	RRT	KI	1242	#Cl	1254	#Cl
2,4,5,3' & 2,3,4,2'	219.0	0.848	2262	40	4	25	4
N.I.	2210.5		2268	41	3	26	4&5
2,4,2',6' & 2,3,6,2',6'	221.5		2286	42	N.I.	27	4&5
2,3,2',3' & 3,4,5	222.0	0.872	2294	-	-	28	4
2,4,6,3',5'			2299	43	5	-	-
2,3,4,5 & 2,5,3',4'	223.5	0.844	2312	44	4	29	4
2,4,3',4' & 2,4,5,2',6'	224.0	0.888	2315	45	4&5	30	5
2,4,6,2',3' & 2,3,5,2',6'	225.0		2317	-	-	31	N.I.
2,3,6,2',5' } 2,3,6,2',4'	225.6	0.900	2335	46	5	32	N.I.
	226.8	0.910	2343	-	-	33	5
2,3,4,4' & 2,3,5,6,2'	227.5	0.913	2348	47	4	34	5
2,4,5,2',5' & 3,4,5,3'	228.5	0.923	2363	48	4	35	5
2,3,5,2',4'	229.0	0.929	2368	49	N.I.	-	
2,4,5,2',4'			2370	50	N.I.	36	5
2,3,6,3',5' & 2,3,3',4'	230.0		2372	51	4	37	4&5
2,3,6,2',3' & 2,3,4,2',6'	231.0	0.946	2390	52	N.I.	38	5
3,4,5,2',4' & 3,4,5,2',5'	233.0	0.959	2400			39	5
2,4,5,3',5'	233.5		2405	-		40	5
2,4,5,2',3' & (2,3,5,6,4')	234.0	0.968	2421	53	N.I.	41	5
2,3,4,2',5' & 2,3,5,2',3'	235.0	0.977	2436	-	-	42	5
3,4,3',4' & 2,4,5,2',4',6'	236.0	0.981	2443	54	N.I.	43	6
3,4,5,2',3'	238.0	0.985	2458	-	-	44	5
N.I.	238.2	1.00	2463	55	N.I.	45	6
2,3,6,3',4'	239.0	1.01	2468	56	N.I.	46	6
3,4,5,2',4',6'	239.5	1.02	2477	-	-	47	N.I.
N.I.	240.0	1.02	2479	-	-	48	"
2,3,6,2',3',6'	240.5		2488	-	-	49	"
2,3,4,2',3'	241.2		2494	-	-	50	"
2,4,5,3',4'	241.5	1.03	2497	57	N.I.	51	"
3,4,5,3',5'	242.5		2505	-	-	52	"
2,3,6,2',4',5'	244.0	1.04	2520	-	-	53	"
3,4,5,3',4'	245.0		2527	-	-	54	"
N.I.	246.0	1.05	2532	58	N.I.	55	"
	246.5		2535	-	-	56	"

Table 4 (Continued)

Identification	Retention Temperature (°C)	RRT	KI	Assigned Peak 1242	#Cl	1254	#Cl
2,3,4,2',4'	249.0	1.08	2573	-	-	57	N.I.
	249.8		2584	59	N.I.	58	"
	250.5		2600			59	"
	251.5		2624			60	"
	253.0	1.11	2635			61	"
	254.5		2656			62	"
	255.0		2660			63	"
	255.6		2663			64	"
2,3,4,5,2',3'	256.5		2670			65	"
	258.5		2676			66	"
2,3,4,5,3',4'	260.0	1.17	2705	60	N.I.	67	"
	260.5		2716			68	"
	262.0		2723			69	"
	263.0	1.19	2737			70	"
			2752				
			2755				
	264.0		2766			71	"
	264.5		2772			72	"
	268.0		2863			73	"
	269.0		2872			74	"
	272.0	1.27	2896			75	"

Table 5 PCBs in Adult Herring Gull Lipid

Method	PCB - Concentration as 1260
Perchlorinated sample	3530 ppm
Non-perchlorinated sample[25]	2800 ppm
WCOT-analyzed sample	3455 ppm

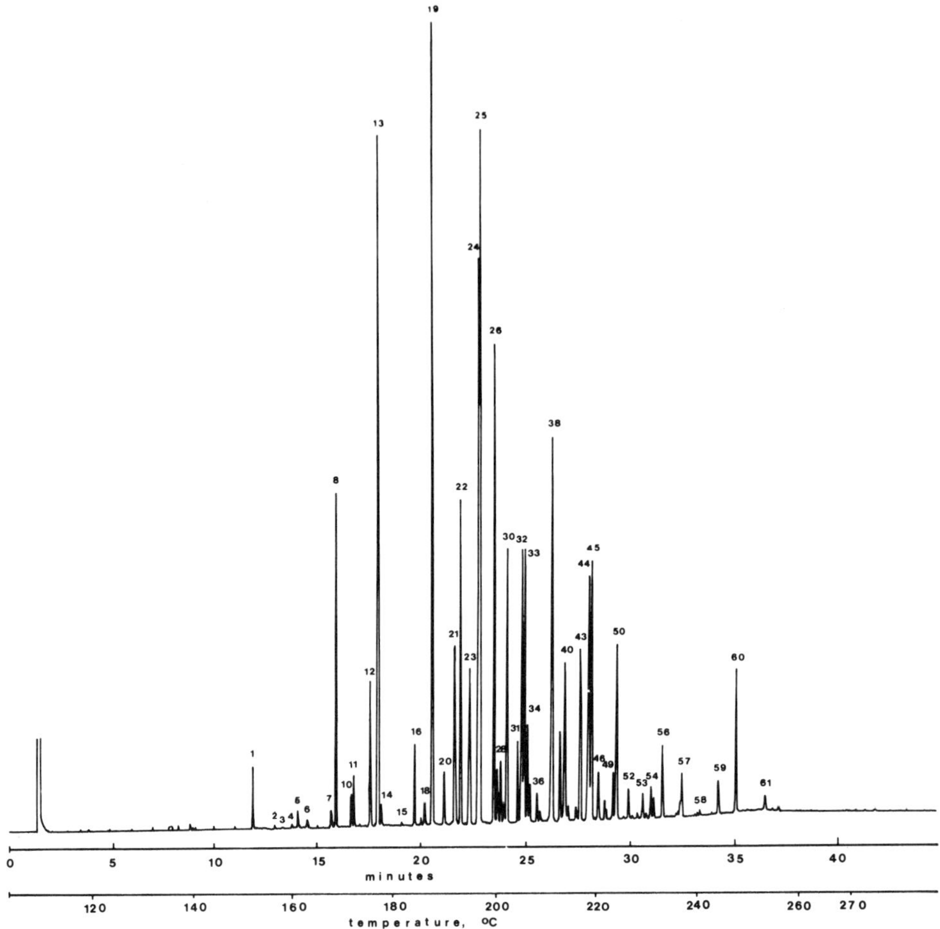

Figure 5. Separation of Aroclor 1242 on a 20 m OV-17 WCOT column. Column temperature programmed from 120-280°C at 4°C/min., 3 min. hold, FID. Peaks are identified in Table 4.

CONCLUSION

This study should contribute some clarity and definition to the problem by providing qualitative and quantitative information, with reasonable analytical precision and accuracy for the determination of many individual isomers contained in Aroclor 1242 and 1254 mixtures.

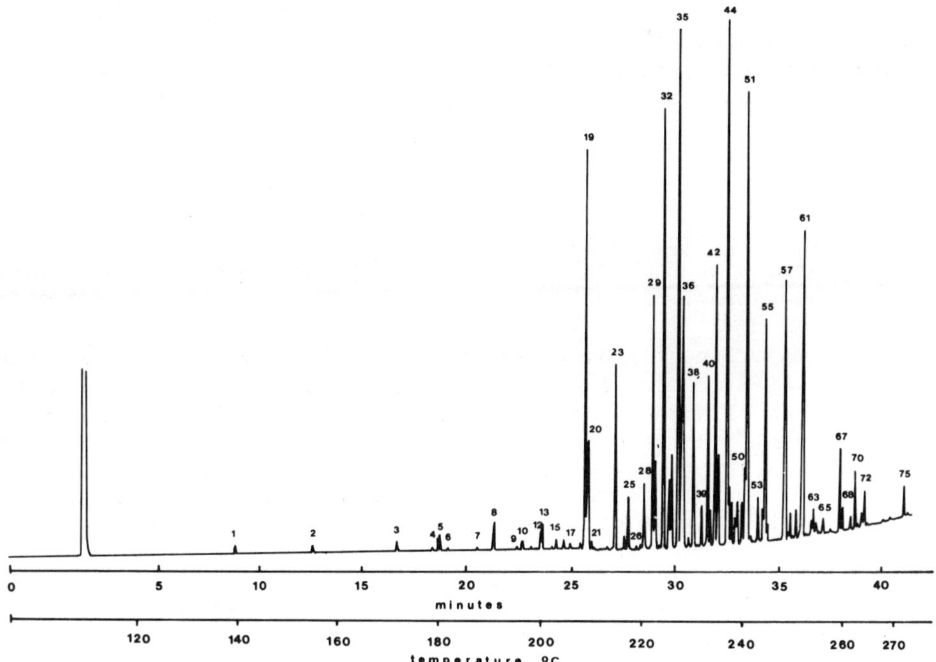

Figure 6. Separation of Aroclor 1254 on a 20 m OV-17 WCOT column. Column temperature programmed from 120-280°C at 4°C/min., 3 min. hold, FID. Peaks are identified in Table 4.

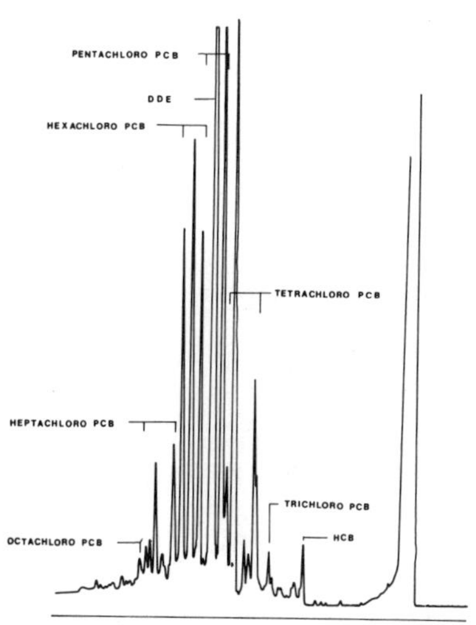

Figure 7. Separation of Aroclor 1254 on a 3 m long 3% OV-101 packed column, temperature programmed from 100-260°C at 6°C/min.

The data should also be beneficial to toxicological, bioaccumulation and biodegradation research studies in assessing toxic threshold effects, dose-response relationships and individual effects of PCB isomers.

SUMMARY

Identification and quantitative analysis of PCB's on WCOT (wall coated open tubular) glass capillary columns were investigated in this study.

The majority of isomers in Aroclors 1242 and 1254 were separated and identified on two glass WCOT columns coated with SP-2100 (methylsilicone phase) and the second column coated with a medium polar OV-17 column. Relative retention indices and response factors with respect to p,p'-DDE for a number of individual isomers are presented. It was found that retention indices could be predicted with a considerable degree of accuracy for programmed temperature runs using the calculated contribution values of ½ RI. These values were correlated with available standard data and confirmed by means of mass spectrometry. The predicted values from mono- to pentachloro- isomers gave excellent fits.

The data should be beneficial to toxicological evaluation of individual isomers, bioaccumulation and biodegradation research in assessing toxic threshold effects, dose-response relationship and individual effects of PCB isomers.

REFERENCES

Ahnoff, M., B. Josefsson, Anal. Let. 6, 1083 (1973).

Berg, O.W., P.L. Diosady, G.A. Recs, Bull. Environ. Contam. Toxicol., 7, 338 (1972).

Chau, A.S.Y., R.C.J. Sampson, Environ. Letters, 8, (2) 89 (1975).

Eichelberger, J.W., L.E. Harris, W.L. Bude, Anal. Chem., 46, (2) 227 (1974).

Fishbein, L., J. Chromatog. 68, 345, (1972).

Gregory, N.L., J.E. Lovelock, Anal. Chem., 33, 45A (1961).

Greichus, Y.A., J.J. Worman, M.A. Pearson, D.J. Call, Bull. Environ. Contam. Toxicol., 11, (2) 113 (1974).

Gustafson, C.G., Environ. Sci. Technol. 4, 814 (1970).

Hennenberg, D., Z. Anal. Chem., 183, 12 (1961).

Holmes, J.C., F.A. Morrell, Appl. Spectrosc., 11, 86 (1957).

Hutzinger, O., S. Safe, V. Zitko, The Chemistry of the PCBs, CRC Press, Cleveland, Ohio (1974).

Jensen, S., G. Sundstrom, Ambio, 3, (2) 70 (1974).

Jensen, S., A.G. Johnels, M. Olsson, G. Otterlind, Ambio Special Report, No. 1, pp. 71-85 (1972).

Jensen, S., M. Olsson, Ambio Special Report, No. 4, pp. 111-123 (1976).

Koeman, J.H., M.C. Ten Noever de Brauw, R.H. de Vos, Nature, 221, 1126 (1969).

Krull, I.S., Recent Advances in PCB Analysis, Residue Rev., 66, 185 (1977).

Krupcik, J., J. Kriz, D. Prusova, P. Suchanek, Z. Cervenka, J. Chromatogr., 142, 797 (1977).

Mieure, P.J., O. Hicks, R.G. Kaley, V.W. Seagal, National Conf. on PCBs, Chicago 1975, Conference Proceedings EPA Office of Toxic Substances, Washington, D.C., p. 84 (1976).

Onuska, F.I., Unpublished Data.

Onuska, F.I., M.E. Comba, J. Chromatog., 126, 133 (1976).

Onuska, F.I., M.E. Comba, J. Chromatog., 119, 385 (1976).

Reynolds, L.M., Bull. Environ. Contam. Toxicol., 4, 128 (1969).

Risebrough, R.W., pp. 5-23 In Chemical Fallout, Berg, G.C., Miller, M.W., Eds., Charles Thomas, Springfield, Ill., (1969).

Rote, J., P.G. Murphy, Bull. Environ. Contam. Toxicol. 6, 377 (1971).

Rote, J.W., W.J. Morris, J. A.O.A.C., 56 (1) 188 (1973).

Sissons, D., D. Welti, J. Chromatogr. 60, 15 (1971).

Stendell, R.C., National Conference on PCBs, Chicago 1975, Conference Proceedings EPA Office of Toxic Substances, Washington, D.C., p. 262 (1976).

Webb, R.G., A.C. McCall, J. A.O.A.C., 55, (4) 747 (1972).

Webb, R.C., A.C. McCall, J. Chromatog. Sci., 11, 366 (1973).

Weil, L., G. Dure, K.E. Quentin, Wasser u. Abwasser Forsch, 7, (6) 227 (1974).

W.H.O. Environmental Health Criteria for PCBs and Terphenyls, Ambio 7, 31-32 (1978).

Zell, M., H.J. Neu, K. Ballschmiter, Chemosphere 2, (3) 69-76 (1977).

Zitko, V., O. Hutzinger, S. Safe, Bull. Environ. Contam. Toxicol. 6, 160 (1971).

AUTOMATIC ANALYSIS OF ORGANIC POLLUTANTS IN WATER VIA GC/MS

David Beggs

Hewlett-Packard Company

1601 California Avenue, Palo Alto, California 94304

The use of gas chromatograph/mass spectrometer (GC/MS) systems for the routine analysis of organic pollutants in water samples has grown significantly during the last few years (Kopfler et al, 1977; Games and Hites, 1977). However, these systems have had several disadvantages; namely, the need for a highly skilled operator and the need for elaborate sample preparation prior to analysis.

The EPA has defined (U.S. EPA, 1977) 114 compounds as chemical indicators of organic pollution in water. These indicators are divided into two groups: 30 purgeable organic compounds and 84 semi-volatile and nonvolatile organic compounds. The latter compounds may be analyzed via liquid/liquid extraction and subsequent direct injection into a GC/MS. The purgeable compounds may be separated from the water substrate by a technique known as vapor stripping (Bellar and Lichtenberg, 1974).

The recent development of microcomputer controlled GC/MS systems allows simplification of operation for routine analysis. The ability to pre-program gas chromatograph and mass spectrometer conditions as well as to format data output, reduces the need for operator attention. The calculator controlled GC/MS can be programmed to automatically tune itself and set up proper gas chromatograph and mass spectrometric conditions. The instrument will accept raw water samples for the analysis of purgeable organic compounds. Subsequent sampling, analysis and data reduction can be carried out under control of the calculator. It is thus possible to make GC/MS analysis of purgeable organic compounds in water semi-automatic. Indeed, the operator need only insert his water sample, and answer four questions asked by the calculator. The sampling process and analysis is completely controlled by the calculator. The analysis and non-purgeable organic compounds can be carried out in this manner, however, the liquid/liquid sampling procedure has not been automated.

EXPERIMENTAL

A Hewlett-Packard 5992A GC/MS system was used for these analyses. A modified purge and trap apparatus was constructed using pneumatic valves to direct the flow of the helium purge and carrier gases. This purge and trap device is similar in concept to that developed by Bellar and Lichtenberg (1974). The valves and trap heater were controlled by the calculator of the GC/MS system. Data was stored on a Hewlett-Packard 18939A Flexibile Disc accessory storage device.

High purity grade helium was used as the purge gas. It was purged at a rate of 40 ml/min for 12 minutes. The purge volume was 200 ml of water. The trap consisted of a glass tube 6.4 mm OD, 2.0 mm ID, 15.3 cm long containing Tenax-GC (60/80 mesh). The trap desorption occurred with a flow of 20 ml/min of helium for 4 minutes at a temperature of 200°C. The GC column was a 180 cm glass column, 6.4 mm OD, 2.0 mm ID packed with Carbopak C (60/80 mesh) coated with 0.2% Carbowax 1500. Column bleed can be reduced by using Carbopak C coated with 0.2% Carbowax 20M. The column was held at room temperature during trap desorption, then rapidly heated to 60°C. During the GC/MS analysis, the column was held at 60°C for 4 minutes then programmed to 160°C at 8°/min. It was then held at 160°C for 14 minutes. The total analysis time was 30 minutes.

RESULTS AND DISCUSSION

The analysis scheme is carried out in the following manner. The operator is queried as to which of the following analysis is needed:

Purgeable Organic Compounds
Acid Extract
Base/Neutral Extract
Pesticides

The calculator then informs the operator of the proper GC column to be used for the particular analysis requested. The operator then is requested to enter the concentration ragne of the sample and the sample identification number. After this is completed, the calculator sets up the proper gas chromatographic and mass spectrometric conditions.

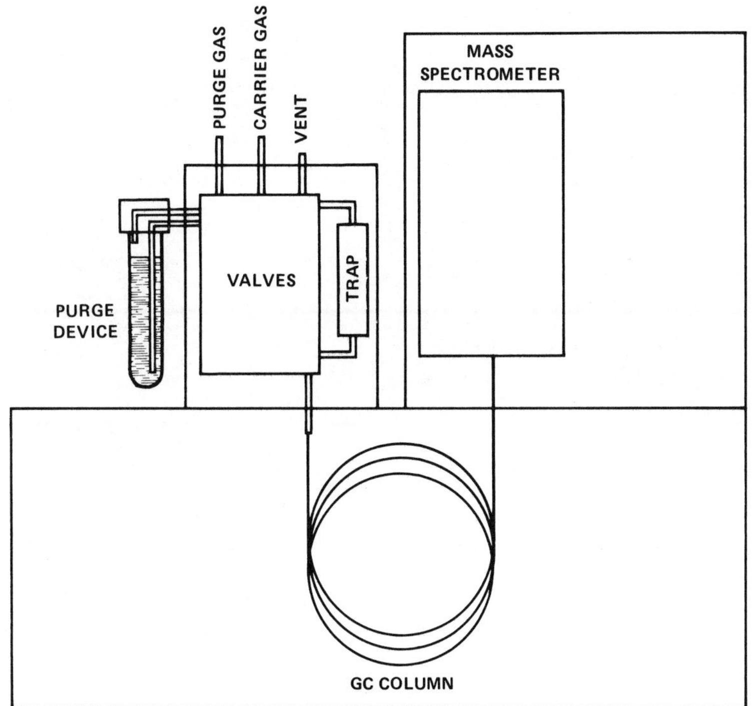

Figure 1. Schematic representation of the analysis system.

If a Purgeable Organic Analysis (POA) has been requested, the calculator will automatically initiate the sampling sequence for this analysis. The sampling sequence consists of a purge cycle and a trap desorb cycle followed by a trap vent cycle. Figure 1 shows a schematic representation of the system.

A stream of helium gas is bubbled through the water sample and then passed through a tube of adsorbent material, Tenax GC. Helium carries the volatile organics out of the water and into the adsorbent trap. The helium flow is then reversed through the trap and shunted into the GC/MS. The trap is heated rapidly to 200°C to drive the purgeable organics into the GC/MS system. The purgeable organic compounds pass through a GC column where they are separated prior to entering the mass spectrometer. A mass spectrum for each compound is then collected and stored on a magnetic disc. A total ion chromatogram is printed which displays the response of the instrument with time. A representative total ion chromatogram is shown in Figure 2.

The chromatogram reveals the presence of several compounds in the sample. Ten ppb of internal standard, 2-bromo-1-chloropropane,

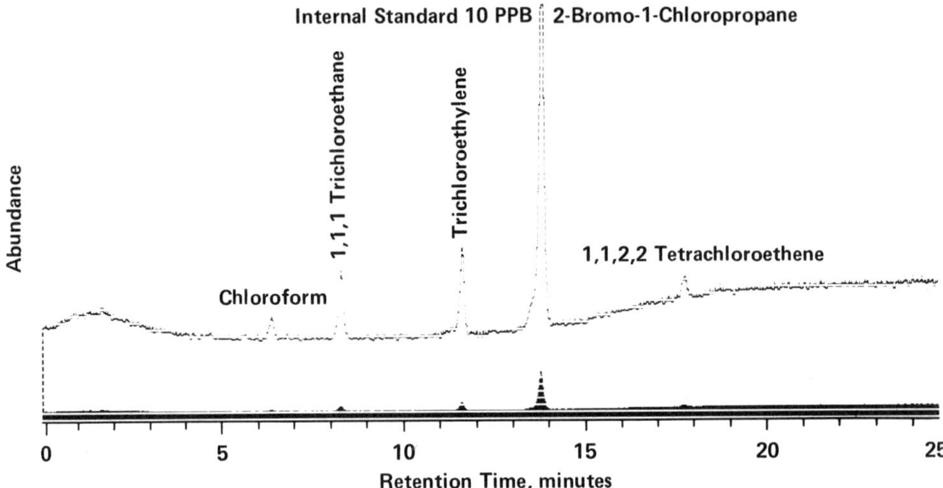

Figure 2. Analysis of a water sample from San Francisco Bay.

has been added to help determine accurate quantitation and relative retention times.

To aid in the identification of the unknowns, the mass spectrum representing the top of each GC peak is stored along with an appropriate background spectrum. The background spectrum is then subtracted from the unknown spectrum and the result can be displayed and then searched against a library of known spectra.

Figure 3 shows the spectrum obtained from the GC peak which occurs at 11.6 minutes. A library search of this spectrum provided the results shown in Figure 4. From these data it can be seen that this peak represents trichloroethylene.

The relative retention time of the unknown compound is then compared with the relative retention time of the compound identified by the library search. The correlation of the retention times is combined with the library search correlation to produce a combined correlation factor which accurately defines the quality of the tentative identification of the unknown compound. If the total correlation factor is above 0.75, the identification can be considered positive.

The internal standards are very helpful in establishing accurate retention times. They also are used to accurately define a quantitative report. The total abundance peak heights for each component are compared to the peak height for the internal standard. A response factor table stored in the calculator memory converts these peak height ratios to actual concentrations.

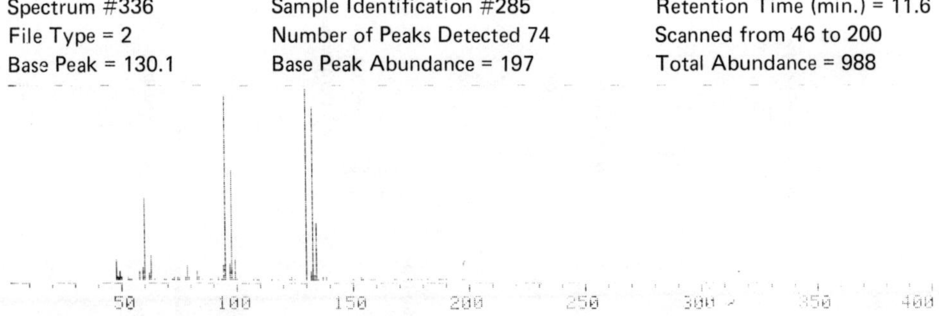

Lower Abundance Cutoff Level = 2.0

MASS	ABUNDANCE	MASS	ABUNDANCE	MASS	ABUNDANCE
47.2	10.7	78.2	7.6	99.2	11.2
49.2	4.6	82.1	5.1	130.1	100.0
52.3	2.0	83.2	2.0	131.1	5.1
57.2	5.1	91.2	2.0	132.1	90.4
59.1	6.6	94.1	8.1	133.0	2.5
60.1	42.6	95.1	96.4	134.0	29.0
61.1	3.6	96.1	9.1	136.0	2.0
62.1	12.7	97.1	57.4		
75.2	2.0				

Figure 3. Spectrum of GC peak located at 6.4 minutes retention time.

10 BEST MATCHES: Library #4

Entry	Similarity Index	Molecular Weight
12	0.9645	130.0 Trichloroethylene
30	0.4285	128.0
3	0.2805	132.0
5	0.2381	132.0
1	0.1034	96.0
27	0.0917	94.0
9	0.0810	112.0
23	0.0529	96.0
8	0.0489	166.0
22	0.0456	98.0

Figure 4. Library search result for spectrum in Figure 3.

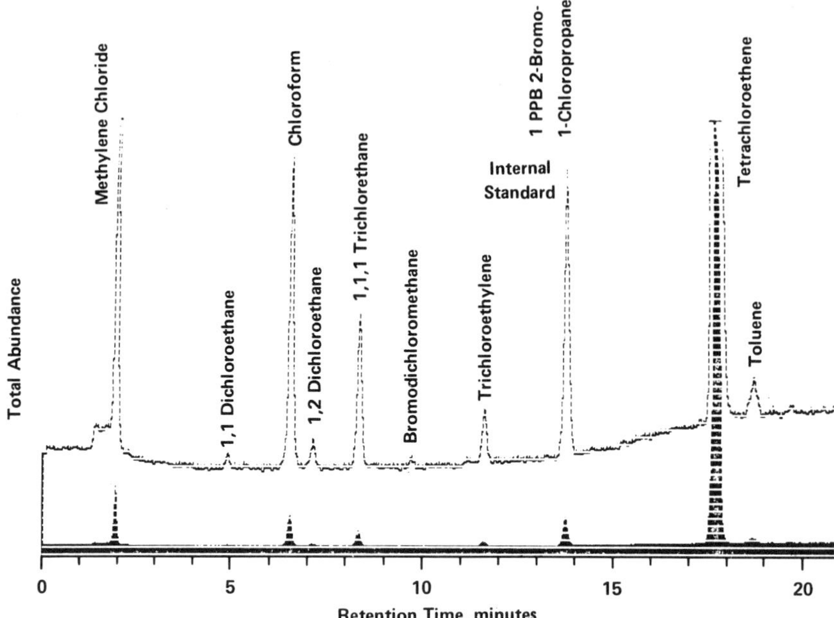

Figure 5. Profile of purgeable organics in industrial effluent sample.

Figure 5 is a profile of the volatile organic compounds emanating from an industrial source. A 1 ppb internal standard was added to this sample and the results indicate significant levels of various chlorinated hydrocarbons. A quantitative analysis was made by comparing the total abundance value at the top of each GC peak with the total abundance value for the internal standard.

Quantitative results are shown in Table 2. This example demonstrates the applicability of this method for monitoring the influent and effluent water quality of industrial sites.

If more sensitivity is desired, the mass spectrometer can be operated in the selected ion mode. This mode increases the apparent sensitivity of the instrument by as much as two orders of magnitude and it is possible to detect compounds in the low parts per trillion range.

In an effort to maximize reuse of reclaimed water, research is being conducted to improve advanced waste treatment facilities so that their effluent may be used as ancilliary water supplies (Water Reclamation Facility, Santa Clara Valley Water District, Palo Alto, CA). Water samples taken prior to and after tertiary treatment were analyzed to determine the efficiency of advanced waste treatment methods. Figures 6 and 7 show the results.

The response factor corrects for differences in purging efficiency, trapping efficiency and responses to the gas chromatograph and mass spectrometer. All of the qualitative and quantitative data is brought together and printed out by the calculator in a Final Report, Table 1.

The same analysis scheme is carried out when any of the other three types of analysis are requested. However, in these cases, the sampling is not controlled by the caluclator. The operator must carry out the liquid/liquid extraction and subsequent concentration. The sample is then injected into the GC/MS by the operator. The GC/MS analysis and data reduction are then controlled and completed by the calculator. The Final Report for these analyses can be combined with the Final Report for the POA to give a complete profile of the organic constituents of a water sample.

Analysis of several different types of water samples were undertaken to demonstrate the versatility of this method. The analysis of water collected from San Francisco Bay, Figure 2, reveals the presence of several compounds in the sample.

The trichloroethane, trichloroethylene, and tetrachloroethene are most likely present due to sewage disposal into the bay. The chloroform results from the chlorination of drinking water which subsequently finds its way into the bay. All of these compounds are at concentrations (<5 ppb) significantly below minimum toxic levels (Fed. Register, February 1977).

Table 1 Final Report

The following compunds were found in Sample 26113 dated 11/26/77.

Name	Spect. #	Ret. Time	Corr.	Conc (ppb)
Chloroform	330	6.4	.984	0.27
1,1,1 Trichloroethane	332	8.2	.971	0.52
Trichloroethylene	336	11.6	.925	0.35
Internal Standard	340	13.8	.941	10.0
1,1,2,2 Tetrachloroethene	354	17.7	.912	0.55

Table 2 Industrial Effluent Quantitative Report

Compound	Concentration (ppb)
Methylene Chloride	1.9
1,1 Dichloroethane	0.06
Chloroform	1.0
1,2 Dichloroethane	0.1
1,1,1 Trichloroethane	0.5
Bromodichloromethane	0.05
Trichloroethylene	0.2
2-Bromo-1-Chloropropane (Int.Std.)	1.0
1,1,2,2 Tetrachloroethene	17.5
Toluene	

The chromatogram shown in Figure 7 was recorded at an increased response factor so that the resulting display is expanded by a factor of 10. After secondary biological treatment and chlorination, there are high levels of chlorinated hydrocarbons. Almost all of these are eliminated or significantly reduced in the tertiary stages of treatment.

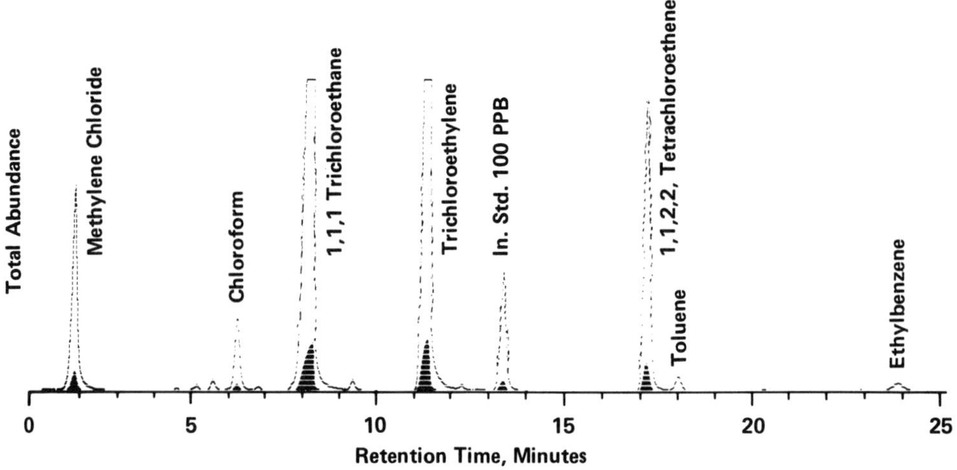

Figure 6. Profile of sewage prior to tertiary treatment.

Total Abundance
Internal Standard 1 PPB
1,1,2,2 Tetrachloroethene
Toluene
Chlorobenzene
0
5
10
15
20
25
Retention Time, minutes

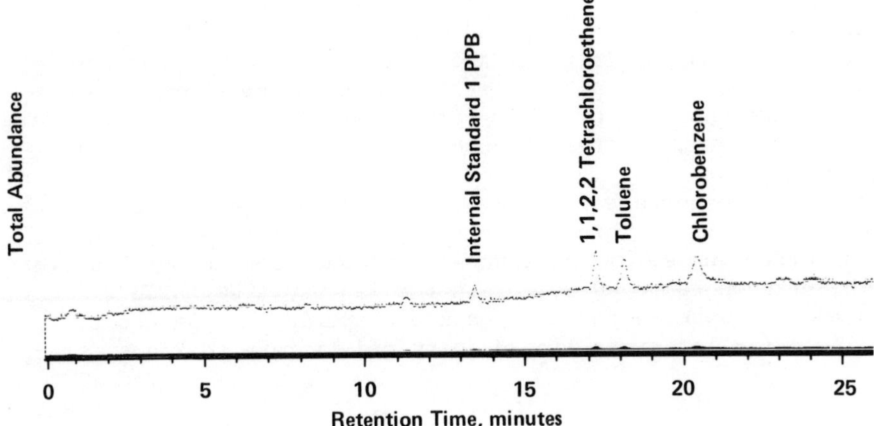

Figure 7. Profile of sewage after tertiary treatment

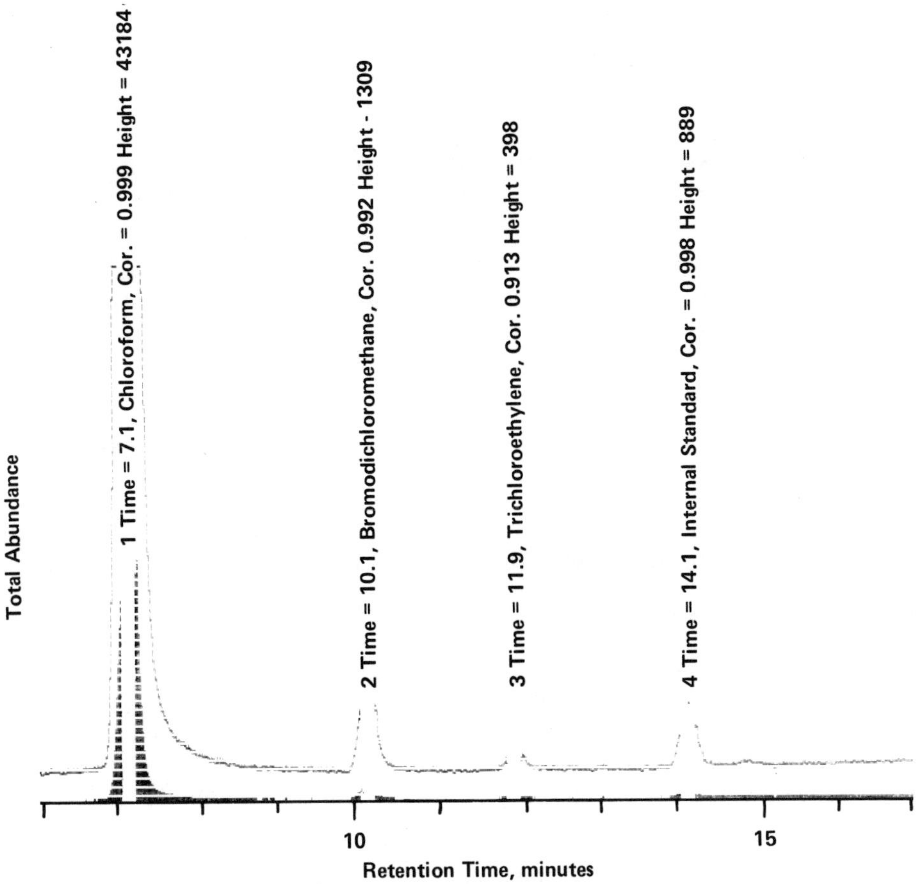

Total Abundance
1 Time = 7.1, Chloroform, Cor. = 0.999 Height = 43184
2 Time = 10.1, Bromodichloromethane, Cor. 0.992 Height - 1309
3 Time = 11.9, Trichloroethylene, Cor. 0.913 Height = 398
4 Time = 14.1, Internal Standard, Cor. = 0.998 Height = 889
10
15
Retention Time, minutes

Figure 8. Real time analysis of drinking water sample

An alternative display method provides qualitative and quantitative results as the compound elutes from the gas chromatogram, Figure 8. Thus the identity and abundance of a GC peak can be printed on top of the peak as it is drawn on the chart paper. This is made possible by a very fast search of a relatively small spectral library. After a successful search is completed the following information is printed on top of the GC peak: compound name, a search correlation factor (how closely the unknown spectrum resembled the library spectrum), the library entry number, the spectrum number of the unknown compound, and the peak height. If the unknown compound is not in the library, "NO MATCH" is displayed. Spectra for all peaks are stored on tape and are available for later off-line display and library searches.

After displaying this information, the printer continues to plot the total abundance chromatogram. Due to the presence of an internal buffer storage device, the top of the GC peak is actually printed after the digital data. Only 1.5 seconds of data collection time is sacrificed to provide the real time search results. For most applications, this is negligible. Figure 8 shows the application of this display method to the analysis of a sample of drinking water.

This is a fairly typical result for drinking water. These compounds are all products of the chlorination process. Chlorinated water contains hypochlorite ions, which react with humic acids to form chloroform and in this case, trichloroethylene. Hypochlorites will also react with any bromine present in the water to form hypobromites, which further react to form brominated products such as bromodichloromethane.

Real time search results were confirmed by the visual comparison of each unknown spectrum to a reference spectrum. The real time search method reduces total analysis time and operator intervention. This results in a greater number of analyses carried out with fewer personnel.

It should be mentioned that if any unusual peaks are present in the chromatogram they can usually be identified through off-line library searches.

Analysis of semi-volatile and nonvolatile components are carried out through the use of a liquid/liquid extraction procedure using methylene chloride (U.S. EPA, 1977). Basic, neutral and acidic fractions are extracted. The basic and neutral fractions can be combined and analyzed together. Figure 9 shows the analysis of a base/neutral fraction from secondary treated sewage. The major components were identified as ortho and meta xylene, and diethyl and dibutyl phthalate.

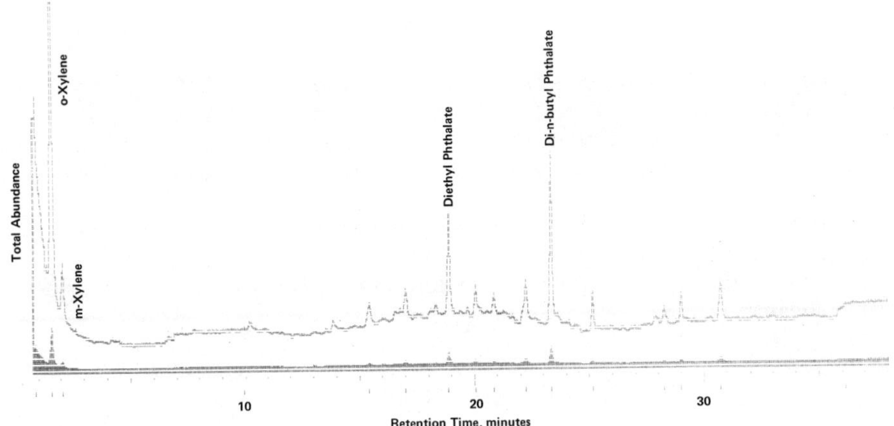

Figure 9. Analysis of base/neutral extract from secondary treated sewage.

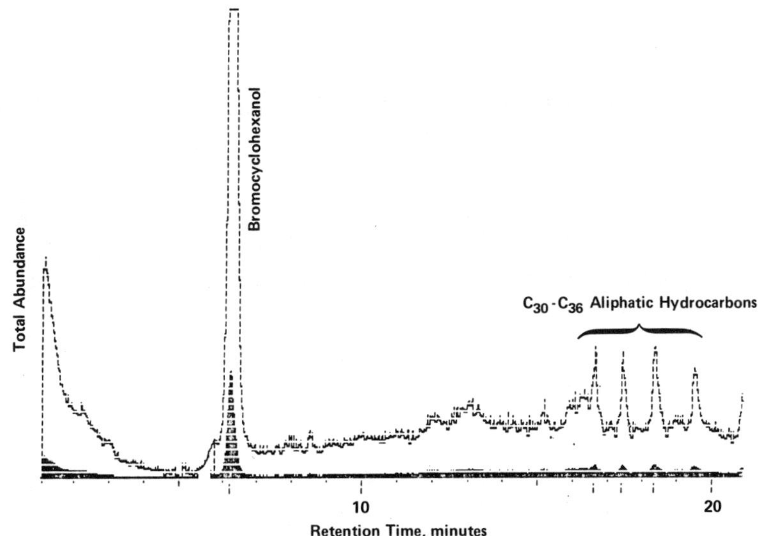

Figure 10. Analysis of acid extract from secondary treated sewage

The acid fraction analysis, Figure 10, resulted in the identification of the C_{30}, C_{32}, C_{34} and C_{36} normal alkanes. Quite surprisingly, a large concentration of bromo-cyclohexanol was detected. This alerted the sewage disposal plant to an unusual source of discharge of this compound, and prevented further contamination by that source.

CONCLUSIONS

These examples demonstrate the utility of a GC/MS system in the routine analysis of organic constituents in water. Positive identification is achieved through the use of a fast, accurate library search routine. Additional confirmation can be achieved by comparison of relative retention times. Accurate quantitation results from the ability to compare responses of contaminants to the response of an internal standard. The high sensitivity of this technique enables positive identification of compounds at levels below 1 ppb.

REFERENCES

Bellar, T.A., and J.J. Lichtenberg. 1974. Determining Volatile Organics at Microgram-per-Litre Levels by Gas Chromatography, Am. Water Works, J., 66, 739.

Games, L., and R.A. Hites. 1977. Composition, Treatment Efficiency, and Environmental Significance of Dye Manufacturing Plant Effluents, Anal. Chem., 49, 1433.

Jolley, R.L., Water Chlorination Environmental Impact and Health Effects (Ann Arbor Science, Ann Arbor, Michigan, 1978), Vol. 1.

Kopfler, F.C., Robert G. Melton, Robert D. Lingg, and W.E. Coleman. 1977. GC/MS Determination of Volatiles for the NORS Survey of Drinking Water, Chapter 6. Identification and Analysis of Organic Pollutants in Water. Ann Arbor Science.

"Pretreatment Standards for Existing and New Sources of Pollution". Fed. Register (Feb. 2, 1977).

Protocol for Sampling and Analysis Procedures for Survey of Industrial Effluents for Priority Pollutants, U.S. EPA, Environmental Monitoring and Support Laboratory, Cincinnati (March 1977). Note: A few of these compounds are actually technical mixtures such as chlodane and PCB 1242.

Water Reclamation Facility, Santa Clara Valley Water District, Palo Alto, CA.

QUANTIFICATION OF POLYCYCLIC AROMATIC HYDROCARBONS IN MARINE ENVIRONMENTAL SAMPLES

W.J. Cretney, P.A. Christensen, B.W. McIntyre and B.R. Fowler

Institute of Ocean Sciences, Patricia Bay

P.O. Box 6000, Sidney, B.C., Canada V8L 4B2

ABSTRACT

The potential for use of perdeuterated polycyclic aromatic hydrocarbons (PAHs) in the analysis of marine sediments and organisms by a method that employs a gas chromatography/mass spectrometer (GC/MS) system operating in the multiple ion monitoring (MIM) mode is examined. Questions concerning chemisorption of PAHs on adsorbents, hydrogen-deuterium exchange, interfering compounds, and the separation of deuterated and protiated PAHs by gas or liquid chromatography are considered and experiments that were performed to obtain answers are described. The evidence indicates that deuterium labelled PAHs are useful carriers and internal standards for their unlabelled forms in the analysis of marine organism and sediment samples.

INTRODUCTION

Does the present rate of influx of PAHs into the marine biosphere have the potential to significantly upset it? Some recent evidence (Blumer and Youngblood, 1975; Youngblood and Blumer, 1975) appears to indicate that PAHs present in soils and recent marine sediments are principally derived from natural grass and forest fires. The implication of this finding would be that the marine biosphere has been subject to contemporary levels of PAHs over geologically long periods of time. Even more recent evidence (Farrington et al, 1977; Hites et al, 1977; Laflamme and Hites, 1978), however, based primarily on the analysis of sediment core samples, indicate that in recent times natural combustion sources of PAH may have become minor to anthropogenic combustion sources, particularly near major industrial centres. Thus, at some sites off the northeast coast of the U.S.A. the PAH burden in the sediments appears to have increased tenfold (Farrington et al, 1977; Hites et al, 1977) over the last century or so. The significance of the anthropogenic input of PAHs on a global scale is not known, although an assessment is in its early stages (Laflamme and Hites, 1978). Also unknown is the biological impact that an increase of PAH levels might have on marine ecosystems.

Some PAHs are potentially carcinogenic, mutagenic and/or teratogenic. There is speculation that marine pollution in general and perhaps PAHs in particular may promote tumor production in some marine animals (Stich et al, 1976; Payne et al, 1978). Other marine animals apparently lack the necessary enzyme systems to transform PAHs into the ultimate carcinogenins or mutagenins and so may be unaffected (Payne, 1977; Vandermeulen, 1978). Some of these animals which are known to be efficient bioaccumulators of hydrocarbons or other lipophilic compounds (Riseborough et al, 1976; Lee, 1977) such as mussels, oysters and scallops, are consumed by humans. The potential exists therefore of a possible direct effect on human health.

Questions regarding the fate and effects of PAHs in the environment will be answered through research, much of which will depend on accurate and precise analytical methods. This paper describes an approach to the analysis of PAHs that has the potential for accurate and precise determination of PAHs in marine sediments and organisms.

EXPERIMENTAL

Materials

Glass-distilled solvents were purchased whenever possible and used without further purification, if found to be of the necessary quality. Ethanol was purchased as the "unmatured spirit" and redistilled until sufficiently pure for use. Water was obtained from a Barnstead Still through a glass transfer tube.

Florisil (60-100 mesh) was washed thoroughly with distilled water and methanol, dried at 60°C in a forced air oven, and calcined at 600°C. The Florisil was then washed again with glass distilled methanol and activated by the method of Howard et al (1966). Potassium hydroxide pellets (ACS grade) were soxhlet-extracted with benzene for four hours and dried at about 90°C for four more hours. Sodium sulfate (ACS grade granular) was treated in the same manner as the potassium hydroxide with an additional heat treatment at 600°C for seven hours. New glassware or glassware which had been washed with soap and water was immersed in potassium dichromate-sulfuric acid solution for three or more hours and rinsed successively with glass distilled water, methanol or acetone, and hexane.

Deuterated PAHs were obtained from Merck, Sharp and Dohme Co. (Montreal, Canada) and used as received. Protiated PAHs were obtained from various suppliers (K & K Laboratories Co., Plainsview, N.Y.; Aldrich Co., Milwaukee, Wisconsin; Eastman Kodak Co., Rochester, N.Y.) and used as received.

Sediment samples and least cisco fish were obtained from the Southern Beaufort Sea. Oysters were purchased locally in Victoria, B.C.

Procedural Blanks

New batches of solvents, adsorbents and other reagents were checked for the presence of contaminants before they were used in an analysis. In addition, analyses of "blank" samples consisting of the internal standards only were performed regularly, i.e., after every five or so analyses.

Extraction of PAHs

The extraction technique was adapted from the method of Howard et al (1966, 1968). Marine organism (20 - 80 g) and sediment (10 - 100 g) samples were treated by stirring for one hour at reflux temperature in a solution of potassium hydroxide (7 g) in ethanol (150 ml) to which had been added an aliquot of a standard solution containing [$^2H_{10}$] phenanthrene, [$^2H_{10}$] pyrene, and [$^2H_{12}$] chrysene, [$^2H_{12}$] benz(a)anthracene, and [$^2H_{12}$] perylene. Then, water (150 ml) was added and the mixture stirred at reflux temperature for an additional thirty minutes. In the analysis of an organism sample, the digest was transferred to a separatory funnel by decanting the entire mixture and rinsing the flask with ethanol (2 x 25 ml) and iso-octane (2 x 100 ml). In the analysis of a sediment sample, after a brief setting period during which the mixture was kept just below the reflux temperature, the hot liquid phase was decanted from the solid phase into a separatory funnel. The solid phase was then washed vigorously with ethanol (2 to 4 x 25 ml) and iso-octane (2 x 100 ml) and the washings were combined with the liquid phase. The contents of the separatory funnel from either sediment or organism samples were vigorously shaken to facilitate partitioning of the hydrocarbons between the organic and aqueous phases. Following separation of the phases, in the analysis of an organism sample, a second partitioning of hydrocarbons from the aqueous phase into a fresh portion (200 ml) of iso-octane was carried out and the combined iso-octane extracts were washed with water (3 x 200 ml). In the analysis of a sediment sample, the fresh portion (200 ml) of iso-octane was used to wash sediment a final time prior to being used in the second partitioning of the hydrocarbons.

A few sediments and organism samples were also analysed without the additon of the labelled standards for comparison purposes. The amount of the standards used in an analysis was varied somewhat in proportion to the perceived amount of analyte PAHs in the sample. Nevertheless, the amount used of a given standard PAH was never less than 2 μg.

Liquid Chromatography

An iso-octane extract was concentrated on a rotary evaporator, applied in a narrow band to a column (4 cm i.d.) of anhydrous sodium sulfate (60 g) over Florisil (30g) and eluted with pentane; iso-octane (1:1) followed by benzene. The benzene eluant was concentrated for the quantification of selected PAHs.

Each newly prepared batch of Florisil was checked to assess its analytical suitability. Several aliquots of a standard mixture containing about 2 μg of each PAH standard compound and hexatriacontane ($\underline{n}C_{36}$) were chromatographed as described. In the case of one aliquot the entire effluent was evaporated, and the residue taken up in carbon disulfide and chromatographed using a gas chromatograph (GC) with a flame ionization detector (FID). The recovery of each PAH standard was calculated assuming 100% recovering of $\underline{n}C_{36}$. In the case of the other aliquots, the pentane:iso-octane and benzene fractions were evaporated and analyzed in a GC/FID separately. The optimum volume of each solvent was determined for the separation of the <u>n</u>-alkane and the recovery of the PAHs.

In the case of a few organism and sediment samples, a portion of the benzene fraction was evaporated to dryness and the residue was re-dissolved in a small amount of hexane for rechromatography using a Waters Associates ACL-401 high performance liquid chromatograph (HPLC) with ultraviolet detector, M-6000 pump, and U6K injector. A 50 cm x 0.26 cm i.d. amino-Sil-X-I column (Perkin Elmer Corp.) was used with hexane as eluant. Two fractions were collected, one corresponding to PAHs with retention times between and including those of phenanthrene and pyrene and the other corresponding to PAHs with retention times between and including those of chrysene and perylene.

Gas Chromatography/Mass Spectrometry

The PAHs were quantified using a Finnigan 9500/3300E GC/MS system with a glass jet separator and a Finnigan PROMIN (programmable multiple ion monitor) with four channels. Chromatographic separation was accomplished using a 1.5 m x 3.2 mm o.d. stainless steel column packed with 2% Dexsil 300 on Chromosorb W(HP), 80-100 mesh or a 3.0 m x 1.8 mm i.d. glass column packed with 2% Dexsil 300 on Chromosorb, W(AW), 100-120 mesh. The carrier gas was helium and its flow rate was 26 ml/min for the former and 20 ml/min for the latter. The temperature program was: 100°C for two minutes then 8°C/min to 300°C, and finally 300°C until elution of all compounds of interest. Ionization of compounds was effected at 15 ev. or 28 ev. The mass spectrometer was standardized daily for optimization of the 219^+ peak of FC - 43 (perfluorotributylamine). The PROMIN was adjusted for nominal masses of 178 and 188 (corresponding to $C_{14}H_{10}$ and $C_{14}{}^2H_{10}$, respectively), 202 and 212 ($C_{16}H_{10}$ and $C_{16}{}^2H_{10}$), 228 and 240 ($C_{18}H_{12}$ and $C_{18}{}^2H_{12}$), and 252 and 264 ($C_{20}H_{12}$ and $C_{20}{}^2H_{12}$). The precise mass settings used in a particular channel were determined using an appropriate mixture of deuterated and protiated standards. The instrument was adjusted to give mass peaks having tops as nearly flat as possible to minimize the effect of mass drift on the isotopic ratio. For these adjustments the mixture was normally introduced through the direct inlet probe. The calibration was subsequently checked from time to time by on column injection of the mixture.

The relative retention times of the protiated and deuterated forms of a given PAH were determined by monitoring their respective molecular ions using the PROMIN as the mixture of the two forms emerged from the GC column.

In the case of samples for which a portion of the PAH fraction was further purified by HPLC, the isotopic ratios, determined before and after the additional purification step, were compared. Mass spectra of the PAHs of interest were also obtained before and after the additional purification step. Obtaining the desired mass spectra was facilitated by visually monitoring, in succession, the molecular ion masses of the deuterated standards on the oscilloscope with the help of the mass marker and scanning when the passage of the standard was indicated.

Isotope Exchange

A typical S. Beaufort Sea sediment sample was soxhlet extracted with benzene-methanol (1:1) for 64 hours. A portion (31 g dry wt.) of the extracted sediment was then carried through the initial steps of the extraction procedure that included the addition of 3-4 μg to each of the five deuterated standards. The alcoholic base treatment was continued for 170 minutes, however. Subsamples, containing about 10 g each of sediment, were withdrawn at regular intervals. Each subsample was set aside until the sediment had settled out and worked up using a modified form of the procedure. After transfer of the supernatant fluid to a separatory funnel, each sediment sample was washed with ethanol (2 x 5 ml) and hexane (4 x 20 ml). The washings were combined with the supernatant fluid from the sediment. After the hydrocarbons had been partitioned, the phases separated, and the hydrocarbon phase washed with water (4 x 40 ml), dried over sodium sulfate and concentrated, measurement in the GC/MS was carried out directly without Florisil separation. Mass spectra were recorded for each of the four PAH peaks in the chromatograms of the extracts from the subsamples withdrawn at 59 minutes and 170 minutes. These mass spectra were compared to those of an untreated standard mixture. The mass spectral scans were one second in duration, 30 amu wide, and were recorded as the PAH peak crested.

A standard mixture of deuterated PAHs was chromatographed on Florisil and the isotopic distribution of the standards before and after chromatography was recorded for comparison as described immediately above.

RESULTS

Recovery of standard PAHs through the procedure was generally found to fall in the 60-90% range with the exception of phenanthrene. Recovery of phenanthrene was poorer presumably because of loss through evaporation and partial elution in the pentane-iso-octane chromatography fraction. When it was properly prepared, the Florisil was found not to chemisorb the PAHs used as standards to any appreciable extent for the amounts of the standards used in the analyses, that is, for amounts greater than about 7×10^{-8} g/g of Florisil. Perylene was observed to be the most readily chemisorbed standard by improperly deactivated Florisil.

The effect of a S. Beaufort sediment sample on five deuterated PAH standards under the conditions of hot base treatment is shown in Table 1. Aside from some minor variations, the relative amounts of the mass peaks remained unchanged from those of the untreated standard. Chromatography on properly prepared Florisil likewise did not cause a significant change in the general distribution of ions in the molecular ion clusters.

The shorter Dexsil 300 column did not possess sufficient resolving power for the unequivocal demonstration of a retention time difference between the labelled and unlabelled PAHs studied. The retention times of the protiated PAHs on the longer Dexsil 300 column, however, were clearly longer than those of their respective

Table 1 The Effect of Sediment and Hot Base on the Isotopic Composition of Perdeuterated PAHs

	PAHs											
	[$^2H_{10}$]Phenanthrene			[$^2H_{10}$]Pyrene			[$^2H_{12}$]Chrysene and [$^2H_{10}$]Benz[a]anthracene			[$^2H_{10}$]Perylene		
	Sampling time (min)			Sampling time (min)			Sampling time (min)			Sampling time (min)		
Nominal Masses	0	59	170	0	59	170	0	59	170	0	59	170
M + 2	0	2	3	2	3	3	2	5	3	2	7	7
M + 1	15	23	20	21	22	18	18	22	23	19	29	30
M[a]	100	100	100	100	100	100	100	100	100	100	100	100
M - 1	16	27	20	20	22	19	12	17	18	10	14	12
M - 2	11	16	12	22	23	21	9	9	10	4	8	8
M - 3	5	9	7	8	15	10	7	9	10	6	13	11
M - 4	23	26	20	23	26	24	31	30	33	23	24	35
M - 5	4	7	5	27[b]	33[b]	32[b]	4	4	8	3	13	12
M - 6	3	4	4	2	7	4	6	5	7	2	8	5

[a]The nominal masses of the molecular ions of the perdeuterated PAHs are 188 for [$^2H_{10}$]phenanthrene ($C_{14}{}^2H_{10}$), 212 for [$^2H_{10}$]pyrene ($C_{16}{}^2H_{10}$), 240 for [$^2H_{12}$]chrysene and [$^2H_{12}$]benz[a]anthracene ($C_{18}{}^2H_{12}$), and 264 for [$^2H_{12}$]perylene ($C_{20}{}^2H_{12}$). [b]There is a large background peak of nominal mass 207; these values are uncorrected.

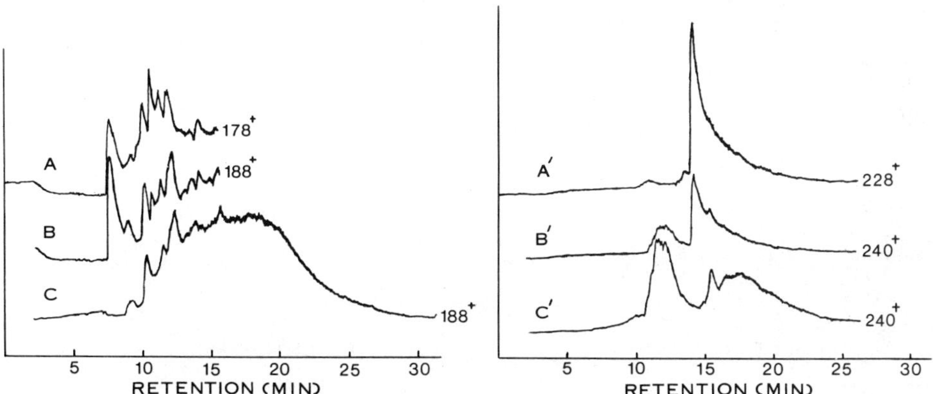

Figure 1. Mass fragmentograms, S. Beaufort Sea sediment. Dexsil 300 column, 1.5 ; IE, 15 ev. A. 178^+ ions, e.g. $C_{14}H_{10}$. B. 188^+ ions, e.g. $C_{14}{}^2H_{10}$, standard added. C. 188^+ ions, no standard added. A'. 228^+ ions, e.g. $C_{18}H_{12}$. B'. 240^+ ions, e.g. $C_{18}{}^2H_{12}$, standards added. C'. 240^+ ions e.g. $C_{18}{}^2H_{12}$, no standards added.

deuterated forms by about 8 seconds. No valley definition attributable to this retention time difference was observable in the PAH peaks in total ion chromatograms. However, since the labelled and unlabelled PAHs were in about equal amounts, the mass spectra of the leading edges of the peaks showed a preponderance of the molecular ions of deuterated forms and the mass spectra of the trailing edges showed a preponderance of the molecular ions of the protiated forms.

The mass fragmentograms of the 188^+ and 240^+ ions that were obtained in the analysis of a typical S. Beaufort Sea sediment sample, to which had been added the standard mixture of deuterated PAHs, are shown in B and B' of Figure 1, respectively. The mass fragmentograms of these same ions are also shown in C and C'. These latter two mass fragmentograms were obtained in the analysis of the same S. Beaufort Sea sample to which the standards had not been added. The difference in the paired fragmentograms, aside from scale, is the absence in C and C' of signals corresponding to the added deuterated PAHs. Comparison of fragmentograms B and C clearly shows that there is no contribution of extraneous ions to the signal from the $[^2H_{10}]$ phenanthrene. Comparison of fragmentograms B' and C' reveals that there is a small extraneous

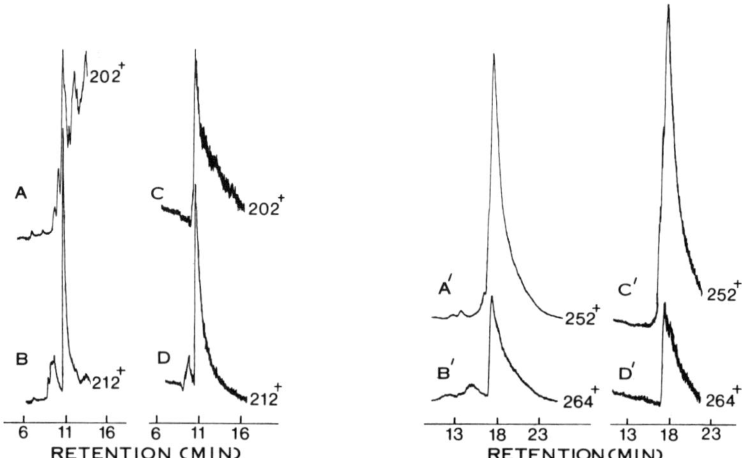

Figure 2. Mass fragmentograms, S. Beaufort Sea sediment. Dexsil 300 column, 1.5 m; IE, 15 ev. A and B. 202^+ and 212^+ ions, e.g. $C_{16}H_{10}$ and $C_{16}{}^2H_{10}$, resp., before separation on the HPLC. C and D. 202^+ and 212^+, resp. after separation on the HPLC. A' and B'. 252^+ and 264^+ ions, e.g. $C_{18}H_{12}$ and $C_{18}{}^2H_{12}$, resp., before separation on the HPLC. C' and D'. 252^+ and 264^+ ions, resp., after separation on the HPLC.

Table 2 The Effect of Additional Purification in a HPLC on Isotopic Ratios Obtained in Sediment Analyses

	Experiment 1			Experiment 2		
Nominal Masses Forming Ratio	R_B[a]	R_A[a]	rsd[b]	R_B[a]	R_A[a,c]	rsd[b]
178/188	0.81	0.97	0.13	1.12	1.81	0.33
202/212	0.38	0.41	0.05	0.68	0.62	0.07
228/240	0.78	0.75	0.03	1.28	0.83	0.30
252/264	4.00	3.70	0.06	5.09	5.10	0.00

[a]R_B is the ratio before use of the HPLC, R_A is the ratio after.

[b]rsd relative standard deviation (coefficient of variation).

[c]R_A was obtained for a different subsample of sediment than was used to obtain R_B and is corrected to compensate for the difference in weights of the two subsamples.

Table 3 The Effect of Additional Purification in a HPLC on Isotopic Ratios Obtained in Analysis of Some Commercial Oysters

	Experiment 1			Experiment 2		
Nominal Masses Forming Ratio	R_B[a]	R_A[a]	rsd[b]	R_B[a]	R_A[a]	rsd[b]
178/188	0.10	0.12	0.13	0.09	0.09	0.00
202/212	0.18	0.23	0.17	0.16	0.15	0.05
228/240	0.23	0.17	0.21	0.17	0.21	0.15
252/264	0.47	0.55	0.11	0.73	0.65	0.07

[a]R_B is the ratio before use of the HPLC, R_A is the ratio after.

[b]rsd relative standard deviation

contribution (ca. 10%) to the signal from $[^2H_{12}]$chrysene and $[^2H_{12}]$ benz(a)anthracene. Comparison of the mass fragmentograms of the 212^+ ions and the 252^+ ions, which are not shown here, reveals that the contribution of extraneous ions to the $[^2H_{10}]$ pyrene and $[^2H_{12}]$perylene signals is about 5%. Analyses of fish and oyster samples with and without the addition of the deuterated standards produced similar results.

The effect on the isotope ratios of the additional purification by chromatography on an amino-Sil-X-I column is demonstrated in Tables 2 and 3. Note that the purification does not change the ratio in a systematic fashion. Instead the change is random. The highest relative standard deviations occur in Experiment 2, Table 2, but in this experiment two different subsamples of a sediment sample were analysed. The ratios R_A were obtained from one subsample and the ratios R_B, from the other. Thus, the relative standard deviations include the between subsample variability in addition to the variation introduced by the additional purification and the GC/MS/PROMIN analysis.

The effect of additional purification on the appearance of the mass fragmentograms is illustrated in Figure 2. In the fragmentograms in Figure 2, which were obtained for a sediment extract, the most noticeable change is the diminution of 202^+ ions arising from compounds eluting more slowly than pyrene (cf A and C). A similar change with purification is observed in the mass fragmentograms (not shown) of the 178^+ ions. In contrast there is very little change in the mass fragmentograms of the 252^+ ions (cf A' and C') and the 228^+ ions. With regard to least cisco fish from the S. Beaufort Sea and commercial oysters, a somewhat different result was obtained. The mass fragmentograms for these organisms all show nearly complete absence of signals from ions from compounds having longer retention times than the analyte compounds after the additional column purification. Prior to the purification several peaks of longer retention time and larger size were present in each mass fragmentogram.

Some mass spectra obtained for a typical S. Beaufort Sea sediment sample are reproduced in Figures 3 and 4. These mass spectra are virtually featureless except for two prominent ion clusters, arising from the molecular ions of the protiated analyte PAHs and the deuterated standard PAHs. Conspicuously absent is the plethora of low mass fragments that was observed in several other mass spectra of unidentified compounds which were scanned in the same GC/MS run. The appearance of the molecular ion clusters of the protiated PAHs differ from those of the deuterated PAHs, since the M-2 peak (loss of 2 hydrogens) in the former appears as an M-4 peak (loss of 2 deuteriums) in the latter. The characteristic M-4 peaks of the deuterated PAHs permit their rapid visual identification as their mass spectra appear on the oscilloscope screen during the GC/MS run.

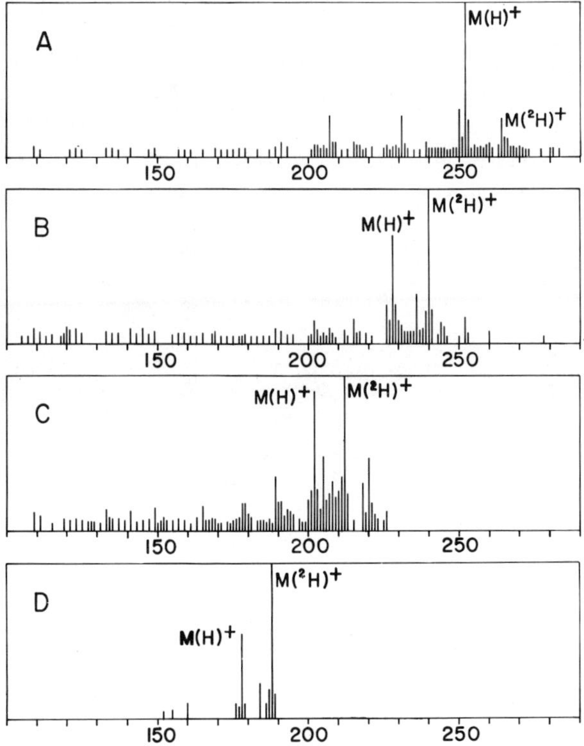

Figure 3. Mass scans obtained during a GC/MS separation of benzene fraction from the Florisil chromatography of a S. Beaufort Sea sediment extract. The 3.0 m glass Dexsil column and an IE of 28 ev. was used. Peaks less than 4% of base peaks are not shown.
A. M(H) = 252, M(^{2}H) = 264. B. M(H) = 228, M(^{2}H) = 240.
C. M(H) = 202, M(^{2}H) = 212. D. M(H) = 178, M(^{2}H) = 188.
This is the only scan for which the contribution of ions of a coeluting compound(s) was subtracted. The mass spectra obtained just before and after the passage of the standard and analyte peak were virtually identical and the subtraction could be done unequivocally.

Because the PAH concentrations in organism samples were typically an order or two less in magnitude than the PAH concentrations in the S. Beaufort Sea sediments, the mass spectra from organism samples are unsatisfactory for positive identification purposes. The signal to noise ratio is too low. Nevertheless, there is some circumstantial evidence to support a conjecture that the analyses of organism samples may be fairly accurate. In Table 4 the absolute responses obtained for the unlabelled analyte PAHs when the analyses were performed in presence and absence of the labelled PAHs are

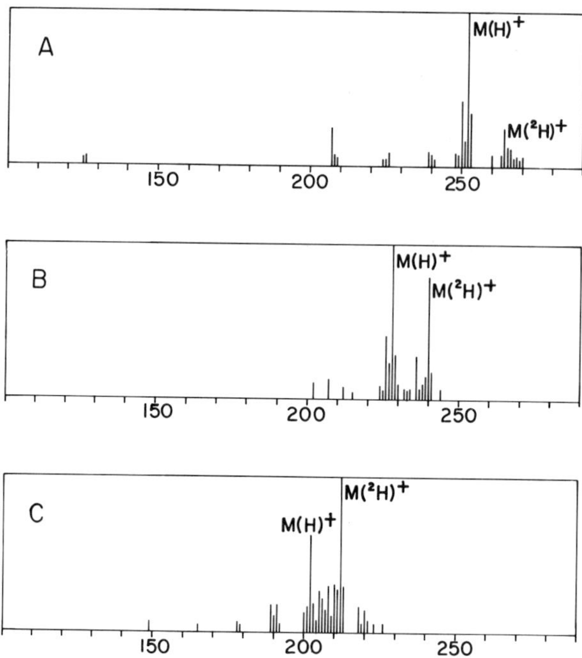

Figure 4. Mass scans obtained during a GC/MS separation of fractions from the HPLC purification of the benzene fraction in Figure 3. The GC/MS operating parameters were the same as used for the scans in Figure 3. Peaks less than 4% of the base peak are not shown.
A. M(H) = 252, M(^{2}H) = 264. Peaks at m/e 207, 208, 209 are present in the background. The small clusters at m/e 225, 240, and 266-270 arise from a coeluting compound(s).
B. M(H) = 228, M(^{2}H) = 240. The peak at m/e 207 is present in the background.
C. M(H) = 202, M(^{2}H) = 212. The clusters at m/e 190, 205, and 220 arise from some coeluting compound(s) whose retention time is slightly longer than that of unlabelled pyrene.

compared. In the absence of added internal standards, the analyte responses are seen to be 5 percent or less of the responses obtained in the presence of the added internal standards. Additional purification using an amino-Sil-X-I column results in complete absence of analyte signals when the internal standards are not added. On the other hand, when the standards are added to the sample, the isotopic ratio remains essentially constant (Table 3) and the absolute response for the analyte does not significantly diminish.

Table 4 Comparison of Analyte Signal Sizes Recorded in the Analysis of Commercial Oysters With and Without the Addition of the Deuterated PAH standards

	Peak heights[a]		
	Following Florisil Separation[b]		Following HPLC Separation[c]
Nominal Masses of Analytes	PAH Standards Present	PAH Standards Absent	PAH Standard Absent
178	231	9	0
202	875	18	0
228	650	5	0
252	n.d.[d]	3	0

[a]In arbitrary units. [b]Regular analytical procedure. [c]Separation in the HPLC was performed in addition to regular procedure. [d]n.d. ≡ no data.

Table 5 Mean Concentrations and Relative Standard Deviations found in the Analysis of S. Beaufort Sea Samples

	Isomeric PAH Groups (Nominal Masses)							
	$C_{14}H_{10}$(178)		$C_{14}H_{10}$(202)		$C_{18}H_{12}$(228)		$C_{20}H_{12}$(252)	
Sample	Conc. (μg/Kg)	rsd	Conc. (μg/Kg)	rsd	Conc. (μg/Kg)	rsd	Conc. (μg/Kg)	rsd
Least Cisco Fish[a]	2.3	0.59	2.4	0.70	4.3	0.19	2.5	0.21
Sediment[b]	93	0.24	49	0.13	157	0.65	348	0.31

[a]Four flesh samples from four different fish were analyzed. The mean concentrations were calculated on a wet weight basis.

[b]Five samples were analyzed, four in duplicate, one in triplicate. The mean concentrations were calculated on a dry weight basis.

Relative standard deviations calculated following the analysis of some S. Beaufort Sea samples are given in Table 5. The least cisco (Oregonus sardinella) fish were chipped from the ice off Pt. Atkinson on the Tuktoyaktuk Peninsula shortly after freeze-up in 1975. Since one flesh sample from each of four different fish was analysed, the relative standard deviation reflects any between sample variability as well as the precision of the method. Common variances and, subsequently, relative standard deviations were calculated for replicate analyses of five different S. Beaufort Sea sediment samples of similar mean concentrations (Lindgren and McElrath, 1969).

Table 6 Comparison of PAH Concentrations found in Near Shore, Benthic Sediments by Different Analytical Methods

		PAH Group Concentrations (μg/Kg[a])			
Publication	Collection Site	$C_{14}H_{10}$	$C_{16}H_{10}$	$C_{18}H_{12}$	$C_{20}H_{12}$
Wong, et al., 1976	S. Beaufort Sea	145 (28-298)[b]	68 (12-103)[b]	207 (30-570)[b]	661 (54-2120)[b]
Giger and Schaffner, 1978	Swiss Lake	370[c]	380[d]	n.d.[h]	410[e]
	Swiss River	230[c]	330[d]	n.d.[h]	370[e]
Hites et al., 1977	Buzzards Bay, Mass.	53[f]	120[d]	160	340
Wong et al., 1976	S. Beaufort Sea	n.d.[h]	41, 52[d]	n.d.[h]	191, 260[g]

[a]Dry weight Basis. [b]Range. [c]Sum of anthracene and phenanthrene. [d]Pyrene. [e]Sum of benzo[a]pyrene, benzo[e]pyrene and perylene. [f]Phenanthrene. [g]Sum of benzo[a]pyrene, perylene and benzo[ghi]-perylene. [h]n.d. ≡ no data.

DISCUSSION

Precise and accurate measurement is the first objective of analytical methodology. Good analysis provides the basis of good interpretation. Isotope dilution analysis techniques have the proven potential to achieve a high degree of accuracy and precision. In the environmental area, isotope dilution analysis by mass spectrometry has been applied with particular success to the analysis of trace elements, particularly lead, in seawater (Chow, 1968; Patterson and Settle, 1975). In the analysis of organic compounds in environmental samples the use of isotopically labelled compounds has been very limited (Wasik and Tsang, 1970; Dunn, 1976; Middleditch and Basile, 1977; Pancirov and Brown, 1977). To the authors' knowledge, no laboratory other than their own routinely analyzes environmental samples by GC/MS/MIM using labelled organic compounds as internal standards and carriers. Nevertheless, there is ample precedence from the pharmacological and biomedical fields for isotope dilution analysis by the GC/MS/MIM method. Some early work was published by Holland et al (1973) and Bertilsson et al (1972).

A key requirement in the successful application of the isotope dilution method is the complete equilibration of the analyte and its labelled form added as the standard and carrier. In the case of elemental analyses, the environmental matrix containing the analyte can generally be reduced to a homogeneous simple ionic medium. Ionic equilibration of an element and its artificial isotope as standard can be reasonably assumed. Unfortunately, in the analysis of organic compounds, complete molecular equilibration is not necessarily assured in the analytical procedure. In the case of the analysis of PAHs in sediments there does not seem to be any way of dissolving the rock and breaking down organic macromolecules that may shield PAH analytes by embracing them in inaccessible interior regions without destroying the PAH analytes. The problem of the unknown influence of macromolecules in the attainment of equilibrium also pertains to the analysis of tissue samples. This problem was not addressed in this study because of its complexity. Therefore, the method requires that the extraction and transfer steps be conducted as quantitatively as possible up to the stage at which the hydrocarbons are partitioned between the isooctane and aqueous ethanol phases and where the equilibration can be assumed to be complete.

Once isotopically labelled internal standards have become equilibrated, several potential advantages inherent in their use can be realized in the analysis of environmental samples. Their chief advantages lie in their close physical and chemical similarity to the compounds being quantified. Thus, analytical results are not necessarily sensitive to losses of analytes through processes such as evaporation, chemisorption, photochemical conversion, and oxidation. The sensitivity of the results to losses through chemical reactions depends on how intimately the isotope label is involved in the mechanism of the reactions. In the case of deuterium labelled PAHs, most of the common reactions that would be expected to take place during the analytical work-up or storage do not intimately involve the deuterium in their mechanisms. An important exception is deuterium-hydrogen exchange, which is acid catalysed and might take place on the surfaces of chromatography adsorbents or sediments. In the method herein described, acidic conditions were assiduously avoided. Moreover, neither the sediment nor the Florisil adsorbent was found to promote the exchange of hydrogen and deuterium. Because of the problem of hydrogen-deuterium exchange, however, deuterated PAHs are not the most ideal internal standards. Carbon-13 labelled PAHs would be preferred. The carbon is not available to isotopic exchange and, further, kinetic isotope effects for carbon are comparatively small for all chemical reactions (Laidler, 1965). Unfortunately carbon-13 labelled PAHs are not available commercially, except by special order.

Since Florisil is used as a chromatography adsorbent, it is important to consider the problem of chemisorption. Florisil is known to chemisorb certain compounds (Snyder, 1963) including, for example, perylene. Florisil prepared by the described method was found to give satisfactory recovery of perylene and the other PAHs used as standards for PAH to adsorbent ratios greater than 2 μg:30 g. This ratio was somewhat arbitrarily chosen and may not represent the minimum usable ratio. The problem of chemisorption is illustrated by the results in Table 4. In the analysis the individual amounts of unlabelled PAHs applied to the Florisil column was of the order of 0.1 μg. Thus, the minimum usable quantity of a PAH standard may be about 1 - 2 μg because the unprotected PAHs in the analysis performed without the addition of the internal standard solution appear to be largely lost. A cautionary note must be added, however, since in the case of the organisms mass spectral confirmation of the analytes was not obtained. Thus, it is conceivable that the internal standards may be acting as carriers for non-PAH compounds. Another consequence of the lack of mass spectral confirmation of the nature of the analytes is that the concentrations determined for organisms must be considered, in the strictest sense, as maximum possible values.

What are the analytes in the analytical method described in this paper, is an important question. Clearly, the analytes are not just the protiated forms of the deuterated PAHs chosen as internal standards. Certain geometrically related isomeric PAHs are not separable by the GC/MS system used. For example, perylene, benzo(a)pyrene and benzo(e)pyrene are inseparable, as are chrysene and benz(a)anthracene. Thus, the method in its present state of development cannot be described as isotope dilution analysis. In the broadest sense, the analytes in this method are simply those compounds which have similar retention times as the standards and molecular ions or fragment ions of the same nominal mass as the standards. In the case of the sediments, the mass spectra (Figures 3 and 4) obtained for the analytes have sufficient quality to indicate that the analytes are most likely non-alkylated, isomeric PAHs. A possible contribution of hetero(O,S,N)-atom containing polymeric aromatic compounds of the same nominal masses cannot be excluded, however. Further identification of the analytes would require techniques such as capillary column chromatography and high resolution mass spectrometry. In the case of the organism samples, there is no mass spectral evidence of the nature of the analytes, but it seems difficult to rationalize mass fragmentograms having <u>only</u> peaks with the retention times of the expected parent PAHs, as were obtained after HPLC separation of oyster extracts, in terms of some unknown compounds.

As mentioned above, certain isomeric PAHs are not separable in the GC/MS system used. Nevertheless, some isomeric PAHs are separable. In Figure 2, mass fragmentograms A' and C' show two small shoulders on the leading edge of the major peak. The longer Dexsil 300 GC column resolves these shoulders into two distinct peaks. The mass spectrum of the peak nearest the major peak indicates it consists of at least one PAH isomeric with perylene. The retention time characteristics of the two small peaks indicate they may be due to benzofluoranthenes. Retention time considerations indicate therefore that the main peak in the mass fragmentograms A' and C' may arise from benzo(a)pyrene, benzo(e)pyrene and perylene, but not from benzofluoranthenes. Similarly, the main peak in the mass fragmentograms A and C in Figure 2 may arise from pyrene, but not from fluoranthene, which elutes before it. The first peak in mass fragmentograms A in Figure 1 may arise from phenanthrene and anthracene. The later peaks appear to arise from fragments of higher molecular weight compounds, some of which seem to be alkalyated PAHs from appearance of their mass spectra. The major peak in mass fragmentogram A' in Figure 1 may arise from chrysene and benz(a)anthracene. In this case the other peaks were considered to be too small to warrant investigation.

Capillary column GC/MS systems would appear to give the most promise for the resolution of the analyte mixtures. Capillary columns have been recently put to good use in the analysis of PAHs in sediments (Giger and Schaffner, 1978; Laflamme and Hites, 1978). Some newly developed stationary phases for packed columns (Janini et al, 1976; Snowdon and Peake, 1978) also show promise of resolving difficult to separate PAHs. A combination of the newer column technologies and recent mass spectral methods of verifying structural assignments of isomeric PAHs, such as mixed charge exchange chemical ionization mass spectrometry (Lee and Hites, 1976) should provide the necessary advance to allow true isotope dilution analyses of PAHs.

An important feature of isotope dilution analyses is the inseparability of the labelled and unlabelled forms of a given compound or element. Thus, the labelled form acts as a carrier for the unlabelled form to assure a common fate for both. Unfortunately, there is ample evidence of the separation of deuterated compounds from their protiated forms by gas chromatography (Van Hook, 1969; Wasik and Tsang, 1970; Jancso and Van Hook, 1974; Middleditch and Basile, 1976) and liquid chromatography (Taneka and Thornton, 1976; Heck et al, 1977). In our study it was clear that the GC retention times of the perdeuterated PAHs were slightly less than those of their unlabelled forms. This result is consistent with the findings of other workers (e.g. Van Hook, 1969). Capillary CG columns would very likely have the necessary resolving power to completely separate fully deuterated and protiated forms of PAHs. The undesirability of this separation, in terms of the isotope dilution aspect of the analysis, may be offset by the possibility of carrying out analyses in GC/FID systems.

Although the separation may be undesirable, its effect on the analytical results should be measurable using appropriate mixtures of labelled and unlabelled PAHs. The separation, if considered undesirable, might be significantly reduced by using carbon-13 labelled PAHs or less heavily labelled deuterated PAHs. Alternatively, liquid crystal GC phases (Janini, 1976), which separate isomeric PAHs efficiently in short packed columns, might be used.

While an analytical method is being developed, the simplest scheme, consistent with desirable precision, accuracy and suitability to the problems at hand, is seldom arrived at initially. In the development of the method described in this paper, it was felt that an additional separation using a HPLC would be required to obtain samples free enough from interfering compounds to be satisfactorily analyzed in the GC/MS system. Since the isotope ratios (Table 2 and 3) remain essentially unchanged by the additional purification, little analytical advantage is gained by the additional step. In the case of the sediment samples, comparison of the mass spectra (Figures 3 and 4) of the analyte-standard mixture before and after using the HPLC shows that the additional purification is redundant rather than ineffective. In fact, a significant purification of the samples is achieved by use of the HPLC as seen in the simplification of some of the mass fragmentograms (cf. A and C in Figure 2) and the reduction of extraneous ions in the analyte mass spectra (Figures 3 and 4). Although our samples may be typical of all marine sediments and organisms, the possibility remains therefore that other samples with different interfering compounds and/or smaller concentrations may greatly benefit from use of a HPLC or some other purification technique. Accordingly, each sample must be approached without prejudice and the method adapted as required.

Considerations of interfering compounds apply more stringently to the analytes than the deuterated standards. In the case of the standards, more can be added to the sample to reduce the amount of interference to a suitable level. Since the deuterated PAHs used as standards contain a small amount of their unlabelled forms, however, addition of more standard to a sample increases the concentration of unlabelled PAH in the sample. Aside from other considerations, there is a practical limit reached when the concentration of added unlabelled PAH in the sample approaches or supersedes that of the analytes.

Since this method has not yet been tested in intercalibration experiments with other laboratories, some other recent analytical results obtained in the analysis of near shore, benthic sediments are given in Table 6 for comparison purposes, although the agreement may be somewhat fortuitous. The last listed results in Table 6 were obtained using the method of Giger and Blumer (1974) to analyze two S. Beaufort Sea sediments.

SUMMARY AND CONCLUSIONS

Perdeuterated PAHs are useful internal standards and carriers for the analysis of marine sediments and organisms. Their use makes for a "rugged" analytical method that is relatively insensitive to an analyst's capability. The method is not lengthy and, although a GC/MS system is required, quite satisfactory work can be done without a computer. The relatively inexpensive PROMIN is necessary, however. The slight difference in retention time of deuterated and protiated PAHs makes the measurement of the isotope ratio from mass spectra impractical.

Although the method is still undergoing development, in its present form quantification of PAHs can be obtained on a group basis. Thus, the method is most ideally suited to survey work where the general level of PAHs in given samples is sought.

ACKNOWLEDGEMENTS

Portions of this work were done under contract serial nos. DSS OST4-0178 (Chemex Labs Ltd., North Vancouver, B.C.) and DSS OSZ5-0066 (Seakem Oceanography Ltd., Sidney, B.C.).

REFERENCES

Bertilsson, L., A.J. Atkinson, Jr., J.R. Althaus, A. Harfast, J.E. Lindgren, and B. Holmstedt. 1972. Quantitative Determination of 5-Hydroxyindole-3-Acetic Acid in Cerebro-spinal Fluid by Gas Chromatography-Mass Spectrometry. Anal. Chem. 44: 1434-1438.

Blumer, M., and W.W. Youngblood. 1975. Polycyclic Aromatic Hydrocarbon in Soils and Recent Sediments. Science 188: 53-55.

Chow, T.J. 1968. Isotope Analysis of Seawater by Mass Spectrometry, Part 1, J. Water Poll. Control Fed. 40: 390-411.

Dunn, B.P. 1976. Techniques for Determination of Benzo(a)pyrene in Marine Organisms and Sediments. Environ. Sci. Technol. 10: 1018-1021.

Farrington, J.W., N.M. Frew, P.M. Gschwend, and B.W. Tripp. 1977. Hydrocarbons in Cores of Northwestern Atlantic Coastal and Continental Margin Sediments. Estuar. Coast. Mar. Sci. 5: 793-808.

Giger, W., and M. Blumer. 1974. Polycyclic Aromatic Hydrocarbons in the Environment: Isolation and Characterization by Chromatography, Visible, Ultraviolet, and Mass Spectrometry. Anal. Chem. 46: 1663-1671.

Giger, W., and C. Schaffner. 1978. Determination of Polycyclic Aromatic Hydrocarbons in the Environment by Glass Capillary Gas Chromatography. Anal. Chem. 50: 243-249.

Heck, H.D., R.L. Simon, and M. Anbar. 1977. Isotopic Fractionation in Thin-Layer Chromatography. J. Chromatog. 133: 281-290.

Hites, R.A., R.L. Laflamme, J.W. Farrington. 1977. Sedimentary Polycyclic Aromatic Hydrocarbons: The Historical Record. Science. 198: 829-831.

Holland, J.F., C.C. Sweeley, R.E. Thrush, R.E. Teets, and M.A. Bieber. 1973. On-Line Computer Controlled Multiple Ion Detection in Combined Gas Chromatography-Mass Spectrometry. Anal. Chem. 45: 308-314.

Howard, J.W., T. Fazio, R.W. White, and B.A. Klimeck. 1968. Extraction and Estimation of Polycyclic Aromatic Hydrocarbons in Total Diet Composites. J. Ass. Offic. Anal. Chem., 51: 122-129.

Howard, J.W., R.T. Teague, Jr., R.W. White, and B.E. Fry, Jr. 1966. Extraction and Estimation of Polycyclic Aromatic Hydrocarbons in Smoked Foods. I. General Method. J. Ass. Offi. Anal. Chem. 49: 595-617.

Jancso, G., and W.A. Van Hook. 1974. Condensed Phase Isotope Effects (Especially Vapor Pressure Isotope Effects). Chem. Rev. 74: 689-750.

Janini, G.M., G.M. Muschik, J.A. Schroer, and W.L. Zielinski, Jr. 1976. Gas-Liquid Chromatographic Evaluation and Gas Chromatography/Mass Spectrometric Application of New High Temperature Liquid Crystal Stationary Phases for Polycyclic Aromatic Hydrocarbon Separations. Anal. Chem. 48: 1879-1883.

Laflamme, R.E., and R.A. Hites. 1978. The Global Distribution of Polycyclic Aromatic Hydrocarbons in Recent Sediments. Geochim. Cosmochim. Acta. 42: 289-303.

Lindgren, B.W., and G.W. Elrath. 1969. "Introduction to Probability and Statistics", p.179. Collier-Macmillan Ltd., London.

Laidler, K.J. 1965. "Chemical Kinetics", pp.97-98. McGraw-Hill Book Co., Toronto.

Lee, M.L., and R.A. Hites. 1976. Mixed Charge Exchange-Chemical Ionization and Mass Spectrometry of Polycyclic Aromatic Hydrocarbons. J. Amer. Chem. Soc. 99: 2008.

Lee, R.F. 1977. Accumulation and Turnover of Petroleum Hydrocarbons in Marine Organisms, Chapter 6. In: D.A. Wolfe (ed.) "Fate and Effects of Petroleum Hydrocarbons in Marine Ecosystems and Organisms". Pergamon Press, New York.

Middleditch, B.S., and B. Basile. 1976. Deuterated Analogs as Internal Standards for the Quantitation of Environmental Alkanes by Gas Chromatography. Anal Lett. 9: 1031-1034.

Pancirov, R.J., and R.A. Brown. 1977. Polynuclear Aromatic Hydrocarbons in Marine Tissues. Environ. Sci. Technol., 11: 989-992.

Patterson, C.C., and D.M. Settle. 1976. The Reduction of Orders of Magnitude Errors in Lead Analyses of Biological Materials and Natural Waters by Evaluating and Controlling the Extent and Sources of Industrial Lead Contamination Introduced During Sample Collecting, Handling, and Analysis, pp. 321-351. In: "Proceedings of the 7th I.M.R. Symposium." NBS Spec. Publ. 422. U.S. Gov. Printing Office, Washington, D.C., U.S.A. 20402.

Payne, J.F. 1977. Mixed Function Oxidases in Marine Organisms in Relation to Petroleum Hydrocarbon Metabolism and Detection. Mar. Poll. Bull. 8: 112-116.

Risebrough, R.W., B.W. De Lappe, and T.T. Schmidt. 1976. Bioaccumulation Factors of Chlorinated Hydrocarbons Between Mussels and Seawater. Mar. Poll. Bull. 7: 225-228.

Snowdon, L.R., and E. Peake. 1978. Gas Chromatography of High Molecular Weight Hydrocarbons with an Inorganic Salt Eutectic Column. Anal. Chem. 50: 379-381.

Snyder, L.R. 1963. Linear Elution Adsorption Chromatography. VI. Deactivated Florisil as Adsorbent. J. Chromatog. 12: 488-509.

Stich, H.F., A.B. Acton, and C.R. Forrester. 1976. Fish Tumors and Sublethal Effects of Pollutants. J. Fish Res. Board Can. 33: 1993-2001.

Tanaka, N., and E.R. Thornton. 1976. Isotope Effects in Hydrophobic Binding Measured by High-Pressure Liquid Chromatography. J. Amer. Chem. Soc. 98: 1617-1619.

Vandermeulen, J.H., and W.R. Penrose. 1978. Absence of Aryl Hydrocarbon Hydroxylase (AHH) in Three Marine Bivalves. J. Fish. Res. Board Can. 35: 643-647.

Van Hook, W.A. 1969. Isotope Separation by Gas Chromatography, pp. 99-118. In: "Isotope Effects in Chemical Processes". Adv. Chem. Ser. No. 89, American Chemical Society, Washington, D.C.

Wasik, S.P., and W. Tsang. 1970. Determination of Trace Amounts of Contaminants in Water by Isotope Dilution Gas Chromatography. Anal. Chem. 42: 1649-1651.

Wong, C.S., W.J. Cretney, R.W. Macdonald, and P. Christensen. 1976. Hydrocarbon Levels in the Marine Environment of the Southern Beaufort Sea. In: "Beaufort Sea Technical Report No.38". Beaufort Sea Project, Institute of Ocean Sciences, Patricia Bay, P.O. Box 6000, Sidney, B.C., Canada V8L 4B2. Unpublished manuscript.

Youngblood, W.W., and M. Blumer. 1975. Polycyclic Aromatic Hydrocarbons in the Environment: Homologous Series in Soils and Recent Marine Sediments. Geochim. Cosmochim. Acta. 39: 1303-1314.

REPRODUCTIVE SUCCESS OF HERRING GULLS AS AN INDICATOR OF GREAT LAKES WATER QUALITY

D.B. Peakall, G.A. Fox, A.P. Gilman, D.J. Hallett and R.J. Norstrom

National Wildlife Research Centre, Canadian Wildlife Service, Ottawa, Ontario, K1A 0E7

"Of all clean birds ye shall eat. But these are they of which ye shall not eat: the eagle, the ossifrage, and the osprey, ...and the pelican, and the gier eagle, and the cormorant."

Deuteronomy 14:11-17

This symposium focuses on analytical chemistry whereas this paper discusses the effects of environmental contaminants on a single species of bird, the Herring Gull (*Larus argentatus*). This can be defended on the grounds that the reason for doing analytical work on the aquatic environment is to further our understanding of the effects chemicals exert on living systems. In a highly technological society man-made compounds are going to escape into the environment and some of them, or their metabolites, are going to persist for appreciable lengths of time. Since we cannot ban all such compounds the challenge for environmental toxicologists is to find out which of these compounds are really causing problems and which are not. The end point of a surveillance programme is not a list of levels of pollutants in various parts of the biota, but a knowledge of the effects that those pollutants are causing in the environment. The problem is very complex because of the large number of species and the large number of pollutants involved.

The information obtained from sampling different parts of an ecosystem are quite different. A comparison of the information obtained from analysis of water, molluscs, and gulls is shown in Table 1. Obviously, measurements of chemicals in many parts of the biota are necessary for a complete picture, but if the number of analyses are restricted, then an overall picture can be obtained from the gull. The gull is so good at concentrating compounds that

Table 1

Comparison of Advantages and Limitations of Analysis of Pollutants in Water, Molluscs and Gulls

Parameter under consideration	Water	Molluscs	
Analytical	Analytical procedures more rapid than for biological samples but levels frequently too low to measure even for pollutants of known importance.	Analytical procedures more complex th higher and therefore more readily de water.	
Time	Covers a single moment in time.	Pollutant level in organism gives integration of pollution over a moderate period of time.	Pollutant level in organism gives integration of pollution over a considerable period of time.
Area	Covers a single water source.	Information specific to a small area as indicator is non-migratory.	Information is integrated over a wide area.
Food-chain	Base.	First step from water.	Head of complex and variable set of food-webs.
Sampling Program	Large number of stations, frequent sampling.	Large number of stations, infrequent sampling.	Small number of stations, infrequent sampling.

contaminants present at only trace levels in other parts of the environment are readily detected. The directive given at the beginning of this article thousands of years ago, is certainly sound advice to anyone living around the Great Lakes today. The 1977 International Joint Commission on Water Quality Objectives shows that the mean limit of detection of ten laboratories currently carrying out determination of all organochlorine and phthalate esters in water in the Great Lakes region was 4 - 6 times greater than the recommended levels.

The validity of the indicator species concept is based on the premise that while nature is incredibly diverse, the underlying biochemistry is remarkably similar. The major energy mechanism of both the gnat and the whale is dependent on the conversion of adenosine triphosphate to adenosine diphosphate. Nevertheless we need to realize the limitations of the concept. The wide range of species response in the case of pesticide-induced eggshell thinning (Peakall, 1975) is a warning not to push the concept too far.

In our work on the Great Lakes we have used the Herring Gull as an indicator species. Only a summary of the reasons for this choice is given here as they have been discussed in detail elsewhere (Gilman et al, 1978a). The Herring Gull eats a wide variety of food and is thus an indicator of overall pollution in the area. The adult Herring Gull is essentially resident within the Great Lakes Basin (Moore, 1976), although there is some movement from lake to lake (Gilman et al, 1977). The Herring Gull nests colonially and thus the entire breeding population of large areas can be counted. Colonial birds are probably the only type of organism for which this information can be readily obtained. Lastly the Herring Gull is widely distributed through the holarctic enabling direct comparisons to be made between work carried out in North America and Europe. Thus the Herring Gull meets the criteria set out by Moore (1966) for a suitable indicator species.

The Canadian Wildlife Service work on the Great Lakes started when poor reproduction and abnormal young were noted in the early 1970's (Gilbertson, 1974). Although attention has been focused on the well-known persistent organochlorines ever since Hickey's work on Lake Michigan more than a decade ago (Hickey et al, 1966, Keith 1966) a wide variety of other compounds has been identified in gulls and their eggs. Our current count is some four hundred man-made compounds in gull material from Lake Ontario. In other papers in this symposium details will be given on mirex and its metabolites, chlorobenzenes, and polynuclear aromatic hydrocarbons. At present the only pollutants for which data is available for a wide geographic area are the common organochlorines. A comparison of levels in gull eggs in various parts of the world is given in Table 2. A trans-Canada comparison for 1974-1975 showed that levels of DDE in eggs from coastal areas were 5 - 10% of Great Lake levels and the levels of PCB's only 1 - 5% (Vermeer and Peakall, 1977).

Table 2 Levels of DDE and PCB's in Herring Gull Eggs

	Year	Residue level (ppm wet weight) DDE	PCB's	Reference
Norway				
(10 colonies)	1969	1.5	1.0	Bjerk and Holt (1971)
	1972	1.6	5.4	
Denmark				
(Baltic colonies)	1972	38	62	Jorgensen and Kraul
(North Sea)	1972	0.2	2	(1974)
Ontario				
Lake Ontario	1975	22	134	Gilman et al (1977)
Lake Superior	1975	21	69	
California*	1973	12	5	Risebrough (priv. comm.)

*Eggs from closely-related glaucous-winged gulls

Most areas are cleaner than the Great Lakes of North America, the only area reporting organochlorine levels comparable to the Great Lakes is that from the Baltic, although analysis of tissue levels of gulls from the sea of Japan (Fujiwara, 1975) indicates that levels of organochlorines are half those found in the Great Lakes.

As part of the studies under the Canada-U.S. Great Lakes Water Quality Agreement, the Canadian Wildlife Service made detailed reproductive studies at four colonies in 1975 (one each on Lakes Superior, Huron, Erie and Ontario) and the productivity (chick survival to 21 days) of the Lake Ontario colony was only a tenth of normal. This confirmed the poor reproduction on Lake Ontario colonies noted previously (Gilbertson, 1974). However, since then, productivity has increased rapidly rising from 0.1 young/pair in 1975 to 0.4 young/pair in 1976 and 1.0 young/pair in 1977. While this altering pattern of reproductive success is encouraging, the shifting base-line hampers research into the causes of the poor reproduction noted in the early 1970's.

Visually, a failing colony, such as those in Lake Ontario in the early to mid-1970's, is quite different from a successful colony. The first abnormality noted occurs while the boat is still several hundreds of meters offshore. The breeding adults leave the island, they wheel around, many settle on the lake well away from the island, a few wheel overhead screaming. On a normal colony the breeding adults remain until one is a few meters away, and some individuals attack the visitors. The use of telemetering eggs has shown that nest attentiveness is lower on the Lake Ontario colonies compared to successful colonies in New Brunswick and that unsuccessful nests are characterized by greater temperature variation and much longer absences of adults from the nest (Fox et al., 1978). As the study continues it is found that the rate of disappearance of eggs is abnormally high in Lake Ontario colonies (nearly 40%) compared to 2 - 10% in other colonies studied, (Gilman et al, 1977). In addition to these losses the percentage of embryos that fail is much higher (35% compared to 6 - 10% in the upper lakes).

The high rate of egg loss, together with the abnormally low nest defence and nest attentiveness, suggests a behavioural change. Embryonic mortality could be caused by toxicants within the egg or by inadequate care, or by a combination of these two factors. In order to separate these two factors, the intrinsic and extrinsic, an egg swap experiment was devised. In theory this is a simple experiment. There are two colonies, one "clean" and one "dirty". Eggs are moved from the clean colony and placed in the dirty colony and vice versa. There are four sets of conditions: clean adults incubating clean eggs, clean adults incubating dirty eggs, dirty adults incubating clean eggs and dirty adults incubating dirty eggs. The experimental design and expected resultant information is shown in Table 3.

In practice all sorts of complications occur. There are, obviously, the effects of transportation on the viability of the eggs. Thus it is necessary to transport control eggs and return them to the colony from which they were removed so that they

Table 3 Overall Design of Egg Exchange Experiments

Adult	Egg	Expected result
Clean	Clean	Normal reproduction
Clean	Dirty	Intrinsic factors only
Dirty	Clean	Extrinsic factors only
Dirty	Dirty	Both intrinsic and extrinsic factors
Incubation	Clean	Normal reproduction
Incubation	Dirty	Intrinsic factors only

correspond to the eggs that were moved from colony to colony. Then nesting occurs at different times in different areas. In all, three egg exchanges, each with somewhat different methodology, have been run. The results are summarized in Table 4. While caution is needed in the interpretaion of the results, it appears that; both intrinsic and extrinsic factors were depressing reproduction in 1975; intrinsic factors were more important than extrinsic in 1976; neither was a factor in 1977. These variable results are consistent with the altering reproductive success pattern over that time period.

The egg exchange experiments examine both intrinsic and extrinsic factors. To examine the effects of the intrinsic factor alone, organochlorine contaminants were extracted from Lake Ontario Herring Gull eggs collected in 1975 and injected into relatively uncontaminated gull eggs in a colony in New Brunswick. Synthetic mixtures of PCB's, DDE, mirex, photomirex and hexachlorobenzene were injected in a similar fashion. All eggs were incubated by their natural parents. Studies showed that the uptake by the embryo of organochlorines injected into the yolk was the same as the uptake of organochlorines when these came from the body burden of the female. No increase in embryonic or chick mortality was observed when compared to the injected controls (Gilman et al, 1978b).

Table 4 Summary of Herring Gull Egg Exchange Experiments

Adult	/ Egg	1975 n	1975 % Hatched	1976 n	1976 % Hatched	1977 n	1977 % Hatched
Clean	/Clean	85	86	-	-	93	68
Clean	/Dirty	41	10	-	-	48	63
Dirty	/Clean	41	7	46	48	58	81
Dirty	/Dirty	49	2	86	37	95	70
Artificial	/Clean	109	60	35	57	-	-
Artificial	/Dirty	86	37	43	23	-	-
Conclusion		Marked intrinsic and extrinsic factors		Marked intrinsic, some extrinsic factors		No effects seen	

Methodology has varied from year-to-year "Clean" colonies were in Prince Edward Island in 1975, and in Lake Huron in 1976-77. The Herring Gull nests two-three weeks earlier on Lake Ontario than in Prince Edward Island. In 1975 the first clutch of eggs in the Lake Ontario colonies were removed and the exchange was between re-laid Lake Ontario eggs and first clutches in Prince Edward Island. The synchronization is better between Lake Ontario and Lake Huron but still involved the use of early eggs from Huron and late eggs from Ontario in 1976 and some storage of eggs in 1977.

The possibility that effects of ingested toxic chemicals may be manifest before the egg is laid is not precluded by these experiments. Such factors as alterations to the genetic material of the embryo, yolk quality and eggshell structure are excluded from consideration in this study.

Analysis of organochlorines have been carried out on Herring Gull eggs collected from the Great Lakes since 1973. The levels are tabulated in Table 5. The changes with time are not completely consistent except from mirex which decreased at all sites. The colony on Muggs Island shows a steady decrease of all residues over the period 1974 to 1977 whereas the trends from the colony in Brother's Island are not clear-cut.

While effects such as increased incidence of abnormal young (Gilbertson et al, 1976) and poor reproductivity (Gilman et al,1977) correlate with high pollutant load, a firm cause and effect relationship remains to be proven.

REFERENCES

Bjerk, J.E., and G. Holt. 1971. Residues of DDE and PCB in eggs from the Herring Gull (*Larus aregentatus*) and the Common Gull (*L. canus*) in Norway. Acta Vet. Scand. 12: 429-441.

Fox, F.A., A.P. Gilman, D.B. Peakall, and F.W. Anderka, 1978. Behavioral abnormalities of Great Lakes Herring Gulls. J. Wildl. Manage. 42(3):477-483, 1978.

Fujiwara, K. 1975. Environmental and food contamination with PCB's in Japan. Sci. Total Environ. 4, 219-247.

Gilbertson, M. 1974. Pollutants in the breeding Herring Gulls in the lower Great Lakes. Canadian Field-Naturalist 88, 273-280.

Gilbertson, M.R., D. Morris, and R.A. Hunter, 1976. Abnormal chicks and PCB residue levels in eggs of colonial birds on the lower Great Lakes. Auk 93: 434-442.

Gilman, A.P., D.B. Peakall, F.A. Fox, D.J. Hallett, and R.J. Norstrom. 1978a. The Herring Gull as a monitor of Great Lakes contamination. Intern. Symp. Patholbiology Environ. Pollut. Storrs, Connecticut. June 1977.

Gilman, A.P., G.A. Fox, D.B. Peakall, S.M. Teeple, T.R. Carroll and G.T. Haymes, 1977. Reproductive parameters and contaminant levels of Great Lakes Herring Gulls. J. Wildl. Manage. 41, 458-468.

Gilman, A.P., D.J. Hallett, G.A. Fox, L.J. Allan, W.J. Learning, and D.B. Peakall, 1978b. Effects of injecting organochlorines into naturally incubated Herring Gull eggs. J. Wildl. Manage. 42(3): 484-493, 1978.

Hickey, J., J.A. Keith, and F.B. Coon, 1966. An exploration of Pesticides in a Lake Michigan Ecosystem. J. Appl. Ecol. 3 (Suppl.): 141-154.

Jørgensen, O.H., and I. Kraul, 1974. Eggshell parameters and residues of PCB and DDE in eggs from Danish Herring Gulls, *Larus a. argentatus.* Ornis Scand. 5, 173-179.

Keith, J.A. 1966. Reproduction in a population of Herring Gulls (*Larus argentatus*) contaminated by DDT. J. Appl. Ecol. 3 (Suppl.): 57-70.

Moore, F.R., 1976. The dynamics of seasonal distribution of Great Lakes herring gulls. Bird banding 47: 141-159.

Moore, N.W., 1966. A pesticide monitoring system with special reference to the selection of indicator species. J. Appl. Ecol. 3 (Suppl.): 261-269.

Peakall, D.B. 1975. Physiological effects of chlorinated hydrocarbons on avian species, pp. 343-360 in R. Hague and V.H. Freed eds. Environmental Dynamics of Pesticides. Plenum Press, N.Y.

Vermeer, K., and D.B. Peakall. 1977. Environmental Contaminants and the Future of Fish-eating Birds in Canada. In Canada's threatened species and habitats. Ed. Musquin, T. and C. Suchal. Ottawa. pp.88-95.

PESTICIDE MONITORING IN THE PRAIRIES OF WESTERN CANADA

Wm. D. Gummer

Water Quality Branch, Fisheries & Environment Canada

1901 Victoria Avenue, Regina, Saskatchewan, S4P 3R4

ABSTRACT

Pesticide monitoring programs conducted by the Water Quality Branch of the Department of Fisheries and the Environment during the period 1971 to 1977 revealed a widespread distribution of 2,4-D; 2,4,5-T; γ-BHC (lindane); and α-BHC as well as a more limited distribution of 2,4-DP (dichloroprop); aldrin; and β-Endosulfan in surface waters of western Canada.

Atmospheric transportation and deposition are the mechanisms believed responsible for the wide distribution of lindane and α-BHC in western Canada. It is speculated that isomerization of lindane to the α-BHC isomer accounts for the abundance of α-BHC. Concentrations of both lindane and α-BHC at times exceeded 0.01 μg/l. The herbicide 2,4-D was observed to be prevalent in the agricultural areas at concentrations above 0.01 μg/l and as high as 4.33 μg/l. In addition to agriculture, industries and municipalities were found to contribute pesticides to the aquatic environment.

INTRODUCTION

Environmental Concern

Public awareness and concern about the quality of the environment in which man coexists with all other forms of life have increased considerably during the last 15 years. Pesticide-induced environmental degradation constitutes part of this concern, the magnitude of which was uncovered in the early sixties by such authors as Carson (1962) and Rudd (1964). Edwards (1973) presents information on the amounts of residues in the environment and lists relevant reviews and symposia on the persistence of pesticides in the environment. The Canadian Wildlife Service (1971) confirms that Canada has not escaped environmental contamination from pesticides, and that nearly all the samples of Canadian wildlife (mammals, fish, fish food, marine invertebrates, migratory and nonmigratory birds) analyzed during a 5-year study contained pesticide residues.

The concentrations of pesticides in water have become a major concern because water is a basic sustenance of life. The health of all life forms can thus be affected by pesticides in water. Because of this and because of the requirement for water of good quality for most human, industrial, agricultural and recreational activities, the Water Quality Branch, Inland Waters Directorate, Environmental Management Service, Fisheries and Environment Canada, conducts a program to measure and assess water quality in Canada.

Environmental Sources and Pathways

Edwards (1973) states that direct and indirect applications of pesticides, agricultural runoff, and industrial discharges are the principal sources of pesticide residues to surface waters. Although it is extremely difficult to quantify, the domestic and municipal use of pesticides undoubtedly contributes significantly to surface water contamination. It is estimated that in 1970-1971, urban home and garden use accounted for about 24% of the pesticide market in Alberta (Alberta Environment Conservation Authority, 1973); this probably typifies domestic use and National scale as well. The unregulated and likely indiscriminant domestic use of pesticides poses a threat to the health of man and the environment.

Pesticides may directly enter the aquatic environment. The direct application of the chemicals DDT and methoxychlor to surface waters has occurred in the Prairies to control mosquito and black fly infestations. Fredeen _et al._ (1970) reported that the Saskatchewan River (North and South) was treated by 3436 kg DDT for the control of black flies. The herbicide 2,4-D (butoxyethylester) is registered in Canada for the control of aquatic weeds, although the extent of aquatic use in western Canada is not known.

Some industrial sources of pesticides in western Canada are the pulp and paper industry, pesticide packaging plants, pesticide manufacturing plants, the food industry, and the seed dressing industry.

Agriculture is the largest single user of pesticides which are essential for high agricultural productivity and for the control of insect-borne diseases. The high agricultural activity in western Canada has resulted in the use of a variety of pesticides which have the potential to seriously degrade the natural environment.

When pesticides are applied in agricultural areas, small amounts drift into the atmosphere and are deposited in remote areas, contaminating both the hydrosphere and the biosphere. Water transport of pesticides is chiefly through surface runoff, which may cause contamination of lower-lying areas and of water bodies (Gerakis and Sticas, 1974).

The dissipation and mobility of pesticides in the environment are contingent upon numerous factors, many of which are illustrated in Figure 1.

Invariably, the aquatic environment is the ultimate recipient of pesticides applied to the land and those released from industry and urban developments. The sources of pesticides to the environment are well known and are controllable. However, the fate of these pesticides in the environment is considerably less known and virtually uncontrollable. Some of the known processes that influence the fate of pesticides are noted in Table 1.

Once in the aquatic environment, pesticides may be volatilized, remain in solution or suspension, precipitate with and as sediment, or be assimilated by biota in the ecosystem. Most pesticides are insoluble or slightly soluble in water and therefore tend to accumulate in bottom sediments, with the result that concentrations in sediments are generally much higher than in the surrounding water. Gerakis and Sticas (1974) report that the accumulation of pesticides in bottom sediments play an important role in the disappearance of pesticides from contaminated water. They support this by stating that studies in major agricultural river basins in California revealed that an average pesticide concentration of 0.1 parts per billion (ppb) to 0.2 ppb in river water may mean that bottom sediments contain 20 ppb to 100 ppb. Bottom sediments are therefore a sink for these toxic substances, posing not just an immediate threat but also a threat to the aquatic environment over time.

Despite recognizing that sediments accumulate pesticides, the Water Quality Branch chose to monitor the water phase in order to better understand the distribution, abundance and movement of pesticides in surface waters of western Canada. In addition, it was felt that any contamination of the aquatic environment would be at least detectable in the water phase, the phase more easily collected and analyzed. The results of monitoring this phase is the topic of this report.

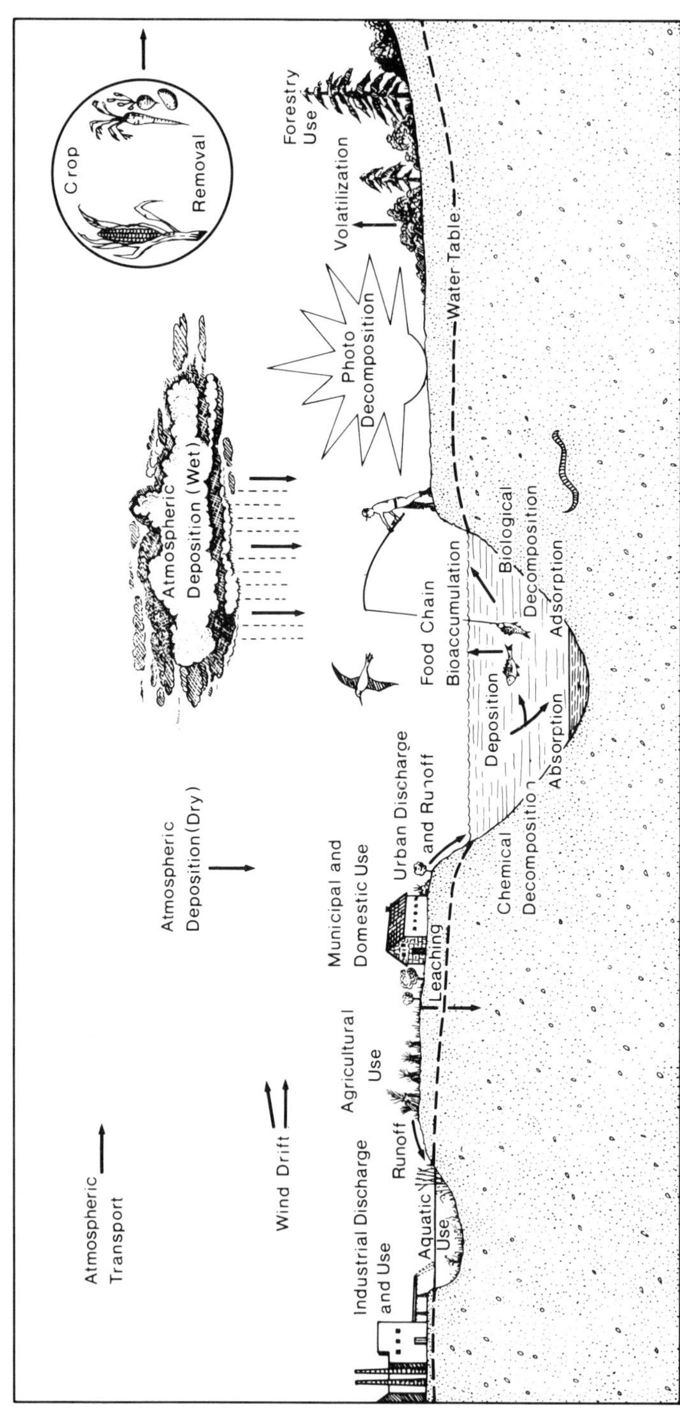

Figure 1. Pesticide Uses and Environmental Pathways

Table 1. Factors Influencing the Fate of Pesticides

Physio-Chemical	Biological
solubility	bioaccumulation
co-distillation	biodegradation
volatility	biouptake
photodegradation	bioavailability
stability	
temperature	Other
active life	
sorption	run-off - precipitation
formulation	management practices
	wind-drift
	erosion

Toxicological Properties

The toxicological effects of pesticides on aquatic life are well documented (U.S. Office of Science and Technology, 1971). Pesticides generally have deleterious effects on aquatic life, wildlife and man. Yet the toxicological properties of the individual pesticides differ considerably. Most pesticides are toxic to most forms of life even at low concentrations, and some are known to be carcinogenic. Certain pesticides, typically the organochlorine pesticides, persist in the aquatic environment and remain a threat over a longer period of time. These pesticides are generally considered to be hazardous to fish and other aquatic life and are slightly to highly accumulative. The Canada Department of Agriculture has placed restrictions and/or suspensions on many of the persistent organochlorine products. Weber (1977) reports that the mobility of this group of pesticides ranges from the relatively nonmobile, nonvolatile chemicals, such as DDT, to the highly mobile, moderately volatile chemicals, such as γ-BHC (lindane).

Weber (1977) also reports that the organophosphorus insecticides, although more toxic to man, are less toxic to fish than the organochlorine group. He states further that organophosphorus insecticides are more soluble, have a lesser tendency to bioaccumulate and are only slightly more mobile than organochlorine pesticides. Their mobility is restricted because of their tendency to complex with soil colloids.

Weber states that carbamate insecticides (carbofuran and Baygon) have toxicological characteristics similar to the organophosphorus insecticides. Carbamate insecticides, however, have not been as fully researched as the others. He says that the chlorophenoxy acid herbicides such as 2,4-D and 2,4,5-T are considerably less toxic than the pesticides above and that these degrade rapidly in the environment. The mobility of this group is less than that of the organochlorine pesticides.

The Pesticide Monitoring Program

In response to environmental concerns the monitoring of pesticides in water commenced in 1971. The initial program was ad hoc, and because of resource limitations, water sampling was done only after spraying activities and primarily only on waters most likely affected. In 1973 a more stable pesticide monitoring program evolved. Thirty-eight strategically selected pesticide sites (Figure 2) were established throughout the prairie region of Canada, with emphasis on monitoring the quality of interjurisdictional waters. The designated pesticide sites were sampled quarterly. In late 1976, monthly sampling for 2,4-D and 2,4,5-T began at the designated interprovincial locations.

In addition to the regular monitoring program, special surveys were initiated that resulted in an increased regional pesticide data base. The pesticide data generated by Branch programs are included in this report. The data derived from other programs jointly financed with other agencies are also included. This report presents the findings of the aforementioned monitoring programs for the 1971-1977 period, which have produced more than 20,000 pieces of pesticide data. Pesticides that have been investigated during this period are shown in Table 2. The report focuses on the five pesticides (2,4-D; 2,4,5-T; 2,4-DP (dichloroprop); lindane; and α-BHC) that are most commonly detected in the surface waters of western Canada.

METHODOLOGY

All water samples collected are defined as "grab samples" and were collected in accordance with Water Quality Branch sampling procedures (Environment Canada, 1973, Water Quality Branch, 1976). Each pesticide datum reflects an instantaneous concentration at a particular place in space and time. The data herein refer to the total dissolved fraction plus that portion associated with the suspended particles.

Table 2. Pesticides and Metabolites Investigated

Pesticide Group	Compounds Investigated		
Organochlorine	aldrin	p,p'-DDT	endrin
	α-BHC	dieldrin	heptachlor
	γ-BHC (lindane)	α-Endosulfan	heptachlor epoxide
	p,p'-DDD	β-Endosulfan	p,p'-methoxychlor
	p,p'-DDE		
Chlorophenoxy acid	2,4-D	2,4-DP	MCPA
	2,4-DB	2,4,5-T	
Organophosphorus	Guthion	malathion	dimethoate
Carbamates	carbofuran	Baygon	

The pesticides monitored were for the most part analyzed in accordance with Water Quality Branch procedures (Environment Canada, 1974).

Water samples for pesticide analyses were collected in acid pre-cleaned 1.2 l glass bottles. At the time of collection, samples for chlorophenoxy acid herbicides and carbamate insecticides were acidified to pH2 with either sulphuric or hydrochloric acid; samples for organophosphorus pesticides were acidified to pH4 with hydrochloric acid. All bottles were tightly capped with an aluminum foil-lined top and stored at 4°C in darkness until analysis.

Organochlorine samples were extracted with pesticide residue grade haxane by mechanical vortexing at room temperature. The extract was concentrated to 1 ml and fractionated using the Florisil column procedure or by the preparative high speed liquid chromatography using a stainless steel column (1.22 m x . 64 cm OD) packed with Porasil "a" (37-75μ). Isocratic elutions of step-wise increasing polarity yielded four fractions which were concentrated and analyzed by isothermal, electron capture gas chromatography.

Chlorophenoxy acid herbicide samples were extrac methylene chloride. The extract was dried using acidic sodium sulfate; the methylene chloride was sequentially replaced with methanol and the solution concentrated to 1 ml. The extracted residues were derivatized to form one of the methyl, 2-chloroethyl or pentafluorobenzyl esters. The derivatives were fractionated and the excess reagent removed by column chromatography with neutral alumina (when required). The concentrate was analyzed by isothermal, electron capture gas chromatography.

Sodium sulphate was added to the water samples requiring carbamate analysis. The sample was then extracted with methylene chloride. The extract was concentrated to 1 ml and analyzed using a nitrogen detector.

Sodium sulphate was added to the water samples requiring organophosphorus pesticide analysis. The sample was then extracted with pesticide residue grade benzene by mechanical vortex stirring at room temperature. The extract was decanted, concentrated and analyzed by isothermal gas chromatography using sulphur and phosphorus specific flame photometric detection.

RESULTS AND DISCUSSION

To facilitate discussion of the pesticide data, the Prairie Provinces have been divided into 14 water quality districts (Figure 2), taking into consideration river basin drainage (Water Survey of Canada, 1974), vegetation formations and the physiographic regions of Canada (Surveys and Mapping Branch, 1973).

The water quality districts are:

I	Mountain
II	Missouri River
III	Western South Saskatchewan River
IV	Red Deer River
V	North Saskatchewan River
VI	Peace-Athabasca
VII	Cold Lake - Lac la Ronge
VIII	Saskatchewan River
IX	Eastern South Saskatchewan River
X	Qu'Appelle River
XI	Souris River
XII	Red River
XIII	Assiniboine River - Red Deer River
XIV	Canadian Shield

The shaded portion of the inset map (Figure 2), land classes 1-3, corresponds closely with the grasslands formation, and the unshaded portion corresponds with the woodland formation of the

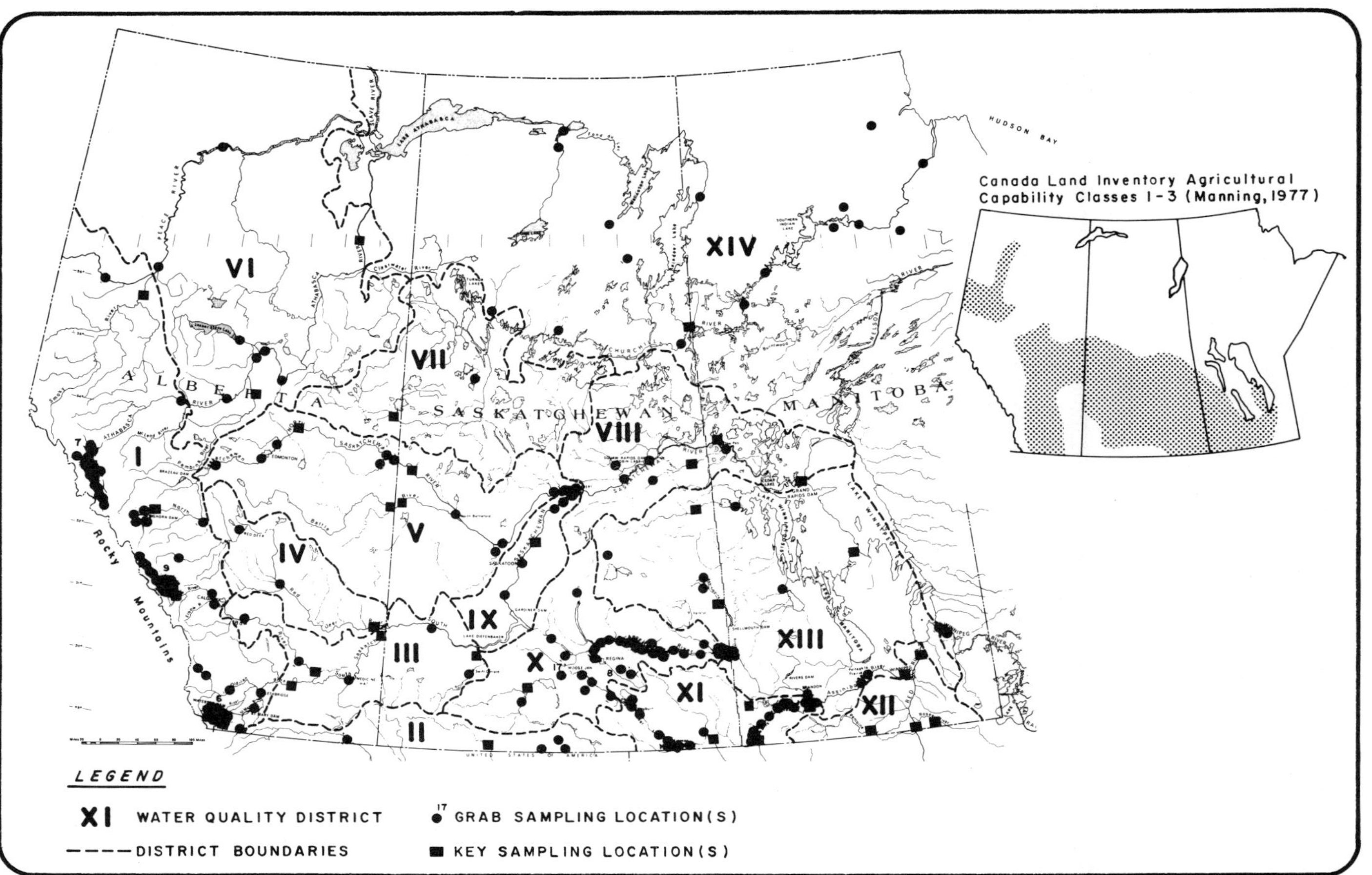

Figure 2. Pesticide Sampling Locations (The insert map reflects the prime grain production areas)

vegetation regions of Canada (Surveys and Mapping Branch, 1973). The land classes 1-3 are considered indicative of the prime grain production areas as well as indicative of the area of greatest pesticide diversity and use. The woodland area (the unshaded area in inset map, Figure 2) on the other hand, hosts a smaller variety of pesticides which are related to the pulp and paper, and the livestock industries; the use of crop pesticides is considerably less and is scattered in this area.

Organochlorine Pesticides

Organochlorines have been used extensively in western Canada. However, their use has diminished considerably over the past few years. The organochlorines and some of their metabolites investigated in surface waters of western Canada are shown in Table 2. These pesticides are presently registered, but their use is restricted and limited.

Approximately 1400 samples were analyzed for 13 organochlorines (Table 2) in the surface waters of western Canada, and only lindane, α-BHC, heptachlor, aldrin and β-Endosulphan were detected. Only seven positive detections out of 1425 tests for heptachlor were observed. These occurred in districts III and X of the grasslands and were between 0.001 μg/l and 0.007 μg/l.

Slightly more than 1400 tests were performed for aldrin and β-Endosulphan. Only two positive detections for aldrin are on record, one in the North Saskatchewan River system (district V) and one in the Qu'Appelle River system (district X) at levels of 0.005 μg/l and 0.002 μg/l, respectively. Beta-Endosulphan was detected only once, in March 1973, at a level of 0.011 μg/l in the Manitoba portion of the Souris River district (XI).

The lack of detection of these organochlorines in surface waters can be explained, for the most part, by the reduction in use over the last few years of these persistent pesticides and most importantly by the fact that these pesticides tend to accumulate in the sediment.

Lindane is registered in Canada for (1) controlling ticks and flies on livestock, (2) seed treatment for wireworm control, (3) controlling infestations of stored logs by the logging industry, and (4) controlling bedbugs by the pest control industry. Additional uses are on nursery stocked trees, as soil treatment, in dog pounds for control of fleas, in farm buildings (wall spraying), and in warehouses where food is not stored.

Table 3 and Figure 3 indicate that of the 14 districts, II, III, IX, X, XI, XII and XIII have the most frequent positive detections of lindane. About 58% of the stations monitored (Table 4) were found to have detectable quantities of lindane at one or more times during the monitoring period. Six of the districts had maximum levels at or above 0.005 μg/l. The highest value, 0.086 μg/l, was recorded in the Qu'Appelle River district (X). Even the Mountain district (I) showed a maximum of 0.007 μg/l. Lindane is not as prevalent as its isomer, α-BHC, in the surface waters of western Canada.

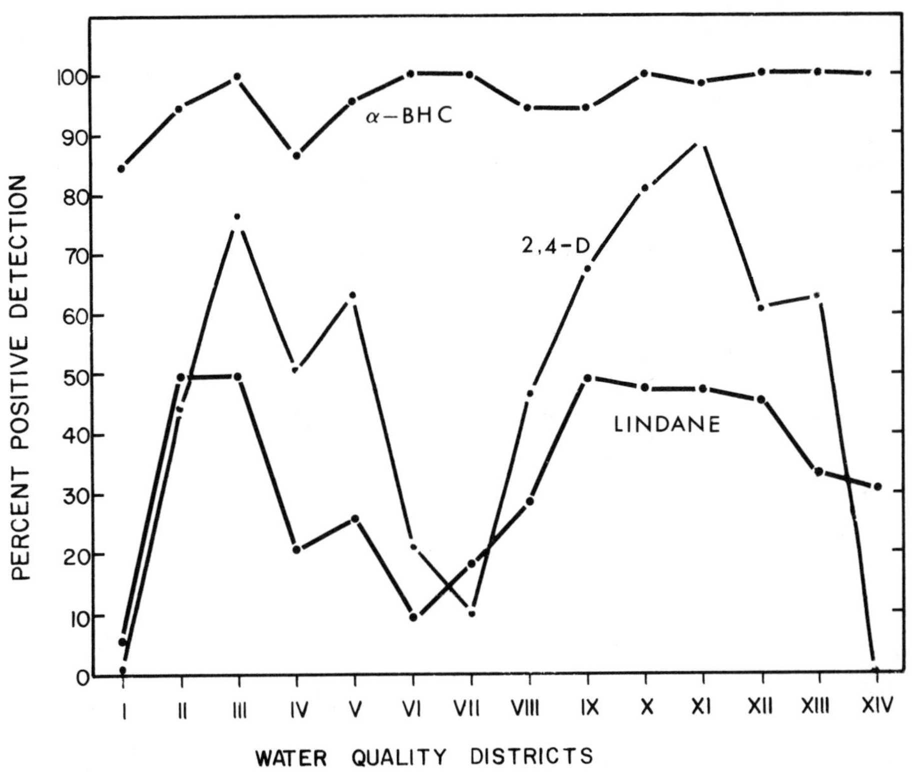

Figure 3: Plot of Percent Positive Detection of α-BHC, Lindane and 2,4-D Against Various Water Quality Districts

Table 3. Lindane Concentrations (μg/l) in Surface Waters of Fourteen Water Quality Districts in Western Canada

Statistics	Water Quality Districts													
	I	II	III	IV	V	VI	VII	VIII	IX	X	XI	XII	XIII	XIV
Minimum	<.001	<.001	<.001	<.001	<.001	<.001	<.001	<.001	<.001	<.001	<.001	<.001	<.001	<.001
Maximum	0.007	0.003	0.002	0.005	0.002	0.002	0.001	0.008	0.010	0.086	0.004	0.003	0.008	0.002
Mean*	<.001	0.001	0.001	<.001	<.001	<.001	<.001	0.001	<.001	<.001	0.001	<.001	0.001	<.001
Median	<.001	<.001	<.001	<.001	<.001	<.001	<.001	<.001	<.001	<.001	0.001	<.001	<.001	<.001
Positive detection(%)	6.1	50.0	50.0	20.7	26.0	9.5	18.2	28.8	49.2	47.9	47.3	45.8	33.3	30.5
Positive detection(No)	38	14	9	16	25	4	4	17	32	90	104	21	41	11
Less than detection(No)	585	14	9	61	71	38	18	42	33	98	116	27	82	25
Total number of tests	623	28	18	77	96	42	22	59	65	188	220	48	123	36

*The computation of the mean assumes 'less than' values as real and equal to the analytical detection limit minus 0.0001 μg/l.

Table 4. Sites with Positive Detection for Selected Pesticides in Each Water Quality District

Water Quality District	Pesticides				
	Lindane	α-BHC	2,4,5-T	2,4-D	2,4-DP
I	27 (78)*	44 (48)	8 (73)	5 (72)	2 (74)
II	3 (6)	5 (5)	1 (6)	5 (6)	1 (4)
III	6 (9)	4 (4)	4 (8)	8 (8)	5 (8)
IV	3 (3)	2 (2)	2 (3)	3 (3)	0 (3)
V	9 (16)	4 (4)	7 (13)	13 (13)	5 (13)
VI	2 (9)	2 (2)	4 (7)	5 (7)	3 (8)
VII	1 (3)	1 (1)	1 (1)	1 (1)	0 (2)
VIII	5 (8)	3 (3)	5 (6)	6 (6)	2 (6)
IX	1 (6)	1 (1)	1 (1)	1 (1)	0 (1)
X	44 (59)	3 (3)	28 (35)	24 (24)	22 (39)
XI	27 (27)	20 (20)	3 (27)	26 (27)	3 (27)
XII	4 (4)	4 (4)	3 (4)	4 (4)	2 (4)
XIII	11 (11)	4 (4)	5 (10)	9 (10)	1 (10)
XIV	6 (19)	1 (1)	0 (4)	0 (4)	0 (16)
Total	149 (258)	98 (102)	72 (198)	110 (186)	46 (215)
Total positive detection (%)	57.8	96.1	36.4	59.1	21.4

*The number preceeding the parentheses is the number of sites with positive detections. The number in parentheses is the total number of sites in the district.

Alpha BHC has few or no insecticidal characteristics (Faust, 1972) and therefore is not an active ingredient in any pesticide product. It is, however, an impurity in the lindane product, but at less than 1%. N. Bostanian (Canada Department of Agriculture, personal communication) states that at one time entomologists viewed BHC as more effective than lindane and therefore BHC was used more extensively. BHC is composed of 60% to 70% alpha, 5% to 6% beta, 12% to 16% gamma, 5% to 7% delta and 2% to 3% epsilon isomers (Canada Department of Agriculture, 1967). J. Stocker (Canada Department of Agriculture, personal communication) confirms that BHC (mixed isomers) was last registered in Canada in 1972, yet α-BHC is observed on a wide geographic scale and in quantities greater than would be expected on the basis of past use and impurity levels in the present lindane products.

As Table 5 shows, more than 80% of the tests performed for α-BHC in all districts were above the analytical detection limit of 0.001 μg/l, with individual district maximums generally in the vicinity of 0.010 μg/l. The maximum level of α-BHC recorded is 0.020 μg/l in district XI. Approximately 92% of the stations monitored (Table 4) and 84% of the tests performed in the remote Mountain district (I) were found to contain α-BHC (Table 5); 35% of the stations monitored and 6.1% of the tests performed were positive for lindane. The Canadian Shield district (XIV) also demonstrates residuals of lindane and α-BHC. Neither lindane nor α-BHC have been used in these districts in significant enough quantities to explain their detections; hence, long range transport and ultimately atmospheric deposition, must be the mechanism by which these organochlorines contaminate the surface waters in these remote districts. Oloffs _et al_. (1972) have suggested that atmospheric transport might, under certain conditions, account for the global redistribution of lindane.

A study or organics in precipitation across Canada was initiated in 1977 by the Water Quality Branch in cooperation with Atmospheric Environment Service of the Canadian Department of Fisheries and Environment. The first results of the study, for July 1977, show precipitation at Edson, Alberta containing 0.074 μg/l α-BHC and 0.022 μg/l lindane. Evidence of organochlorine contamination of the atmosphere has been reported elsewhere. Tarrant and Tatton (1968) report that organochlorines are frequently present in rainwater in the British Isles. In England, rainwater collected from two sites near London contained from 0.015 μg/l to 0.040 μg/l α-BHC and from 0.020 μg/l to 0.155 μg/l lindane (Morrison, 1972). Widespread detection of lindane and α-BHC has been reported in surface waters of Israel at levels as high as 0.04 μg/l and 0.119 μg/l, respectively (Kahanovitch and Lahau, 1974). In this last example the author suggests atmospheric pollution as one mechanism responsible for the surface water contamination.

Table 5. Alpha BHC Concentrations (μg/1) in Surface Waters of Fourteen Water Quality Districts in Western Canada

Statistics	Water Quality Districts													
	I	II	III	IV	V	VI	VII	VIII	IX	X	XI	XII	XIII	XIV
Minimum	<.001	<.001	0.005	<.001	<.001	0.001	0.002	<.001	<.001	0.002	<.001	0.002	0.002	0.003
Maximum	0.010	0.020	0.009	0.010	0.009	0.008	0.005	0.007	0.010	0.010	0.010	0.006	0.010	0.006
Mean*	0.003	0.009	0.007	0.004	0.004	0.005	0.004	0.003	0.006	0.005	0.006	0.004	0.004	0.004
Median	0.002	0.007	0.008	0.005	0.004	0.005	0.005	0.005	0.006	0.005	0.006	0.004	0.004	0.004
Positive detection(%)	84.6	94.1	100	86.6	95.5	100	100	94.1	94.1	100	98.2	100	100	100
Positive detection(No)	55	16	4	13	21	8	7	16	16	9	55	18	19	5
Less than detection(No)	10	1	-	2	1	-	-	1	1	-	1	-	0	0
Total number of tests	65	17	4	15	22	8	7	17	17	9	56	18	19	5

*The computation of the mean assumes 'less than' values as real and equal to the analytical detection limit minus 0.0001 μg/l.

Characteristics such as volatility, metabolic rate, and solubility may account for the generally fewer detections of lindane than α-BHC. The reported isomerization of lindane to α-BHC in the environment (Sethanathan, 1973, Benezet and Matsumura, 1973) may partially explain the higher than expected levels of α-BHC observed in the surface waters monitored.

Although agriculture is the predominant source of lindane and α-BHC contamination of freshwaters in western Canada, urban centres and industry (packaging and manufacturing plants) are additional potential sources. During a 2,4-D survey of the Red River (discussed in the subsequent section) during July 1977, the effluents from the City of Winnipeg's North End Pollution Control Centre and South End Pollution Control Centre were found to contain lindane at concentrations of 0.10 μg/l and 0.08 μg/l, respectively.

Lindane is generally more toxic than DDT and is known to bioaccumulate in the food chain. It has been recommended (Great Lakes Water Quality Board, 1974) that lindane in water should not exceed 0.01 μg/l for the protection of aquatic life. Tests for lindane and α-BHC conducted in 1976 on fish and wildlife in Alberta have been reported (Alberta Department of Agriculture, 1977, unpublished) as 0.734 ppm, and 0.150 ppm to 0.282 ppm, respectively. Although there is little information on the toxicity of α-BHC, this isomer is recognized as being more persistent, but less toxic, than the gamma isomer.

Chlorophenoxy Acid Herbicides

In western Canada, MCPA, 2,4-D and 2,4-DP are widely used on cereal crops for the control of broadleaf weeds; 2,4,5-T and 2,4-D are used for brush control in pastures. The chlorophenoxy acid herbicides, 2,4-D and MCPA account for more than 90% of the total herbicides used (Grover, 1974). Agricultural use patterns for the period 1972 to 1976 (F. Holm, Canada Department of Agriculture, personal communication) show a decline in quantities used of the 2,4-D formulation since 1972 in Alberta and Saskatchewan, whereas in Manitoba there has been a slight increase (Figure 4). The quantities of MCPA used have increased in Alberta and decreased in Saskatchewan and Manitoba during this five-year period. Saskatchewan uses about three times as much 2,4-D than Manitoba and about twice as much as Alberta.

Both 2,4-D esters and amines are used in the Prairie provinces. In 1976, in Saskatchewan the 2,4-D formulation applied contained about 27% amine salt; in Alberta it contained about 33% amine salt; and in Manitoba it contained about 81% amine salt. Since 2,4-D (free acid) is a breakdown product of the salts and esters and is

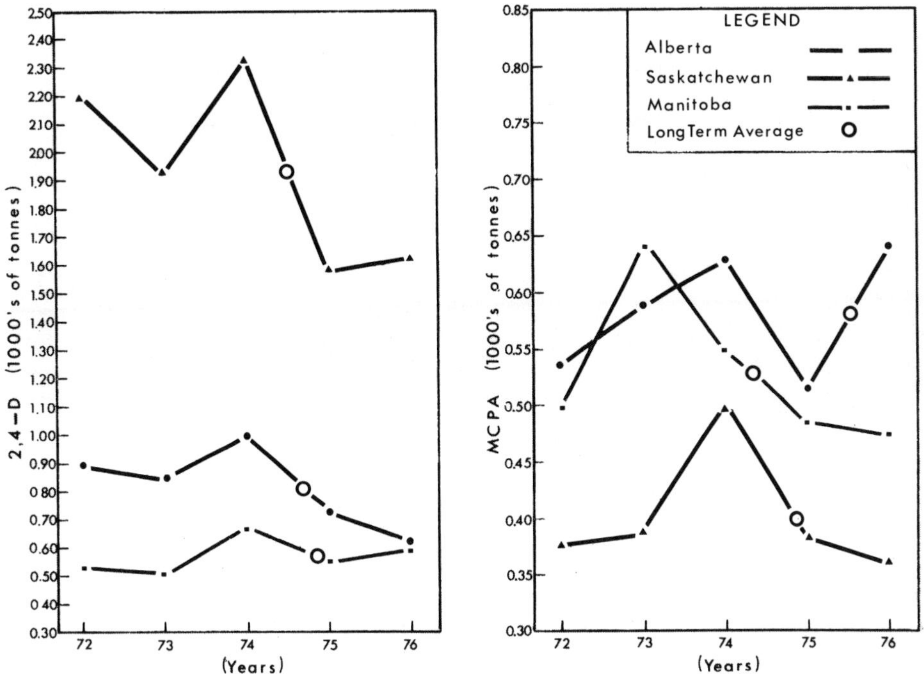

Figure 4: Use Patterns of 2,4-D and MCPA in the Prairie Provinces

biologically and chemically more stable (Paris and Lewis, 1973), its fate in the aquatic environment is of concern.

Positive concentrations of 2,4-D have been frequently observed in most of the 14 water quality districts (see Table 6 and Figure 3). The highest percentages of positive detection were found in those districts in and overlapping the agricultural areas of Saskatchewan. The remote areas I and XIV showed 1.1% and 0.0% positive detection 2,4-D, respectively. This is contrary to the observation for lindane and α-BHC which are prevalent in the surface waters of these areas.

The lack of 2,4-D in the remote surface waters can be explained in part by the relatively short persistence of 2,4-D, the limited use of this herbicide in the remote districts and possibly the volatility (atmospheric mobility) of 2,4-D formulations. Grover (1974) reports that air monitoring near Regina, Saskatchewan to determine herbicidal drift detected esters of 2,4-D whereas residues of the amine salt went undetected. He states that the predominant form of 2,4-D in the air is the high volatile butyl

Table 6. 2,4-D Concentrations (μg/1) in Surface Waters of Fourteen Water Quality Districts in Western Canada

Statistics	Water Quality Districts													
	I	II	III	IV	V	VI	VII	VIII	IX	X	XI	XII	XIII	XIV
Minimum	<.004	<.004	<.004	<.004	<.004	<.004	<.004	<.004	<.004	<.004	<.004	<.004	<.004	<.004
Maximum	0.025	0.080	0.106	0.095	0.248	0.090	0.010	0.228	0.334	4.33	0.60	3.4	0.07	<.004
Mean*	<.004	0.020	0.054	0.013	0.005	0.010	0.004	0.015	0.035	0.235	0.038	0.101	0.016	<.004
Median	<.004	<.004	0.061	0.004	0.016	<.004	<.004	<.004	0.019	0.040	0.02	0.010	0.008	<.004
Positive detection(%)	1.1	44.7	76.9	50.9	63.1	20.9	10.0	46.2	67.2	80.6	89.0	60.4	62.9	0.0
Positive detection(No)	6	17	10	30	53	9	2	24	37	54	187	32	73	-
Less than detection(No)	548	21	3	29	31	34	18	28	18	13	23	21	43	22
Total number of tests	554	38	13	59	84	43	20	52	55	67	210	53	116	22

*The computation of the mean assumes 'less than' values as real and equal to the analytical detection limit minus 0.0001 μg/l.

ester. The significance of this observation is that the mechanism of atmospheric contamination of surface waters appears to be dependent to some extent on the herbicide formulation. In Manitoba which uses large quantities of the nonvolatile amine salt, contamination derived from the atmosphere is primarily from direct non-target application of the herbicide and from drift within short distances of herbicide application. In Alberta and Saskatchewan where 2,4-D ester formulations are used extensively, atmospheric contamination of surface waters probably results from direct non-target application, drift, and long range transport within the degradation life time of 2,4-D. Although atmospheric transportation helps account for the geographic distribution of 2,4-D, the rapid breakdown of 2,4-D probably precludes anything more than periodic detection of 2,4-D in surface waters that are remotely situated (as waters in districts I and XIV) from the areas of herbicide use.

In the rural areas, slight seasonal variations in 2,4-D concentrations are discernible from the detailed data, with the higher levels generally detected during the period from May to September which corresponds to the normal period of 2,4-D application. Occasional high levels were observed in the winter periods. This is not surprising, considering that the degradation of 2,4-D is inhibited by colder temperatures and lower dissolved oxygen (Paris and Lewis, 1973). Throughout the entire monitoring period 1971 to 1977, 59.1% of the stations sampled had detectable 2,4-D on one or more occasions. Of the total number of 2,4-D tests (1386) performed, 38.5% were positive. Excluding those districts remote from direct agricultural influence (I, VI, XII and XIV), 69.2% of the tests performed had detectable 2,4-D.

It is readily apparent from the results of the monitoring programs that there is a significant contribution of 2,4-D to the aquatic environment via municipal and industrial effluents. The highest 2,4-D values on record occurred at sites below urban centres. A maximum level of 4.33 μg/l was recorded in Wascana Creek below the City of Regina, Saskatchewan (in district X) during January 1974; in February of 1977, a maximum of 3.4 μg/l was recorded in the Red River near Selkirk, Manitoba below the City of Winnipeg (in district XII). Because both the winter 2,4-D maximum and the magnitude of the 2,4-D concentration were unprecedented in the Red River, the Water Quality Branch conducted an investigation to determine the primary source of 2,4-D.

The results of the investigation determined a major source of 2,4-D to the Red River to be the effluent from the City of Winnipeg's North End Pollution Control Centre (Chacko 1977, unpublished). A 24-hour composite sample collected of the effluent, during April 27th and 28th, 1977, had a 2,4-D concentration of 30.5 μg/l. Chacko suggests that this effluent may have also resulted in the February

1977 maximum. He further speculates that the maximum level which corresponds with the herbicide packaging activities in Winnipeg, is likely more than just a coincidence. There are two packaging plants within Winnipeg that have the potential to contaminate water received at the North End Pollution Control Centre for treatment. It is interesting to note that the effluent from the City of Winnipeg's South End Pollution Control Centre had a 24-hour composite concentration of less than the analytical detection limit of 0.004 μg/l for the April 27-28th, 1977 time period.

The Water Quality Branch monitoring program detected a more subtle increase in 2,4-D during late 1976 and early 1977 in the North Saskatchewan River at the Alberta-Saskatchewan border. A survey to locate the cause of this increase was conducted in January 1977 by the Alberta Department of Environment. The survey revealed 2,4-D levels as high as 4.35 milligrams/liter in the effluent from a Uniroyal Limited development located a few kilometers east of Edmonton, Alberta (P. Grant, Alberta Department of Environment, personal communication). The river monitoring site at which the increase in the North Saskatchewan River was first detected is approximately 280 km downstream of this effluent.

Most pesticides have a strong tendency to be adsorbed on suspended matter, on sediments and other surfaces. Hurlbert (1975) has reported that following an application of 112 kg/ha of 2,4-D, the concentration of 2,4-D in water dropped below 1 μg/l in eight hours. Four days after treatment, bottom sediment had 2,4-D residues ranging from 950 μg/kg to 56,000 μg/kg; ten months after treatment sediment had 2,4-D residues ranging from 240 μg/kg to 58,000 μg/kg. Contrary to these earlier findings, one to two days after applying 148 cc/ha to 590 cc/ha of 2,4-D (low volatile ester) the 2,4-D concentrations in the water phase were 10 μg/l to 15 μg/l and no 2,4-D was observed in bottom sediment (J. Patterson, Canadian Wildlife Service, personal communication). In work conducted by Chacko (1977), 2,4-D was not found in Red River bottom sediment which suggests that 2,4-D was in solution or associated with the suspended sediments and thus held in suspension throughout the course of the Red River studied. Flow rate, river depth and other flow-related characteristics may play an important role in the transport and distribution of 2,4-D in river systems.

Zepp _et al_. (1975) report that the fate of 2,4-D esters is of interest to the U.S. Environmental Protection Agency because of their toxicity to fish and other types of wildlife. J. Patterson (Canadian Wildlife Service, personal communication) indicates that preliminary information from studies now in progress shows that the application of 2,4-D (low volatile ester) and MCPA to wetlands at a rate of 148 cc/hectare to 590 cc/hectare has reduced the herbivore and carnivore invertebrate communities 30% to 40%, and 80%, respectively. The detritus communities were unaffected or possibly slightly enhanced. Patterson suggested that 2,4-D levels in the order of 1.0 μg/l in surface water may pose a threat to the submerged macrophytic and invertebrate communities, thereby threatening food availability in duckling habitat. The 2,4-D LD_{50} for fish has been reported as 250 μg/l, with a 2,4-D longevity in half-life of one to four weeks (Weber, 1977).

As shown in Figure 4, the use of MCPA has declined in Manitoba and Saskatchewan, but increased slightly in Alberta. More than 400 tests have been performed for MCPA in surface waters of the districts. However, detections above the analytical detection limit of 0.2 μg/l have gone unrecorded. The preliminary work of Patterson (Canadian Wildlife Service, personal communication) has revealed that MCPA migrates rapidly from the water to the bottom sediments. This phenomenon may help explain why MCPA has gone undetected when monitoring the water phase. In addition, the high detection limit (1000 times higher than that for 2,4-D) may have precluded an adequate evaluation of MCPA in the surface waters during the monitoring period.

The ester and amine compounds of the herbicide 2,4,5-T are used regionally. Statistics on how much has been, and is used, are not available; however, quantities are known to be relatively small compared to 2,4-D quantities. Nevertheless, 48.3% of the tests performed in the Qu'Appelle River district (X) and 22.2% of the tests performed in the North Saskatchewan River district (V) were positive for 2,4,5-T (Table 7). All other districts had less than 15% positive detections. The maximum value recorded is 3.12 μg/l in district X. Of the total number of sites monitored (198), 36.4% had at least one positive detection. Weber (1977) reports the LD_{50} for 2,4,5-T to fish to be 500 μg/l and the persistency (half-life) to be 1 to 12 weeks. Kenaga (1975) reviewed the stability of esters of 2,4,5-T in the aquatic environment, finding them to be rapidly hydrolyzed in most waters with the exception of those that are acidic.

The herbicide 2,4,5-T does not appear to be problematic on a wide scale, but in areas of high use it could affect the aquatic habitat of local wetlands and sloughs. Monitoring by the Branch has not focused on these aquatic ecosystems.

Dichloroprop, or 2,4-DP, is not used in as large quantities as 2,4-D or MCPA. Most often it is used in conjunction with 2,4-D rather than alone for the control of broadleaf weeds in wheat and barley crops.

Dichloroprop is not as prevalent as 2,4-D in the surface waters of western Canada (Table 8). Twenty-one percent of the stations monitored recorded positive detections one or more times,

Table 7. 2,4,5-T Concentrations (μg/1) in Surface Waters of Fourteen Water Quality Districts in Western Canada

Statistics	Water Quality Districts													
	I	II	III	IV	V	VI	VII	VIII	IX	X	XI	XII	XIII	XIV
Minimum	<.002	<.002	<.002	<.002	<.002	<.002	<.002	<.002	<.002	<.002	<.002	0.002	<.002	<.002
Maximum	0.014	0.008	0.002	0.015	0.020	0.04	0.002	0.071	0.013	3.12	0.071	<.002	0.020	<.002
Mean*	<.002	0.002	<.002	<.002	0.003	<.002	<.002	0.004	<.002	0.090	<.002	0.003	0.003	<.002
Median	<.002	<.002	<.002	<.002	<.002	<.002	<.002	<.002	<.002	<.002	<.002	<.002	<.002	<.002
Positive detection(%)	2.0	2.6	7.7	5.9	22.2	9.3	5.0	13.2	5.3	48.3	3.5	13.2	6.7	0.0
Positive detection(No)	11	1	1	5	20	4	1	7	3	42	7	7	8	-
Less than detection(No)	541	37	12	79	70	39	19	46	53	45	190	46	110	22
Total number of tests	552	38	13	84	90	43	20	53	56	87	197	53	118	22

*The computation of the mean assumes 'less than' values as real and equal to the analytical detection limit minus 0.0001 μg/1.

and only 4.8% of 1455 tests were above the analytical detection limit of 0.004 μg/l. The maximum level recorded in surface waters of western Canada was 7.15 μg/l in district X (Table 8). This same district accounted for 59% of the total positive detections. Dichloroprop is more persistent than 2,4-D (J. Holm, Saskatchewan Department of Agriculture, personal communication, 1977) and is harmful to pollinating insects such as bees and bumblebees (Makkula, 1975).

Organophosphorus Insecticides

These insecticides do not constitute part of the Branch routine monitoring programs. Approximately 120 tests were performed during the summers of 1973 and 1974 for each of Guthion, malathion and dimethoate. Grab samples for these organophosphorus insecticides were collected from surface waters located within and contiguous to areas sprayed with these chemicals. Examination of surface waters for organophosphorus insecticides has not been performed since 1974. As previously mentioned, they degrade rapidly in the environment and, as a result, have gone undetected.

Carbamate Insecticides

Significant amounts of carbamate insecticides are not used in the Prairies; consequently, there has been limited monitoring of these insecticides in surface waters. As a result of a potential outbreak of the Western Encephalomyelitis virus in Manitoba during the summer of 1975, Baygon was applied aerially to control the adult mosquito. Twenty-five samples for Baygon analysis were collected and analyzed from surface waters in and around the City of Winnipeg, Manitoba. Of the total number of samples analyzed, 50% were positive for Baygon (Gummer, 1975). The highest level, 36.4 μg/l, was recorded in Sturgeon Creek in the City of Winnipeg about 12 hours after aerial application. Four samples from the La Salle River had levels in excess of 17.5 μg/l.

The biodegradation of Baygon is quite rapid, particularly in waters with high bacterial activity and elevated temperatures. The colder water temperatures (15°C) that occurred during the latter days of aerial application may have slightly retarded the otherwise rapid degradation.

Carbofuran is used in western Canada for the control of grasshoppers. Monitoring of this carbamate has been limited, amounting to about 60 samples collected during 1974-1975. All the tests conducted revealed no concentrations at or above the analytical detection limit of 1.0 μg/l. In 1974, the Branch participated in

Table 8. 2,4-DP Concentrations (μg/1) in Surface Waters of Fourteen Water Quality Districts in Western Canada

Statistics	Water Quality Districts													
	I	II	III	IV	V	VI	VII	VIII	IX	X	XI	XII	XIII	XIV
Minimum	<.004	<.004	<.004	<.004	<.004	<.004	<.004	<.004	<.004	<.004	<.004	<.004	<.004	<.004
Maximum	0.016	0.028	<.004	0.007	0.033	0.044	<.004	0.038	0.028	7.15	0.027	0.131	0.030	<.004
Mean*	<.004	0.005	<.004	<.004	0.005	<.004	<.004	0.005	<.004	0.233	<.004	0.009	0.004	<.004
Median	<.004	<.004	<.004	<.004	<.004	<.004	<.004	<.004	<.004	<.004	<.004	<.004	<.004	<.004
Positive detection(%)	0.4	6.6	0.0	1.7	8.6	6.5	0.0	6.1	7.4	35.3	1.8	5.1	0.8	0.0
Positive detection(No)	2	2	-	1	6	3	-	3	4	42	4	3	1	-
Less than detection(No)	563	28	15	57	64	43	16	46	50	77	223	56	117	29
Total number of tests	565	30	15	58	70	46	16	49	54	119	227	59	118	29

*The computation of the mean assumes 'less than' values as real and equal to the analytical detection limit minus 0.0001 μg/1.

a special study with Agriculture Canada to investigate the residual carbofuran levels in an aquatic environment (a slough) aerially sprayed with carbofuran. The results of the study indicated no identifiable residues of carbofuran or hydroxy-carbofuran in soil, water, or bottom muds immediately after spraying, or at 1, 7 and 16 days post treatment (Ormrod, 1976). This experiment suggests rapid degradation of carbofuran and further suggests that routine monitoring in western Canada would not detect residual carbofuran levels in surface waters.

SUMMARY

The monitoring of surface waters in western Canada has revealed that the chlorophenoxy acid herbicides 2,4-D; 2,4-DP; and 2,4,5-T as well as the organochlorines lindane and α-BHC are frequently detected. Lindane has been found on occasion to exceed the limit of 0.01 µg/l prescribed by the Great Lakes Water Quality Board (1974) for the protection of aquatic life. The herbicide 2,4-D has been found in water at levels in excess of 1.0 µg/l, which may be detrimental to the invertebrate community.

The highest concentrations of 2,4-D; 2,4-DP; 2,4,5-T; and lindane were recorded in the Qu'Appelle River district (X), which is probably reflective of the more intense agricultural activity in this district. Sampling sites below large urban and industrial centres such as Winnipeg, Regina and Edmonton confirm that substantial quantities of pesticides are introduced to the environment via municipal and industrial effluents. No seasonal patterns were discernible at these locations, whereas the high levels in the rural regions were generally observed to correspond to the May to September herbicide application period.

Alpha BHC commonly occurs in waters throughout all of the districts at levels in excess of 0.001 µg/l. Although α-BHC is not presently viewed as a problem to the aquatic community, because of its prevalence the need exists to define the toxicological properties of this substance more fully. Its prevalence is thought to result from the extensive past agricultural use of technical-grade BHC, impurities in presently used lindane products and the isomerization of γ-BHC to α-BHC. The wide geographic dispersion of both lindane and α-BHC beyond the areas of use, suggests that atmospheric transport and deposition are key factors in the environmental distribution of these pesticides. The data suggest that the phenoxyacid herbicides are less volatile and degraded faster than the organochlorines and therefore are not found as extensively.

Carbamate insecticides are not routinely monitored in western Canada. Baygon monitoring during and after an emergency spray program in Manitoba showed levels as high as 36.4 μg/l 12 hours after application. A special study investigating residual levels of carbofuran in an aquatic environment suggests that carbofuran degradation is rapid and that residual levels would therefore not likely be detected in a monitoring program.

Sediment (benthic) investigations for pesticides have not been conducted by the Branch as part of the routine monitoring programs. However, the Branch has undertaken steps to increase its activities in monitoring for pesticides in suspended and deposited sediments, and aquatic organisms.

ACKNOWLEDGEMENTS

The water quality data on which this report is based have been derived from Water Quality Branch projects and other projects jointly funded with Parks Canada and the Prairie Provinces Water Board. Additionally, the author received information from many individuals with Agriculture Canada and the agriculture departments of Saskatchewan, Manitoba and Alberta.

The conscientiousness and support of the analytical staff at the Water Quality Branch Laboratory, in the analysis of pesticide samples is gratefully acknowledged. Special thanks are to J. Temple for his cartographic assistance. Thanks are also extended to G. Bergson and the staff of the Water Quality Branch who assisted in researching and compiling pertinent information. The editorial assistance and constructive comments and criticism provided by R. McNeely and K. Reid are greatly appreciated.

REFERENCES

Alberta Department of Agriculture. 1977 Unpublished. "Summary of Pesticide Residue Analyses for the Period December 1, 1975 to November 30, 1976".

Alberta Environment Conservation Authority. 1973. Public Hearings into the Use of Pesticides and Herbicides in Alberta Bulletin No. 3.

Benezet, H.J., and F. Matsumura. 1973. Isomerization of γ-BHC to α-BHC in the environment. Nature, Vol. 243.

Canada Department of Agriculture, BHC, Plant Products Division. May 1967.

Canadian Wildlife Service. 1971. Pesticides and Wildlife No. R66-4071. Information Canada.

Carson, R. 1962. The Silent Spring. Greenwish, Connecticut: Fawcett Publications Inc.

Chacko, V.T. "The Content and Distribution Pattern of 2,4-D in the Red River". Inland Waters Directorate, Fisheries and Environment Canada, 1977. Unpublished.

Edwards, C.A. 1973. Persistent Pesticides in the Environment. Cleveland, Ohio: CRC Press.

Environment Canada. 1973. Instructions for Taking and Shipping Water Samples for Physical and Chemical Analyses. 2nd ed. Inland Waters Directorate, Water Quality Branch.

Environment Canada. 1974. Analytical Methods Manual. Inland Waters Directorate, Water Quality Branch.

Faust, S.D. 1972. Fate of Organic Pesticides in the Aquatic Environment. Advances in Chemistry Series III, American Chemical Society.

Fredeen, F.J.H., J.G. Saha and L.M. Royer. 1970. Residues of DDT, DDE, DDD in fish in the Saskatchewan River after using DDT as blackfly larvicide for twenty years. J. Fish. Res. Board Can., Vol. 28, No. 1.

Gerakis, P.A., and A. G. Sticas. 1974. The presence and cycling of pesticides in the ecosphere. Residue Rev., Vol. 52.

Great Lakes Water Quality Board. 1974. Great Lakes Water Quality, Appendix A. International Joint Commission.

Grover, R. 1974. Herbicide Entry into the Atmospheric Environment; Chemistry in Canada. Vol. 26:7.

Gummer, W.D. 1975. Unpublished. "The Aerial Application of the Insecticide Baygon and the Receiving Water Environment Winnipeg, Manitoba". Environment Canada.

Hurlbert, S.H. 1975. Secondary effects of pesticides on aquatic systems. Residue Rev., Vol. 57.

Inland Waters Directorate. 1974. Analytical Methods Manual. Environment Canada.

Kahanovitch, Y., and N. Lahau. August 1974. Occurrence of pesticides in selected water sources in Israel. Environ. Sci. Technol., Vol. 59.

Kenaga, E.E. 1975. The evaluation of the safety of 2,4,5-T to birds in areas treated for vegetation control. Residue Rev., Vol. 59.

Manning, E.W. and J.D. McCraig. 1977. Agricultural Land and Urban Centres. Fisheries and Environment Canada. Ottawa.

Makkula, M. 1975. Regulation of pesticides in Finland. Residue Rev., Vol. 59.

Morrison, F.O. 1972. A review of the use and place of lindane in the protection of stored products from the ravages of insect pests. Residue Rev., Vol. 41.

Oloffs, P.C., L.J. Albright and S.Y. Szeto. 1972. Fate and behavior of five chlorinated hydrocarbons in three natural waters. Can. J. Microbiol., Vol. 18.

Ormrod, S.W. 1976. Unpublished. Residue Monitoring of Carbofuran as Applied for Grasshopper Control in Western Canada. Agriculture Canada File 834. 352R4.

Paris, D.F., and David L. Lewis. 1973. Chemical and microbial degradation of ten selected pesticides in aquatic systems. Residue Rev., Vol. 45.

Rudd, R.L. 1964. Pesticides and Living Landscape. Binghamton, New York: The University of Wisconsin Press Jail-Balleu Press Inc.

Sethunathan, N. 1973. Microbial degradation of insecticides in flooded soil and in anaerobic cultures. Residue Rev., Vol. 47.

Surveys and Mapping Branch. 1973. The National Atlas of Canada. 4th ed. Department of Energy, Mines and Resources. Ottawa.

Tarrant, K.R. and J.D.G. Tatton. 1968. Organochlorine pesticides in rainwater in the British Isles. Nature, Lond., 219.

United States Office of Science and Technology. 1971. Ecological effects of pesticides on non-target species. Executive Office of the President, Stock No. 4106-0029.

Water Quality Branch. 1976. "Handbook for the Collection, Preservation and Shipping of Water Samples". Unpublished, Inland Waters Directorate, Western and Northern Region.

Water Survey of Canada. 1974. Surface Water Data Reference Index. Environment Canada.

Weber, J.B. 1977. The pesticide scoreboard. Environ. Sci. Technol., Vol. 11, No. 8.

Zepp, R.G., N. Lee Wolfe, John A. Gordon and George L. Baughman. 1975. Dynamics of 2,4-D esters in surface waters. Environ. Sci. Technol., Vol. 9.

IMPORTANCE OF PARTICULATE MATTER ON THE LOAD OF HYDROCARBONS OF MOTORWAY RUNOFF AND SECONDARY EFFLUENTS

Fritz Zürcher, Markus Thüer* and James A. Davis

Federal Institute for Water Resources and

Water Pollution Control,

CH-8600 Dübendorf, Switzerland

ABSTRACT

Motorway runoff and secondary effluents contribute substantially to the anthropogenic input of hydrocarbons to natural waters. Hydrocarbons from both wastewaters are discharged primarily in association with suspended solids. Model weathering experiments and chemical analyses indicate that adsorbed hydrocarbons are negligible compared to oil-particulate agglomerates. The agglomeration mechanism produces particulates with high hydrocarbon concentrations in mineral suspensions, motorway runoff and secondary effluents. Hydrocarbons associated with motorway runoff particulates probably originate from motor oil exhuasted by automobiles.

INTRODUCTION

Increasing amounts of hydrocarbons are present in anthropogenic wastewaters which enter natural aquatic systems. Estimates for Swiss natural waters (Stumm and Thüer, 1973) implicate urban drainage and municipal wastewater as major sources. Although frequently causing serious local water quality impairments, accidental spillages contribute in a negligible amount to the total anthropogenic input of hydrocarbons to natural waters. The major contribution of the two land sources on the hydrocarbon pollution of coastal waters was reported by Storrs (1973).

* Present Address: Ciba-Geigy AG, Environmental Technology Dept.
CH-4000 Basel, Switzerland.

Motorway runoff contributes substantially to the hydrocarbon load of urban drainages (Dauber et al, 1978). The relation between traffic and the composition of motorway runoff was demonstrated earlier by Shaheen (1975). In contrast to motorway stormwater runoff, secondary treatment plants continuously feed large amounts of hydrocarbons into receiving waters. Van Vleet and Quinn (1977) found that approximately 95% of the hydrocarbons in secondary effluents occurred in association with suspended solids. Thüer and Stumm (1977) have shown that suspended mineral particulates can play an important role in the transport of hydrocarbons. These conclusions were supported by model experiments (Zürcher and Thüer 1978). Therefore, both the amount and the distribution mode are significant factors in evaluating the ecological impact of hydrocarbons entering surface waters.

The present study was undertaken to measure the relative input of petroleum hydrocarbons to natural water by motorway runoffs and secondary effluents and to determine the distribution of hydrocarbons in these polluting sources. A treatment plant which received no industrial waste discharges was selected for the study of secondary effluents. The origin of petroleum hydrocarbons in each source will be discussed. The investigation of motorway runoff was part of a larger study by Dauber et al (1978).

EXPERIMENTAL

Model experiments. No.2 fuel oil was added to a batch suspension of kaolinite (10 g/1) in 10^{-3} n NaCl solution. Experiments were performed under two different conditions, first without rupture of the oil film (dissolution) and second with high turbulence, breaking the oil film into droplets (dispersion). Water and mineral phases were analyzed separately in order to quantify and characterize the hydrocarbon distribution. Experimental details are described elsewhere (Thüer, 1975).

Motorway stormwater runoff. Hydrocarbons were measured within an extended study by Dauber et al (1978) to estimate the contribution of motorway runoff to the pollution of receiving waters. The samples were preserved in glass bottles with nitric acid (0.05 n) at 5°C until analyzed in the laboratory. Chemical analyses were conducted according to new methods recommended by the Swiss Federal Department of the Interior (not yet published).

Secondary effluent. A small treatment plant serving 1230 inhabitants and receiving no industrial waste discharges was selected for study. Daily cumulative samples of a secondary effluent were collected during one week and preserved with dilute nitric acid acid (0.05 n) at 5°C before analysis. Hydrocarbon concentrations were measured in dissolved and particulate phases. An extended chemical characterization of the effluent is given by Conrad et al (1978).

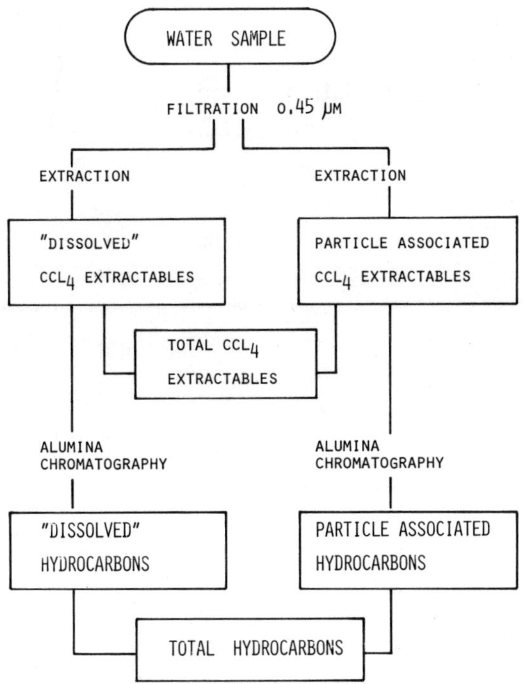

Figure 1. Flow diagram for analysis of total hydrocarbons in water samples.

Analytical. The analytical scheme is shown in Figure 1. Samples were first filtered through 0.45 μm cellulose nitrate filters (Sartorius SM 11306). The water and particle phases were subsequently extracted with CCl_4, and the extracts were percolated through a column (0.4 mm i.d. x 50 mm) of activated alumina (alkaline Al_2O_3, 0% H_2O) to remove hydrocarbons. The extracts were quantified before and after percolation by infrared spectrometry (ir). The sum of absorption from wave numbers 2925 and 2960 cm^{-1} was compared with the corresponding sum of a "Simard" standard (Simard et al, 1951). The detection limit was 100 μg/g for hydrocarbons extracted from the aqueous phase and 20 μg/g for extraction of dry particulates. Total hydrocarbons and total CCl_4 extractables were calculated by summing the values for "dis solved" and particle-associated fractions. The separation of dissolved and particulate phases before extraction by CCl_4 is an important step in the analytical scheme. With this procedure, the total hydrocarbon recovery for street runoff samples was increased by more than 50% as compared to direct extraction of unfiltered samples. This is probably due to reduced solvent contact of particulates in aquatic suspension. The yields of extractions were above 90% for filtrates and above 95% for dried particulates.

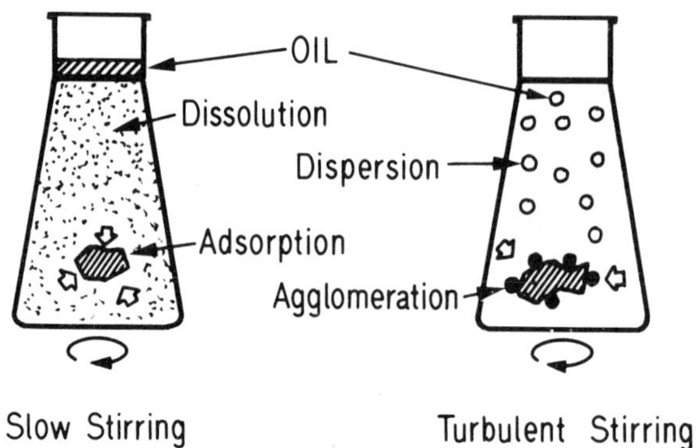

Figure 2. Distribution of hydrocarbons from No. 2 fuel oil during contact with an aquatic clay suspension.

Gas chromatography (gc) was performed on a Carlo Erba gas chromatograph, equipped with glass capillary column SE 52 (20 m) which was provided by K. Grob. The Grob splitless injection technique was used (Grob and Grob, 1974). The column temperature was programmed at 3°/min. from 50°to 250°C. Peak identifications were based on retention times.

RESULTS

Laboratory experiments. The modes of hydrocarbon distribution used in laboratory experiments are shown in Figure 2. By varying the energy input one may examine systems with an oil film at the surface or with dispersed oil droplets. Low molecular weight hydrocarbons of No. 2 fuel oil, especially alkylated benzenes and naphthalenes are dissolved in the water phase. Higher molecular weight hydrocarbons are preferably adsorbed at the solid/liquid or air/water interfaces because of their greater hydrophobic nature. Dispersed droplets readily form agglomerates with suspended solids (Thüer and Stumm, 1977). After such simulated weathering conditions the hydrocarbon fractions were specified and quantified (see Table 1). In comparing experiments with and without turbulence, we find that the amount of hydrocarbons associated with the particulate phase generally was 100 times greater when agglomeration occurred. Adsorbed hydrocarbons (400 μg/g dry weight) were primarily long-chain aliphatic molecules. In contrast, ir and gc analyses of the hydrocarbon composition of oil-kaolinite agglomerates were very similar to the original No. 2 fuel oil (Zürcher and Thüer, 1978). Adsorption was not limited by the slow stirring rate since kinetic experiments have shown that equilibrium is attained (Thüer, 1975).

Table 1 Specific Hydrocarbon Distribution in the Aquatic Clay Suspension Exposed to No. 2 Fuel Oil Under Laboratory Conditions

ARTIFICIAL WATER WITH 10 G/L SUSPENDED CLAY *	METHOD OF ANALYSIS **	HYDROCARBON TYPE	HYDROCARBON CONTENT	
			DRY WEIGHT MG/G	WATER SAMPLE MG/L
DISSOLVED IN WATER	IR, GC, MS	AROMATIC, ONE- AND TWO RING	–	5
ADSORBED ON KAOLINITE	IR, GC	ALIPHATIC, STRAIGHT CHAIN & BRANCHED (MW 220-420)	0.4	4
AGGLOMERATED WITH KAOLINITE (TURBULENCE)	IR, GC	PARAFFINIC, NAPHTHENIC & AROMATIC (ORIGINAL MIXTURE) (MW 120-300)	20 - 100	200 - 1000

* FOR DETAILS SEE EXPERIMENTAL

** IR: INFRARED SPECTROMETRY, GC: HIGH RESOLUTION CAPILLARY GAS CHROMATOGRAPHY, MS: MASS SPECTROMETRY

Motorway stormwater runoff. Measurements of organic pollutants and lead in 70 motorway runoff events are summarized by concentration distribution in Figure 3. The cumulative relative frequencies indicate log-normal distributions. Therefore the geometrical mean can be considered as the most characteristic mean value (Dauber et al, 1978). The 95% value for hydrocarbons does not exceed the Swiss standard of 10 mg/1 for discharge to receiving water.

Table 2 summarizes the measured concentrations of organic constituents. CCl_4 extractables and hydrocarbons are predominantly associated with the particulate phase. Percentages of total organic carbon, hydrocarbons and CCl_4 extractables in filtrates were 34%, 5%, 4%, respectively. Since the dissolved fractions may contain particulates of colloidal size, these portions represent the upper limit for dissolved compounds. A number of samples which were extracted after centrifugation showed the same distribution as found with filtration. Thus, the distribution of CCl_4 extractables and hydrocarbons is not an artifact of the filtration method.

The concentrations of pollutants associated with particulate matter in motorway runoff is given in Table 2 (dry weight). The concentrations of hydrocarbons is 1000 times larger than background values for unpolluted sediments (Unger, 1971). Table 2 also shows calculated mass loads of pollutants per car kilometer.

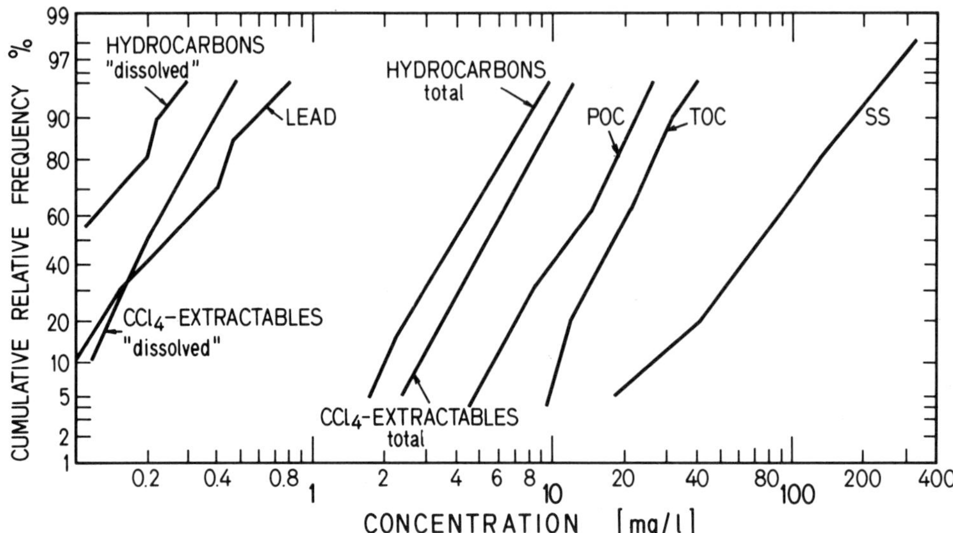

Figure 3. Cumulative relative frequencies of measured runoff concentrations of organic constituents and lead.

Table 2 Mean Concentrations, Dry Weight Portions and Specific Mass Loads of Organic Constituents in Motorway Storm-water Runoff

Polluant	Concentrations: Geometrical mean mg/l	Concentrations: Flow weighted mean mg/l	Mean dry weight portions in particulate matter mg/g	Specific mass loads mg/car km *
CCl_4 extractables, particle associated	5.2	6.0	75	14.7
CCl_4 extratables, total	5.5	6.2		15.4
Hydrocarbons, particle associated	3.7	4.4	60	10.5
Hydrocarbons, total	3.9	4.6		10.9
Total organic carbon (TOC)**	18.5	20		47
Particular carbons (TOC-DOC)**	10.7	13.4	175	31
Total suspended solids**	73	98		240
Lead**	0.26	3	4.3	0.7

* According to automatic measurements: 9.6 million car kilometers during 7 months period, on the 1.7-km length of roadway studied

** "Dauber et al (1978)"

Table 3 Distribution and Specific Load of Hydrocarbons and CCl_4 Extractables in a Domestic Secondary Effluent

POLLUTANT	MEAN CONCENTRATION MG/L	MEAN DRY WEIGHT PORTIONS IN SUSPENDED SOLIDS MG/G	MEAN MASS LOADS *) MG/HAB.-DAY
CCL_4 EXTRACTABLES, PARTICLE ASSOCIATED	1.0	200	463
CCL_4 EXTRACTABLES, TOTAL	1.2		520
HYDROCARBONS, PARTICLE ASSOCIATED	0.6	100	252
HYDROCARBONS, TOTAL	0.6		256
SUSPENDED SOLIDS **	7		2930
PARTICULATE ORGANIC CARBON **		380	1100
LEAD **	0.001		0.4

* NUMBER OF INHABITANTS: 1230

** "CONRAD (1973)"

Secondary effluent. The mean concentrations and the distribution of organic constituents in the secondary effluent are shown in Table 3. Mean dry weight portions related to suspended solids, and mean mass loads per capita, are included. Again the dissolved fractions are small in comparison to total concentrations (7% of organic carbon, 11% of CCl_4 extractables, 2% of hydrocarbons). Similar results were observed by Van Vleet and Quinn (1977) who found only about 5% of hydrocarbons in the soluble form.

The specific hydrocarbon load determined for the secondary effluent (0.25 g/capita per day) is low compared to values from Storre (1973) (8 g/capita per day). This discrepancy can be explained by the absence of any industrial raw-water input to the waste-water plant used for our study.

DISCUSSION

In order to understand the transport and distribution of hydrocarbons in natural aquatic systems, the physio-chemical form of possible input sources must be known. Dispersed oil fractions act as transition state before agglomeration and dissolution. Dissolved hydrocarbons may eventually enter the atmosphere by transport across the air/water interface (McAuliffe, 1977). However, adsorbed and agglomerated hydrocarbons are most frequently found in natural aquatic systems as persistent fractions. Numerous studies of sediments from rivers (Wenzlow, 1975; Hellmann, 1977) and lakes (Unger, 1971; Giger et al, 1974; Wakeham, 1977; Hellmann, 1977) have shown that hydrocarbons accumulate in the sediments following anthropogenic inputs. Since the association of hydrocarbons with particulates is controlled primarily by agglomeration, sediments with finer particles can accumulate a higher concentration of hydrocarbons on a dry weight basis. Thus the suspended solids of the Streu River show higher hydrocarbon concentrations than the underlying river sediments (Table 4). This observation also applies to particulates in secondary effluents which have an average smaller diameter as compared to those in sewage sludge. The results summarized in Table 4 also demonstrate the analytical advantage of phase separation before quantification of hydrocarbons by utilizing the natural enrichment on particulates, which results in higher extraction yields.

The high concentration of hydrocarbons (20 - 120 mg/g) in particulates from secondary effluent and motorway runoff demonstrate the pollution potential of these sources for receiving waters. The bulk of petroleum hydrocarbons entering natural waters are apparently associated with particulates (95%). Infrared and

Table 4 Hydrocarbon Concentrations of Particulate Matter in Aquatic Environments

	Total hydrocarbons dry weight (mg/g)	Anal. method	Investigator
Surface sediments lake Constance	0.2 - 0.3	ir	Unger, 1971
Sediment river Streu	0.03 ± 0.04	gc	Wenzlow, 1975
Suspended solids river Streu	1 - 3	gc	Wenzlow, 1975
Surface sediments lake Washington	1.6	gc	Wakeham, 1976
Street dust	1 - 9	ir	Hellmann, 1977
Sewage sludge	2 - 6	ir	Hellmann, 1977
Secondary effluent particulates	100 ± 20	ir	this study
Motorway runoff	60 ± 40	ir	this study
Laboratory experiments	100 ± 20	ir	Thüer, 1975

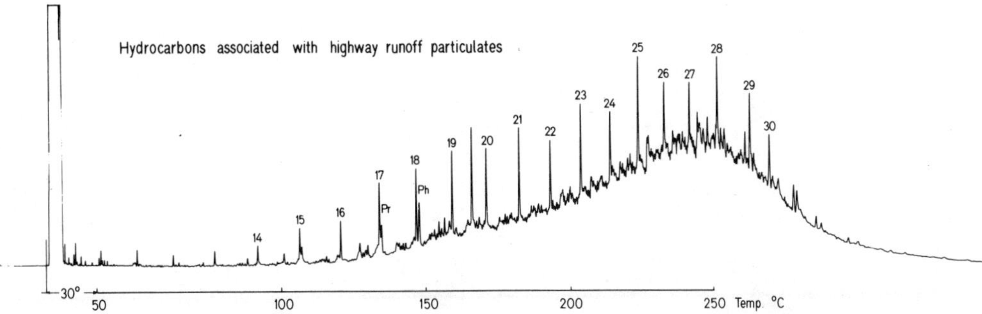

Figure 5. Gas-chromatographic separation of hydrocarbons associated with motorway runoff particulates. Numbers referring to chain length of corresponding n-alkanes, Pristane and Phytane indicated as Pr and Ph. Column: SE 52 coated glass capillary (20 m x 0.32 mm i.d.) Conditions: 1.5 µl, splitless injection, carrier gas pressure H_2: 6 psig (415 mbar), 3°/min. from 50 to 250°C.

gas-chromatographic analyses evidence that the composition of these hydrocarbons in secondary effluents and motorway runoff is identical. A typical gas chromatogram for particulated hydrocarbons in motorway runoff is shown in Figure 5. Finally, in agreement with laboratory experiments, we can conclude that the dominant associated species are oil-particle agglomerates (Zürcher and Thüer, 1978). For motorway runoff particulates of mineral type, the adsorbed hydrocarbons probably contribute less than 1% to total hydrocarbons, since the maximum adsorption of hydrocarbons observed on kaolinite crorespond to a concentration of 0.4 mg/g. The mean hydrocarbon concentration of secondary effluent particulates related to dry weight reached 10% (100 mg/g, Table 3). The adsorption capacity of organic particulates is not well known, but probably cannot account for the high concentration observed in particulates from secondary effluents.

In addition, we have attempted to identify the origin of hydrocarbons in motorway stormwater runoff. Automobile exhaust must be viewed as the primary hydrocarbon source for street runoff since lead is well correlated with hydrocarbons (Figure 6). Total CCl_4 extractables also correlate well with total hydrocarbons (Figure 6.). This correspondence (correlation coefficient 0.98) is typical for time- and climate-independent sources. Therefore, atmospheric sources can be neglected as contributors to motorway runoff. Also, rainwater samples have shown low concentrations of hydrocarbons. It is likely that various automotive sources of hydrocarbons have to be considred, e.g. burned fuel, exhausted motor oil, leakage of

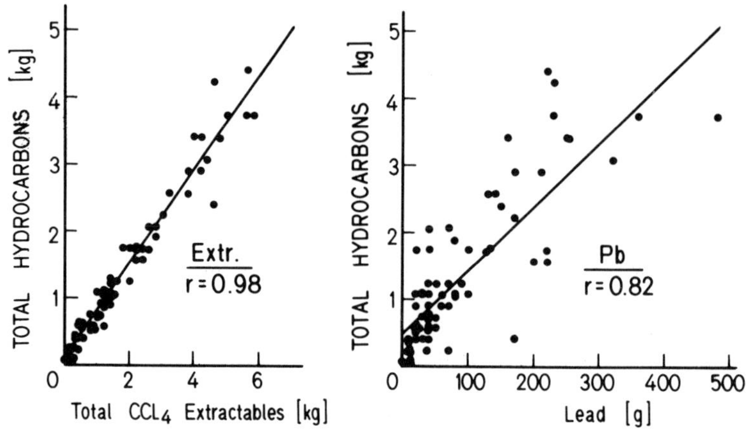

Figure 6. Correlation of the loads of hydrocarbons with CCl_4 extractables and lead, measured in the motorway runoff.

lubricants, and erosion from asphalt pavement and tires. Traffic-related hydrocarbon emissions have been reported by Shaheen (1975) as independent of the pavement, asphalt or cement. We therefore conclude that the asphalt pavement of the motorway used in this study did not greatly influence the total hdyrocarbon mass load. This assumption is supported by the lack of a significant correlation between total unsoluble matter (eroded pavement) and hydrocarbons.

The bulk of hydrocarbons in exhaust particulates analyzed by Boyer and Laitinen (1975) include several unresolved peaks for compounds greater than C_{22} and a medium retention time equivalent to that of n-C_{29}. In our study, the hydrocarbons associated with highway runoff particulates showed a similar unresolved complex mixture above C_{22} and a medium retention time of n-C_{28} (Figure 5). In addition, these compounds have an ir-spectrum very similar to altered motor oil (Kägler, 1969). Thus, automobile exhaust is also implicated as the primary source of hydrocarbons by gc- and ir- analyses.

The origin of hydrocarbons in municipal wastewater is difficult to determine. Since this study was made during a dry-weather period (no precipitation during the study), the contribution of street runoff to wastewater influent was limited. The hydrocarbon/lead ratio found in the wastewater influent was 10 times higher than observed for motorway runoff. Thus, it is possible that sources other than stormwater runoff contribute to the hydrocarbons entering the treatment plant, e.g. cleaning solvents and lubricating oil wastes from small automotive industries. However, the hydrocarbon/lead ratio in the influent may also be affected by differing removal processes that occur during transport to the plant. Further study is needed to determine the source of hydrocarbons in municipal waste water.

CONCLUSIONS

Filtration of water samples followed by separate extraction of the aqueous and particulate phases results in higher analytical yields for total hydrocarbons. Hydrocarbon concentrations in motorway runoff and domestic secondary effluent were 10 to 100 times larger than background concentrations of surface and rain-waters. The hydrocarbons were discharged almost exclusively associated with particulates (95%). Thus, the ecological impact of hydrocarbons in anthropogenic wastewaters will be largely determined by the availability of the particle-associated fraction. This fraction is important since the concentrations in motorway runoff and secondary effluent are 100 to 1000 times larger than those found in recent lake sediments. Based on the results of model weathering experiments and chemical analyses (ir, gc), we conclude that the hydrocarbons are associated with particulates by an agglomeration-predominant mechanism rather than an adsorption mechanism.

Because of their composition and correlation with lead, hydrocarbons in motorway runoff originate primarily from motor oil exhausted by automobiles. Estimates of the hydrocarbon portion of particulate organic carbon were 33% for motorway runoff and 23% for secondary effluent.

ACKNOWLEDGEMENT

We are indebted to L. Dauber and B. Novak for useful advice and technical support on motorway runoff. We thank Franziska Lutz, W. Staudenmann and D. Widmer for their assistance on hydrocarbon analysis. We are grateful to the collaborators of the analytical laboratories of the EAWAG who determined lead and carbon and to Ch. Schaffner for gas chromatographic analyses of secondary effluents. The writer has greatly benefited from discussions with Th. Conrad, J. Zobrist, S. Wakeham and W. Giger.

This work was supported in part by the Swiss Department of Commerce (Commission of the European Communities, Project COST 64b). We are indebted to Prof. Werner Stumm who supported our activities.

REFERENCES

Boyer, K.W., and H.A. Laitinen, 1975. Automobile exhaust particulates; properties of environmental significance. Env. Sci. Technol. 9: 457-469.

Conrad, Th., et al, 1978. Unpublished results, EAWAG, Dübendorf, Switzerland.

Dauber, L., B. Novak, J. Zobrist and F. Zürcher, 1978. Pollutants in motorway stormwater runoff, In: Road Drainage, Symposium organized by the Swiss Federal Office of Highways and Rivers, Berne, edited by Organization for Economic Co-operation and Development, Paris.

Giger, W., M. Reinhard, C. Schaffner, and W. Stumm, 1974. Petroleum-derived and indigenous hydrocarbons in recent sediments of Lake Zug, Switzerland, Env. Sci. Technol. 8: 454-455.

Grob, K. and K. Grob Jr., 1974. Isothermal analysis on capillary columns without stream splitting. The role of the solvent, J. Chromatogr. 94: 53-64

Hellmann, H., 1977. Zur Belastung von Sedimenten, Klärschlämmen, Luft- und Strassenstaub durch organische Stoffe, insbesondere Kohlenwasserstoffe und Phosphate, Vom Wasser 48: 225-246

Kägler, S.H., 1969. Neue Mineralölanalyse, Verlag A. Huthig, pp. 310.

McAuliffe, C.D., 1977. Evaporation and solution of C_2 to C_{10} hydrocarbons from crude oils on the sea surface, In: Fate and Effects of Petroleum in Marine Ecosystems and Organisms (ed. D.A. Wolfe), Pergamon Press, New York, pp. 363-372.

Shaheen, D.G., 1975. Contribution of urban roadway usage to water pollution, Environmental Protection Technology Series, Washington, D.C.

Simard, R.G., I. Hasegawa, W. Bandaruk, and C.E. Headington, 1951. Infrared spectrophotometric determination of oils and phenol in water, Anal. Chem. 23: 1384-1387.

Storrs, P.N., 1973. Petroleum inputs to the marine environment from land sources, Background papers, workshop on petroleum in marine environment, Ocean Affairs Board, National Academy of Sciences, Washington, D.C.

Stumm, W., and M. Thüer, 1973. Der Beitrag des Autos zur Gewässerverschmutzung - Quellen, Anteile und Auswirkungen, Auto-Mensch-Umwelt 16: 209-222.

Thüer, M., 1975. Tramsportmöglichkeiten von Oel in schwebestoffhaltigem Wasser, PhD. Thesis 5628, ETH Zürich.

Thüer, M. and W. Stumm, 1977. Sedimentation of dispersed oil in surface waters, In: Progress Water Technology, Pergamon Press, 9: 183-194.

Unger, U., 1971. Untersuchungen uber die Verunreinigung des Bodensees durch Mineralöl, Gas- und Wasserfach 112: 256-261.

Van Vleet, E.S., and J.G. Quinn, 1977. Input and fate of petroleum hydrocarbons entering the Providence River and Upper Narragansett Bay from waste water effluents, Env. Sci. Technol. 12: 1086-1092.

Wakeham, S.G., 1977. Hydrocarbon budgets for Lake Washington, Limnol. Oceanogr. 22: 952-957.

Wenzlow, B., 1975. Kohlenwasserstoffe in Oberflächengewässern am Beispiel der Fränkischen Saale und der Streu, PhD. Thesis, TH Aachen, West Germany.

Zürcher, F. and M. Thüer, 1978. Rapid weathering processes of fuel oil in natural waters - analyses and interpretations, Env. Sci. Technol. 12: 838-843.

ORGANOCHLORINES IN PRECIPITATION IN THE GREAT LAKES REGION

W.M.J. Strachan, H. Huneault, W.M. Schertzer and

F.C. Elder

Canada Centre For Inland Waters

P.O. Box 5050, Burlington, Ontario L7R 4A6

ABSTRACT

Atmospheric precipitation in the form of snow (1976) and rain (1976 and 1977) was collected from around the Canadian side of the Great Lakes as well as inland. All sites were remote from any nearby industrial or urban contamination. Rain samples were collected with a 0.36 m^2 stainless steel funnel and were event related. Snow was obtained from the accumulated columns during the month of February, 1976. A total of 81 rain and 17 snow samples were examined.

Quantifiable levels of PCBs, α-BHC, lindane, DDT residues, endosulfan, dieldrin and methoxychlor were all observed in most samples; HCB and chlordane were found infrequently; heptachlor, heptachlor epoxide, aldrin, endrin, and mirex were not detected. The observed concentrations of these substances were generally the same in both 1976 and 1977 rainfall, except that α-BHC appeared to be increased during 1977. Levels in snowmelt were considerably less than for rainfall, except for PCBs for which they were approximately the same.

An examination of air movements preceding several of the events failed to indicate any particular region(s) or source(s) for any of these contaminants. Indications were that most of the substances were associated with particulate deposition during a rainfall.

INTRODUCTION

Organochlorine compounds (OCs) have been reported for Great Lakes region samples as far back as 1958 (Breindenbach et al, 1960) when waters from Lakes Superior, Michigan and Erie were sampled for nine organochlorine pesticides. Since then, determinations in various parts of the aquatic ecosystem have demonstrated the widespread occurrence of these and other OCs and PCBs. The 1976 report of the IJC-Water Quality Board (IJC, 1977a) presented a long list of chlorinated and non-chlorinated persistent organic compounds which have been observed in the Lake Ontario system. While the presence of these chemicals can be understood for the populous industrial and agricultural southern Ontario, the same cannot be said for the two northern lakes. Many of these substances have, however, also been recorded for both Lakes Superior and Huron (IJC, 1977b). This widespread occurrence, coupled with an inability to explain this fact on population grounds alone, suggests a broadly based source such as precipitation. Calculations that the annual loadings to the Great Lakes, based on the assumption of ten nanogram/litre of a chemical might be present in the rainfall of the region, indicates the potential of this input mechanism. Such calculations, along with estimated PCB loadings from municipal sources are given in Table 1.

The seriousness of the atmospheric problem was indicated in 1975 by Sanderson and Frank (PCB Task Force Report, 1976, p.66) and by ourselves, both of whom observed rainfall concentrations of 10 - 100 ng/ℓ. Subsequent to the commencement of these studies, Murphy and Rzeszutko (1978) and Swain (1978) both published results which tended to confirm atmospheric input as a major contaminant source.

Table 1 Theoretical Loadings by 10 ng/ℓ in Rainfall

Lake	Atmospheric Loading		Municipal
	To Lake	To Basin	PCB Loading †
	(kg/annum)		
Superior	630	1600	27
Huron + Georgian Bay	450	1450	25
Erie	220	870	453
Ontario	170	800	317

†loadings from known Canadian municipal sources (PCB Task Force, 1976) and adjusted for total population using data from IJC 1969 and 1977b. Factors: Ontario 1.6; Erie 8.9; Huron + Georgian Bay 2.3; Superior 4.5.

Initial results of this study have been presented by Strachan and Huneault (1978) dealing with the 1976 environmental samples. The work reported here includes the 1977 rain data and some experiments on the involvement of particulates in the precipitation of a given rainfall event.

EXPERIMENTAL

Most of the experimental details have been presented in Strachan and Huneault (1978). Rain samples were collected by means of 0.36 m^2 samplers during the period May - November in both years; snow samples were integrated columns collected during February 1976 after most of the season's snow had fallen. Locations for all samplings, shown in the accompanying map (Figure 1), were chosen for their remoteness from immediate potential pollution sources. The station at CCIW was an exception.

Four of the seven 1976 rain stations peripheral to the lakes were moved inland in 1977 in order to examine for the deposition of PCBs and OCs unaffected by the immediate Great Lakes environment. The stations at Batchawana Bay, CCIW and Picton were retained for reference. Additionally, all 1977 stations were equipped with tipping bucket rain gauges (0.25 mm sensitivity) where the 1976 stations depended upon nearby meteorological stations.

Analyses were according to procedures of the methods of the Inland Waters Directorate (IWD, 1974). Substances analyzed for were: PCBs, HCB, dieldrin, aldrin, endrin, α-BHC, lindane, DDT residues (p,p'-DDT, o,p'-DDT, p,p'-DDD and p,p'-DDE), methoxychlor, mirex, chlordane, endosulfan, heptachlor and heptachlor epoxide. Quantitation limits (10 - 20% precision) were 1 - 2 nanogram/litre of whole water except for PCBs where the limit was 20 nanogram/litre. Mirex, aldrin, heptachlor and heptachlor epoxide were not observed; endrin, γ-chlordane and HCB were detected on several occasions but quantitation and confirmation were not usually possible. The remaining substances were each observed in at least 29% of the rain. PCBs were found in virtually every sample. All data presented here pertain to quantified and confirmed observations based upon three different gas chromatographic columns.

In addition to the survey aspects outlined above, an attempt was made to ascertain the role of particulates in wetfall deposition of PCBs and OCs. A larger (3.0 m^2) sampler was used to collect subsamples of three different rain events. A tipping bucket rain gauge was used in conjunction with these samples.

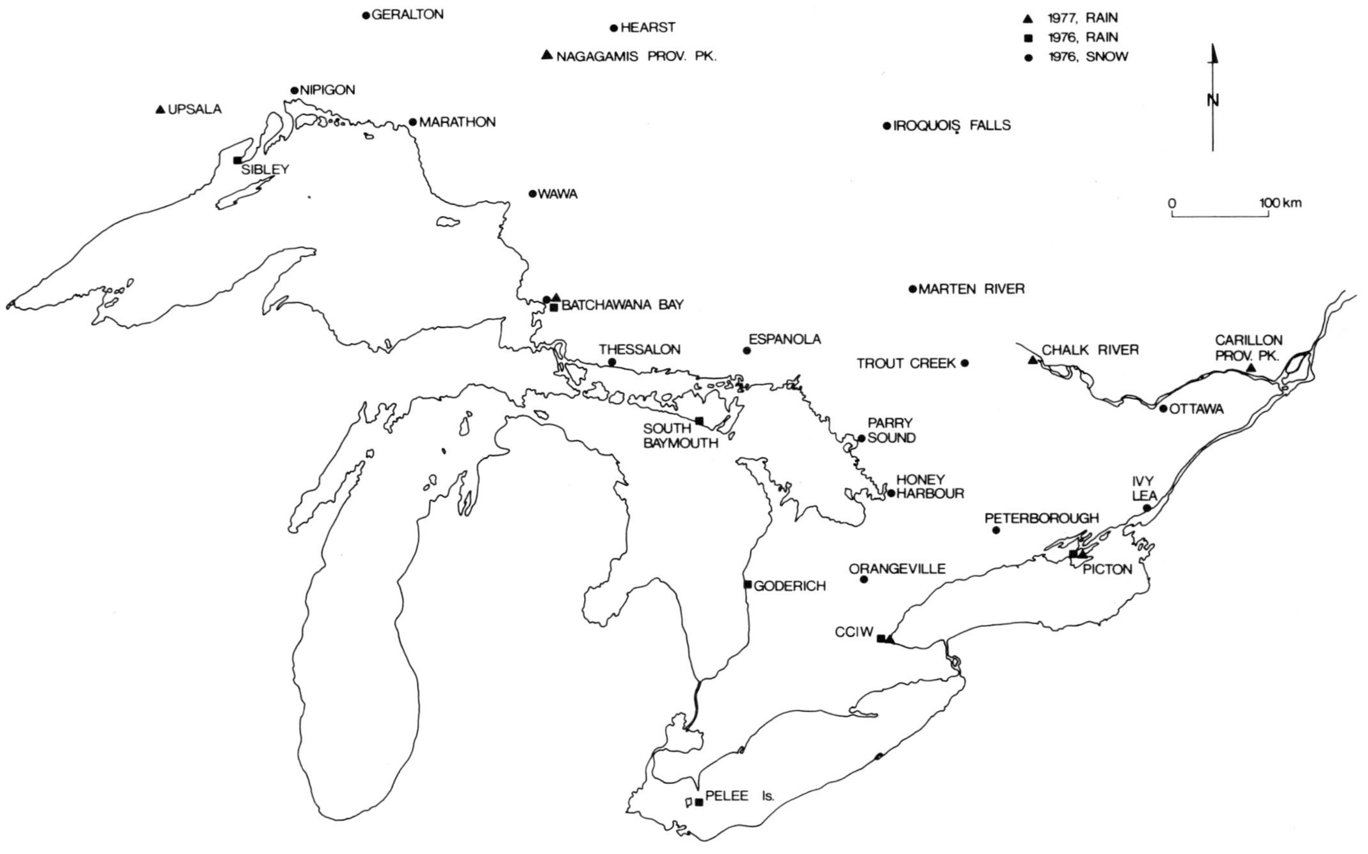

Figure 1. Locations of sampling ports

RESULTS

A total of 50 rain samples were collected during 1976 and 31 during 1977; 17 snowfall samples were also obtained. On average, the rainfalls of largely single events collected once per month represented 14% of the total rainfall for any area and snowfall averaged approximately 30% of the total annual precipitation. Analytical results for the substances in these samples are presented in Table 2.

Snowmelt values are considerably reduced from levels in rain, with the possible exception of PCBs and DDT residues. This may be the result of either decreased usage or lowered volatility druing winter. The "constant" PCB and DDT residue levels may reflect continuous input to the atmosphere from sources far removed from the colder northern hemisphere. The observation that most of the DDT residues is represented by pp'-DDT is indicative of fresh application since this isomer is readily converted to the other "DDT residue" forms in the environment. Since DDT is banned for most usages in Canada and the United States, this observation lends weight to the theory of global transport of this material.

A second noteworthy aspect of Table 2 is the considerable similarity between rain values of 1976 and 1977, despite the fact that the locations of many of the stations are considerably different in the two years. This would seem to indicate the widespread, long range nature of atmospheric transport of these pollutants. It is also of concern that the levels of PCBs, lindane and α-BHC (and their loadings) are as high as the 10 ng/ℓ concentration used to calculate the loadings of Table 1. Methoxychlor would also appear to be approaching that level.

Table 2 Mean Concentrations of PCBs and OCs in Precipitation

	Snow	Rain	
	1975/6	1976	1977
No. of Samples	17	50	31
	ng/ℓ		
PCBs	29	21	28
Lindane	tr	5	9
α-BHC	1	12	28
DDT residues*	1	3	2
α-Endosulfan	0	2	1
β-Endosulfan	0	5	4
Dieldrin	2	1	1
Methoxychlor	2	8	6

*76% pp'-DDT, 9% pp'-DDE, 9% op'-DDT, 6% pp'-DDD

Table 3 Regional Precipitation

	Batchawana 1976	Batchawana 1977	CCIW + Picton 1976	CCIW + Picton 1977	Inland 1977
No. of Samples	8	4	16	10	17
	ng/ℓ				
PCBs	44	14	43	22	22
Lindane	6	12	5	8	9
α-BHC	1	40	19	23	28
DDT residues	0	0	6	3	tr
α-Endosulfan	tr	tr	5	2	tr
β-Endosulfan	2	1	12	4	2
Dieldrin	1	tr	2	1	tr
Methoxychlor	2	2	9	9	5

In Table 3, the rain data are presented for the three stations which were common to both the 1976 and 1977 programmes as well as for the four 1977 inland stations.

Batchawana on the shore of Lake Superior is located in an area with little of man's influence; CCIW and Picton are on Lake Ontario, an area of high industrialization, urbanization and intensive agriculture. Most of the "inland" stations are remote from allochthonous influence and geographically resemble Batchawana more than CCIW and Picton. For DDT residues, endosulfan, dieldrin and methoxychlor in both years, there is an apparent difference between the two types of areas. Inland and Batchawana samples are quite similar with collectively their levels averaging only 30% of those observed for CCIW plus Picton. For PCBs, lindane and α-BHC, the differences between years were more pronounced than for the other OCs and were greater than any regional difference. If any pattern is indicated, it is a downward trend in PCBs and an upward one for lindane and α-BHC.

Table 4 presents the rain data as a function of season (spring = May and June; summer = July, August and September; autumn = October and November).

Methoxychlor shows a pattern in which spring values are substantially higher than those during the rest of the year (even the relatively high summer values arise from July samples). A similar but less pronounced behaviour is found for DDT residues. The increase in a α-BHC and lindane levels between 1976 and 1977 and which are noted in Table 1, is seen to be generally observed regardless of season. Endosulfan in rain is found to be strictly a spring and summer phenomenon. Dieldrin levels seem devoid of seasonal influence.

Table 4 Seasonal Precipitation

	Spring		Summer		Autumn	
	1976	1977	1976	1977	1976	1977
No. of samples	12	10	26	17	10	4
	ng/ℓ					
PCBs	23	31	17	30	27	11
Lindane	4	13	5	8	7	5
α-BHC	0	30	18	27	11	29
DDT residues	7	2	2	1	2	1
α-Endosulfan	1	2	3	1	0	0
β-Endosulfan	4	6	8	3	0	0
Dieldrin	1	1	1	1	2	2
Methoxychlor	13	10	8	5	3	1

In Figure 2, graphic data are presented which illustrate that the bulk of the PCBs and OCs are deposited early in the rainfall. Those substances not shown in the Figure were not observed at concentrations sufficiently above the quantification levels to indicate any pattern.

DISCUSSION

The ambiguity of the regional and seasonal behaviour patterns of the pollutants plus their uneven distribution through the course of a rainfall makes the use of mean values for loading estimates suspect. Despite such limitations, the observed concentrations indicate very substantial loadings - equal to the tributary loadings (IJC, 1977b) in the case of Lakes Superior and Huron (1740 and 760 kg/annum, respectively). Indeed, much of the tributary loadings may themselves be derived from atmospheric precipitation. Data for similar comparisons in the lower Great Lakes are in preparation (IJC, 1978) but it may be that the proportion of atmospheric input is not as great there as in the upper lakes. Nevertheless, it is apparent that precipitation is a very important factor in the PCB and OC budgets in both the lakes region and inland. Indeed, it is likely to be a factor in areas far removed from the study area.

Eight clear-cut examples of single events, including the three in the sub-event study, are being examined as to movement of the air mass involved (at several altitudes) for the 48 hours preceding the event. This approach is a current study but preliminary indications support the contention that these substances are generally contaminants in the atmosphere and that their sources are not localized.

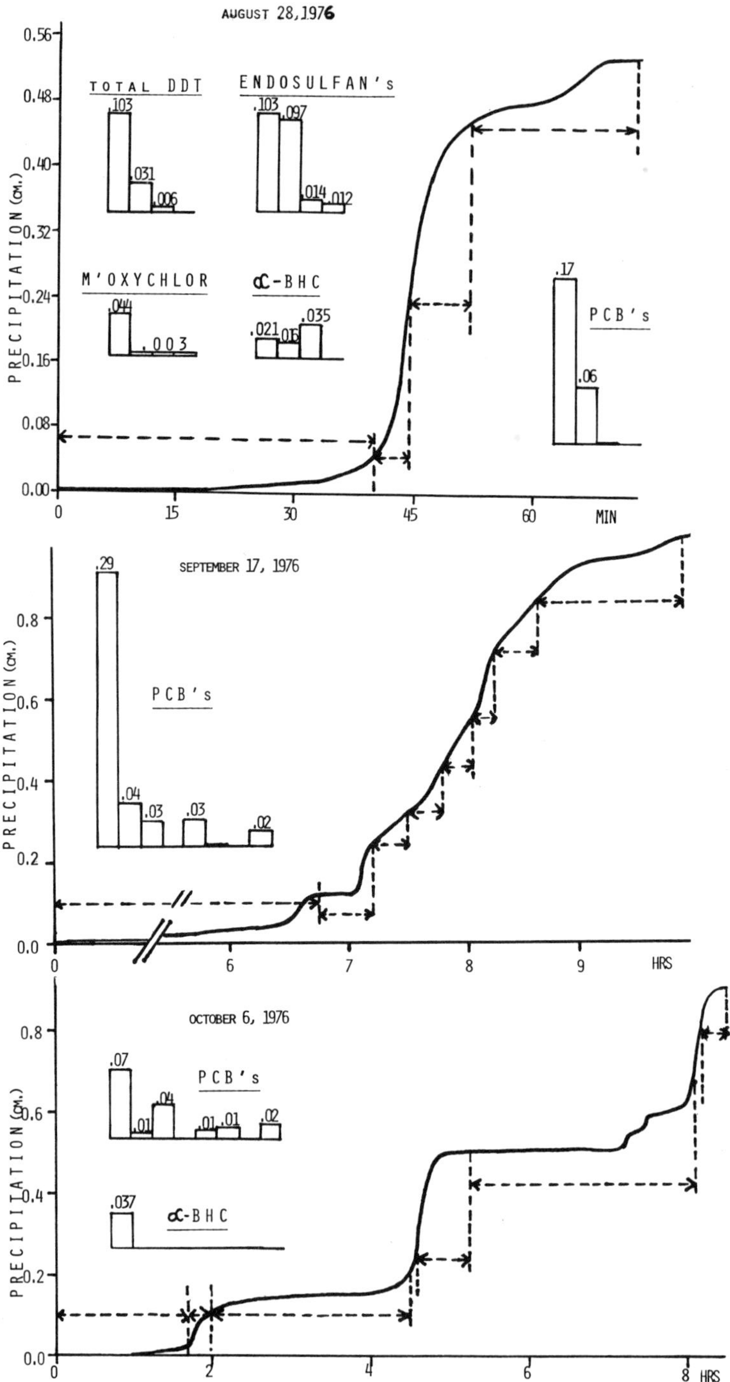

Figure 2. C.C.I.W. Precipitation

ACKNOWLEDGEMENTS

The authors would like to thank R.C.J. Sampson, J.A. Coburn and I.A. Valdmanis of the Water Quality Branch (Ontario Region) for the analyses.

REFERENCES

Breidenbach, A.W., C.G. Gunnerson, F.K. Kawahara, J.J. Lichtenberg, and R.S. Green. 1967. Chlorinated hydrocarbon pesticides in major river basins, 1957-65. U.S. Public Health Service, Public Health Report 82, 139-156.

I.J.C. 1969. Pollution of Lake Erie, Lake Ontario and the International Section of the St. Lawrence River. Vol. 2: Lake Erie; and, Vol. 3: Lake Ontario and the International Section of the St. Lawrence River. Reports by the Water Pollution Board. pp. 316 and 329, respectively.

I.J.C. 1977a. Great Lakes Water Quality Board Annual Report. 1976. Appendix E: Status report on the persistent toxic pollutants in the Lake Ontario basin. p.95.

I.J.C. 1977b. The Waters of Lake Huron and Superior. Vol II: Lake Huron, Georgian Bay and the North Channel; and, Vol. III: Lake Superior. Reports of the Upper Lakes Reference Group. pp. 743 and 575, respectively.

I.J.C. 1978. Environmental Management Strategy for the Great Lakes System. Final report of the Pollution from Land Use Activities Reference Group. pp.115.

I.W.D. 1974. Analytical Methods Manual and subsequent amendments to procedures therein. Inland Waters Directorate, Water Quality Branch, Canadian Dept. of Fisheries and Environment.

Murphy, T.J., and Rzeszutko. 1978. Precipitation inputs of PCBs to Lake Michigan. J. Great Lakes Res. 3, 305-312.

PCB Task Force. 1976. Background to the regulation of polychlorinated biphenyls (PCB) in Canada. Report of the PCB Task Force to the Enviornmental Contaminants Committee of Environment Canada and Health and Welfare Canada. Technical Report 76-1. p.169.

Strachan, W.M.J., and Huneault. 1978. Polychlorinated biphenyls and organochlorine pesticides in Great Lakes precipitation. J. Great Lakes Research (in press).

Swain, W.R. 1978. Chlorinated residues in fish, water and precipitation from the vicinity of Isle Royale, Lake Superior. J. Great Lakes Res. (in press).

POLYCYCLIC AROMATIC HYDROCARBONS IN THE MARINE ENVIRONMENT: GULF OF MAINE SEDIMENTS AND NOVA SCOTIA SOILS

R.A. Hites, R.E. Laflamme, and J.G. Windsor, Jr.

Department of Chemical Engineering

Massachusetts Institute of Technology

Cambridge, Massachusetts, 02139

Polycyclic aromatic hydrocarbons (PAH) occur widely in the lithosphere. They have been reported in various soils (Youngblood and Blumer, 1975; Laflamme and Hites, 1978), in marine sediments (Youngblood and Blumer, 1975; Hites et al, 1977), and in urban limnic sediments in the United States (Hites and Biemann, 1975; Wakeham, 1977) and in Europe (Grimmer and Böhnke, 1975; Muller et al, 1977; Giger and Schaffner, 1978). The concentrations of PAH range from less than 100 ppb for abyssal plain sediments to more than 100,000 ppb for sediments from highly urbanized areas. Much of the past work on the organic geochemistry of PAH has centered on understanding their sources, and there now seems to be a consesus (Youngblood and Blumer, 1975; Laflamme and Hites, 1978; Hites et al, 1977) that most (but not all) PAH in the sedimentary environment are due to combustion processes. However, little information is available on the mode(s) by which PAH are transported from the combustion source to the sediment. Existing data indicate high PAH levels at those locations which are close to centers of intense human activity and very low PAH levels at sites remote from anthropogenic influence. This is, however, only a qualitative observation which needs to be supplemented by a quantitative study. This paper represents such a study.

We have selected the Gulf of Maine and Nova Scotia for detailed study (see Figure 1). This selection was based on the expectation that the industrialized New England area would be the major regional source of combustion-generated PAH. Since the prevailing wind directions for these latitudes are predominantly southwesterly (Owens and Bowen, 1977), we would have a known, up-wind, semi-point

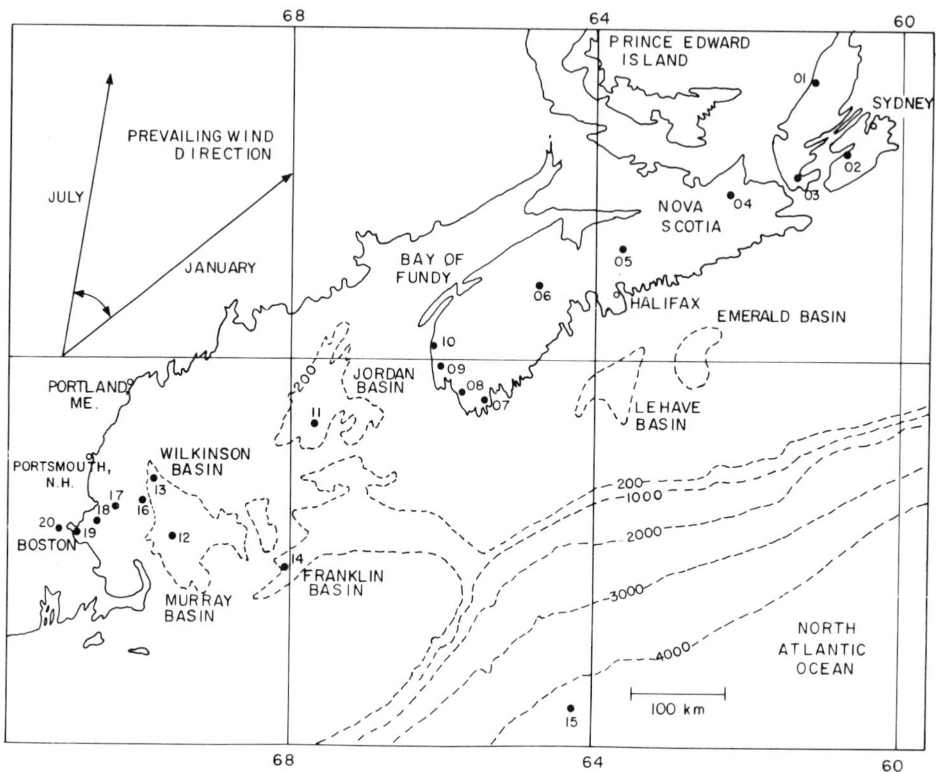

Figure 1. Gulf of Maine, Massachusetts Bay, and Nova Scotia sample locations. Approximate populations: Metropolitan Boston, MA, 4,000,000; Portland, ME, 60,000; Portsmouth, NH, 30,000; Nova Scotia, 770,000; Halifax-Dartmouth, NS, 270,000; Sydney, NS, 60,000. The latitude line is 44°N

source input of PAH into this geographical area. Therefore, the sedimentary PAH concentration distribution over this region should be indicative of PAH transport mechanism(s).

EXPERIMENTAL METHODS

The sample locations are shown in Figure 1. The soil samples from Nova Scotia (number 01 to 10) were obtained by MIT in October 1976. In each case, we attempted to minimize local contamination effects, such as highway runoff. None of our soil sample sites seemed to be effected by previous forest fires. The Gulf of Maine (numbers 11 to 15) and the deep-ocean sediment samples (numbers 21 to 23) were obtained by the Woods Hole Oceanographic Institution.

The Massachusetts Bay sediment samples (numbers 16 to 19) were taken with a Smith-MacIntyre grab sampler from the R/V Edgerton (MIT); the Charles River sediment sample (number 20) was from the MIT collection.

All sediment and soil samples were stored in acid-washed, solvent-rinsed, glass jars with foil-lined lids; they were frozen upon collection and kept frozen until extracted. Soxhlet extraction of the wet sediment, solvent partitioning, and silicic acid chromatography yielded a PAH fraction that was qualitatively and quantitatively analyzed by gas chromatographic mass spectrometry. All of these procedures are detailed elsewhere (Laflamme and Hites, 1978).

RESULTS AND DISCUSSION

Nova Scotia Soil Samples. In all of these samples, the relative distribution of individual PAH was highly indicative of combustion-generation, a subject which has been discussed previously (Youngblood and Blumer, 1975; Laflamme and Hites, 1978; Hites et al, 1977). The PAH levels measured at sites 01 to 10 exhibited a rather random, site-to-site variability which we attribute to three factors: (a) variations in soil porosity; (b) variations in lipophilic content of the surface layers; and (c) variations in the humic substances content of the sub-surface layers. Combinations of these three factors have apparently caused the large PAH variations observed in these soil samples. In fact, these site specific variations are so great that they mask variations due to deposition of PAH originating in New England. We, therefore, conclude that these soil samples cannot be used to determine the source of the PAH. Because of their variability, the Nova Scotia data have all been combined to give a representative picture of this location. The median total PAH level in Nova Scotia is 50 ppb (see Table 1), a value which is quite low when compared to sample sites closer to Boston (see below).

Massachusetts Bay Sediment Samples. Samples 16 to 20 represent a transect starting in Metropolitan Boston (Charles River, sample 20) and extending out into Massachusetts Bay in a northeasterly direction for about 94 km (samples 16 to 18). A PAH pattern which is indicative of combustion sources was observed for samples 16 to 20; even in the Boston Harbor (sample 19) there was little indication of petroleum derived PAH. (We do not suggest that there is no petroleum impact on Boston Harbor but that PAH from combustion sources far exceed those from spilled oil). The total PAH abundance (see Table 1) in these samples ranged from 120,000 to 160 ppb and decreased rapidly as a function of distance from Boston. In fact, this decrease fits a semi-logarithmic function quite well as shown in Figure 2; the correlation coefficient

Table 1
Total PAH Concentrations (PPB, Dry Weight Basis) at Various Sites

Location	Site No. (see Fig. 1)	Total PAH Conc. (ppb)
Nova Scotia	01-10	50[a]
Gulf of Maine	11	500
	12	540
	13	870
	14	200
Deep Ocean	15	160
Massachusetts Bay	16	160
	17	830
	18	3,400
	19	8,500
Charles River, Boston	20	120,000
Continental Slope	21[b]	120
Deep Ocean	22[c]	97
	23[d]	18

a. median of all Nova Scotia samples
b. located at 38.9° N, 71.8° W
c. located at 32.4° N, 70.2° W
d. located at 30.0° N, 60.0° W

is -0.958 which is significant at the 99% confidence level. These data certainly indicate that Boston is the source of the PAH observed in these Massachusetts Bay sediments. Furthermore, the rate at which PAH concentrations decrease, as a function of distance from Boston, is quite high; we observe about one order of magnitude less PAH for each 40 km distance from Boston.

Gulf of Maine Basins. Samples 11 to 14 were from four basins in the Gulf of Maine. The total PAH concentrations ranged from 870 to 200 ppb (see Table 1), and only a combustion PAH pattern was observed. Although these samples were more distant from the presumed PAH source (Boston) than the Massachusetts Bay samples, the PAH values were disproportionately high. That is, these PAH concentrations do not fit on the line for the Massachusetts Bay samples shown in Figure 2. We think that these higher values occur because the basins in the Gulf of Maine are areas of unusually high sediment accumulation and act as sinks for fine grained sediments (Shepard, 1963) which contain relatively high levels of organic matter scoured from the surrounding area. The effect can be seen by comparing the sample taken at the edge of the Wilkinson basin (no. 16, 160 ppb total PAH) with that taken in the basin

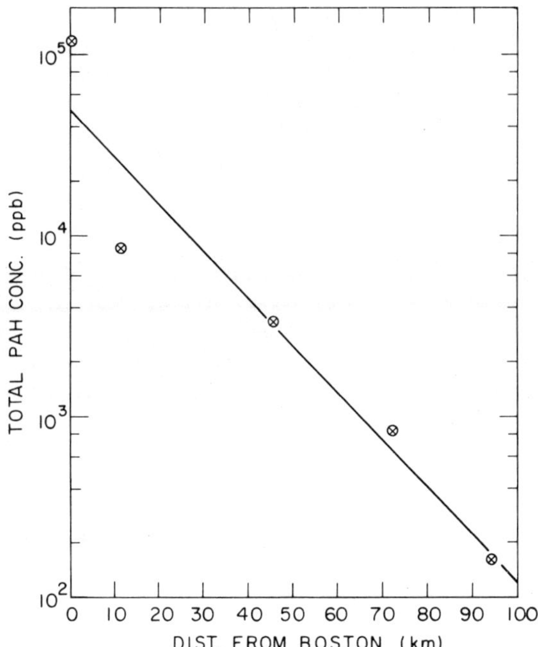

Figure 2. Total PAH concentrations vs. distance from Boston for Massachusetts Bay samples.

(no. 13, 870 ppb). The same effect is noted in the other basins as well, although it is less pronounced.

Deep Ocean Samples. The four samples taken off of the continental shelf (no. 15, 21-23) also gave PAH patterns indicative of combustion sources. The total PAH levels in these samples were 18 to 160 ppb (see Table 1). It is interesting to note that the lowest value (18 ppb) was measured in the sample which was the most remote from the North American continent. Based on the rapid fall off near Boston, these total PAH levels were also unusually high; that is they would not be predicted by extrapolation of the near-shore data presented in Figure 2. For example, the total PAH concentrations at sampling sites 15 and 16 (see Figure 1) are the same despite the fact that site 16 is within 94 km of Boston while site 15 is in the deep ocean and is 620 km from Boston.

Mechanism of PAH Transport. Based on the above measurements of PAH levels, we suggest the following scenario for the transport of PAH. The various fuels which are burned in the metropolitan Boston area produce airborn particulate matter (soot and fly ash) on which polycyclic aromatic hydrocarbons are adsorbed. These particles are transported (toward the northeast) by the prevailing wind for distances which are a strong function of the particle's diameter. We suggest that the long range airborn transport of small particles accounts for the low PAH levels in Nova Scotia soils and in the deep ocean sediments. Remember that the median level in Nova Scotia was only 50 ppb, and the deep ocean values ranged from 18 to 160 ppb.

Larger airborn particles will settle back onto Boston; precipitation then washes them from the streets and buildings. The PAH in this urban run-off eventually accumulate in local sinks such as the Charles River and Boston Harbor. We suggest that these highly contaminated sediments are slowly transported by resuspension and currents to sea-ward locations where the sediments once again accumulate in basins or the deep ocean. The rapid decrease in PAH to a level of 160 ppb within 94 km of Boston (see Figure 2) indicates that this transport mode is a rather short range effect. Certainly sediment transport is necessary to explain the PAH levels observed in the Gulf of Maine basins. In addition, the more near-shore deep ocean sites (no. 15 and 21) may well have PAH contributions due to both sediment resuspension and airborn transport.

ACKNOWLEDGEMENTS

We are grateful to John W. Farrington (WHOI) for helpful discussions and for some sediment samples and to the National Science Foundation for support (grant number OCE 77-20252).

REFERENCES

Giger W., and C. Schaffner. 1978. Determination of polycyclic aromatic hydrocarbons in the environment by glass capillary gas chromatography. Anal. Chem. 50, 243-249.

Grimmer, G., and H. Böhnke. 1975. Profile analysis of polycyclic aromatic hydrocarbons and metal content in sediment layers of a lake. Cancer lett. 1, 75-84.

Hites,R.A., R.E. Laflamme, and J.W. Farrington. 1977. Polycyclic aromatic hydrocarbons in recent sediments: the historical record. Science 198, 829-831.

Hites, R.A., and W.G. Biemann. 1975. Identification of specific organic compounds in a highly anoxic sediment by GC/MS and HRMS. Advan. Chem. Ser. 147, 188-201.

Laflamme, R.E., and R.A. Hites. 1978. The global distribution of polycyclic aromatic hydrocarbons. Geochim. Cosmochim. Acta 42, 289-303.

Muller G., G. Grimmer, and H. Böhnke. 1977. Sedimentary record of heavy metals and polycyclic aromatic hydrocarbons in Lake Constance. Naturwissen. 64, 427-431.

Owens, E.H., and A.J. Bowen. 1977. Coastal environments of the Maritime Provinces. Maritime Sed. 13, 1-31.

Shepard, F.P. 1963. Submarine Geology, 2nd edition, Harper and Rowe, New York, N.Y.

Wakeham, S.G. 1977. Synchronous fluorescence spectroscopy and its application to indigenous and petroleum-derived hydrocarbons in lacustrine sediments. Env. Sci. Technol. 11, 272-276.

Youngblood, W.W., and M. Blumer. 1975. Polycyclic aromatic hydrocarbons in the environment: homologous series in soils and recent marine sediments. Geochim. Cosmochim. Acta 39, 1303-1314.

HALOGENATED HYDROCARBONS IN DUTCH WATER SAMPLES OVER THE YEARS 1969 - 1977

Ronald C.C. Wegman and Peter A. Greve

National Institute of Public Health

P.O. Box 1, Bilthoven, The Netherlands

ABSTRACT

The results of a surveillance program with respect to hexachlorobenzene, α-, β- and γ-hexachlorocyclohexane, heptachlor, heptachlorepoxide, dieldrin, endrin, o.p'-DDT, p.p'-DDT, p.p'-DDE, TDE, α- and β-endosulfan, polychlorobifenyls and the parameter "Extractable Organic Chlorine" (EOCl) in the Dutch aquatic environment are presented.

In the period 1969 - 1977, 1,826 water samples were taken at 99 sampling sites.

The highest concentrations of halogenated hydrocarbons were found in the River Rhine and its tributaries.

A sampling trip by boat made along the River Rhine from Rheinfelden in Switzerland to Rotterdam in the Netherlands proved that the source of the α-, β- and γ-HCH-contamination was located between Basel (Switzerland) and Karlsruhe (Federal Republic of Germany).

INTRODUCTION

Since 1969, after the endosulfan wave in the River Rhine (Greve and Wit, 1971), a systematic long-term investigation with regard to the presence of pesticides and related substances in the Dutch aquatic environment has been carried out.

The sampling program in the Dutch aquatic environment comprised surface water, drinking water prepared from surface water, ground water and rain water. In view of the scope of the project, not all sites were sampled at the same time, but, except for a few fixed sites of permanent interest (Rivers Rhine, Meuse, etc), the sampling sites were changed every year so that after nine years all parts of the Netherlands were screened.

The sampling plans were established in cooperation with the Head Inspectorate for Environmental Hygiene of the Dutch Department of Public Health and Environmental Hygiene.

In 1974 a sampling trip by boat was made along the River Rhine from Rheinfelden in Switzerland to Rotterdam in the Netherlands in order to determine the exact source of the organochlorine discharges.

EXPERIMENTAL

In the period 1969 - 1977, 1,826 samples were taken at 99 sampling sites (18 sites for drinking water prepared from surface waters, three sites for groundwater, one site for rainwater, and 77 sites for surface water). Surface water samples were collected by means of a bail ("grab samples") from an approximate one meter depth and analysed inclusive of the silt.

During the sampling trip along the River Rhine, every 15 km a water sample was taken in the middle of the stream by means of pumping for 15 minutes. The location of the sampling sites of the boat trip are given in Figure 1.

METHODS

Organochlorine Compounds

1000 ml of water, including the silt, were extracted successively with 200, 100 and 100 ml of petroleum ether (boiling range 40 - 60°C). The combined extracts were dried over anhydrous sodium sulphate and concentrated to about 5 ml in a Kuderna-Danish apparatus. The last few millilitres of solvent were evaporated to exactly 1 ml by a gentle stream of nitrogen at room temperature. The concentrated extract was brought on a microcolumn containing 2.00 g of basic alumina W-200, activity Super I, Woelm, activated for 16 hours at 150°C and then deactivated with 11% water (11 g water + 89 g alumina). The first elution was carried out with 5 ml of petroleum ether to give "Eluate A". The receiving tube was changed and a second elution was carried out with 10 ml of an

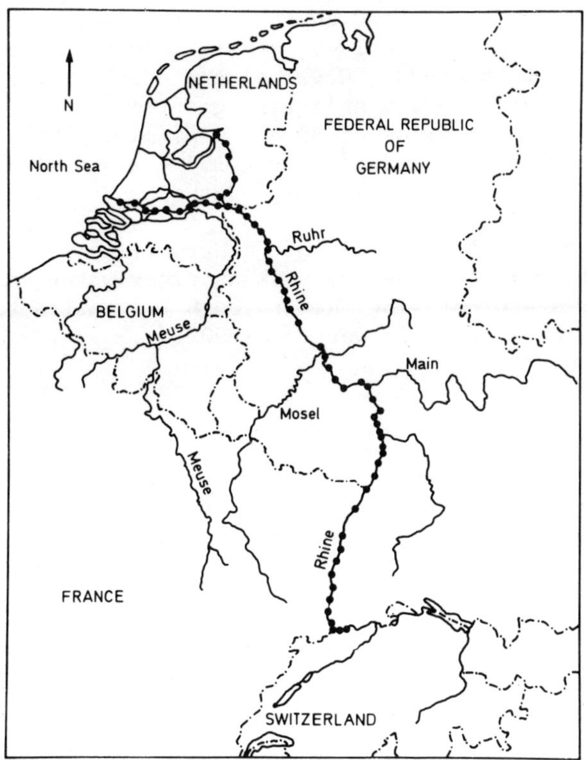

Figure 1. Sampling sites, Rhine trip by boat.

ethyl ether/petroleum ether 20/80 (v/v) mixture to give "Eluate B". The eluates were concentrated to exactly 1 ml by a gentle stream of nitrogen at room temperature. Eluate A contains: HCB, α- and γ-HCH, heptachlorepoxide (± 10%), p.p'-DDE, o.p'-DDT, TDE, p.p'-DDT, telodrin, isodrin, aldrin and heptachlor. Eluate B contains β-HCH, heptachlorepoxide (± 90%), dieldrin and endrin.

For the determination of α- and β-endosulfan a microcolumn was used containing 2.00 g of silicagel, Fisher S 661, 60-200 mesh, activated for 2 - 3 hours at 140°C. The first elution was carried out with 8 ml of a hexane/toluene mixture 80/20 (v/v). The second elution was carried out with 8 ml hexane/toluene mixture 40/60 (v/v) and 8 ml of toluene. In the second eluate α- and β-endosulfan were present. 1 μl of the concentrated eluates was injected in the gas chromatograph. The gas chromatographic conditions were:

1) A Varian Model 1800 equipped with a tritium electron capture detector, a pyrex column 180 cm x 0.3 cm I.D. packed with Chromosorb W-HP (80-100 mesh) coated with 5% OV-120 + 5% OV-17 (4+1) and a nitrogen carrier gas flow of 40 ml/min. Temperatures were: injection port 205°C; oven 190°C; detector 200°C.

2) A Perkin-Elmer Model F 22 equipped with a ^{63}Ni electron capture detector, a capillary column, pyrex, 40 m x 0.35 mm I.D. coated dynamically with SE-30 (GC grade) and a helium carrier gas flow of 2 - 3 ml/min, helium splitting gas flow of 0 - 60 ml/min, nitrogen purge gas flow of 80 - 100 ml/min. Temperatures were: injection port 215°C; oven 155 - 225°C at 3°C/min with a linear temperature programmer, detector 250°C.

The practical lower limit of detectability was 0.01 μg/1. Recovery data were obtained by spiking river water samples with the pesticides and carrying them through the entire analytical procedure: the recoveries were over 90%.

Polychlorinated Biphenyls

1 ml of a concentrated petroleum ether extract of 1000 ml of water was brought on a microcolumn containing basic alumina (see "Organochlorine Compounds") and eluted with 13.5 ml of petroleum ether. The eluate was concentrated to exactly 1 ml by a gentle stream of nitrogen at room temperature and brought on a column containing 5.00 g of activated silicagel. The elution was carried out with 40 ml of petroleum ether. The eluate was concentrated to a volume of 1 ml and brought in a chlorination tube (Sovirel No. 611.53).

After addition of 0.5 ml of chloroform the solution was concentrated by a gentle stream of nitrogen at room temperature to approximately 0.5 ml. After addition of 1.0 ml of chloroform the procedure was repeated. 0.3 ml of antimony pentachloride was added and the tube was immediately closed and heated for 3 hours in an oilbath at a temperature of 200°C. 5 μl of the solution were injected into the gas chromatograph. The decachlorobiphenyl formed was determined and recalculated into Aroclor 1260 by multiplication by the ratio of the molecular weights, 0.72.

The gaschromatographic conditions were:

A Varian Model 1800 equipped with a tritium electron capture detector, a pyrex column 0.7 m x 2 mm I.D., packed with a 1:4 mixture of 3% OV-17 and 3% OV-210, both on Chromosorb W-HP (80-100 mesh). Nitrogen carrier gas flow was 30 ml/min. Temperatures were: injection port 220°C; oven 205°C; detector 210°C. The practical detection limit is 0.1 μg Aroclor 1260/1.

Extractable Organic Chlorine (EOCl)

1000 ml of water, including silt, were extracted successively with 200, 100 and 100 ml of petroleum ether (boiling range 40-60°C).

Table 1 Concentrations (in μg/1 = ppb) of Organochlorine Pesticides and Related Substances in the River Meuse near Eysden. I = maximum value II = median value

Period	HCB		α-HCH		γ-HCH		dieldrin		endosulfan (α+β)		DDT-complex		Aroclor 1260		EOCl	
	I	II	I	II	I	II	I	II	I	II	I	II	I	II	I	II
1969	not determined				0.18	0.02	-	-	0.09	-	0.08	-				
1970	0.04	-	0.03	-	0.06	0.02	0.01	-	0.03	-	-	-	not determined			
1971	0.05	0.01	0.01	0.01	0.03	0.01	0.03	-	-	-	0.10	-				
1972	0.03	0.01	0.07	0.01	0.07	0.02	0.01	-	-	-	0.13	-				
1973	0.29	0.01	0.02	0.01	0.05	0.01	-	-	0.01	-	0.05	-	not determined		58	2.3
1974	0.05	0.01	0.02	0.01	0.04	0.02									8.7	1.9
1975	0.02	-	0.01	0.01	0.03	0.02	not determined								11	3.6
1976	0.01	-	0.01	-	0.05	0.03							1.2	<0.10		
1977	0.01	-	0.02	0.01	0.06	0.03							0.24	0.17	14	4.8

"-" = non detectable

Table 2 Concentrations (in μg/1 = ppb) of Organochlorine Pesticides and Related Substances in the River Rhine at Kilometre 865 (Lobith) I = maximum value II = median value

Period	HCB		α-HCH		γ-HCH		dieldrin		endosulfan (α+β)		DDT-complex		Aroclor 1260		EOCl	
	I	II	I	II	I	II	I	II	I	II	I	II	I	II	I	II
1969	not determined				0.24	0.18	0.04	-	0.81	0.24	0.31	-				
1970	0.39	0.08	0.26	0.14	0.16	0.08	0.04	-	0.40	0.03	0.26	-	not determined			
1971	0.52	0.14	0.48	0.16	0.34	0.10	0.06	-	0.25	-	0.21	-				
1972	0.37	0.13	0.57	0.16	0.28	0.11	0.02	-	0.03	-	0.23	-				
1973	0.55	0.08	0.45	0.19	0.42	0.12	0.02	-	0.10	-	0.12	0.01	not determined		40	23
1974	0.39	0.10	0.60	0.22	0.33	0.13	0.05	-	0.02	-	0.05	-			55	18
1975	0.21	0.06	0.21	0.06	0.14	0.04	0.02	-	0.02	-	0.02	-			27	11
1976	0.39	0.12	0.12	0.02	0.06	0.02	-	-	0.03	-	-	-	0.73	0.34	31	17
1977	1.2	0.06	0.07	0.02	0.09	0.02	-	-	-	-	-	-	0.47	0.21	31	12

"-" = non detectable

The combined extracts were dried over anhydrous sodium sulphate and concentrated to about 5 ml in a Kuderna-Danish apparatus. After addition of 0.5 ml of n-hexadecane the last few millilitres of solvent were evaporated to exactly 1 ml by a gentle stream of nitrogen at room temperature; 5 μl, corresponding to 5 ml of water, were then injected into the furnace of the microcoulometer.

The optimal working conditions of the microcoulometer were:

Furnace temperature: inlet zone 350°C, combustion and outlet zone 850°C; gas flow: argon 50 ml/min, oxygen 140 ml/min; bias potential 265 mV; range-ohms 500; injection rate 0.2 μl/sec.

The practical detection limit is 1 μg cl/l. The volatility of the organochlorine compounds has a great influence on their contribution to EOCl (Wegman and Greve, 1977). Recovery experiments by spiking over water samples with various organochlorine compounds having different vapour pressures gave for dichlorobenzenes 45%, for trichlorobenzenes 70%, and for hexachlorobenzenes 94% recovery. Tetrachloroethylene for example does not contribute to the parameter.

RESULTS

Over 10,000 data were collected in the monitoring program for the years 1969-1977. As an example the results of two sampling sites of permanent interest, the River Meuse near Eysden (borderline with Belgium) and the River Rhine near Lobith (borderline with Germany) are summarized in Tables 1 and 2. Samples were taken weekly near Lobith and monthly near Eysden.

In view of the low frequency of occurrence and the low concentrations found, the concentrations of β-HCH, aldrin, heptachlor, heptachlorepoxide and endrin are not given in the Tables. The River Rhine was studied in more detail than the other surface waters; the concentration of HCB, α- and γ-HCH, EOCl and PCB's are given as a function of time in Figures 2 to 5.

The concentrations of organochlorine compounds in drinking, ground and rain water were very low and are not given here.

In Table 3 the results of a sampling trip by boat along the borders of the Rhine are summarized. Also in this Table the low concentrations of organochlorine compounds are not mentioned. PCB's were not determined.

The geographical distribution of α-, β- and γ-HCH and EOCl in the River Rhine from Rheinfelden to Rotterdam in the Netherlands are illustrated in Figures 6 and 7.

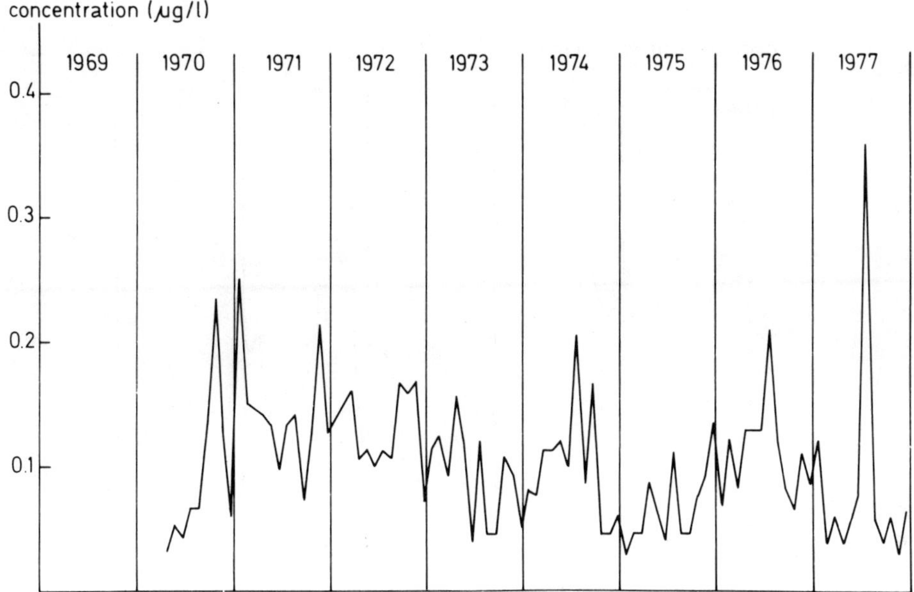

Figure 2. Concentration of HCB in the River Rhine at Rhine-Km 865 (Lobith)

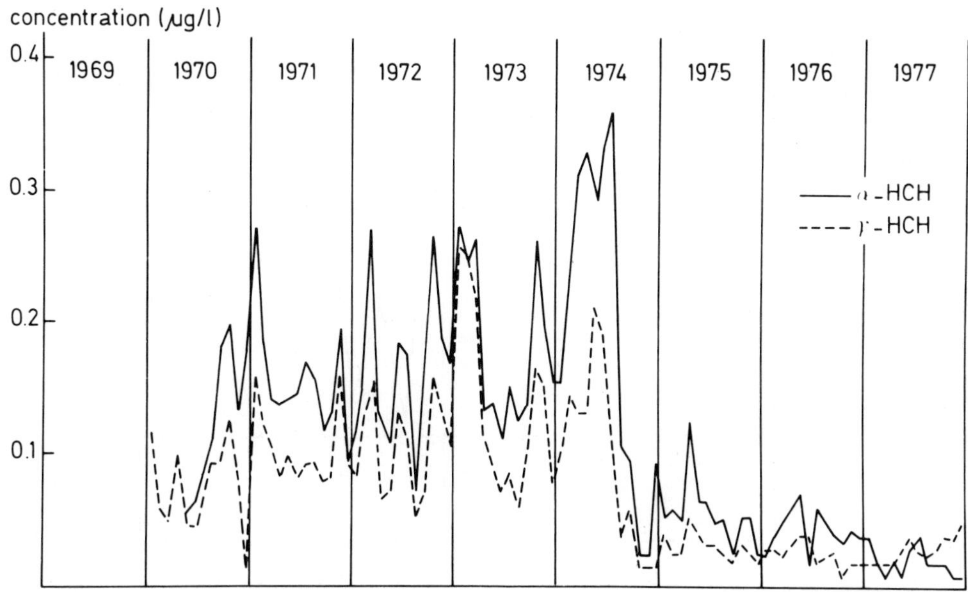

Figure 3. Concentration of α- and γ-HCH in the River Rhine at Rhine-Km 865 (Lobith)

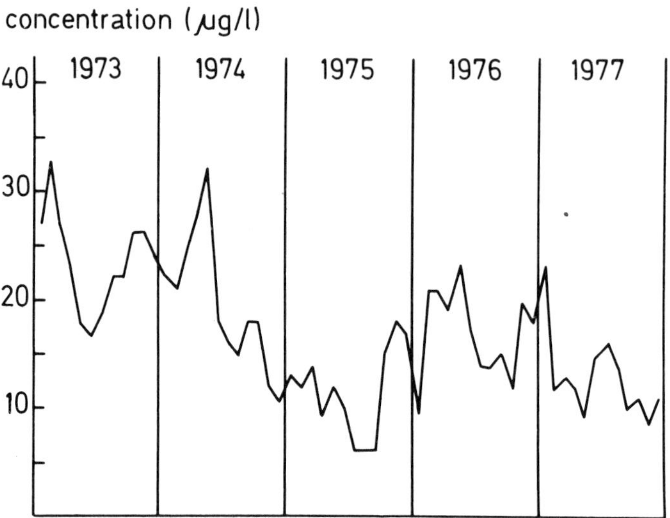

Figure 4. Concentration of EOCl in the River Rhine at Rhine-Km 865 (Lobith)

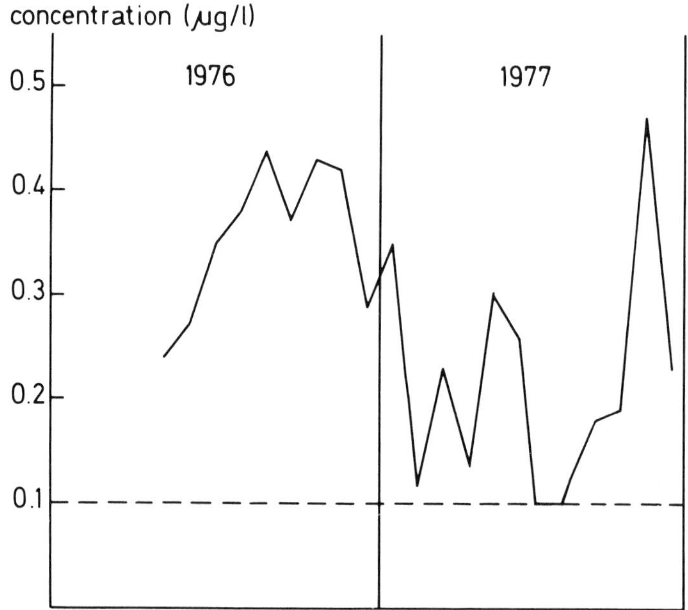

Figure 5. Concentration of PCB's (Calculated as Aroclor 1260) in the River Rhine at Rhine-km 865 (Lobith)

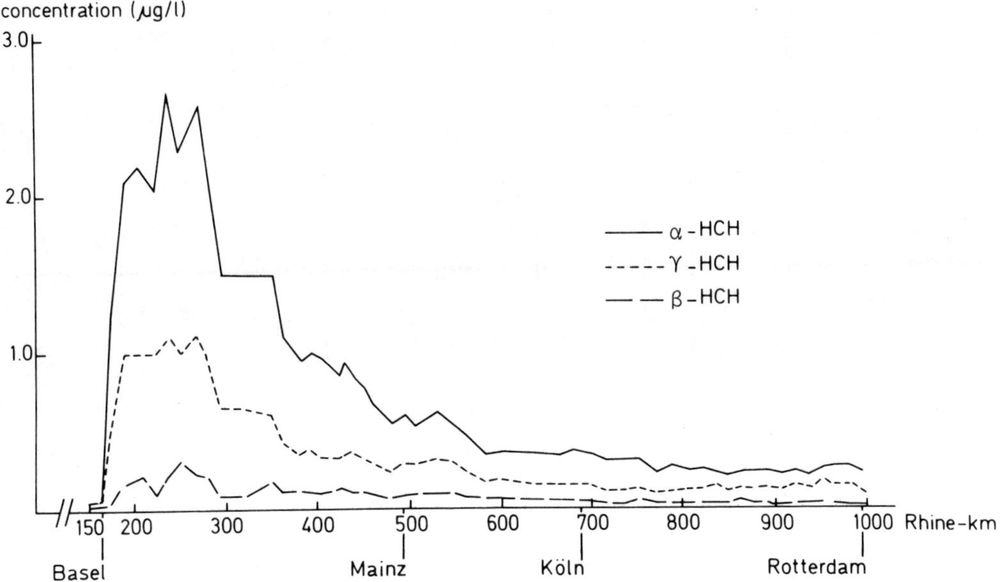

Figure 6. Concentration of α-, β- and γ-HCH in the River Rhine

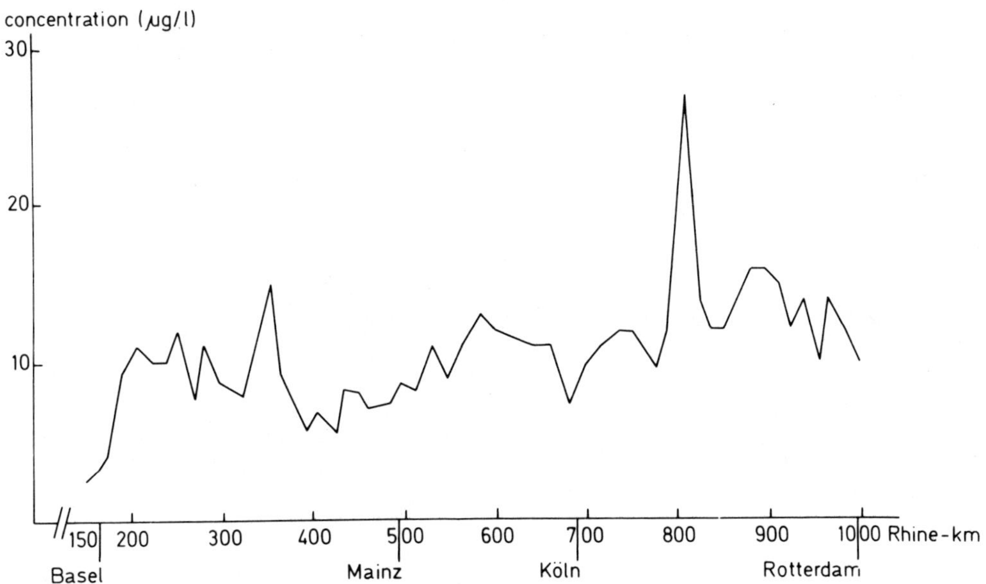

Figure 7. Concentration of EOCl in the River Rhine

Table 3 Concentrations (in μg/1) of Organochlorine Pesticides and EOCL in the River Rhine from Rheinfelden in Switzerland to Rotterdam in the Netherlands, June/ July, 1974

Sampling site	Rhine-km*	HCB	α-HCH	β-HCH	γ-HCH	EOCl
Rheinfelden	147	-	0.02	0.02	0.02	2.6
Basel (boundary)	173	0.04	1.2	0.15	0.47	9.5
Bellingen	190	0.02	2.1	0.22	1.0	11
Artzenheim	235	0.04	2.7	0.30	1.1	11
Strasbourg	294	0.03	1.5	0.09	0.64	8.0
Karlsruhe	362	0.05	1.1	0.13	0.42	6.9
Ludwigshaven	425	0.03	0.86	0.10	0.33	5.6
Mainz	496	0.04	0.60	0.08	0.29	9.1
Leverkusen	698	0.03	0.33	0.05	0.15	9.7
Duisburg	785	0.04	0.27	0.03	0.11	12
Lobith (boundary)	865	0.06	0.24	0.05	0.13	14
Rotterdam	996	0.01	0.23	-	0.10	10

* Rhine-km 0 = Constance "-" = non detectable

DISCUSSION

The highest concentrations of organochlorine pesticides and related substances are found in the River Rhine and its tributaries. The concentrations in the River Meuse are much lower, whereas the other surface waters contain still lower or not detectable concentrations of these compounds. From the C_6-compounds HCB, α- and γ-HCH are nearly always present in the Rhine water samples. Since July 1974 the concentrations of α- and γ-HCH in the River Rhine have decreased considerably.

From the results of the sampling trip by boat it can be concluded that the source of the α-, β- and γ-HCH-contamination is located between Basel (Switzerland) and Karlsruhe (Federal Republic of Germany).

The high concentrations of EOCl in the Rhinewater samples cannot be explained by the concentrations found for HCB, α- and γ-HCH and PCB's alone as is confirmed by other investigations (Van de Meent, 1976; Sontheimer, 1975).

REFERENCES

Greve, P.A., and S.L. Wit. 1971. J. Water Poll.Contr.Fed., 43, 2338-2348.

Sontheimer, N. 1975. 5. Arbeitstagung der I.A.W.R., Amsterdam.

Van de Meent, W. 1976. Proceedings Colloquium on the Analysis of Organic Micropollutants in Water, Voorburg.

Wegman, R.C.C., and P.A. Greve. 1977. Sci.Total Environ., 7, 235-245.

A MODEL SYSTEM FOR THE INVESTIGATION OF METHODS OF HERBICIDE SPECIATION

Robert D. Guy and Deoraj R. Narine

Trace Analysis Research Centre, Dept. of Chemistry

Dalhousie University, Halifax, Nova Scotia, B3H 4J1

Organic chemicals in the aquatic environment may exist in a number of chemical forms; for example, the chemical may be in true solution, adsorbed onto inorganic colloids (Haque, 1975), associated with soluble and particulate humic substance (Haque, 1975), dissolved in lipid coatings on particulates (Hartung, 1975; Hance, 1969), and incorporated into the biotic material. The relative distribution of an organic chemical among the different forms will depend on the properties of the compounds (charge, polar functional groups, size etc.), the composition of the particulates, and the ultimate relationship between chemical speciation and bioavailability.

There is ample evidence in the literature to indicate that chemical speciation plays an important role in biological toxicity. The formation of complexes between toxic metal ions and organic ligands reduces the toxicity of metal to algae (Hutchinson and Stokes, 1975; Steeman Nielsen and Wium-Andersen, 1970) but enhances organic toxicity to bacteria and fungi (Foye, 1977). One of the prime thrusts in agricultural research was the investigation into the relationship between herbicidal activity and herbicide-colloid interactions. Since persistence, transport and biological activity of the organic chemicals will depend on the chemical form, it is important to develop procedures suitable for the speciation of organic substances in the aquatic environment.

To date most research in chemical speciation in the aquatic environment has been concerned with toxic metals (Florence and Batley, 1977). One of the principal lessons to be learnt from the metal speciation research is the extreme complexity of the problem - if one attempts to develop analytical methods using real samples one soon finds that the interpretation of results is difficult. An alternative approach is to develop a working chemical model in which the interactions between the chemicals of interest and the matrix components are well characterized; such a model system can then be used to test proposed speciation schemes and will serve as a frame of reference for the interpretation of analytical results obtained on real samples. The purpose of this paper is to present a model system suitable for the testing of speciation schemes for herbicides.

BACKGROUND

The aquatic system can be divided into four compartments: (1) sediment, (2) suspended particulates, (3) soluble salts and organic compounds, and (4) biota. A viable model system must contain individual components from each of these compartments. The purpose of this section is to describe the properties of each component selected for the model.

Algae and bacteria are the lowest trophic levels in the aquatic ecosystem. An alga, <u>Selenastrum capricornutum</u>, (UTEX 1648), was selected as a biological monitor because:

1. algae are easy to culture in the laboratory;
2. growth experiments involve 1000-1,000,000 algae per flask, hence, better statistics than other monitors;
3. <u>Selenastrum capricornutum</u> is easy to identify under a microscope;
4. short-term effects can be readily monitored by following the algae growth. Unicellular algae grow rapidly and "doubling times" of six hours are feasible permitting experimental runs of 96 hours.

Pesticides (of which herbicides are a subgroup) have been classified into several classes (Weber, 1972):

1. charged species, such as the cationic bipyridyl herbicides (paraquat and diquat) and the salts of weak acids (2,4-D) which may be anionic in alkaline waters;
2. polar compounds such as the s-triazines (simazine, atrazine etc.) and carbamates;
3. nonpolar compounds such as the chlorinated hydrocarbons.

The herbicides in the model system are chosen from groups 1 and 2. Charged species and polar molecules will be more soluble in water and have a better chance of being transported into the aquatic environment. Paraquat and atrazine have been included in the model.

A cationic dye, methylene blue, was selected as a model herbicide for several reasons:

1. the positive charge and aromatic ring systems are very similar to the bipyridyl systems;

2. methylene blue is very easy to analyze for in model systems; for example, using molecular fluorescence concentrations as low as 1 ng/ml can be detected;

3. the chemistry of methylene blue has been exhaustively studies by geochemists for use in determining the cation exchange capacities of sediments and suspended material (Hang and Brindley, 1970) and by biochemists as a model for studying photosynthesis (Frankowiak and Rabinowitch, 1966).

Clays and humic substances have been suggested as prime agents for fixing herbicides in the environment (Green, 1974; Bailey and White, 1970). Clays can interact with the herbicides by ion exchange (both cation and anion exchange), chemisorption (reactions between herbicides and clay surface groups such as hydroxyl, oxygen, and adsorbed water), and physical adsorption (van der Waal's forces). Terce and Calvert (1978) have compared the adsorption behaviour for a number of herbicides and noted extensive adsorption for the s-triazines at low pH values. The ionic herbicides paraquat and methylene blue can be expected to undergo ion exchange reactions at all pH values, whereas, the polar molecule atrazine will interact only by chemisorption/ physisorption mechanisms at normal natural water pH levels.

Humic substances are large, polydisperse polyelectrolytes that contain acidic functional groups, especially carboxylic and phenolic acids. The ionic herbicides can interact with the humic substances via ion exchange reactions and polar molecules via H-bonding and dipole-dipole interactions (Schnitzer and Khan, 1972). To simulate the sediment and particulate compartments clay (bentonite at levels of 0-10 mg/l) and dissolved humic acid (0-20 mg/l) are included in the model.

This paper describes the interactions between the individual components (herbicide-clay, herbicide-HA, herbicide-algae), illustrates the behaviour in complicated systems (herbicide-clay/HA-algae), and suggests that a speciation scheme based on size fractionation may be feasible.

EXPERIMENTAL

Chemicals. Bentonite (Fisher Scientific) was saturated with potassium ion using standard methods. The cation exchange capacity at pH 6 was found to be 51±2 meq/100 grams (using the procedure of Chapman (1965).) Suspensions of the homoionic clay were prepared by suspending 5 grams of clay in distilled water in a one litre graduated cylinder; after a 24-hour settling period the top 900 ml was removed and kept suspended by a magnetic stirrer. The clay content of the suspensions were accurately determined by weighing aliquots collected on a 0.2 μm Nuclepore membrane filter. A typical suspension contained 4.1 mg/ml of clay.

Humic acid (aldrich) was treated with 3M HNO_3 to remove heavy metals. Total acidity of the humic acid (Schnitzer and Khan, 1972) was determined to be 6.0±0.1 meq/g and the total carboxyl content (Schnitzer and Khan, 1972) was found to be 2.1±0.02 meq/g. Humic acid was dissolved in distilled water (with 3 pettets KOH), filtered through a 0.2 μm Nuclepore membrane filter, and the concentration determined by weight of undissolved material on the filter.

Methylene blue (Matheson) was used as the chloride hydrate salt. The purity was checked using thin layer chromatography. Chromatograms run on silica gel (Merck) developed by a 1:9 glacial acetic acid/methanol eluent indicated no Azure A, B, or C to be present. Methyl viologen hydrate (Aldrich) was used for the herbicide paraquat. All other chemicals used in the study were reagent grade.

<u>Characterization of organic-particulate interactions</u>. The distribution of methylene blue and paraquat between solution and clay colloid was determined by three methods: 1. a titration method in which the herbicide was slowly added to a clay suspension (~20 mg/50 ml) and the herbicide in solution monitored by visible spectrometry after centrifugation; 2. a bulk adsorption method in which known amounts of methylene blue or paraquat were added to a clay suspension and allowed to equilibrate for 24 hours. After centrifugation the equilibrium concentration of herbicide in solution was determined by spectrometry and the amount adsorbed calculated by difference; 3. a dialysis bag method in which a clay suspension was placed in a Spectrapor 2 dialysis bag and the herbicide solution in an outer solution. After equilibration the amount of free herbicide in solution was determined by visible spectrometry. The advantages of methods 1 and 3 are that the clay is in contact with low concentrations of herbicide and, therefore, interactions between clay and organic micelles are minimized.

The distribution of herbicide between solution and humic acid was determined using the dialysis bag procedure described above. In all cases the distribution was studied at ionic strengths of 0, .01, 0.1 and 0.5 (set using potassium nitrate). The pH of the solutions was 6.0±0.1.

<u>Algal assays</u>. Algae growth experiments were done in 250 ml erlenmeyer flasks containing 200 ml of solution. The flasks were placed in a light box consisting of four fluorescent tubes (cool white, 35 watts). The flasks were aerated with water-saturated air at a rate of 0.5 litres/minute. The algal growth medium was taken from Standard Methods for Water and Wastewater Control (1977). The nutrient and lighting conditions were such that the chlorophyll content doubled every 6.7±0.2 hours.

A typical growth experiment consisted of twelve flasks (six samples in duplicate) containing herbicide and algae or herbicide-algae-clay/HA suspension to study individual interactions. Algae growth was monitored using the chlorophyll fluorescence. Ten millilitres of algal suspension was centrifuged in the presence of one gram of magnesium carbonate powder, the algal mat was extracted with ten millilitres of ACS methanol for 30 minutes. The fluorescence (λ_{exc} 421 nm; λ_{em} 667 nm) was measured on a scanning Aminco-Bowman spectrofluorimeter.

RESULTS AND DISCUSSION

Herbicide-algae interactions. Algae growth can be expressed mathematically by the function:

$$N = N_o \; 2^{(t/t_d)} \tag{1}$$

where N_o is the initial number of algae present, t is the time since the start of the experiment, and t_d is the "doubling time". A herbicide can affect the algae in a number of ways; for example, there could be no effect (N_o, t_d same as the control), the herbicide may kill a fraction of the initial amount of algae present but follow normal growth (N_o decreases, t_d same as the control), or the herbicide may change the growth rate ($t_d \neq t_d$ control).

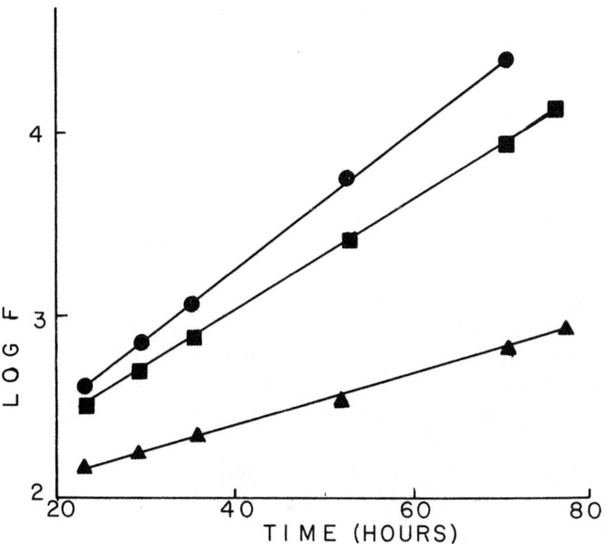

Figure 1. Algae growth curves for S. capricornutum in the presence of atrazine. ● control ■ 0.39 μM ▲ 1.55 μM

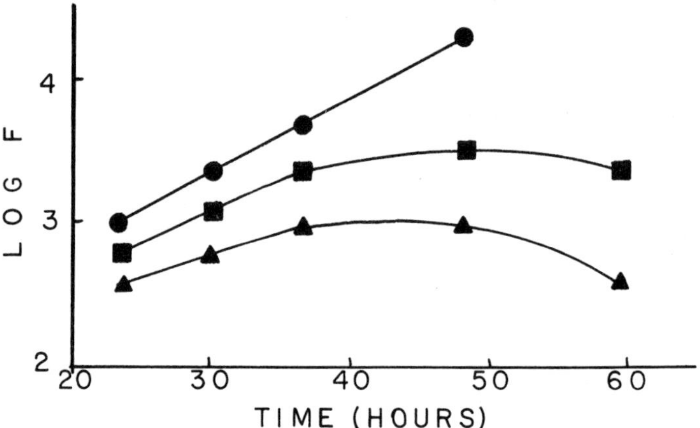

Figure 2. Algae growth curve for S. capricornutum in the presence of paraquat. ● control ■ 1.3 μM ▲ 6.7 μM

Algae growth curves in the presence of atrazine and paraquat are given in Figures 1 and 2, respectively. The herbicide atrazine decreases the growth rate constant and over the range investigated (up to 20 μM) no effect on N_o was observed. Paraquat exhibited a different behaviour. The paraquat decreased N_o but for the first 40 hours a normal growth rate was observed; after 40 hours a limiting growth region occurred. The limiting growth region was not a result of depleted nutrient levels because respiking the solutions with nutrient did not prevent algae death. Algae from the flasks containing paraquat were centrifuged and used to spike paraquat free nutrient; a lag period of 30 hours was followed by normal growth.

The methylene blue-algae growth experiments were complicated by the photodecomposition of methylene blue in the light box. A number of workers (Bonneau and Pereyre, 1975; Zugel et al, 1972) have reported that methylene blue photobleaches in the presence of reducing agents such as EDTA and also an oxidative self-destruction reaction occurs (the latter is a result of an excited triplet MB molecule reacting with a ground state singlet MB molecule). Under the continuous illumination-aeration conditions of our growth experiments the methylene blue was found to decompose by first-order kinetics with a half-life of 9.2±0.3 hours. To overcome the effect of the photodecomposition the methylene blue solutions were respiked periodically to give an average concentration over the whole experiment. The algae growth curves in the presence of increasing amounts of methylene blue are given in Figure 3. The methylene blue affects the algae in a manner similar to atrazine - no initial decrease in N_o but a decrease

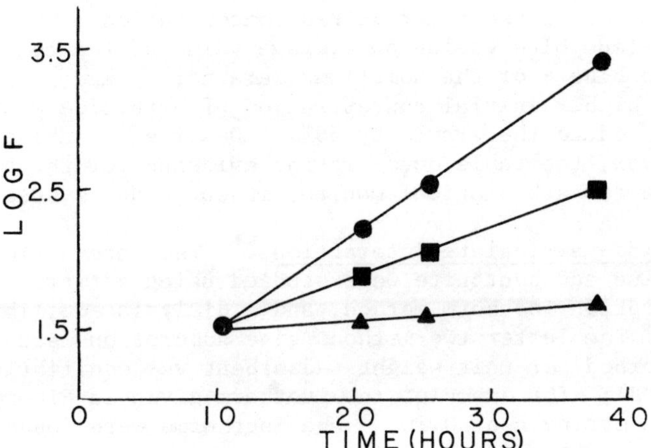

Figure 3. Algae growth curve for S. capricornutum in the presence of methylene blue. ● control ■ 0.08 μM ▲ 1.16 μM

in t_d as the concentration of methylene blue increases. The decrease in algae growth was not a result of the absorption of light by the methylene blue, that is, methylene blue has a broad absorption maximum at 667 nm that overlaps the red light absorption of chlorophyll. If a growth experiment is performed in which the algae flask is surrounded by a solution of methylene blue in a beaker, normal growth is observed. (Additional experiments indicate that methylene blue must absorb light for the substance to toxic indicating that a photolytic mechanism must be operative. This effect is currently under investigation and will be reported in a subsequent publication.)

A comparison of the three herbicides is presented in Table 1. The data in Table 1 must be interpreted with caution. The growth rates for the algae are different at different levels of the herbicide hence the 50% growth values will change with time. The 48 hour time limit was chosen to be in the paraquat "limiting growth" region - a shorter time interval would increase the paraquat concentration necessary to decrease growth by 50% and a longer time

Table 1. A Comparison of Herbicide Toxicity

Compound	50% Growth (48 h) (μM)	t_d
MB	0.088	decreases
atrazine	0.39	decreases
paraquat	1.16	no change

interval would decrease the required concentration. The respiking of the methylene blue yields an average value of the concentration of methylene blue - of the solutions were not spiked at periodic intervals a higher initial concentration of herbicide would be required to reduce the growth by 50%. Despite the above limitations, however, the table does provide evidence for the suitability of the algae for a biological monitor in our model system.

Herbicide-particulate interactions. The interactions between methylene blue and bentonite were studied using a titration method, a bulk adsorption isotherm method, and a dialysis equilibration procedure. The latter two methods give adsorption isotherms (amount adsorbed per unit weight adsorbent vs. equilibrium concentration of MB); for example, the isotherm given in Figure 4 is for an ionic strength of 0.10. The isotherms were found to obey the Langmuir equation:

$$X = \frac{X_m \cdot B \cdot C}{1 + B \cdot C} \tag{2}$$

where X is the amount adsorbed per unit weight of the adsorbent, X_m is the limiting adsorption value, B is a constant, and C is the equilibrium concentration of the methylene blue. Figure 5 presents the titration results for methylene blue onto bentonite at different ionic strengths. The curves are presented for the titration of 3.21 mg of clay and the intercept for I = 0.1 yields a value of X_m of 70 ± 2 meq/100 grams clay. The curves in Figure 5 indicate that as the concentration of counter ions increases, the

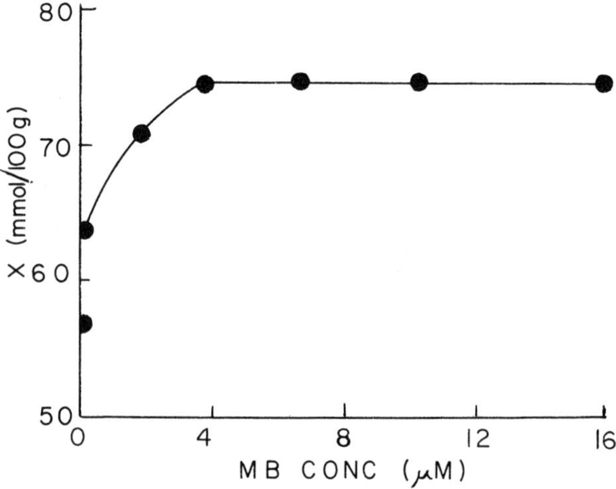

Figure 4. Adsorption isotherm for methylene blue onto bentonite at an ionic strength of 0.10.

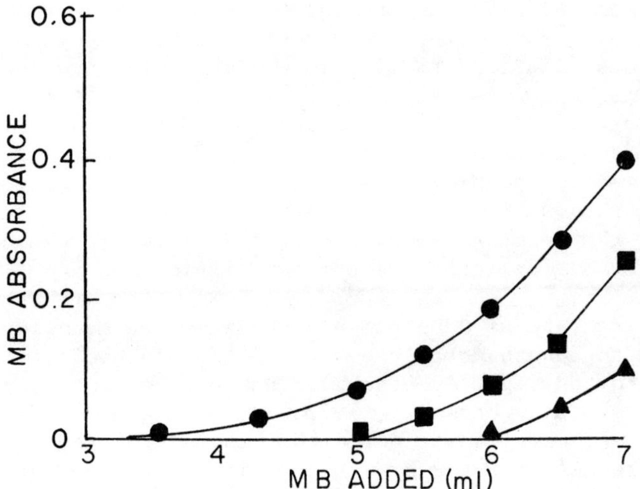

Figure 5. Titration experiment for bentonite and methylene blue at ionic strength of ● 0.5 ■ 0.1 ▲ 0

adsorption maximum decreases because of the ion exchange competition:

$$\text{CLAY-K} + \text{MB}^+ \rightleftharpoons \text{CLAY-MB} + \text{K}^+$$

A Comparison of the adsorption maxima determined for methylene blue onto the clay be the three methods is given in Table 2. The table indicates that the three methods provide similar results.

Two aspects of the ion exchange reaction are important: does the amount of potassium exchanged equal the amount of methylene blue adsorbed? and is the reaction reversible? An experiment in which the amount of potassium ion released (determined by atomic absorption) was found to be 51 meq/100 g clay compared to 77 meq/100 g clay of adsorbed methylene blue. The 51 meq/100 g

Table 2. MB-Clay Interaction

	X_m (mm01/100 g)		
I	Titration	Dialysis	Bulk Adsorption
0.00	79 ± 2	77 ± 3	76 ± 6
0.01	75 ± 2	77 ± 1	76 ± 5
0.10	70 ± 2	75 ± 0.5	71 ± 5
0.50	64 ± 3	70 ± 2	69 ± 3

of potassium released was equivalent to the cation exchange capacity of the clay. When methylene blue-saturated clay was equilibrated with various solutions ranging from 0.1 M potassium ion to 0.1 M aluminum ion solutions at pH 3 less than 1.5% of the methylene blue eas exchanged. These two experiments indicate that:

1. methylene blue is adsorbed by an ion exchange mechanism in addition to a chemisorption mechanism;
2. the adsorbed methylene blue is fixed to the clay surface and the exchange reaction is not reversible.

A third type of adsorption noted in which methylene blue was attracted to adsorb methylene blue fixed onto the clay surface. This type of aggregation behaviour has been extensively reported in the literature (Braswell, 1972; Bergman and O'Konski, 1963). In our experiments it was noted that the aggregated methylene blue could be readily removed by washing with distilled water and caused a shift in the visible spectrum of the methylene blue to shorter wavelengths.

The interaction between paraquat and clay is illustrated in the titration curves shown in Figure 6. The amount of paraquat adsorbed at various ionic strengths is given in Table 3. Exchange and desorption studies similar to the methylene blue study indicated that the paraquat was adsorbed at values greater than the cation exchange capacity of the clay and the paraquat was not desorbed by solutions with an ionic strength of 0.50.

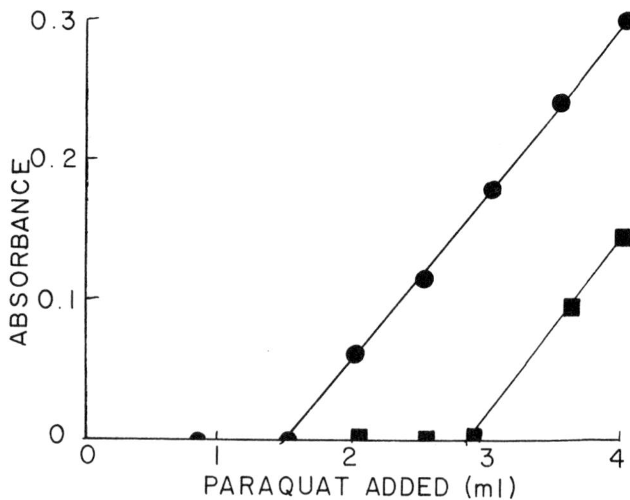

Figure 6. Titration experiment for paraquat and bentonite at various ionic strengths ● 0.5 ■ 0

Table 3. Paraquat-Clay Interactions

Ionic Strength	X_m (mmols/100 g)
0.0	41.7
0.1	31.4
0.5	24.0
"sea water"	32.8

Humic substances are large, polydisperse systems of electrolytes with molecular weights ranging from 2000 to greater than 100,000. The large size of the humic molecule permits a separation of herbicide bound to humic acid from the herbicide in true solution using a dialysis membrane. Experiments using methylene blue gave the adsorption isotherm shown in Figure 7. The isotherm obeys the Freundlich relationship:

$$X = k\ C^{1/n} \tag{3}$$

where for humic acid and methylene blue k = 1.05 and n = 1.85. Unlike the methylene blue-clay interaction the methylene blue-humic acid system does not give a limiting sorption value - the system can be best described by giving the amount of herbicide bound and free herbicide present in a solution of known humic acid concentration and total methylene blue concentration.

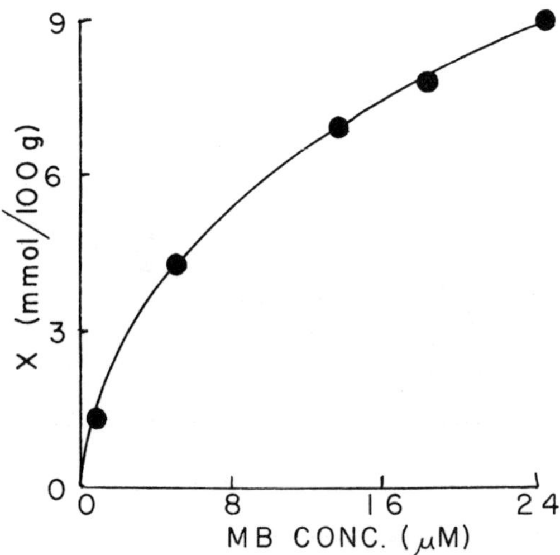

Figure 7. Adsorption isotherm for methylene blue onto humic acid.

Table 4. Herbicide-Colloid Interactions

Colloid	Amount Bound (%)*		
	MB	Paraquat	Atrazine
1.0 ppm clay	64	33	0
2.5 ppm HA	13	8	0

* assuming 1.25 M solution of herbicide

The interaction of atrazine with humic substances and clay has been described in the literature (Harris, 1962). In the pH range of natural waters (6 < pH < 8) atrazine is not adsorbed by bentonite (Harris, 1962) and approximately 3 μmoles/gram HA are bound (Schnitzer and Khan, 1972). Table 4 is a summary of the herbicide-colloid interactions.

Herbicide-Algae-Particulate Studies. The values in Table 4 indicate that clay and humic acid at levels similar to the amounts present in lakes and rivers can tie up significant quantities of herbicides. If the bound herbicide is non-toxic then one could expect methylene blue and paraquat to be non-toxic in the presence of clay and humic acid whereas atrazine should not be affected by the colloids. To test these conclusions algae growth studies were undertaken to investigate mixed systems containing colloid, algae and herbicide. The algae growth curves for atrazine-clay, paraquat-clay, and methylene blue-humic acid are given in Figures 8, 9 and 10, respectively. The atrazine-clay-algae system illustrates the valid

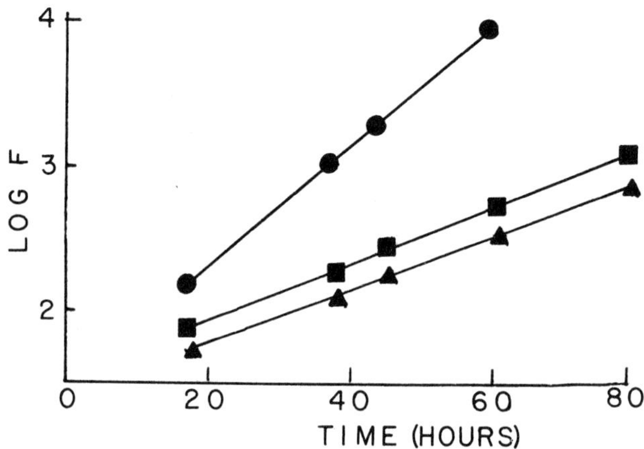

Figure 8. Algae growth curves in the presence of atrazine and bentonite. ● control ■ 4.9 ppm clay and 1.8 μM atrazine ▲ 1.8 μM atrazine

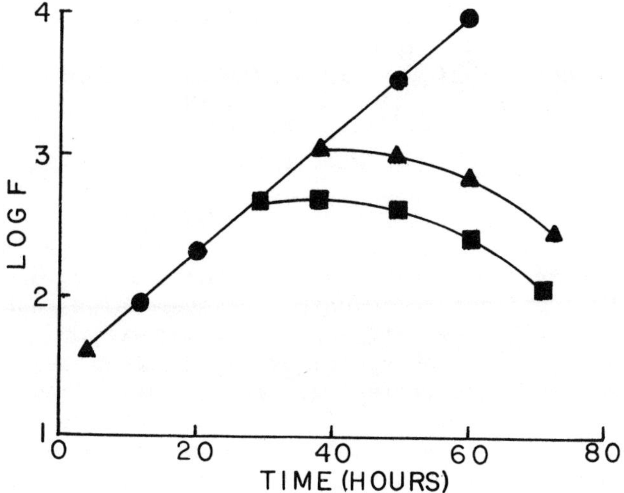

Figure 9. Algae growth curves in the presence of clay and 3.89 M paraquat. ● control and 7.5 ppm clay ▲ 3.75 ppm clay ■ 1.9 ppm clay

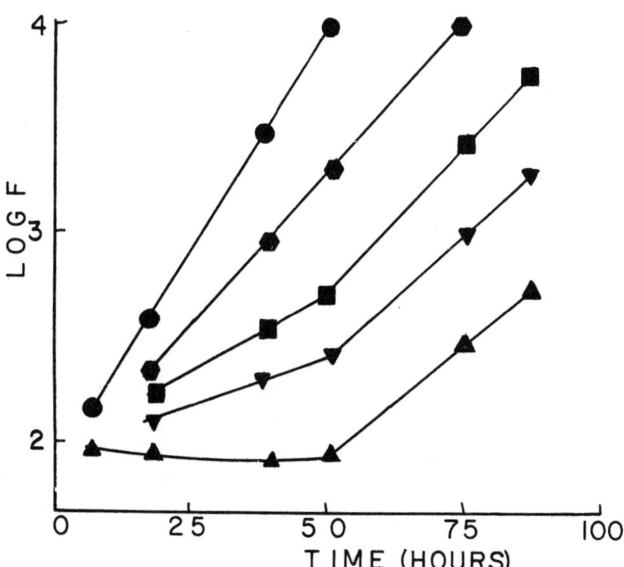

Figure 10. Algae growth curves in the presence of 2.5 ppm humic acid and methylene blue. ● control ⬢ 200 ng/ml MB ■ 300 ng/ml MB ▼400 ng/ml BM ▲500 ng/ml MB

conclusion that the atrazine-clay interaction is not extensive enough to reduce the atrazine toxicity. Similar results were obtained for the atrazine-humic acid-algae system. The effect of clay on the paraquat solution shows that as the concentration of clay increases the free paraquat concentration decreases. A 7.5 ppm solution of clay reduced the toxicity to nil over a 70 hour growth experiment. Similar results were obtained for a paraquat-humic acid-algae system.

The MB-clay-algae system showed that the clay bound all the methylene blue present up to dark blue suspensions and no toxicity was observed until free methylene blue was present. The humic acid methylene blue-algae system provided some interesting results. In the experiment illustrated in Figure 10 the methylene blue was not spiked at periodic intervals and the concentrations are the initial MB concentration. A 500 ng/ml methylene blue solution should become nontoxic after about 36 hours (that is, after four half lives) but in each case the toxicity is reduced by the humic acid (any solution above 100 ng/ml MB completely kills the algae in the absence of humic acid) but normal growth rates are not reached until a 50 hour time period has elapsed. It is evident, therefore, that the herbicide-particulate interactions can have a complex role in the aquatic environment - the particulates can reduce the short-term toxicity but may also extend herbicide effect to longer time periods by trapping the herbicide and releasing it slowly into the water environment as the free herbicide is degraded by bacteria or photolysis.

The methylene blue-humic acid-algae system was selected to test the possibility of using dialysis as a speciation procedure. The humic acid-methylene blue complex passes through a 0.2 μ membrane filter and would meet the common definition of "soluble" species. The dialysis membrane used in the study has a pore size of 19 nm and allows the free passage of methylene blue (half life for dialysis is about 6 hours) but does not pass the humic acid-MB complex. The free methylene blue was monitored via molecular fluorescence spectrometry a preliminary experiment was done: a 50 ppm solution of humic acid was ultrafiltered through a 12000-14000 molecular weight dialysis membrane. The ultrafiltrate was used to prepare a methylene blue standard (100 ng/ml MB, 50 ppm HA ultrafiltrate). The fluorescence of the ultrafiltrate solution was identical to a 100 ng/ml MB standard solution. This is strong evidence for dialysis of free MB only because in the presence of humic acid the MB fluorescence is suppressed by quenching. Table 5 presents the results of an algae growth run comparing MB_{total}, MB_{free}, and doubling time of the MB and MB-HA solutions. The free MB was determined by dialysis separation. A one-to-one correspondence between free MB concentration and the doubling time is not present in the MB-HA systems - a MB-HA solution appears to have a higher methylene blue concentration than the MB

Table 5. MB Speciation - Effect on Algae

MB_{total} (ng/ml)	MB_{free} (ng/ml)*	t_d (hours)
0	0	6.7
100	100	8.8
200	200	13.2
300	300	kill
500	500	kill
100	6	7.0
200	29	8.3
300	76	10.3
400	113	18.4

*MB_{free} was determined by dialysis

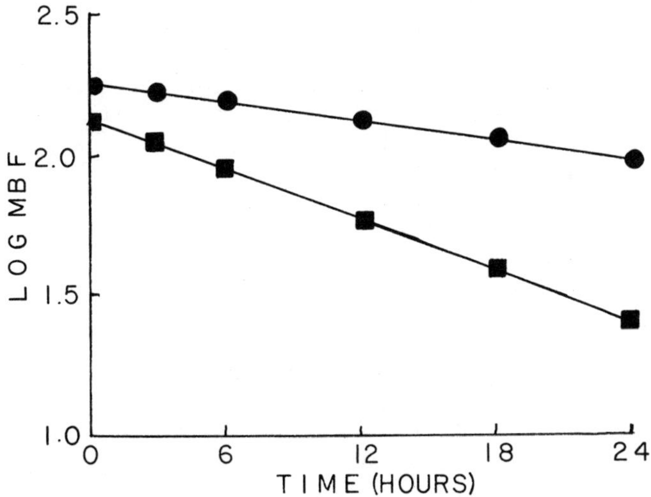

Figure 11. Decomposition of methylene blue as a function of time ● 400 ng/ml MB and 2.5 ppm HA ■ 100 ng/ml MB no HA. (In the presence of humic acid the free MB was determined after dialysis separation).

solutions without humic acid. The reason for the difference is illustrated in Figure 11. An experiment was completed that compared the photolysis rate of free methylene blue at 100 ng/ml to the rate of photolysis of a solution containing 400 ng/ml MB and 2.5 ppm humic acid that gave a free methylene blue concentration (by dialysis) of 113 ng/ml. The MB-HA system tends to buffer the free MB concentration:

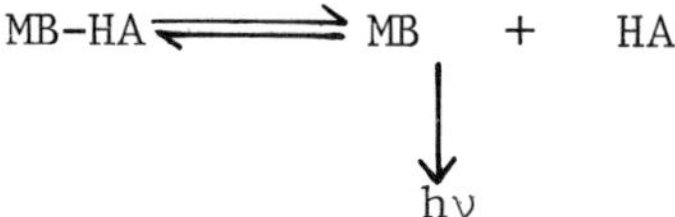

decomposition products

As the free MB photolyzes the complexation equilibrium shifts to replenish the free MB - the result is that a toxic MB concentration is present for a longer period of time (the half life of free MB is 9.3 hours and the half life of MB-HA is 23 hours.)

CONCLUSIONS

A study of the model system indicates that:

1. a clay-HA-algae system contains essential components for herbicide studies;
2. algae are a good biological monitor for herbicides but are limited by variable sensitivity to other organic molecules;
3. dialysis and ultrafiltration may serve as a speciation method based on the size of the species;
4. methylene blue proved to be toxic to the algae;
5. the humic acid-methylene blue system illustrates that particulates may tie up the herbicides but prolong the toxic effects by slowly releasing them into the aquatic environment.

REFERENCES

Bailey, G.W. and L.L. White, 1970. Adsorption, desorption and movement of pesticides in soil. Residue Reviews, 32, 29 (1970).

Bergman, K. and C.T. O'Konski, 1963. A spectroscopic study of methylene blue monomer, dimer, and complexes with montmorillonite. J. Phys. Chem., 67, 2169 (1963).

Bonneau, Roland and Josette Pereyre, 1975. Mechanism of photoreduction of thiazine dyes by EDTA. Photochem. Photobiol., 21, 173 (1975).

Braswell, E.H. 1972. Equilibrium sedimentation studies of the aggregation of methylene blue. J. Phys. Chem., 76, 4026 (1972).

Chapman, H.D., 1965. Methods of Soil Analysis, Chapter 57. Edited by C.A. Black, Amer. Soc. Agronomy.

Florence, T.M. and G.E. Batley, 1977. Chemical forms of trace metals in natural waters with special reference to copper, lead, cadmium and zinc. Talanta, 24, 151 (1977).

Foye, W.O., 1977. Antimicrobial activities of mineral elements, in Microorganisms and Minerals, edited by E.D. Weinberg, Marcel Dekker, New York, 1977. page 387.

Frackowiak, D. and E. Rabinowitch, 1966. Methylene blue-ferrous iron reaction in a two phase system. J. Phys. Chem., 70, 3012, (1966).

Green, R.E., 1974. Pesticide-Clay-Water Interactions, in Pesticides in Soil and Water., Soil Science Soc. America, Madison, Wisconsin. page 3.

Hance, R.J., 1969. Adsorption of linuron, atrazine, and EPTC by model aliphatic adsorbents and soil organic preparations. Weed Res., 9, 103 (1969).

Hang, P. Thi, and G.W. Brindley, 1970. Methylene blue adsorption by clay minerals. Determination of surface areas and cation exchange capacities. Clay and Clay Minerals, 18, 203 (1970).

Haque, Rizwanthul, 1975. Role of adsorption in studying the dynamics of pesticides in a soil environment, in Environmental Dynamics of Pesticides, ed. by R. Haque and V.H. Freed, Plenum, New York, 1975. p.97.

Harris, C.I., 1962. Adsorption and desorption of herbicides by soil. Ph.D. Thesis, Purdue University.

Hartung, Rolf, 1975. Accumulation of chemicals in the hydrosphere, in Environmental Dynamics of Pesticides, ed. by R. Haque and V.H. Freed, Plenum, New York, 1975. p.185.

Hutchinson, T.C. and P.M. Stokes, 1975. Heavy metal toxicity and algal bioassays, in Water Quality Parameters, ed. by S. Barabas, ASTM STP 573, Philadelphia, 1975. p.320.

Nielsen, E. Steeman and S. Wium-Anderson, 1970. Copper ions as a poison in sea and in fresh water. Marine Biol., 6, 93 (1970).

Schnitzer, M. and S.U. Khan, 1972. Humic Substances in the Environment, Marcel Dekker, New York, 1972.

Standard Methods for the Examination of Water and Wastewater, 14 Edition, APHA-AWWA-WPCF, 1975. p.748.

Terce, Martine and R. Calvet, 1978. Adsorption of several herbicides by montmorillonite, kaolinite and illite clays. Chemosphere, 4, 365 (1978).

Weber, J.B. 1972. Interactions of organic pesticides with particulate matter in aquatic and soil systems, in Fate of Organic Pesticides in the Aquatic Environment, Ed. by R.F. Gould. Amer. Chem. Soc., Washington, D.C., page 55.

Zugel, M., Th. Forster and H.E.A. Kramer, 1972. Sensitized photooxygenation according to type I mechanism (radical mechanism). Part III. Continuous Illumination. Photochem. Photobiol., 15, 33 (1972).

ANTHROPOGENIC C_1 AND C_2 HALOCARBONS: POTENTIAL APPLICATION AS COASTAL WATER-MASS TRACERS

George R. Helz

Department of Chemistry, University of Maryland

College Park, Maryland, 20742, U.S.A.

INTRODUCTION

Chemical water-mass tracers are often employed to obtain knowledge of water movements and mixing rates. In coastal waters, this knowledge is essential for wise planning of municipal and industrial development. Tracers sometimes also can be used to establish the legal responsibility for a pollution episode or spill.

Virtually any natural or anthropogenic substance which meets the following criteria would be useful as a tracer. It should be readily detectable without background interferences even after dilution to levels far below the concentration at its source. It should have relatively few, easily characterizable sources in the region of interest. Finally, it should either mix conservatively or else change with time in a simple, predictable manner.

Halogenated halocarbons are an obvious candidate for water-mass tracers both because they may be detected easily at very low levels and because natural sources which would contribute to background are quite rare. Unfortunately, the highly non-polar members of this class, such as halogenated pesticides, are not very suitable because interactions with suspended particles and living organisms lead to irregular changes in concentration with time. On the other hand the potential of fluorochloromethanes as global oceanic tracers was first pointed out by Lovelock et al (1973) and is under continued development (Hammer et al, 1978). There are an additional half dozen or so volatile chloro- and bromocarbons which are probably common in coastal waters near population centers and which may prove useful as near-shore tracers. Because of concern about trihalomethanes in drinking waters, the analytical methods for these compounds have become highly developed and the necessary equipment is very widely available. Detection limits are of the order of 1 nM (nanomole per litre).

SOURCES

Figure 1 presents estimated environmental release rates of the major industrially produced C_1 and C_2 halocarbons in the United States according to Nelson and Van Duuren (1975). A more recent critical review of release rates for only the halogenated methanes is also available (National Academy of Sciences, 1978). The values in Figure 1 may be precise to no better than a factor of 2, but they illustrate the kinds of compounds which are likely to enter the environment from anthropogenic sources.

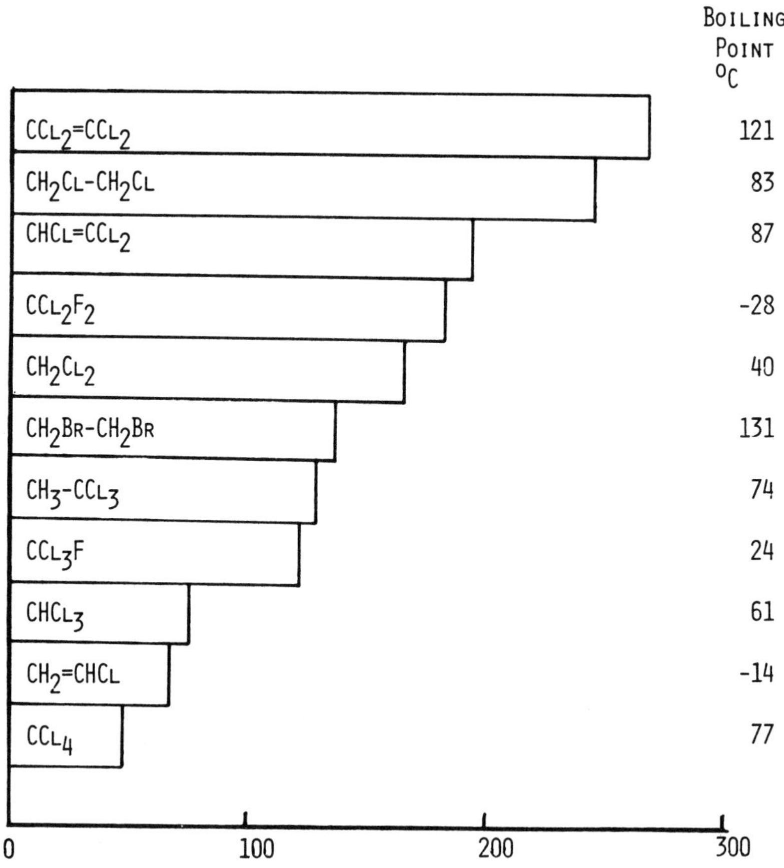

Figure 1. Estimated annual rates of release of volatile halocarbons into the environment in the United States (Date from Nelson and Van Duuren, 1975).

A large variety of conceivable sources exists. However, Helz and Hsu (1978) concluded that near population centers the important, strong sources would probably be: (a) sewage treatment plants which receive wastes from firms engaged in such activities as dry cleaning, degreasing, paint stripping, etc.; (b) waste heat dischargers (power plants, steel mills, oil refineries) which chlorinate their cooling waters to prevent biofouling in their heat exchangers; and (c) major manufacturers, shippers or users of industrially produced halocarbons. Compared to sources such as these which can produce halocarbon concentrations well above 10 nM, Helz and Hsu (1978) concluded that processes such as washout or diffusive transfer from the atmosphere and natural biosynthesis would be of minor importance in populated coastal regions.

The synthesis of a considerable variety of halogenated organic compounds by chlorination of sewage has been shown by Jolley (1974) and Glaze and Henderson (1975). However, the data of Bellar et al (1974) and Helz and Hsu (see Table 1) suggest that most of the C_1 and C_2 halocarbons in the effluents of large treatment plants may be present prior to chlorination. This contamination probably represents discharge solvents.

The production of trihalomethanes by direct chlorination of coastal waters has been reported and discussed by Helz et al (1978 a,b), Helz and Hsu (1978) and Carpenter and Smith (1978). Volatile halocarbons other than trihalomethanes are not formed at levels above 10 nM. The total chlorine to total bromine ratio in the trihalomethane products depends upon salinity. Figure 2 indicates

Table 1 Halocarbons in Sewage Treatment Plant Effluents (Values in nanomoles per litre)

Compound	1	2	3	4
CH_2Cl_2	34	40	775	47
$CHCl_3$	59	84	413	106
CCl_4	-	-	186 (combined with CH_3CCl_3)	14 (combined with CH_3CCl_3)
CH_3CCl_3	67	63		
$CHClCCl_2$	65	75	230	106
CCl_2CCl_2	23	25	369	50

1. Cincinnati, OH (Bellar et al, 1974) before chlorination.
2. Same plant as column 1 after chlorination.
3. Baltimore, MD (Helz and Hsu, 1978) before chlorination in winter time; air temperature slightly below 0^oC.
4. Same plant as column 3 before chlorination in spring time; air temperature approximately 25^oC.

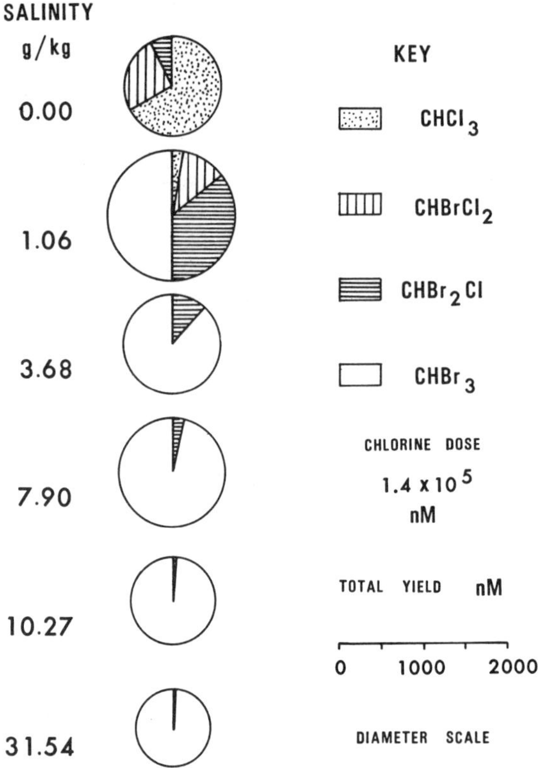

Figure 2. Trihalomethanes produced by chlorinating selected estuarine waters. Diameters indicate total yield and shading indicates product distribution.

that at very low salinities, chloroform is the major product (as usually observed in drinking waters) whereas above a salinity of about 2 g/kg, bromoform is the chief product. Bromoform is produced through oxidation by chlorine of the abundant Br^- in seawater. In laboratory experiments, molar yields of the order of 1 - 5% of the theoretical limit for substitution reactions were obtained by Helz and Hsu; Carpenter and Smith obtained yields of 4 - 8%.

The importance of factories and port facilities as halocarbon sources probably varies extremely widely. Table 2 presents some frequency of occurrence data from Ewing et al (1977) on volatile halocarbons in waters near industrial sites. The previously mentioned National Academy of Sciences report (1978) assembled unpublished data from the U.S. Environmental Protection Agency for halomethanes in industrial effluents. As much as 1.55 x 10^6 nM of CH_2Cl_2 was found in the effluent of one electrical manufacturer; $CHCl_3$ and CCl_4 were found in excess of 10^4 nM at various plants.

Table 2 Volatile Compounds Found in Waters Near Industrial Sites[1]

Compound	Frequency (%)
CCl_2CCl_2	38
CH_2ClCH_2Cl	26
$CHClCCl_2$	43
CH_2Cl_2	16
CH_2BrCH_2Br	1
CH_3CCl_3	9
CCl_4	3
$CHCl_3$	87
$CHCl_2Br$	12
$CHClBr_2$	5
$CHBr_3$	3

[1]Data from Ewing et al. (1977) based on analysis of 204 water samples collected near industrial sites in the United States.

SINKS

As stated in the introduction, a useful tracer must either behave conservatively or else change with time in a well-established manner. In Figure 3, the data of Helz and Hsu (1978) are presented for 5 halocarbons in the 1 - 2 m deep waters of Back River, Md., which receives the effluent from Baltimore's principal sewage treatment plant. In Back River, the effluent mixes with saline Chesapeake Bay water, so a plot of halocarbon concentration vs. salinity (or specific conductance at 25°C) will be linear if mixing is conservative. However, as seen in Figure 3, the concentration of each compound diminishes more rapidly than would be expected for conservative mixing; this results in the concave-up distribution curves. It is interesting that these data were obtained in winter after the estuary had been covered with 10 - 40 cm of ice for more than a month. When sampling was repeated in spring, after the ice disappeared, halocarbon levels dropped below detection limits very rapidly.

The general processes which might lead to non-conservative behaviour, as observed in Back River, are the following: photolysis, hydrolysis, adsorption, biological degradation and volatilization to the atmosphere. Although the available information is still somewhat limited, it is probable that the first four processes are

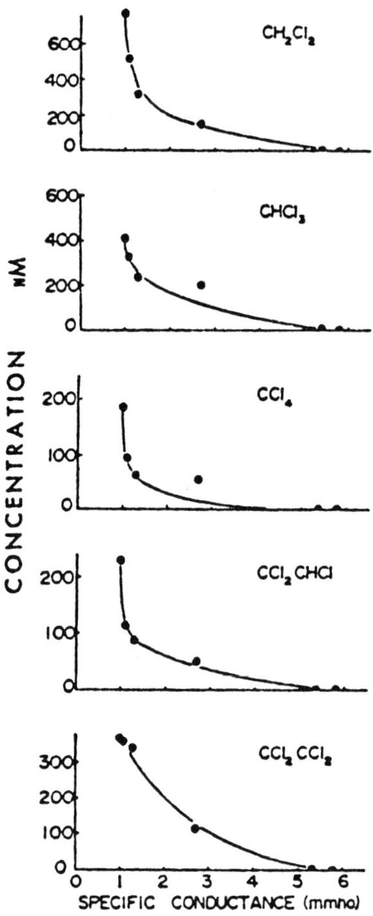

Figure 3. Decrease of halocarbons in a seaward direction in Back River, Maryland. The estuary was covered with ice when these samples were taken.

of minor importance for the C_1 and C_2 halocarbons. For example, Dilling et al (1975) measured the persistance of five C_1 and C_2 chlorocarbons which had been sealed in quartz tubes and exposed to sunlight. After one year, the amounts remaining ranged from 65% for CH_2Cl_2 to about 25% for CCl_2CHCl and CCl_2CCl_2. Zafiriou (1975) estimated that 50% conversion of CH_3Cl to CH_3OH in seawater would take 2.5 years at 20°C and 14 years at 10°C. Thus abiotic chemical degradation of the C_1 and C_2 halocarbons, by either photolysis or hydrolysis, appears to be fairly slow.

Similarly, it is likely that adsorption onto settling particles is a fairly ineffective removal process. Helz and Hsu (1978) found trihalomethanes to be very weakly adsorbed on such substances as montmorillonite, ferric hydroxide and humic acid; Diachenko (Ph.D. thesis in preparation, University of Maryland) has further substantiated these findings. Karickhoff et al (1978) have measured the partition coefficients between water and sediment for a wide range of compounds. They find that the partition coefficient between water and the organic matter fraction of fine-grained sediment is about 0.6 of the octanol-water partition coefficient. For C_1 and C_2 halocarbons, octanol-water partition coefficients are mostly in the range, 10 - 100. From these figures it can be estimated that if a water contained 10 mg/L particulate organic carbon (a reasonable value for many natural waters), it would carry less than 0.1% of its C_1 and C_2 halocarbon load on organic-laden particles.

The extent to which biological degradation of C_1 and C_2 halocarbons will contribute to non-conservative behaviour remains to be established. Helz and Hsu (1978) speculated that biological degradation might contribute to the loss of halocarbons that they observed under ice-cover conditions (Figure 3). However, the general opinion appears to be that these compounds are rather resistant to biological attack (Pearson and McConnell, 1975). More information is needed on this problem.

The only removal mechanism for C_1 and C_2 halocarbons which has been demonstrated to be fast is volatilization. In stirred beakers in the laboratory, Dilling et al (1975) and Dilling (1977) have measured half lives due to volatilization of 30 ± 15 minutes for a large number of chlorocarbons. Helz and Hsu (1978) calculated half lives due to volatilization of less than 10 hours for the compounds in Figure 2 in 1 - 2 m deep Back River. These results explained the very rapid halocarbon loss that they observed in the absence of ice-cover.

DISCUSSION

From the preceeding discussion, it can be concluded that strong sources of C_1 and C_2 halocarbons are probably common in coastal regions near population centers. It remains to be seen whether the halocarbon concentration of municipal and industrial wastewaters are sufficiently stable for tracer applications. However, heat dischargers such as power plants represent an ideal tracer source. In those estuarine waters where fouling by hardshell organisms, such as barnacles or mussels, is a problem, continuous chlorination at a constant level is often practiced. Even when application of chlorine is made only periodically, the schedule and dosages are often established in advance and documented in plant records. Thus the source is predictable.

Power plants are particularly important because of the vast volumes of water they circulate. In the northern Chesapeake Bay, one of the largest estuaries in the United States, power plants currently circulate a volume of water equivalent to 10% of the freshwater throughput.

Although a detailed understanding of the fate of C_1 and C_2 halocarbons remains to be achieved, it appears that dilution and volatilization are the main mechanisms which control concentrations of these compounds in receiving waters. The principles which control volatilization have been discussed by Liss and Slater (1974), Mackay and Leinonen (1975) and others. Making use of this information, the rate of concentration change for a halocarbon with time can be presented as follows:

$$\frac{dC}{dt} = -\frac{k}{L}C - D'(t)C \qquad (1)$$

where C = concentration, t = time, k = volatilization rate constant with units of length/time, L = mixing depth and D'(t) = the instantaneous dilution rate. If at time t = 0, a parcel of water contains a concentration, C_A^o, of halocarbon A, then at some later time, by integration of equation 1:

$$\ln\frac{C_A}{C_A^o} = -k_A\frac{t}{L} - D(t) \qquad (2)$$

If the halocarbon source contains more than one halocarbon, which is often the case with the sources described in this paper, then the dilution function, D(t), can be eliminated by considering halocarbon ratios:

$$\ln\frac{C_A}{C_B} = \ln\frac{C_A^o}{C_B^o} - (k_A - k_B)\frac{t}{L} \qquad (3)$$

Thus if C_A^o/C_B^o can be determined for a source and if $(k_A - k_B)$ can be estimated by methods described in the literature (e.g. Liss and Slater, 1974) then t/L can be computed for any parcel of water in a mixing system from a measurement of C_A/C_B. Knowing t/L, the dilution function in equation 2 can be evaluated for the sample; multiple determinations will allow this function to be mapped. If the mixing depth, L, is known from hydrographic data, then the "age" of the water parcel (i.e. time since t = 0) can be computed. Such an application is not possible with conservative tracers. Undoubtedly, hydrographers and physical oceanographers will be able to devise other, more sophisticated applications than these.

CONCLUSION

Strong sources of C_1 and C_2 halocarbons are probably relatively common in the coastal waters of populated regions. Concentrations in excess of 10^4 nM have been reported from some industrial sources. Analytical methodology for these compounds is now highly developed with detection limits often around 1 nM. The necessary equipment has become widely available. At present it appears that dilution and volatilization are the main processes which control the concentration of the volatile halocarbons after discharge. It appears that C_1 and C_2 halocarbons may prove useful as water-mass tracers for coastal waters.

REFERENCES

Bellar, T.A., J.J. Lichtenberg, and R.C. Kroner. 1974. The Occurrence of Organohalides in Chlorinated Drinking Water. J. Am. Water Works Assoc., 66, 703-6.

Carpenter, J.H., and C.A. Smith. 1978. Reactions in Chlorinated Sea Water. In: Water Chlorination: Environmental Impact and Health Effects, Vol. 2 (R.L. Jolley, D.H. Hamilton and H. Gorchev, Eds.) Ann Arbor Science, p. 195-208.

Dilling, W.L., N.B. Tefertiller, and G.J. Kallos. 1975. Evaporation Rates and Reactivities of Methylene Chloride, Chloroform, 1,1,1-Trichloroethane, Trichloroethylene, Tetrachloroethylene and Other Chlorinated Compounds in Dilute Aqueous Solutions. Environ. Sci. & Technol., 9, 833-8.

Dilling, W.L. 1977. Interphase Transfer Processes. II. Evaporation Rates of Chloro-Methanes, Ethanes, Ethylenes, Propanes, and Propylenes from Dilute Aqueous Solutions. Comparisons with Theoretical Predictions. Environ. Sci. & Technol., 11, 405-9.

Ewing, B.B., E.S. Chian, J.C. Cook, C.A. Evans, P.K. Hopke, and E.G. Perkins. 1977. Monitoring to Detect Previously Unrecognized Pollutants in Surface Waters. U.S. Environ. Protection Agency EPA-560-6-77-015, 77 pp.

Glaze, W.H. and J.E. Henderson IV. 1975. Formation of Organochlorine Compounds from the Chlorination of a Municipal Secondary Effluent. J. Wat. Pollut. Control Fed., 47, 2511-15

Hammer, P.M., J.M. Hayes, W. Jenkins, and R. Gagosian. 1978. Freon-11 Measurements in North Atlantic Water Columns. Trans. Am. Geophys. Union. 59, p.307 (Abst.).

Helz, G.R., and R.Y. Hsu. 1978. Volatile Chloro- and Bromocarbons in Coastal Waters. Limnol. and Oceanogr. In Press.

Helz, G.R., R. Sugam, and R.Y. Hsu. 1978a. Chlorine Degradation and Halocarbon Production in Estuarine Waters. In: Water Chlorination: Environmental Impact and Health Effects. Vol 2 (R.L. Jolley, D.H. Hamilton and H. Gorchev, Eds.) Ann Arbor Science, p. 209-222.

Helz, G.R., R.Y. Hsu, and R.M. Block. 1978b. Bromoform Production by Oxidative Biocides in Marine Waters. In: R.G. Rice, J.A. Cotruro and M.E. Browning (Eds.), Ozone-Chlorine Dioxide Oxidation Products of Organic Materials. International Ozone Institute. p.68-76.

Jolley, R.L. 1974. Determination of Chlorine-Containing Organics in Chlorinated Sewage Effluents by Coupled ^{36}Cl Tracer-High Resolution Chromatography. Environ. Lett., 7, 321-340.

Lovelock, J.E., R.J. Maggs, and R.J. Wade. 1973. Halogenated Hydrocarbons In and Over the Atlantic, Nature 241, 194-6.

Liss, P.S., and P.G. Slater. 1974. Flux of Gases Across the Air-Sea Interface, Nature 242, 181-4.

Mackay, D., and P.J. Leinonen. 1975. Rate of Evaporation of Low-Solubility Contaminants from Water Bodies to Atmosphere. Environ. Sci. Technol. 9, 1178-1180.

National Academy of Sciences. 1978. Nonfluorinated Halomethanes in the Environment. Washington, D.C. 297 pp.

Nelson, N., and B. Van Durren. 1975. Final Report of NSF Workshop Panel to Select Organic Compounds Hazardous to the Environment. New York Univ. Medical Center, 78 pp.

Pearson, C.R., And G. McConnell. 1975. Chlorinated C_1 and C_2 Hydrocarbons in the Marine Environment. Proc. Roy. Soc. Lond. 189, 305-32.

Zafiriou, O.C. 1975. Reaction of Methyl Halides with Seawater and Marine Aerosols. J. Mar. Res. 33, 75-81

PREDICTION OF VOLATILIZATION RATES OF CHEMICALS IN WATER

James H. Smith and David C. Bomberger

SRI International

Menlo Park, California 94025

INTRODUCTION

Transport of chemicals from water bodies and wastewater treatment facilities to the atmosphere by volatilization (evaporation) can be an important environmental pathway for certain chemicals. Liss and Slater (1974) and Mackay and Leinonen (1975) assumed a two-film model in which the rates of diffusion in air and in water control the rate of transfer of a chemical across the interface between air and water. The volatilization rate constant for solute S, k_v^S, may be expressed as:

$$k_v^S = \frac{A}{V}\left\{\frac{1}{K_L^S} + \frac{RT}{H_c^S K_G^S}\right\}^{-1} \tag{1}$$

where:

k_v^S	Overall mass transfer coefficient (hr^{-1})
A	Interfacial area (cm^2)
V	Liquid volume (cm^3)
H_c	Henry's law constant (torr litre $mole^{-1}$)
K_L	Liquid film mass transfer coefficient (cm hr^{-1})
K_G	Gas film mass transfer coefficient (cm hr^{-1})
R	Gas constant (litre torr K^{-1} $mole^{-1}$)
T	Temperature (K)

To use equation (1) to estimate k_v^S, it is necessary to measure or estimate the mass transfer coefficients and the Henry's law constant for each solute of interest. However, it is difficult to relate the mass transfer coefficients determined in the laboratory to those in a natural water body and there is no convenient way to measure the mass transfer coefficients directly in the environment.

The procedure that we have used for estimating k_v^S (Smith et al, 1977, 1978) is based on the observation by Tsivoglou (1965, 1967) that the ratio of k_v^S and the oxygen reaeration rate constant, k_v^0, is a constant for a range of turbulence conditions. Both k_v^S for a specific substance and k_v^0 can be measured simultaneously in the laboratory, and k_v^S/k_v^0 can then be calculated.

Representative oxygen reaeration rate constants are available for several different types of water bodies and typical wastewater treatment facilities. Therefore, multiplying a measured value of k_v^S/k_v^0 by the oxygen reaeration rate constant for a specific water body gives the value of the volatilization rate constant of the chemical in the water body or facility.

Mass transfer in the liquid phase is rate controlling when the value of H_c is greater than about 1000 torr mole litre^{-1}. Then, since

$$K_L = D/\delta \tag{2}$$

where δ is the liquid film thickness, equation (1) reduces to

$$K_v^S/k_v^0 = D^S/D^0 \tag{3}$$

where D^S and D^0 are the diffusion coefficients of solute and oxygen in water. Tsivoglou (1967) has shown that the liquid phase diffusion coefficient is often inversely proportional to the molecular diameter. The equation (3) reduces to

$$K_v^S/k_v^0 = d^0/d^S \tag{4}$$

where

where d^0 and d^S are the molecular diameters of O_2 and the substance S. Either equation (3) or equation (4) can be used to estimate k_v^S/k_v^0 for high volatility compounds at the low turbulence levels likely to be found in natural water bodies. If a value for D^S (in water) cannot be found ($D^0 = 2.1 \times 10^{-5}$ cm sec^{-1}, Tsivoglou, 1967), the molecular diameter can be estimated from the critical volume (Present, 1958), which can be estimated by methods reviewed by Reid and Sherwood (1967) (an accepted value for d^0 is 2.98 A).

For compounds with a Henry's law constant less than 10^3 torr mole litre^{-1}), so that vapor phase mass transport resistance is large, the ratio k_v^S/k_v^0 does not reduce to a simple form. It may be easier (that is, more economical) to measure the ratio in the laboratory than to estimate the various transport coefficients and H_c^S, which are needed to estimate k_v^S/k_v^0.

EXPERIMENTAL PROCEDURES

Volatilization rates were measured using the method described by Hill et al (1976). Solutions of the solute were prepared in deionized, distilled water. For low volatility compounds, 1 litre of the solution was placed in a 2-litre beaker equipped with a stirring bar. The solution was sparged with nitrogen to remove most of the dissolved oxygen. For highly volatile compounds, the solute was added to previously sparged water. The measurements of k_v^S and k_v^0 for benzo[b]thiophene at low stirring rates (k_v^0 <0.5 hr^{-1}) were obtained from solutions in 1-litre Erlenmeyer flasks.

At the start of the experiment, the concentration of the solute was measured by either liquid or gas chromatography. The oxygen concentration was measured with a Delta model 2110 dissolved oxygen analyzer. Successive solute and dissolved oxygen concentrations were made at regular time intervals.

The solute concentration data were fit to equation (5),

$$\ln[S_t] = k_v^S t + \ln[S_0] \tag{5}$$

using the linear least squares routines in hand-held calculators. The dissolved oxygen concentration data were fit to equation (6) in the same way,

$$\ln[O_2]_t = k_v^0 t + \ln[O_2]_{sat} - \ln([O_2]_{sat} - [O_2]_o) \tag{6}$$

where $[O_2]_{sat}$ is the oxygen saturation concentration.

RESULTS

The k_v^S/k_v^0 values predicted for all substrates studied in our laboratories are given in Tables 1 and 2. The data for chloroform, Figure 1, are typical of the data for carbon tetrachloride and 1,1-dichloroethane. The data for benzo[b]thiophene are shown in Figure 2.

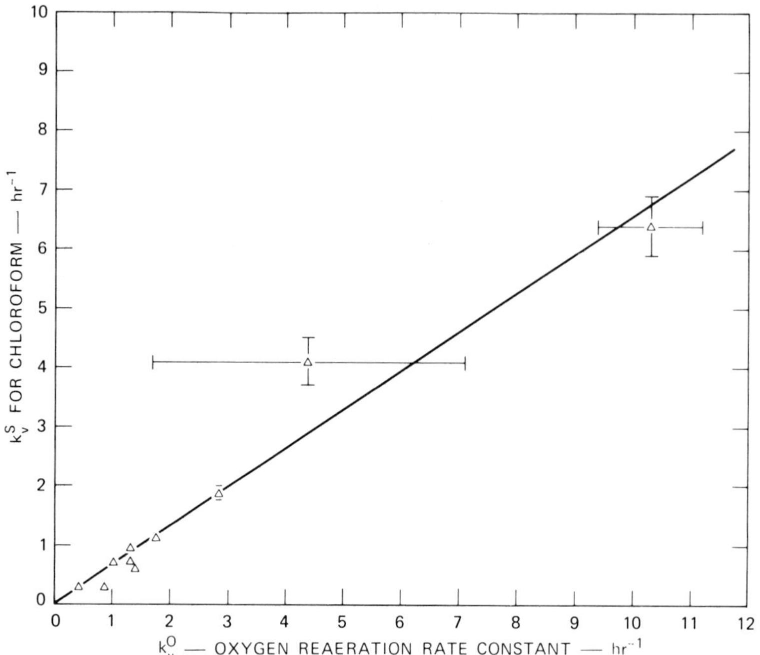

Figure 1. Volatilization rate data for chloroform

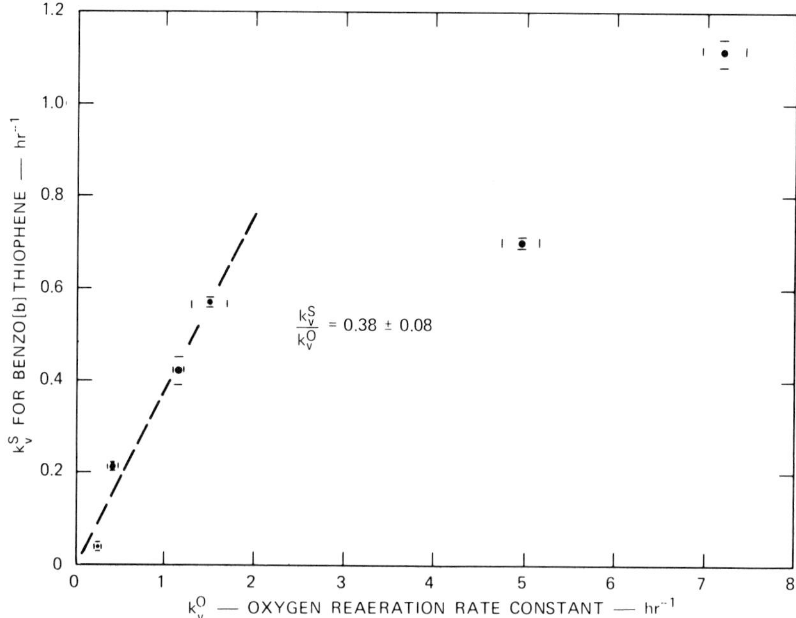

Figure 2. Volatilization rate data for benzo[b]thiophene

Table 1 Volatilization Rate Data for High Volatility Compounds

Compound	H_c [a] (torr liter mole^{-1})	k_v^S/k_v^0 Predicted d^0/d^S	k_v^S/k_v^0 Predicted $(d^0/d^S)^{1/2}$	k_v^S/k_v^0 Measured
Vinyl Chloride	14,000	0.49	0.70	1.0 ±0.6 [b]
Chloroform	2,700	0.40	0.63	0.66±0.11
Carbon tetrachloride	16,200	0.42	0.65	0.79±0.19
1,1-Dichloroethane	4,400	0.44	0.66	0.71±0.11
Oxygen	59,000	1.00	1.00	1.00
Benzo[b]thiophene	210	0.38	0.62	0.38±0.08
Dibenzothiophene	332	0.33	0.57	0.14

[a] All values, except that for oxygen, were estimated from vapor pressure and solubility data by the method of Mackay and Wolkolf (1973).

[b] Hill et al. (1976).

CONCLUSIONS

The data shown in Table 1 for the three high volatility chlorinated compounds indicate that the value of k_v^S/k_v^0, predicted from equation (3), which is based on film theory (Liss and Slater, 1974; Mackay and Leinonen, 1975), does not agree with the values measured in our laboratory. However, our data agree, within experimental error, with predictions based on equation (7).

$$k_v^S/k_v = (d^0/d^S)^n \tag{7}$$

where n = 0.5, which is the form predicted by penetration theory (Higbie, 1935) and by surface renewal theory (Danckwerts, 1951, 1955). Additional work is under way in our laboratories to further define the mass transport mechanism of volatile compounds. However, film theory remains a useful conceptual framework for describing the volatilization process.

The plot of k_v^S versus k_v^0 for benzo[b]thiophene (Figure 2) is not a straight line, and data for dibenzothiophene are similar. These results suggest that both terms in equation (1) are of the same magnitude, which is also reflected in the values of H_c.

The measured values of k_v^S/k_v^0 for the low volatility compounds, Table 2, are significantly lower than the values predicted by equations (4) (0.27 to 0.43) or equation (7) (0.52 to 0.66). Also, the values of H_c are all low. These observations suggest that the volatilization rates of these compounds are limited by gas phase mass transport resistance.

Table 2 Volatilization Rate Data for Low Volatility Compounds

Compound	H_c [a] torr liter mole^{-1}	Measured k_v^S/k_v^0
p-Cresol	6.5	0.0014
Benz[a]anthracene	0.10	0.0004
Benzo[a]pyrene	1.0	0.0036
Quinoline	0.19	0.0025
Benzo[f]quinoline	0.072	0.0002
Carbazole	120	0.0001
Dibenzo[c,g]carbazole	0.0036	0.0001
Methyl parathion	59	0.0001
Mirex	144	0.063

[a] All H_c values were estimated from pressure and solubility data by the method of Mackay and Wolkoff (1973).

The theory and measurements discussed above have been applied to modelling the loss of most of the compounds listed in Tables 1 and 2 from natural water bodies such as lakes and streams (Smith et al, 1977, 1978). The same methodology is also useful for predicting volatilization rates of organic compounds from wastewater treatment operations such as dissolved air flotation, settling, evaporation ponds, and activated sludge treatment (Smith and Bomberger, 1978) as well as cooling towers.

These studies suggest that volatilization can be an important mechanism for transfer of many pollutants from water bodies and wastewater treatment facilities to the atmosphere. It may be the major environmental process of nonpolar chlorinated compounds and nonpolar aliphatic and aromatic hydrocarbons that have a molecular weight less than about 150. Estimates of volatilization rate constants can also be made for less volatile compounds ($H_c < 10^3$ torr litre mole^{-1}), provided laboratory measurements of k_v^S/k_v^0 are available. For these compounds, volatilization is not important unless processes such as photolysis or biodegradation are very slow.

ACKNOWLEDGEMENTS

We thank Ms. Barbara Kingsley and Messrs. Daniel Haynes and Mark Zinnecker for their assistance in carrying out the laboratory measurements. A portion of this work was supported by the U.S. Environmental Protection Agency under contract No. 68-03-2227.

REFERENCES

Danckwerts, P.V., (1951), Ind. Eng. Chem. 43, 1460.

Danckwerts, P.V., (1955), J. AIChE, 1, 456.

Hill IV, J., et al (1976), "Dynamic Behavior of Vinyl Chloride in Aquatic Ecosystems", U.S. Environmental Protection Agency, EPA-600/13-76-001.

Higbie, R., (1935), Trans. AIChE, 31, 365.

Liss, P.S. and P.G. Slater, (1974), Nature, 247, 181-184.

Present, R.D., (1958), Kinetic Theory of Gases (McGraw-Hill, New York).

Mackay, D., and P.J. Leinonen, (1975), Environ. Sci. Tech., 9, 1178-1180.

Mackay, D., and A.W. Wolkoff, (1973), Environ. Sci. Tech., 7, 611-614.

Reid, R.C., and T.K. Sherwood, (1966), The Properties of Gases and Liquids, 2nd ed. (McGraw-Hill, New York).

Smith, J.H., and D.C. Bomberger, (1978), "Prediction of Volatilization Rates of Chemicals in Water", paper presented at the AIChE 85th National Meeting, June 4-8, 1978.

Smith, J.H. et al, (1977), "Environmental Pathways of Selected Chemicals in Freshwater Systems. Part 1: Background and Experimental Procedures", EPA report No. EPA-600/7-77-113; (1978) "Part 2: Laboratory Results", EPA report No.EPA-600/7-78-074.

Tsivoglou, E.C., (1965), J. Water Poll. Control Fed., 37, 1343-1362; (1967), "Measurement of Stream Reaeration", U.S. Department of the Interior, Washington, D.C., June 1967.

CHLORINATED PHENOLIC COMPOUNDS: FORMATION, DETECTION, AND TOXICITY IN DRINKING WATER

Phanibhushan B. Joshipura* and Peter N. Keliher

Chemistry Department, Villanova University

Villanova, Pa. 19085, U.S.A.

Chlorination is extensively practiced in wastewater treatment to disinfect the effluent prior to discharge, particularly where the receiving water may be used for recreational purposes or as a source of potable water. Recently, it has been demonstrated that industrial effluents, particularly those subjected to biological treatment may require disinfection to meet current receiving water bacterial standards. Zillick (1972) has pointed out that the chlorination of municipal wastewater increases its toxicity to aquatic life. One of the major problems associated with disinfection of water supplies by chlorination is that the organoleptic properties of the chlorinated water may be increased. This malodorous water often is produced by a reaction between free available chlorine and trace concentrations of organic compounds, especially phenol and its derivatives.

Phenols are widely distributed by man and, at trace levels, by nature. They are readily detected by taste and odor, particularly when the unsubstituted phenol is chlorinated in public water supplies. It should be noted that, in the context of this paper, the terms "chlorophenols" and "chlorinated phenolic compounds" refer only to those phenolic compounds which are directly formed by reaction between free available chlorine and phenol. The important industrial pollutant, pentachlorophenol (PCP) is not formed in this way and will, therefore, not be considered further here. We have discussed the determination of pentachlorophenol elsewhere (Fountaine et al, 1975, 1976).

* Present address: Naval Supply Center Chemistry Laboratory Portsmouth, Virginia 23703, U.S.A.

The nature of the compound or compounds which gives rise to "chlorophenolic" taste and odors resulting from the reaction between aqueous chlorine and phenolic compounds has been investigated repeatedly (Adams, 1931; Ettinger and Ruchoft, 1951; Todd, 1942). Little was known with certainty, however, about the chemical reaction involved in the production or destruction of chlorophenolic taste and odor until the appearance of a significant paper by Burttschell and co-workers (1959). These workers proposed a reaction scheme to account for the production and subsequent elimination of chlorophenolic taste and odors in water supplies.

According to Burttschell et al, the chlorination of phenol proceeds by stepwise substitution of the 2,4, and 6 protons of the aromatic ring. Initially phenol is chlorinated to form either 2- or 4- chlorophenol. The 2-chlorophenol is chlorinated to form 2,4- and/or 2,6- dichlorophenol while 4-chlorophenol forms 2,4-dichlorophenol. Both 2,4- and 2,6- dichlorophenols are chlorinated to form, 2,4,6- trichlorophenol. The 2,4,6-trichlorophenol reacts with aqueous chlorine to form a mixture of non-phenolic oxidation products.

Burttschell and co-workers (1959) also determined threshold odor concentrations (organoleptic properties) for chlorophenols. The compounds with the strongest organoleptic properties were 2-chlorophenol, 2,4-dichlorophenol, and 2,6-dichlorophenol; these were detectable at concentrations as low as 2 to 3 parts-per-billion (ng/ml). 4-chlorophenol had an organoleptic value of 250 ppb. By contrast, phenol and 2,4,6-trichlorophenol were observed to have organoleptic values above 1 part-per-million.

Lee and Morris (1962) studied the kinetics of reactions between chlorine and phenolic compounds. They stated that the rate of reaction of aqueous chlorine and phenol or the chlorophenols obeys a neo-second-order rate expression where the rate of reaction is proportional to the product of formal concentration of aqueous chlorine and phenolic compound. The rates of these reactions are highly pH dependent with the maximum rate occurring, depending upon the compound being chlorinated, in the neutral or slightly alkaline pH range. Lee and Morris concluded that the chlorination of phenol bearing waters should be conducted with maximum possible free chlorine in the pH range 7 to 8.

An important constituent of natural waters which has been reported to alter the kinetics of the chlorination of phenol is ammonia. Weil and Morris (1949) indicated that for equal initial molar concentrations of ammonia and phenol at pH 8 and at 25°C, ammonia is chlorinated to form monochloroamine about one thousand times faster than phenol is chlorinated to form a monochlorophenol. Therefore, little chlorophenol would be expected in the presence of excess ammonia. Burttschell and his co-workers, however, found that if sufficient time was allowed for the reaction to proceed (several days to a week), the same chlorophenols are formed when chloroamine reacts with phenol as were found in a considerably shorter period of time when the chlorination of phenol was conducted in the absence of ammonia. This extremely slow formation of chlorophenols in the presence of ammonia may account for some of the problems found in water supply practice with taste and odor development after the water has left the distribution plant, particularly in the dead ends of the distribution system (Lee and Morris, 1962).

Murphy and his co-workers (1975) obtained confirmation of the rapid chlorination of phenol by free available chlorine by comparing the infrared spectra of ether extracted material to standard reference scans (Sadtler, 1967) of phenol, 2-chlorophenol, 2,6-dichlorophenol, and 2,4,6-trichlorophenol. Murphy and his co-workers also indicated that with long reaction times or where high chlorine concentrations were employed, a brown-black precipitate was formed. They presumed that the brown-black precipitate might consist of some oxidized phenolic products. However, no attempt was made to purify the precipitate or to characterize it further.

Grimley and Gordon (1973) reported that at lower pH's the major products of chlorination of a dilute phenolic aqueous solution were 2-chlorophenol and 4-chlorophenol. Very small amounts of 2,4-dichlorophenol are formed at lower pH's. Lee and Morris (1962) had observed that in the reaction mixture 2,4-dichlorophenol predominates only at or above pH 9. They recommended that the chlorination of phenol containing waters should be carried out at pH 7. At this pH, initially concentrations of 2-chlorophenol and 4-chlorophenol predominate.

No specific method is available for the rapid quantitative determination of different chlorinated phenolics at very low concentration levels. Hence, in the ASTM procedure, all chlorinated phenols are determined in terms of 2,4-dichlorophenol. The ASTM procedure for chlorinated phenolic compounds is a modification of the standard procedure for phenolic compounds (American Society for Testing and Materials, 1970) in which the phenolic compound is reacted with the colorimetric reagent, 4-aminoantipyrine (4-AAP). Since different chlorinated phenols have different molar absorptivities, determination of the "total" concentration of the chlorinated phenols based on the 2,4-dichlorophenol calibration curve can result in an obvious error. This error will, of course be particularly great when the percentage of 2,4-dichlorophenol in the sample is small.

In a recent publication (Fountaine et al, 1974), we had described a novel and unique instrumental system for the determination of total phenolic compounds. We had referred to the system as an "Ultraviolet Ratio Spectrophotometer". The principle of the instrument is based upon the bathochromic shift effect which occurs when phenolic compounds are made basic. The instrument utilizes two hollow cathode lamps (sources normally used in atomic absorption spectrometry) as spectral sources and the system is set up such that one lamp is in the "pH dependent" region for a phenolic compound while the other lamp is in the "pH independent" region. By proper adjustment of the light output from the two lamps, it is possible to determine phenolic compounds at the low ppb level. In our 1974 publication, we performed a comparison study between the ultraviolet ratio spectrophotometric system (UVRS) and the standard ASTM (4-AAP) method. When analyses of real samples were performed, typical results using the UVRS were ca. 30 - 50 percent higher than when the ASTM procedure was employed. This is because the colorimetric reagent, 4-AAP, does not react with para-cresol, an important industrial phenolic pollutant.

In this study, a direct comparison was made between the 4-AAP method and the UVRS for the determination of chlorinated phenolic compounds. The procedure used for the UVRS was experimentally identical to the procedure described in our 1974 publication. The procedure used for the 4-AAP method included extraction of samples with petroleum ether(b.p. range 30-60°C) followed by extraction with 0.5 M ammonium hydroxide solution. The ammonium hydroxide extracts containing chlorophenolic compounds were used for the color development with 4-AAP.

Figures 1 and 2 show the calibration curves of different chlorophenols (and phenol itself) under two different sets of conditions. Figure 1 shows the calibration curves of chlorophenols when the instrument was optimized for phenol itself. Optimization is accomplished by selecting, via a small monochromator in the instrument, the wavelength from the "pH dependent" hollow cathode lamp which will give the best slope for the selected phenolic compound. As seen in Figure 1, other phenolic compounds will have different slopes under these conditions, showing responses usually less than but sometimes greater than the optimized phenolic compound.

Figure 2 shows the calibration curves of different phenolic compounds when the instrument was optimized for 2,4-dichlorophenol by changing the "pH dependent" wavelength. Once again, different phenolic compounds are observed to have different slopes. These different slopes are due to their differences in bathochromic shifts, differences in wavelengths of maximum absorption, and differences in their molar absorptivities at the "pH dependent" wavelengths. Also obtained were a similar set of calibration curves using the standard ASTM method. The different slopes of the calibration curves of different chlorophenols are due, in the ASTM procedure, to the differences in molar absorptivities formed between chlorophenols and 4-AAP.

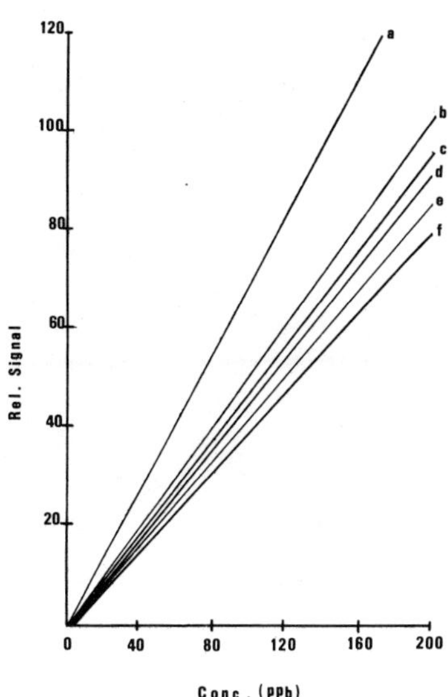

Figure 1. Response curves for some chlorinated phenolic compounds at original wavelength (290 nm) optimized for phenol. (a) 2-chlorophenol, (b) phenol, (c) 3-chlorophenol, (d) 2,4,6-trichlorophenol, (e) 2,4-dichlorophenol, (f) 4-chlorophenol.

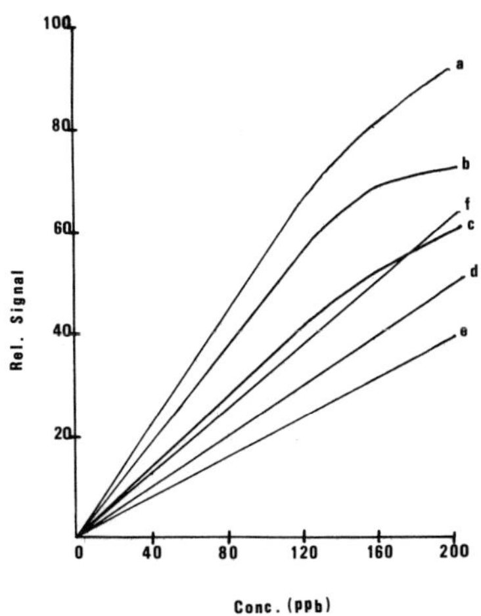

Figure 2. Response curves of some chlorinated phenolic compounds at 2,4-dichlorophenol optimized wavelength (310 nm). (a) 2,4,6-trichlorophenol, (b) 4-chlorophenol, (c) 2,4-dichlorophenol, (d) 3-chlorophenol, (e) phenol, (f) 2-chlorophenol.

Table 1 Comparison of Errors in Interpretation of Chlorophenol Concentration by Both Methods

Compounds	ASTM Method	UVRS Method[1]	UVRS Method[2]
2,4-dichlorophenol	0%	0%	0%
2-chlorophenol	-6.6%	+60%	0%
4-chlorophenol	-70%	-10%	+40%
2,4,6-trichlorophenol	-25%	+10%	+56%

1. Optimized for phenol, see text
2. Optimized for 2,4-dichlorophenol, see text

Table 1 compares the errors in interpreting concentrations of different chlorophenols in terms of 2,4-dichlorophenol. As noted previously, in practical situations when chlorination of water containing phenol is carried out, the major products are 2-chlorophenol and 4-chlorophenol with a very small amount of 2,4-dichlorophenol initially. The formation of 2,4,6-trichlorophenol is negligible. It is evident from Table 1 that the standard method is going to give a small concentration of chlorinated phenols due to the negative errors for both 2-chlorophenol (-6.6%) and 4-chlorophenol (-70%). The UVRS, on the other hand, when optimized for 2,4-dichlorophenol, will give a higher concentration of chlorinated phenols because of the error for 4-chlorophenol (+40%). This partially explains why the UVRS gives higher concentration levels than the standard 4-AAP method, but it is clear that of the two methods, the 2,4-dichlorophenol optimized UVRS method should be closer to the actual concentration levels than is the standard method. Also evident from Table 1 is how it is easy and efficient to make the instrument relatively specific for chlorinated phenols by simple wavelength adjustment of the "pH dependent" hollow cathode lamp.

Table 2 shows a series of measurements taken over a period of time for the determination of chlorinated phenols in Philadelphia drinking water. All samples were collected from the same location and, in all cases, the tap was allowed to run for at least 20 minutes before sample collection.

The measurements were made with both methods - 4-AAP and UVRS. Since the concentration levels of chlorophenols in drinking water were very low, readings were obtained in the vicinity of the detection limits by both methods.

Table 2 Chlorinated Phenol Concentration Levels in Drinking Water

Sample Number	Date Collected	ASTM Method	UVRS Method[1]
1	2/26/75	5 ppb	6 ppb
2	2/27/75	4 ppb	6 ppb
3	2/28/75	<4 ppb[2]	4 ppb
4	3/3/75	4 ppb	5 ppb
5	3/4/75	5 ppb	6 ppb
6	3/5/75	<4 ppb	<4 ppb
7	3/6/75	4 ppb	5 ppb
8	3/7/75	<4 ppb	<4 ppb

1. UVRS optimized for 2,4-dichlorophenol
2. Less than 4 ppb; 4 ppb is about the detection limit.

ACKNOWLEDGEMENTS

We thank B.R. Keliher for her assistance with the figures.

DEDICATION

This paper is dedicated, with respect and affection, to the memory of Mr. B.U. Joshipura.

REFERENCES

Adams, B.A. 1931. Water Works Eng., 33, 387.

American Society for Testing and Materials, Method D-1783-70. 1970 Philadelphia, Pa.

Burttschell, R.H., A.A. Rosen, F.M. Middleton, and M.B. Ettinger. 1959. J. Am. Water Works Assoc., 51, 205.

Ettinger, M.B., and C.C. Ruchoft. 1951. J. Am. Water Works Assoc., 43, 561.

Fountaine, J.E., P.B. Joshipura, P.N. Keliher, and J.D. Johnson. 1974. Anal Chem., 46, 62.

Fountaine, J.E., P.B. Joshipura, P.N. Keliher, and J.D. Johnson. 1975. Anal. Chem., 47, 157.

Fountaine, J.E., P.B. Joshipura, and P.N. Keliher. 1976. Water Research, 10, 185.

Grimley, E., and G. Gordon. 1973. J. Phys. Chem., 77, 973.

Lee, C.F., and J.C. Morris. 1962. Int. J. Air Water Pollut., 6, 419.

Murphy, K.L. R. Zaloum, and D. Fulford. 1975. Water Research, 9, 389.

Todd, A.R. 1942. J. Am. Water Works Assoc., 34, 1805.

Weil, I., and J.C. Morris. 1949. J. Am. Chem. Soc., 71, 1664.

Zillick, J.A. 1972. J. Water Pollut. Control. Fed., 44, 212.

SCREENING SOUTH AFRICAN DRINKING-WATERS FOR ORGANIC SUBSTANCES HAZARDOUS TO HEALTH

J.F.J. van Rensburg, Senior Chief Research Officer and

A.J. Hassett, Chief Technical Officer

National Institute for Water Research of the Council for Scientific and Industrial Research, P.O. Box 395 Pretoria, 0001, South Africa

INTRODUCTION

When screening water supplies for organic pollutants which may be hazardous to health, the choice of analytical techniques is determined by their adaptability to routine use. Important considerations are simplicity, use of automated or semi-automated equipment and economic factors such as high sample throughput per man hour and low sampling and sample preparation costs.

The list of 11 compounds designated by a committee of experts (WHO, 1975) as being "abundantly present" in water and classified as "very toxic" was modified in other publications, e.g., Harrison et al (1974)and Symons et al (1975) and adapted for South African conditions (Table 1). Lower hazard limits were based upon limits set elsewhere (EPA, 1975; WHO, 1971) and available toxicological data.

A preliminary screening programme was introduced to estimate the health risks posed by South African drinking-waters by determining the possible presence of these compounds.

TABLE 1: "Very toxic" compounds analysed for initial screening programme

Organic compounds	Lower hazard limit (μg/ℓ)
Volatile halogenated hydrocarbons (VHH)	
Trihalomethanes (chloroform, bromoform, bromodichloromethane, dibromochloromethane) and tetrachloromethane	1 per compound
Chlorinated hydrocarbons	
Lindane, chlordane, dieldrin, hexachlorobutadiene, hexachlorobenzene, PCB's	0,1 per compound
Dichlorobenzene	1
Chlorophenols	
Di-, tri-, tetra- and pentachlorophenols	1 per compound
Polynuclear aromatic hydrocarbons (PNA)	
Benz(a)anthracene (BAN), benzo(b) and (k) fluoranthene, (BbFL, BkFL) benzo(a)pyrene (BaP), dibenz(a,h) anthracene (DBaHAN), indeno(1,2,3-cd) pyrene (IPN), Anthracene (AN), phenanthrene (PN) and fluoranthene (FL)	0,1 per compound

ANALYTICAL METHODS

All glassware was thoroughly cleaned with detergent, distilled water and acetone and baked for 8 hours at 200 °C (Bevenue *et al*, 1971).

Sampling and extraction

Clean 750 mℓ beer bottles, containing about 100 mg ascorbic acid (Kissinger and Fritz, 1976) and sealed with crown corks lined with aluminium foil, were used for sampling. These were completely filled with the sample, resealed with aluminium-lined crown corks, air-freighted to the laboratories and stored at 3 °C until extraction.

The VHH were extracted by means of the syringe technique (van Rensburg *et al*, 1978a) and labelled 'Extract A'.

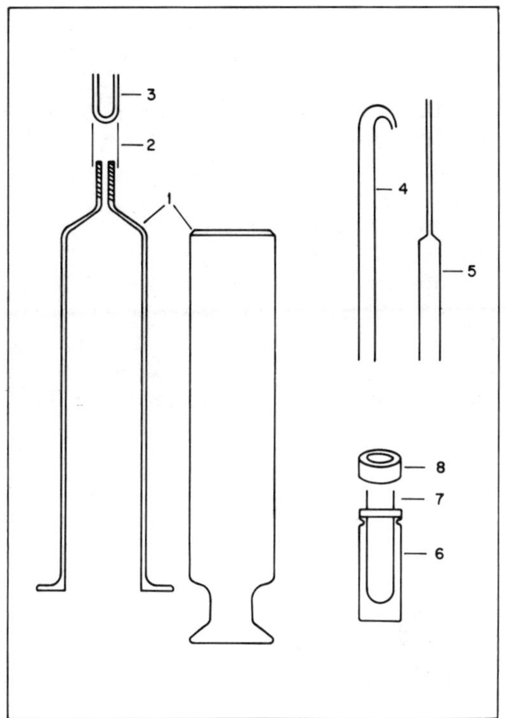

Fig. 1: Apparatus for the syringe extraction technique

The glass tube (5, Fig. 1) was fitted to the all-glass syringe (1) by means of the PTFE connector (2) and a 10 mℓ water sample drawn into the syringe from the beer bottle. A small amount of air (1 to 2 mℓ) was also drawn into the syringe to facilitate shaking. With the syringe tip uppermost, the glass tube was removed, 250 μℓ hexane (BDH extra pure grade), containing dibromo-ethane as internal standard, was added through the tip of the syringe, the tip stoppered (3) and the mixture shaken for 3 minutes in a tumbling machine. After separation of the two phases the transfer tube (4) was fitted to the syringe tip and the syringe clamped above a labjack, with the labjack supporting the plunger. The extract was then carried over to the glass insert (7) by raising the syringe plunger by means of the labjack. The insert was sealed in the 2 mℓ serum vial (6) normally used on the gas chromatographic automatic sampler. A further 10 mℓ sample was discarded, the pH of the remaining sample in the beer bottle adjusted to pH 10 to 11 and extracted with 15 mℓ benzene/hexane (1:9) containing aldrin as internal standard, for 3 minutes in a tumbling machine.

After separation of the phases, 200 μℓ of the organic phase (extract B) was placed into an insert and sealed in a serum vial for the gas chromatographic determination of the chlorinated hydrocarbons. The rest of the organic phase was drawn off the sample with a Pasteur pipette and concentrated by evaporation to 50 μℓ for the determination of the PNA's by MPD-TLC.

The water sample was re-extracted with 5 mℓ benzene, which was discarded, the sample acidified to pH <2 with sulphuric acid and then extracted as before with 15 mℓ benzene containing hexachlorobenzene as internal standard (extract C). A 200 μℓ aliquot (extract C) was transferred to an insert, 20 μℓ Trizil Z (trimethylsilyl imidazole in ultra-pure pyridine (0,25 ℓ/ℓ, Pierce Chemical Co.) was added and the insert sealed in a 2 mℓ serum vial. The silyl derivatives of the chlorophenols thus formed were analysed by gas chromatography. Confirmation of the chlorophenols was obtained by concentrating extract C to 50 μℓ and analysed by MPD-TLC on NH_3-buffered silica gel plates. Visualization was obtained by $AgNO_3$-reagent (van Rensburg *et al*, 1978(b)).

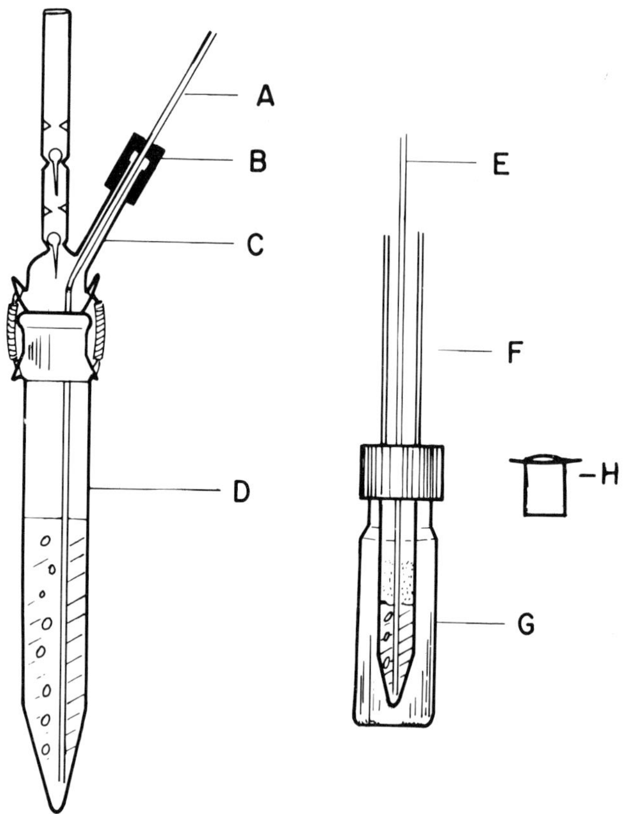

Fig. 2: Concentration apparatus
2.1 Micro-Snyder evaporator
2.2 Micro-evaporator

Concentration by evaporation

A two-stage evaporation procedure was used (Fig. 2), consisting of a modified micro-Snyder evaporator (Fig. 2.1) (Kontes, Vineland, N.J) and a micro-evaporator (Fig. 2.2).

The micro-Snyder was modified by the addition of a glass side tube (C), 6 x 3 x 20 mm, through which a 1,6 x 0,5 mm stainless steel tube was sealed, with a PTFE slip-on sleeve as illustrated. Evaporation was achieved by bubbling pure nitrogen or air through A at the rate of about 50 mℓ/min and at a temperature just below the boiling point of the solvent. The micro-evaporator was made from a standard 1 mℓ Reactivial (Pierce Chemical Co., Rockford, Il.). A glass tube (F) was sealed onto the lid of the Reactivial with a PTFE sleeve (H) which replaced the conventional rubber seal.

The bulk of the extract was concentrated to 0,5 mℓ in the modified micro-Snyder system, then carried over to the micro-evaporator for final concentration to 50 μℓ.

Gas chromatography

A Hewlett Packard 5710A g.c. with an ECD (Ni-63) detector and a 5671A auto sampler linked to a 3352 data system was used. Two glass columns, 3 m x 2,5 mm i.d., were used, one packed with 0,15 (kg/kg) XF 1150 on Chromosorb W AW (80-100 mesh) at 90 °C for the VHH and the other packed with 0,04 OV 101/0,06 OV 210 (kg/kg) on Chromosorb W HP (80-100 mesh) at 200 °C for the chlorinated hydrocarbons and at 180 °C for the chlorophenol derivatives.

Thin layer chromatography

Multiple progressive thin layer chromatography (MPD-TLC) (van Rensburg, 1978) was employed with self-built apparatus (Fig. 3) which allows development for progressively longer intervals according to $T_r = a.r$, where a is the development time for the first interval and T_r the development time for the rth interval. Plates were dried between development times by blowing air over the plate in such a way that the air pressure inside the developing chamber forced the developing solvent to below the lower limit of the chromatographic material. For the PNA, caffeine impregnated silica gel plates were used in a 7-stage development procedure with benzene/hexane (0,55 ℓ/ℓ). Visualization of spots by fluorescence occurs at a wavelength of 350 nm.

Standard silica gel TLC plates (Merck) were impregnated with caffeine by TLC-development with a solution prepared by dissolving 15 g caffeine in 105 mℓ chloroform and 45 mℓ methanol. The plates were air dried and the last traces of solvent removed by heating in an oven at 120 °C for 20 to 30 minutes.

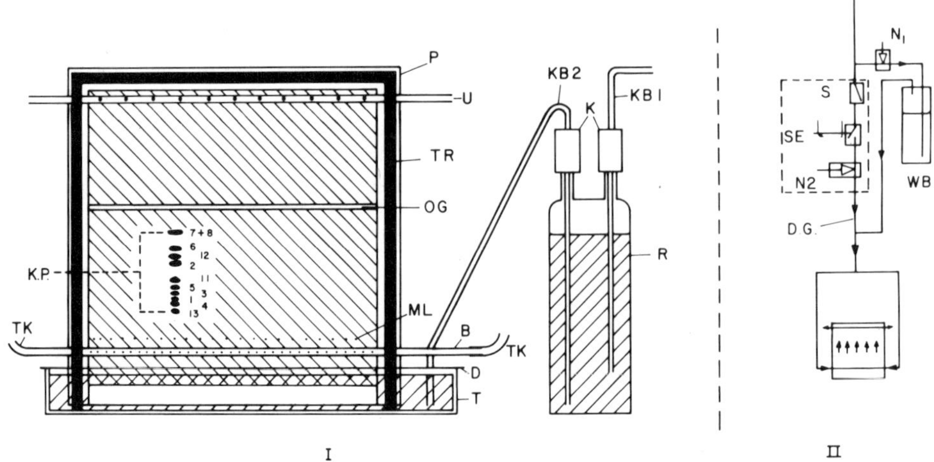

Fig. 3: Instrumentation for MPD-TLC

Before application of the samples, 10 mm of the silica gel was removed from the top and sides and 5 mm from the bottom of the plate (P, Fig. 3) and the development line (OG) scratched out, 110 mm from the bottom of the plate. Samples were applied on a line (ML) 35 mm from the bottom of the plate at 10 mm intervals. An external standard containing fluoranthene and benzo(ghi)perylene and two calibration mixtures were interspersed among the samples.

A sandwich development chamber was prepared by placing the PTFE frame (TR) and a clean glass plate on the TLC plate and clamping them together. The development chamber was completed by placing the sandwich in the trough (T) and coupling the gas connection with PTFE slip-on sleeves (TK). In the 7-stage progressive development procedure, provision was made for drying periods of 5 minutes between each step. During the drying stages the programmable switching valve (SE) permitted the flow of gas (controlled by needle valve (N_2)) through the development chamber. The drying gas was admitted to the chamber through 0,5 mm holes in the 3,2 mm stainless steel tube (B) and removed via a similar tube (U) through 1 mm holes. The gas flow was adjusted so that the gas pressure inside the chamber forced the developing solvent below the chromatographic material.

In the development cycle this gas stream was switched off. A residual gas stream through WB, controlled by needle valve (N_1), provided a saturated atmosphere in the development chamber. After completion of the preset number of development steps, the drying gas remained switched on. The level of the developing solvent in the trough (T) was kept constant by syphoning solvent from the reservoir (R) and the height of the solvent was adjusted by the position of KB1.

The plates were scrutinized with 350 nm UV light, PNA's identified by their fluorescent colours and positions on the plate and quantified by comparison of spot intensities with those of calibration mixtures.

Scanning with a 350 nm UV light with alternate use of 442 nm narrow-band transmission and 520 nm cut-in (U_4) filters, obtained more accurate results (5 to 10%).

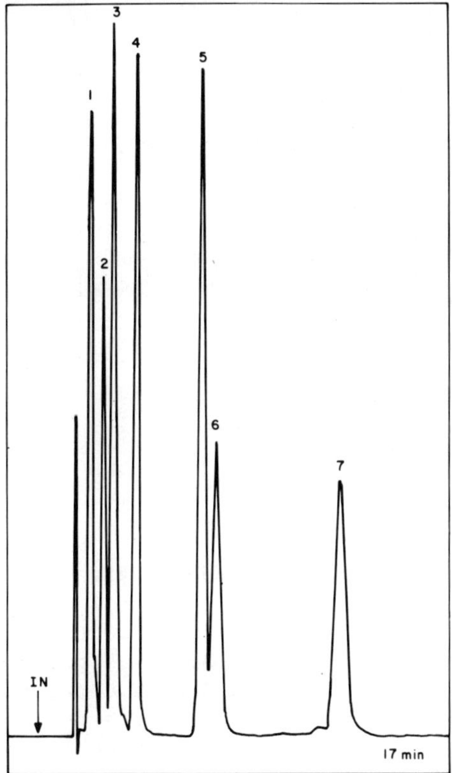

Fig. 4: Chromatogram of volatile halogenated hydrocarbons

1. Tetrachloromethane
2. Chloroform
3. Tetrachlorethylene
4. Bromodichloromethane
5. Dibromochloromethane
6. Dibromethylene (internal standard)
7. Bromoform

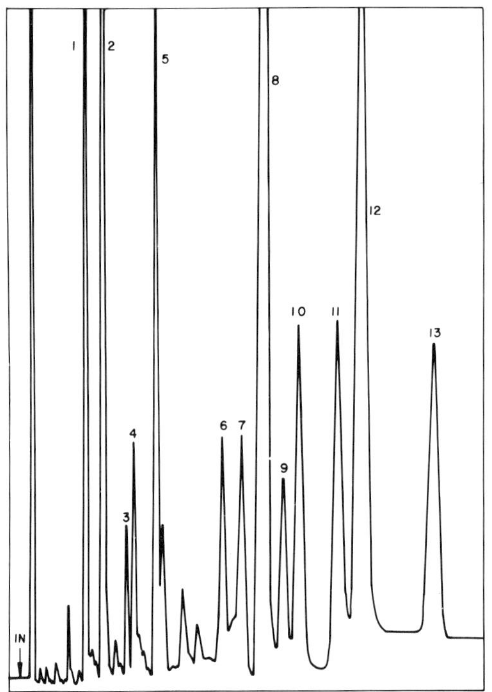

Fig. 5: Chromatogram of chlorinated hydrocarbons.
1. Hexachlorobenzene
2. Lindane
3,4,6,7: Chlordane
5. Aldrin
8. PP'-DDE
9,12: TDE
10. Dieldrin
11. Endrin
13. PP'-DDT

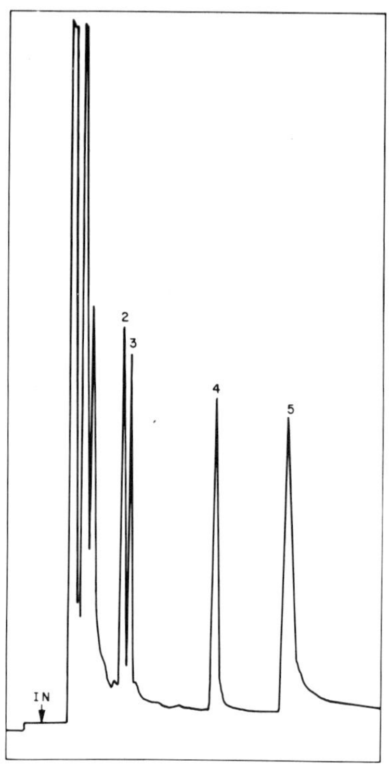

Fig. 6: Chromatogram of Trisil Z derivatives of chlorophenols.
1. Trisil Z
2. 2,4,6 trichlorophenol
3. 2,4,5 trichlorophenol
4. 2,4,6 tribromophenol
(internal standard)
5. Pentachlorophenol

TABLE 2: Analysis of PNA's by MPD-TLC

Compound	Activity	R_b-value*	Fluorescence	Detection limit (μg)	
				Visual	Instrument
1. BaAn	+	89	Blue-green	0,01	0,01 (442)
2. BbFL	++	68	Blue-white	<0,001	<0,001 (442)
3. BkFL	-	71	Blue	-	-
4. BaP	+++	63	Blue	0,005	0,002 (442)
5. DBahAN	+	55	Purple	0,01	0,01 (442)
6. IPN	+	49	Yellow	<0,001	<0,001 (U4)
7. AN	?	113	Blue-purple	0,02	0,002 (442)
8. PN	?	113	Blue-purple	0,02	0,002 (442)
9. FL	-	100	Green-white	0,001	0,002 (442)

*R_b-value: R_f(compound) x 100/R_f(fluoranthene)

RESULTS AND DISCUSSION

Standard gas chromatograms are shown in Figures 4 to 6 and the relevant data from the MPD-TLC system are given in Table 2.

The methods selected were found to be simple and can be introduced into a small laboratory with a relatively small budget, with the exception of the non-essential automated g.c. apparatus.

The extraction techniques gave recoveries of >70 per cent with overall method reproducibilities of <20 per cent standard deviation with the exception of pentachlorophenol which gave a standard deviation of 40 per cent. The use of only two g.c. columns simplifies column changeover, with less down-time and column breakage.

The MPD-TLC technique, although requiring a higher capital outlay through the construction or purchase of a programmable gas flow valve, represents an improvement upon the conventional two-dimensional technique of Borneff and Kunte (1969), commonly in use in Europe, by distinguishing the carcinogens from the non-carcinogens and by having a faster sample throughput (8 to 10 samples per plate every two hours, unattended).

An average of 8 to 10 samples per day can be analysed by two laboratory assistants; the limitation is the concentration of extract B for the PNA. Complete results, when required, can be obtained within 5 hours.

The Trisil Z method was preferred to the acetic anhydride method (Rudling, 1970) because of the difficulty in obtaining pure acetic anhydride, while the diazomethane method (Thompson, 1974) involved the risk of handling highly hazardous materials, thus being impractical for routine work.

The use of the micro-Snyder evaporator resulted from a report by Webb (1975) who used the Kuderna-Danish evaporator for quantities greater than 5 mℓ and the micro-Snyder evaporator for quantities from 5 to 0,3 mℓ . This combination gave the most reproducible results and a recovery of 90 to 95 per cent. The bubbling of N_2 through the solvent allowed for better control of the boiling and reflux conditions eliminated bumping and other problems associated with boiling stones.

The micro refluxer, developed by Dünges (1973), whereby microlitre quantities can be heated under reflux conditions, was used to design the micro evaporator. The two key principles of continuous renewal of the liquid surface and evaporation under reflux conditions were obtained by the bubbling nitrogen, with the glass and stainless steel tubes acting as a reflux condenser. Recoveries of 86 to 99 per cent were obtained for the compounds of interest.

Results from water samples taken in South Africa indicated an excellent water quality: No PNAs were found above the minimum limits with only traces of fluoranthene (<0,05 μg/ℓ). VHH occurred up to a maximum of 40 μg/ℓ, which was below the limit of 100 μg/ℓ suggested by the EPA (1978), and chlorophenols within a range 1 to 10 μg/ℓ were present in some cases.

Only one water source serving several small communities near Pretoria, contained chlorinated hydrocarbons (BHC-isomers) within the range 0,1 to 1 μg/ℓ.

This preliminary screening programme has now been succeeded by a more sophisticated one, involving capillary column g.c.-m.s. techniques.

For this purpose the beer bottle samples were replaced by extraction of 2 ℓ water samples with XAD-resin for g.c.-m.s. work, since the liquid-liquid extraction procedure produced too many contaminant peaks.

The chlorinated compounds are, however, still analysed using a separate sample, as outlined above, but the syringe extraction technique is used throughout, requiring only about 50 mℓ of sample. Derivatization of the chlorophenols is now obtained through acetyl chloride (Hassett *et al* , 1978), providing a better stability and reproducibility than the Trisil Z method, especially for pentachlorophenol.

The list of compounds could be expanded to include "moderately toxic" compounds (WHO, 1975), but these were excluded in the primary screening programme because of the increased analytical difficulties involved.

ACKNOWLEDGEMENT

This paper was presented with the permission of the Director of the National Institute for Water Research.

REFERENCES

BEVENUE, A., KELLEY, T.W. and HYLIN, J.W. (1971) J. Chromat. 54, 71-76.

BORNEFF, J. and KUNTE, H. (1969) Arch. Hyg. Bakt. 153, 220-229.

DÜNGES, W. (1973) Anal. Chem. 45, 963-964.

ENVIRONMENTAL PROTECTION AGENCY (EPA) (1975) National interim drinking water regulations. EPA Federal Register, 24th December. Washington, D.C.

ENVIRONMENTAL PROTECTION AGENCY (EPA) (1978) Control of organic chemical contaminants in drinking water. EPA Federal Register, 9th February. Washington, D.C.

HARRISON, R.M., PERRY, R. and WELLINGS, R.A. (1975) Water Research, 9, 331-346.

HASSETT, A.J., VAN RENSBURG, J.F.J. and VAN ROSSUM, P.G. (1978) An improved method for the routine gas chromatographic determination of several halogenated organic compounds. To be submitted for publication in Bull. Environ. Contam. and Toxicol.

KISSINGER, L.D. and FRITZ, J.S. (1976) J. Am. Wat. Wks Ass. 68, 435.

RUDLING, LARS. (1970) Water Research, 4, 533-537.

SYMONS, J.M., BELLAR, T.A., CARSWELL, J.K., DEMAREO, J., KROPP, K.R., ROBECK, G.G., SEEGER, D.R., SLOCUM, C.J., SMITH, B.L. and STEVENS, A.A. National organics reconnaissance survey for halogenated organics in drinking water, EPA., Cincinnati.

THOMPSON, J.F. (editor) (1974) Analysis of pesticide residues in human and environmental samples, prepared by Pesticides and Toxic Substances Effects Laboratory, National Environmental Research Centre, US Environmental Protection Agency, Research Triangle Park, NC27711.

VAN RENSBURG, J.F.J. (1978) A routine multiple progressive thin layer chromatographic technique for polynuclear aromatic hydrocarbons. Submitted for publication in Bull. Environ. Contam. and Toxicol.

VAN RENSBURG, J.F.J., VAN HUYSSTEEN, J.J. and HASSETT, A.J. (1978a) Water Research, 12, 127-131.

VAN RENSBURG, J.F.J., KUHN, A. and VAN ROSSUM, P. (1978b) A multiple progressive thin layer technique for the determination of chlorophenols. Submitted for publication in Bull. Environ. Contam. and Toxicol.

WEBB, R.G. (1975) National Environmental Research Centre, Office of Research and Development, U.S. Environmental Protection Agency, Corvallis, Oregon 97330, Document no. EPA-660/4-75-003 (June).

WORLD HEALTH ORGANIZATION (WHO) (1971) International standards for drinking water, 3rd ed., WHO, Geneva.

WORLD HEALTH ORGANIZATION (WHO) (1975) Health effects relating to direct and indirect reuse of wastewater for human consumption. Technical Paper no. 7, WHO, Geneva.

TRIHALOMETHANE LEVELS AT SELECTED WATER UTILITIES IN KENTUCKY, U.S.A.

G.D. Allgeier, R.L. Mullins Jr., and D.A. Wilding

University of Louisville, Louisville, Kentucky, 40208

J.S. Zogorski

Indiana University, Bloomington, Indiana, 47401

S.A. Hubbs

Louisville Water Company, Louisville, Kentucky, 40202

INTRODUCTION

For almost half a century, Americans have relied confidently upon their domestic water supply, assuming that drinking water was free of harmful contaminants. In November 1974, the U.S. Environmental Protection Agency (EPA) announced officially that trace quantities of 66 organic chemicals were identified in the New Orleans drinking water supply. Several of the compounds identified were suspected carcinogens. Partially in response to the events surrounding New Orleans, Congress passed legislation, Section 1442a-9 of the Safe Drinking Water Act, that directed EPA to conduct a comprehensive study of public water supplies and drinking water sources to determine the nature, extent, sources and means of control of contamination by chemicals or other substances suspected of being carcinogenic. In complying with this mandate by Congress, EPA initiated an extensive program to answer the questions raised by Congress concerning suspected carcinogens in drinking water.

The occurrence of trace organics in drinking waters throughout the United States was initially monitored by EPA's National Organics Reconnaissance Study (NORS) (Symons et al, 1975). Two of the major findings of NORS were: (a) small quantities of trace organics were present in all of the 80 municipal water supplies sampled and (b) the presence of trihalomethanes (THM) in drinking waters resulted directly from the chlorination of the water supply. Since the completion of NORS, a number of other state and federal sampling programs have been initiated or completed to further define the nature and extent of organic contamination in drinking waters. As of July 1978, a total of over 700 specific organic compounds have been identified in various drinking water supplies. The majority of these organic compounds have not been examined for their potential chronic health effects. (Anon., 1975a; Anon, 1975b).

As might be expected, the results of EPA's studies raised serious questions about the overall quality of drinking water and further illustrated the need for a better delineation between substances found in drinking waters and associated health effects. While the presence of trace amounts of organic substances in drinking water does not create any acute threat to public health or welfare, more research into the possible long-term health effects is certainly warranted. The health effects of substances identified in drinking water are currently being investigated (Tardiff, 1978; Burton and Cornhill, 1977; Cantor and McCabe, 1978; Loper et al, 1977). Numerous research projects have also been initiated to evaluate the effectiveness of various treatment processes and in-plant treatment modifications for the removal of trihalomethanes (Anon., 1976; Boettger et al, 1977; Harms and Looyenga, 1977; Hoehn et al, 1977; Hubbs, 1976; Love, 1975; Lovett and McMullen, 1977; O'Connor et al, 1977; Hubbs et al, 1977; Kinmen and Rickbaugh, 1976; Love, 1976; Rook, 1976; Smith et al, 1977; Symons, 1976; Zogorski and Wilding, 1977; Zogorski et al, 1978).

Interim primary drinking water regulations for the control of organic chemical contaminants were proposed recently by EPA (Anon., 1978). Among other standards, a maximum concentration level for total trihalomethanes (TTHM) was established at 100 μg/l (yearly average). This drinking water standard would initially be applicable only to community water systems serving greater than 75,000 people and which add a disinfectant to the water in any part of the treatment process. Compliance at small utilities may be required at a later date.

FORMATION AND SIGNIFICANCE OF TRIHALOMETHANES

One of the significant findings of NORS was that chlorine reacts with certain organic materials to produce chloroform and other related organic byproducts. This reaction has been occurring probably since chlorination was first practiced at water treatment plants. In general terms, the reaction may be simplified as:

Chlorine + Precursors ⟶ Chloroform + Other Trihalomethanes

Trihalomethanes are derivatives of methane containing three halogen atoms. Chloroform ($CHCl_3$) and bromodichloromethane ($CH\ Cl_2Br$) are commonly the most dominant trihalomethanes found in drinking waters. Dibromochloromethane ($CHClBr_2$), bromoform ($CHBr_3$) and dichloroiodo methane ($CHCl_2I$) are also frequently detected in chlorinated drinking waters. The precursors which react with chlorine to form chloroform and other trihalomethanes have not been identified specifically but are believed to be humic substances, a group of naturally occurring organic compounds resulting from the degradation of vegetation.

The occurrence of trihalomethanes in drinking waters was first reported by Rook (1974). Since this date, chloroform has been found in detectable amounts in nearly every water supply sampled, with the exception of those water works facilities that do not use chlorine anywhere within their treatment and distribution system.

A recently released National Cancer Institute (NCI) report indicates that chloroform causes cancer in rats and mice under laboratory test conditions (Anon., 1975c). In response to the NCI report, the U.S. Food and Drug Administration has banned the use of chloroform in human drugs, cosmetics and food packaging. The carcinogenicity and other physiological effects of trihalomethanes other than chloroform have not been reported to date. However, these compounds may also be classified as health hazards based on the structural similarity of these compounds to chloroform.

Reports from several epidemiological studies also suggest a link between human health and the level of trihalomethanes in drinking water. Recently, Cantor and McCabe (1978) presented a review of available epidemiological evidence for trihalomethanes. It was noted that a statistically significant difference in the incidence of human cancer exists (especially for males) between drinking waters with and without trihalomethanes present (i.e. with and without chlorination).

In summary, the occurrence of trihalomethanes in chlorinated drinking waters is widespread with at least one of these compounds being a known animal carcinogen. Steps to reduce man's exposure to the trihalomethanes, especially chloroform, appears prudent and warranted provided they do not increase the risk of microbial contamination or other detrimental effects.

DESCRIPTION AND SCOPE OF THIS INVESTIGATION

The investigations described herein were begun in the summer of 1975 and were jointly supported by the Kentucky Water Resources Research Institute and the Louisville Water Company. A total of fifteen public drinking water utilities were selected for an investigation of their trihalomethane levels and causative factors. Collectively, these treatment facilities covered a wide range of water sources, treatment techniques, treatment chemicals and production capacity. The population served by the fifteen plants selected for study represented a significant portion (nearly 40%) of Kentucky's population served by municipal water systems.

The Louisville Water Company is the largest water utility in Kentucky and delivers daily about 130 million gallons of water to the Louisville metropolitan area. More detailed studies of trihalomethane levels were conducted at this utility by the research team, in comparison to the statewide survey described above. Sampling programs at Louisville included: (a) yearly variation of trihalomethanes in tap water; (b) yearly variation of trihalomethane levels in six process waters; (c) seasonal variation of precursors; and (d) seasonal variation of trihalomethane levels in the distribution system. A detailed description of this work has been reported elsewhere (Zogorski and Wilding, 1977).

This paper will describe the major research findings from both the fifteen plant survey and Louisville's monitoring program. Several correlations are presented, comparing finished water trihalomethane levels with various environmental parameters.

Fifteen Plant Survey

During the summer of 1977, water samples were collected at each of the chosen water utilities for laboratory analysis of trihalomethane levels. Three plants were revisited in the winter of 1977. Field measurements and observations were also made and included: pH value, temperature, chlorine residual, chemical dosages, points of chemical addition, treatment processes and scheme, and other factors associated with the formation of haloforms.

Figure 1 locates each of the fifteen water utilities on a Kentucky State map to show their relative positions. Since water treatment plants within a certain watershed are likely to have chemically similar raw water sources, their trihalomethane levels may also be similar given equivalent treatment conditions. Each of the utilities sampled is listed in Table 1 along with its respective average production, raw water source and treatment scheme.

Monitoring at Louisville Water Company

Approximately 750,000 residents in the Louisville metropolitan area are served by a 240 MGD capacity treatment facility, with a second 60 MGD facility recently constructed. The primary facility,

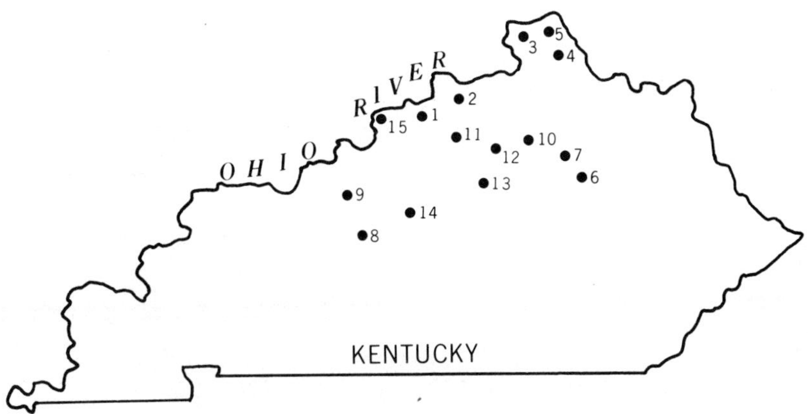

Figure 1. Location of water utilities selected for trihalomethane investigation

the Crescent Hill treatment plant, was originally constructed in 1877. Filtration and coagulation processes were added in the early 1900's and precipitive softening was added in the 1950's, thereby completing the treatment scheme as it exists today with pre-sedimentation, coagulation, softening, filtration, fluoridation and disinfection. Typical chemical additions for the various treatment operations are shown in Table 2. These data are approximations of average conditions and vary substantially depending upon the quality of the Ohio River. Further details concerning the treatment of water at the Crescent Hill plant have been described elsewhere (Hubbs, 1976; Hubbs et al, 1977; Smith et al, 1977; Zogorski and Wilding, 1977; Zogorski et al, 1978).

Trihalomethane sampling programs at the Crescent Hill plant were initiated by this research team in the summer of 1976 and were continued through the fall of 1977. Thereafter, personnel at the Louisville Water Company assumed responsibility for monitoring trihalomethane levels on a regular schedule.

ANALYSIS OF TRIHALOMETHANES

A modified Bellar procedure was used for the analysis of trihalomethanes in this investigation (Bellar et al, 1974). In this method, volatile organics are concentrated via gas stripping and then separated and quantified via gas chromatography.

A model 5710A Hewlett Packard gas chromatograph was used for the analysis of trihalomethanes in drinking water. This chromatograph was equipped with dual-flame ionization detectors and temper-

Table 1
Water Treatment Plants Sampled for Trihalomethanes as Part of the Fifteen Plant Survey

PLANT NUMBER	PLANT SIZE (MGD)	RAW WATER SOURCE	TREATMENT SCHEME[A]
1	1.7	Ohio River Alluvium	IRO, FLO, DIS
2	0.9	Ohio River Alluvium	FLO, DIS
3	9.0	Medium River, Tributary of Ohio River	PRE, COA, SED, FIL, TAS[C], FLO, DIS
4	9.5	Ohio River	PRE, COA, SED, FIL, FLO, DIS
5	9.0	Ohio River	PRE, COA, SED, FIL, TAS, DIS
6	2.5	Long, Narrow Shallow Impoundment	PRE, COA, SED, FIL, TAS, FIL, DIS
7	1.2	Medium Size Creek	PRE, COA, SED, FIL, TAS[B]. FLO, DIS
8	1.1	Small, Stream-Fed Lake	PRE, COA, SED, FIL, TAS, FLO, DIS
9	2.0	Ohio River Alluvium	PRE, COA, SED, FIL, FLO, DIS
10	1.5	Medium Size Spring Feeding Small Lake	PRE[C], COA, SED, FIL, TAS[C], FLO, DIS
11	1.7	Creek-Fed Medium Size Lake	PRE, COA, SED, FIL, TAS, FLO, DIS
12	6.0	Highly Turbid, Medium Size River	PRE, COA, SED, FIL, TAS[C], FLO, DIS
13	1.0	Highly Turbid, Medium Size River	PRE, COA, SED, FIL, TASC, FLO, DIS
14	3.3	Large Lake	PRE, COA[C], SED, FIL, TAS[C], FLO, DIS
15	130.0	Ohio River	PRE, COA, SED, FIL, SOF[C], TAS[C], FLO, DIS

A - PRE (Prechlorination), COA (Coagulation), SED (Sedimentation), FIL (Rapid Sand Filtration), SOF (Softening), TAS (Taste & Odor Control), IRO (Iron Removal), FLO (Fluoridation), DIS (Post-chlorination)

B - 24" of GAC with 6" sand in filters

C - Occasionally Used

Table 2
Typical Dosages of Chemicals at Crescent Hill Water Treatment Plant, Louisville, Kentucky

TREATMENT PROCESS	CHEMICALS	DOSAGE (LBS/MG)	REMARKS
Pre-sedimentation	chlorine	40	once per month for about 24 hours
Chemical Coagulation	chlorine	35	continuous addition
	alum	100	high turbidity water
	alum	35	low turbidity water
	polymer	12	high turbidity water
	polymer	4	low turbidity water
Precipitive Softening	lime	120	hardness > 140 mg/l as $CaCO_3$
	soda ash	80	
	carbon dioxide	50	
	lime	60	hardness ≤ 140 mg/l as $CaCO_3$
Filtration (anthracite/sand)	-	-	2 gpm/ft^2
Post-chlorination	chlorine	6	free residual = 2.0 mg/l
Fluoridation	fluoride	6	residual = 1.0 mg/l

ature programming. Gases used with this instrument include air, hydrogen, and nitrogen, all of which were purchased in the highest purity available from local dealers. Two identical six-foot long, 1/8 inch diameter, glass columns were used. These columns contained 80 x 100 mesh Chromosorb 101 packing. The unit was operated in a differential mode (A-B mode) and chromatograms were recorded by a SRG pen recorder.

Since small quantities of trihalomethanes are present in water supplies, a concentraton step was used prior to analysis. Concentration was accomplished with a model LSC-1 Liquid Concentrator (Tekmar Co., Cincinnati, Ohio). A 25 ml purging cell was built and used instead of the 5 ml unit supplied with LSC-1. After preliminary experimentation with the LSC and GC units, standard conditions for processing water samples for trihalomethatne levels were established as follows:

LSC-1 Concentrator

sample volume	= 25 ml
purge time	= 10 min
purge setting	= 40
desorb time	= 5 min
desorb temp.	= 180°C
desorb flowrate	= 28 ml/min

GC Unit

attenuation	= 64-128 (typically)
mode	= A-B (differential)
injection port temp.	= 200°C
initial column temp.	= 25°C
detector temp.	= 250°C
temperature program	= 120°C - 180°C at 8°C/min (typically)
hold times @ 120°C	= 0 min
hold time @ 180°C	= 0 min

Peaks on chromatograms were identified by their individual retention time, as well as by their retention time from the chloroform peak. Standards were prepared daily and were analyzed in a manner identical to water samples. Each standard was prepared in an ethanol-water solution and typically contained chloroform, bromodichloromethane and chlorodibromomethane. The concentration of each compound in the standard was varied from season to season to closely reflect the content of the compounds in tap water.

All samples to be analyzed for trihalomethanes were collected in 60 or 120 ml serum bottles and were sealed with teflon coated septa. Prior to sealing, about 0.1 g of sodium thiosulfate was added to remove any free residual chlorine. Samples were stored in a refrigerator until processing was commenced.

The reproducibility of the LSC-GC system was determined on four separate dates. The results of this evaluation for chloroform are shown in Table 3. Especially noteworthy are the coefficients of variation for chloroform, which were 5.8, 4.5, 5.8, and 2.3% on 8/20/76, 12/2/76, 1/29/78 and 2/8/78, respectively. Similar results were evident for other trihalomethanes. In general, data collected to date indicate that the analytical system used for trihalomethane analyses was very reproducible, especially for the lower concentration range.

Table 3
Experimental Data Illustrating the Reproducibility of the LSC-GC System for Analysis of Trihalomethanes

DATE	REPLICATE No.	CONCENTRATION (μg/l)	REMARKS
8/20/76	1	103	
8/20/76	2	112	mean = 109 μg/l
8/20/76	3	119	SD = 6.3 μg/l
8/20/76	4	108	CV = 5.8%
8/20/76	5	105	
12/2/76	1	24.3	
12/2/76	2	25.7	mean = 24.5 μg/l
12/2/76	3	22.8	SD = 1.1 μg/l
12/2/76	4	25.1	CV = 4.5%
12/2/76	5	24.6	
1/29/78	1	275	
1/29/78	2	302	mean = 303 μg/l
1/29/78	3	302	SD = 17.6 μg/l
1/29/78	4	319	CV = 5.8%
1/29/78	5	317	
2/8/78	1	38.4	
2/8/78	2	36.6	mean = 38 μg/l
2/8/78	3	37.9	SD = 0.86 μg/l
2/8/78	4	38.4	CV = 2.3%
2/8/78	5	38.8	

SD = standard deviation
CV = coefficient of variation

RESULTS AND DISCUSSION

Results of laboratory and field measurements completed in conjunction with the fifteen plant survey are listed in Tables 4 and 5. All data were collected within a five-day period in August, 1977. A majority of the information listed in Table 4 is considered typical of water treatment practices in Kentucky. It is noteworthy that considerably lower total chlorine dosages were applied at facilities using ground water supplies (plants 1,

Table 4 Listing of Field Data Collected in Association with the Fifteen Plant Survey

	Chemical Dosages (mg/l)					Physical and Chemical Analyses[a]			
Plant Number	Chlorine Pre	Chlorine Post	Chlorine Total	Alum	Lime	pH (units)	Temp(°C)	Chlorine Residual (mg/l)	NVTOC (mg/l)
1	-	--	1.8	--	--	7.7	19	1.2	2.0
2	-	--	0.7	--	--	7.2	17	0.7	---
3	3.6	1.4	5.0	22.7	8.5	7.2	25	2.7	---
4	5.3[b]	2.1	7.4	42.4	35.2	7.9	26	>3.0	3.2
5	4.7	0.5	5.2	18.0	8.3	7.6	26	1.9	1.7
6	4.8	1.4	6.2	26.4	14.4	7.9	24	1.4	5.4
7	10.0	2.5	12.5	40.0	15.0	7.6	27	2.6	5.1
8	4.2	1.1	5.3	37.2	17.0	7.1	27	1.6	4.4
9	2.7	1.8	4.5	4.5	39.0	8.2	17	0.6	2.1
10	2.4	1.6	2.0[c]	24.3	18.2	7.3	27	1.4	3.0
11	10.6	1.4	12.0	42.4	21.2	7.7	21	>3.0	7.5
12	4.7	1.6	6.3	d	d	7.2	26	2.3	7.1
13	4.0	1.9	5.9	78.0	18.0	7.5	25	1.9	5.0
14	3.9	10.2	14.1	d	d	7.9	29	1.4	5.4
15	4.2	1.0	5.2	12.0	14.4	8.5	18	1.4	4.3

a - Conducted on finished water.
b - Partial ClO_2 dosages.
c - Pre-chlorination rarely used with post-chlorination, average value given.
d - This chemical is used occasionally.

2,9 and 10), in comparison to plants using surface water supplies. Mean total chlorine dosages for ground water and surface water plants were 2.2 and 7.7 mg/l, respectively. These data are not surprising, considering the higher chlorine demand normally associated with surface water supplies due to presence of suspended solids, dissolved and colloidal organic matter, reduced inorganic ions, and microorganisms. It is also evident in Table 4 that the mean concentration of non-volatile total organic carbon (NVTOC) in ground water sources was only 2.4 mg/l, contrasted to a mean for surface water supplies of 4.9 mg/l.

The dependence of total trihalomethane levels in finished water at the 15 plants investigated (Table 5) with chlorine demand and NVTOC is illustrated in Figures 2 and 3 respectively. The chlorine demand shown in Figure 2 was calculated as the numerical difference between the total chlorine dosed within the plant and the chlorine residual leaving the clearwell (i.e. residual present in finished water). Each data point in Figures 2 and 3 represents and individual water treatment facility.

As indicated earlier, the concentration of NVTOC represents an approximation of the precursors associated with the formation

Table 5 Concentration of Trihalomethanes in Finished Water at Fifteen Water Treatment Plants in Kentucky[a]

	Concentration of Trihalomethanes (THM)				
Plant Number	$CHCl_3$ (µg/l)	$CHCl_2Br$ (µg/l)	$CHClBr_2$ (µg/l)	TOTAL THM (µg/l)	TOTAL THM (µmol/l)
1	TR	TR	TR	< 15	< 0.1
2	TR	TR	TR	< 15	< 0.1
3	55	5	ND	60	0.5
4	45	15	TR	60-65	0.5
5	40	10	TR	50-55	0.4
6	130	10	ND	140	1.2
7	50	15	TR	65	0.5
8	60	10	TR	70-75	0.6
9	TR	TR	TR	< 15	< 0.1
10	20	15	TR	35-40	0.3
11	85	10	ND	95	0.8
12	75	15	ND	90	0.7
13	40	10	ND	50	0.4
14	100	10	ND	110	0.9
15	80	10	TR	90-100	0.8

a - In summer of 1977, ND = not detected, TR = trace (<5 ug/l).

Table 6 Seasonal Variation of Chloroform at Selected Water Treatment Plants in Kentucky

PLANT NUMBER	Concentration of Chloroform, ug/l Summer (8/77)	Winter (1/78)	Percent Reduction in Chloroform Levels
6	130	10	92
12	75	4	95
15	83	13	84

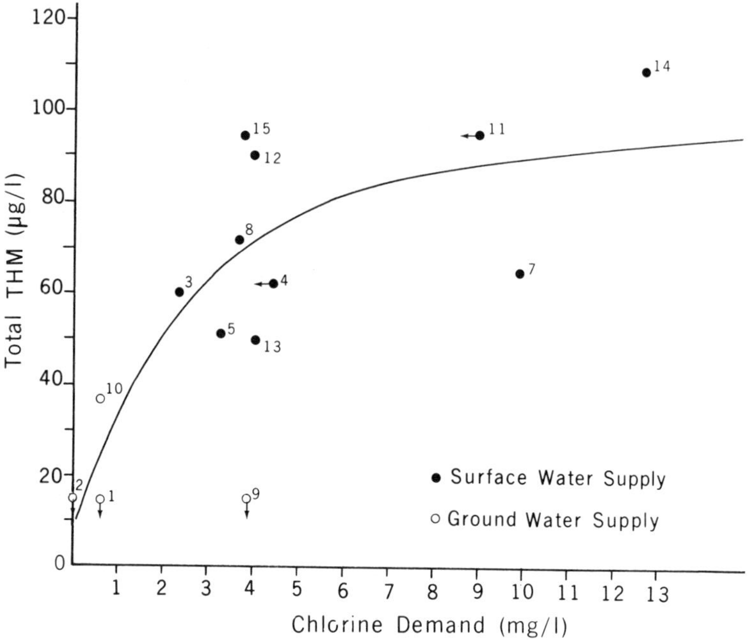

Figure 2. Total THM levels as a function of chlorine demand and water supply

of trihalomethanes. A modest positive correlation (r = +0.70) was evident between total trihalomethane levels and NVTOC for the 13 plants on which NVTOC analyses were completed (Figure 3). Total trihalomethane concentrations were also found to increase with increased chlorine demand, although a plateauing effect appeared evident at chlorine demands exceeding 10 mg/l (Figure 2).

Both laboratory and field surveys have shown that higher levels of trihalomethanes are found at water treatment plants using a surface water supply (for reasons previously described) in comparison to water supplied from aquifers (Symons et al, 1975; Harms and Looyenga, 1977; Hoehn et al, 1977; Hubbs et al, 1977; Kinmen and Rickbaugh, 1976; Symons, 1976; Zogorski et al, 1978). Indeed, this was the case for the fifteen plants surveyed in this investigation. Table 5 lists the finished water trihalomethane data for the fifteen plants. The average concentration of total trihalomethanes at the four plants using ground water sources was 0.15 μmol/l, whereas a similar value for the surface water plants was 0.66 μmol/l. This 4.1 to 1.0 ratio is consistent with the previous statement that most water utilities served by surface water supplies (streams, lakes, impoundments, and so forth) will have higher levels of trihalomethanes in finished water. As shown in Table 5, five of the eleven surface water plants had instantaneous total trihalomethane levels near or above the proposed federal drinking water standards of 100 μg/l.

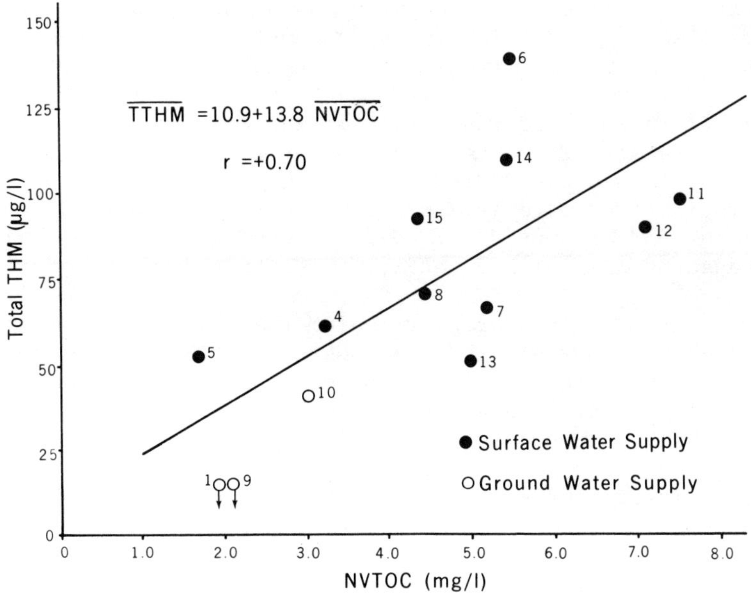

Figure 3. Total THM levels as a function of non-volatile total organic carbon for the fifteen plant survey

Three of the surface water plants were revisited in January, 1978. Water samples were collected at plants 6, 12 and 15 and subsequent trihalomethane analyses were completed. The level of trihalomethanes found in finished water in January was much lower than concentrations detected in August. A comparison of chloroform data for the two sampling dates is given in Table 6. As shown, winter chloroform concentrations were on the average only about 10 percent of similar readings found during the summer survey. The lower levels of trihalomethanes detected in the winter survey are presumed to have resulted primarily from lower ambient water temperatures. Elevated water temperatures have been shown to enhance the formation of trihalomethanes (Kinmen and Rickbaugh, 1976; Zogorski and Wilding, 1977; Zogorski, et al, 1978).

LOUISVILLE WATER COMPANY

Whereas the fifteen plant survey provided data on the variation of trihalomethanes in finished water as influenced by source-treatment effects and spacial variations, a more detailed study was completed over a fifteen month period at the Louisville Water

Company to provide temporal information on trihalomethane levels. As part of this latter program, process and drinking water samples were collected at random intervals and analyzed for levels of trihalomethanes. Trihalomethane surveys through the Crescent Hill plant showed that on the average, 35-40% of the total trihalomethane levels in finished water was formed in the chemical coagulation basins as a result of pre-chlorination at the head of this treatment step. It was also found that the high pH levels present during precipitive softening further enhanced the formation of trihalomethanes and accounted for 40-50% of the total trihalomethane levels measured in finished water.

Review of monthly average trihalomethane data for Louisville's distribution system indicated a strong seasonal variation (Figure 4). This seasonal pattern closely resembles Louisville's annual water temperature profile. That is, when water temperatures of the Ohio River were high, the concentrations of chloroform and bromodichloromethane were high. Conversely, low levels of trihalomethanes were recorded during winter months when water temperatures were lowest. Least-squares linear regression analyses were computed. The independent variable was the monthly average water temperature of the Ohio River (°F), while the dependent variables were the monthly average levels of trihalomethanes in Louisville's distribution system (μg/l).

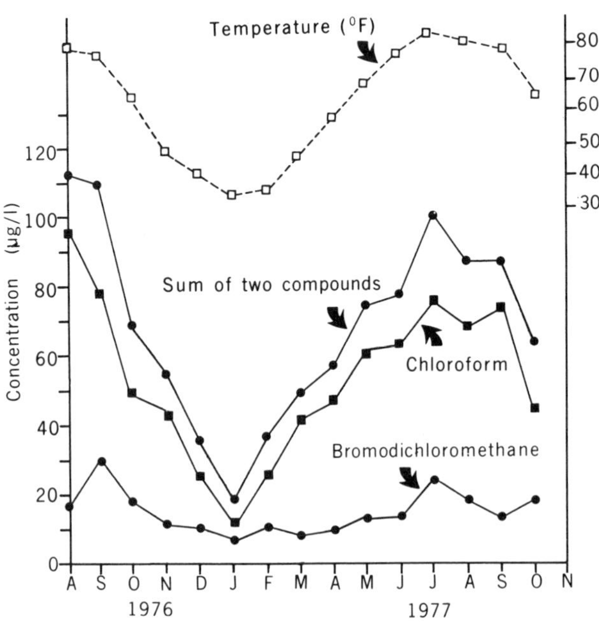

Figure 4. Monthly variation of two trihalomethanes within Louisville's distribution system.

Resulting expressions were as follows:

$$\overline{CHCl_3} = -26.9 + 1.37 \times \overline{TEMP} \qquad R = +0.94$$

$$\overline{CHCl_2Br} = -1.6 + 0.28 \times \overline{TEMP} \qquad R = 0.78$$

$$\overline{TTHM} = -28.5 + 1.64 \times \overline{TEMP} \qquad R = 0.93$$

where $\overline{CHCl_3}$, $\overline{CHCl_2Br}$ and $\overline{TTHM}$ are the monthly average concentrations of chloroform, bromodichloromethane and total trihalomethanes (estimated as sum of $CHCl_3$ and $CHCl_2Br$), and TEMP is the monthly mean water temperature. The concentrations of chloroform and total trihalomethanes were high correlated with water temperature whereas a moderate correlation was evident for bromodichloromethane. Laboratory experiments completed in the authors' laboratory indicated that temperature was, in fact, the major environmental factor responsible for the annual variation in trihalomethanes illustrated in Figure 4.

CONCLUSIONS

Based upon the findings of this investigation, the following specific conclusions are made:

1. The GC-LSC methodology for measuring trihalomethanes was reproducible and completely satisfactory for concentrations in excess of 5 μg/l.
2. Water utilities which use ground water as their water supply contained lower levels of trihalomethanes, whereas surface water supplies containing large amounts of color, turbidity and organic matter recorded the highest concentration of trihalomethanes.
3. The concentrations of chloroform and total trihalomethanes in drinking water were associated with the chlorine demand and the non-volatile total organic carbon concentration of the water supply.
4. Levels of trihalomethanes detected during the summer survey at 15 water treatment plants were found to average nearly ten times higher than readings recorded from the winter survey.
5. A spacial variation of trihalomethane levels was found through Louisville's Crescent Hill treatment plant. On the average, about 35-40% of the total trihalomethane content of finished water formed during the chemical coagulation process. The concentration of total trihalomethanes was further increased during precipitive softening due to the elevated pH values used in this treatment step and the continued presence of free chlorine. About 40-50% of the total trihalomethanes in finished water originated within the softening basins.

6. Chloroform and bromodichloromethane were the predominant trihalomethanes routinely identified in Louisville's drinking water.
7. Concentrations of trihalomethanes in Louisville tap water were at a maximum during the warmer summer months and at a minimum during the cooler winter months. Monthly chloroform and total trihalomethane levels were closely associated with monthly water temperatures.

ACKNOWLEDGEMENTS

The authors express their sincere gratitude to the water utilities which participated in this survey. The financial support provided jointly by the Louisville Water Company and the Kentucky Water Resources Research Institute is gratefully acknowledged.

REFERENCES

Anon., "Assessment of Health Risk from Organics in Drinking Water", Hazardous Materials Advisory Committee, Science Advisory Board, U.S. Environmental Protection Agency, Washington, D.C., 1975a.

Anon., "Preliminary Assessment of Suspected Carcinogens in Drinking Water - Interim Report to Congress", U.S. Environmental Protection Agency, Washington, D.C., 1975b.

Anon., "Report on the Carcinogenesia Bioassay of Chloroform", National Cancer Institute, Bethesda, Maryland, 1975c.

Anon., "EPA Proposal on Control Options for Organic Chemical Contaminants", Federal Register, Volume 4, pp. 28991-28998. 1976.

Anon., "Interim Primary Drinking Water Regulations for Control of Organic Chemical Contaminants in Drinking Water, Federal Register, Volume 43, No. 28, pp. 5756-5780. 1978.

Bellar, T.A., J.J. Lichtenberg, and R.C. Kromer. 1974. "The Occurrence of Organohalides in Chlorinated Drinking Water", Journal American Water Works Association, Volume 66, pp. 703-706.

Boettger, J., A. Hess, E. Shervin, J. Coyle, and P. Joshipura. 1977. "Trace Organic Removal by Activated Carbon and Polymeric Adsorbents for Potable Water", proceedings of American Water Works Conference, Anaheim, California.

Burton, A.C., and J.F. Cornhill. 1977. "Correlation of Cancer Death Rates with Altitude and with the Quality of Water Supply of the 100 Largest Cities in the United States", Journal of Toxicology and Environmental Health, Volume 3, pp 465-478.

Cantor, K.P., and L.J. McCabe. 1978. "The Epidemiologic Approach to the Evaluation of Chemicals in Drinking Water", proceedings of the Annual American Water Works Conference, Atlantic City, New Jersey.

Harms, L.L., and R.W. Looyenga. 1977. "Chlorination Adjustments to Reduce Chloroform Formation", Journal of American Water Works Association, Volume 69, pp. 258-263.

Hoehn, R.C., R.P. Goode, C.W. Randall, and P.T.B. Shaffer. 1977. "Chlorination and Water Treatment for Minimizing Trihalomethanes in Drinking Water", proceedings of the Water Chlorination: Environmental Impact and Health Effects Conference, Gatlinburg, Tennessee.

Hubbs, S.A. 1976. "The Oxidation of Haloforms and Haloform Precursors Utilizing Ozone", International Ozone Institute Workshop of Ozone and Chlorine Dioxide Oxidation Products, Cincinnati, Ohio.

Hubbs, S.A., J.S. Zogorski, D.A. Wilding, A.N. Arbuckle, T.L. Kockert, G.D. Allgeier, and R.L. Mullins, Jr. 1977. "Trihalomethane Reductions at the Louisville Water Company", proceedings of Water Chlorination: Environmental Impact and Health Effects Conference.

Kinmen, R.N., and J. Rickbaugh. 1976. "Study of In-Plant Modifications for the Removal of Trace Organics from Cincinnati Drinking Water", Unpublished report to Cincinnati Water Utility, Cincinnati, Ohio.

Loper, J.C., D.R. Land, and C.C. Smith. 1977. "Mutagenicity of Complex Mixtures from Drinking Water", proceedings of the Water Chlorination: Environmental Impact and Health Effects Conference, Gatlinburg, Tennessee.

Love, O.T., Jr. 1976. "Treatment for the Prevention or Removal of Trihalomethanes in Drinking Water - Appendix 3 - Interim Treatment Guide for the Control of Chloroform and Other Trihalomethanes", U.S. Environmental Protection Agency, Cincinnati, Ohio.

Love, O.T. 1975. "Treatment of Drinking Water for Prevention and Removal of Halogenated Organic Compounds", proceedings of the American Water Works Association Conference, Minneapolis, Minnesota.

Lovett, D.R., and L.D. McMullen. 1977. "Formation of Trihalomethanes in a Lime Softening Water Plant", proceedings of American Water Works Association Conference, Anaheim, California.

O'Connor, J.T., D. Badorek, and J.R. Popalisky. 1977. "Design, Construction and Operation of Pilot Plant for Removal of Organics from Missouri River Water", proceedings of Annual American Water Works Association Conference, Anaheim, California.

Rook, J.J. 1974. "Formation of Haloforms During Chlorination of Natural Waters:, Journal Water Treatment and Examination, Vol. 23, pp. 234-243.

Rook, J.J. 1976. "Haloforms in Drinking Water", Journal American Water Works Association, Volume 68, pp. 168-172.

Smith, D.R., J.S. Zogorski, D.A. Wilding, A. A. Arbuckle. 1977. "An Evaluation of Granular Activated Carbons and Synthetic Resin for Reducing the Content of Trihalomethanes in Drinking Water", proceedings of American Water Works Association Conference, Anaheim, California.

Symons, J.M. 1976. "Interim Treatment Guide for the Control of Chloroform and Other Trihalomethanes", United States Environmental Protection Agency, Cincinnati, Ohio.

Symons, J.M., T.A. Bellar, J.K. Carswell. J. Demarco. K.L. Kropp, G.G. Robeck, D.R. Seeger, C.J. Slocum, B.I.Smith and A.A. Stevens. 1975. "National Organics Reconnaissance Survey for Halogenated Organics", Journal of American Water Works Association, Volume 67, pp. 634-647.

Tardiff, R.G. 1978. "Toxicity Evaluation and Risk Determination of Chemicals in Drinking Water: An Overview and Selected Examples", Proceedings of International Symposium on the Analysis of Hydrocarbons and Halogenated Hydrocarbons, Hamilton, Ontario.

Zogorski, J.S., G.D. Allgeier, and R.L. Mullins, Jr. 1978. "Removal of Chloroform fron Drinking Water", final report to Kentucky Water Resources Research Institute, Lexington, Kentucky.

Zogorski, J.S., S.A. Hubbs, D.A. Wilding, A.N. Arbuckly, G.D. Allgeier, and R.L. Mullins, Jr. 1978. "Modifying Water Treatment Practices for the Removal of Chloroform and Other Trihalomethanes - A Case Study" accepted for publication in proceedings of Third World Congress on Water Resources, Sao Paulo, Brazil.

Zogorski, J.S., and D.A. Wilding. 1977. An Investigation of the Occurrence and Removal of Trace Organics in Louisville Water Company's Drinking Water Supply and Process Waters, final report to Louisville Water Company, Louisville, Kentucky, 1978.

FORMATION OF TRIHALOMETHANES DURING CHLORINATION AND DETERMINATION OF HALOGENATED HYDROCARBONS IN DRINKING WATER

Masashi Kajino, and Masakazu Yagi

Water Exam. Lab., Osaka Municipal Water Works Bureau

1-Hamacho Higashiyodogawa-ku, Osaka 533, Japan

INTRODUCTION

In 1974, Rook reported that chloroform and bromine-containing trihalomethanes (THM) were present in chlorinated water at much higher concentrations than in raw water. It was proposed that these trihalomethanes were produced by action of chlorine on humic substances comprising natural organic color in raw water.

Since it was felt necessary to survey the chloroform contamination of Osaka city tap waters, a suitable analytical method was required to monitor chloroform content. Direct aqueous injection method (Nicholson and Meresz, 1975), solvent extraction method (Richard and Junk, 1977), and headspace analysis (Dow Chemical, 1972) were studied.

From the comparison of the methods (Kajino, 1977), headspace analysis was found to be the most reliable method, and was also proved to be preferable from the point of view that gas injection minimizes the deterioration of an electron capture detector. By using headspace analysis, chloroform and other halogenated hydrocarbons in Osaka city tap waters were determined at μg/l levels.

Various kinds of industrial wastes discharged into the Yodo river were chlorinated to investigate which substances were precursors to chloroform. The result of this experiment showed that industrial wastes had little or no effect on chloroform formation (Kajino, 1977).

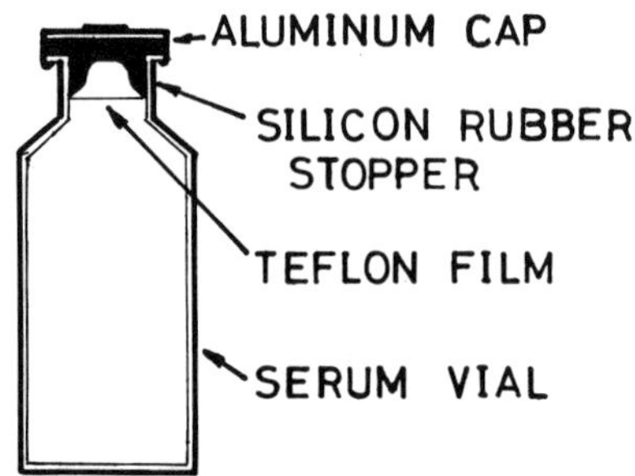

Figure 1. Serum vial for analysis and sample transportation

This article has two objectives. One is to describe the analytical conditions of headspace analysis in detail. The other is to discuss the reaction between humic acids and free and combined chlorine.

TRANSPORTATION AND PRESERVATION OF SAMPLES

Samples of Osaka city tap water were taken in a 100 ml serum vial with no headspace (Figure 1). The vial was sealed with a sheet of teflon film and a silicon rubber stopper and covered with an aluminum cap crimped at the neck of the vial. When the vials were transported upside-down, neither loss of the volatile compounds in the sample water nor contamination from the surroundings could be observed. The chloroform concentration was unchanged for a week, when the vial was stored upside-down in a refrigerator held below 5°C (Figure 2.)

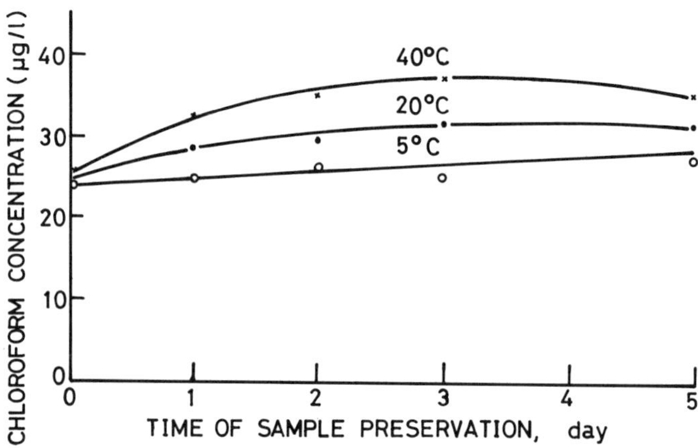

Figure 2. Chloroform concentration change after sampling

PREPARATION OF STANDARD SOLUTION

Standard stock solutions were prepared by adding 0.2 g of the compound to be determined to a 20 ml volumetric flask almost full of methanol. The mixture was then diluted to the mark with methanol. One ml of this solution was again diluted to 100 ml with methanol. One μl to 100 μl of this diluted methanol solution was injected directly into the aqueous phase in a 100 ml partially filled volumetric flask and diluted to the mark with organic-free distilled water, which gave concentrations between 1 to 100 μg/l.

HEADSPACE GAS CHROMATOGRAPHIC ANALYSIS

A 10 ml aliquot of the sample transported in the 100 ml vial was pipetted into a 13 ml vial (Figure 3.) The vial was sealed by the same method as described in the previous section, and kept for an hour in a water bath held at 25±0.1°C. A 0.5 ml aliquot of the upper gas was injected into a gas chromatograph. Gas chromatography was carried out on a Shimadzu GC-4CM-FE with a 63 Ni electron capture detector using a glass 3 m x 3 mm i.d. column containing 20% DC-550+ 20%SF-96 (2+8) on 60-80 mesh Chromosorb W (AW-DMCS). The flow rate of the carrier gas N_2 was 50 ml/min. The column temperature was isothermal at 80°C, and the detector and the injection port temperature was 150°C.

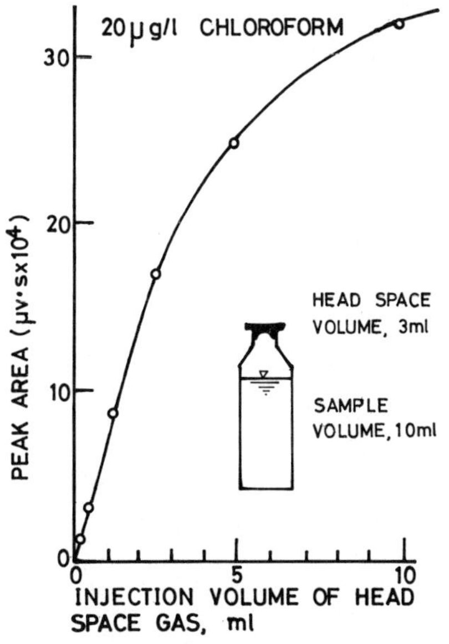

Figure 3. Relationship between peak area and injection volume of head space gas

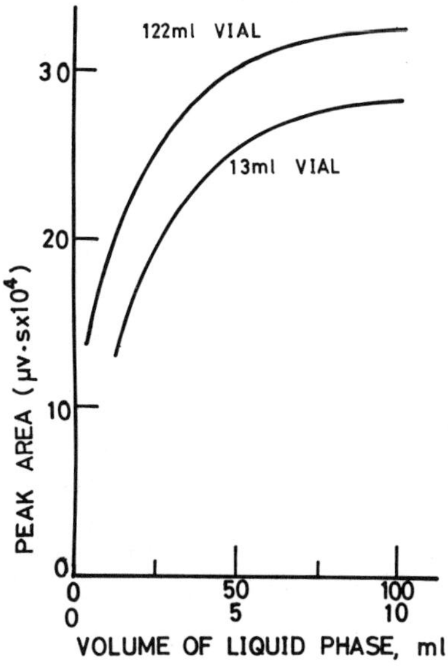

Figure 4. Effect of sample volume on chloroform concentration in gas phase

Figure 3 shows the relationship between peak area and injection volume, 20 μg/l chloroform standard solution being used. Peak area increased linearly with the injection volume up to 2 or 3 ml. By increasing the injection volume, determination at lower concentrations was found to be possible. Being able to change the injection volume within a wide range (μl to ml), according to the sample, is one of the advantages of this headspace analysis.

The suitable ratio of the volume of liquid to that of the vial was investigated with two types of vials, that is, 13 ml and 122 ml. In each vial, the peak area became constant when the ratio was higher than about 0.6 (Figure 4). From this observation, the liquid volume in the 13 ml vial was determined to be 10 ml.

Water temperature significantly influenced the partition coefficient between gas and liquid phase (Figure 5). Thus water temperature must be held constant. Before injection to a gas chromatograph, samples were kept in a water bath rigidly held at 25°C ± 0.1°C. It can be seen from this figure that at 25°C, the chloroform concentration in the headspace gas is about one-tenth to that of the liquid in equilibrium.

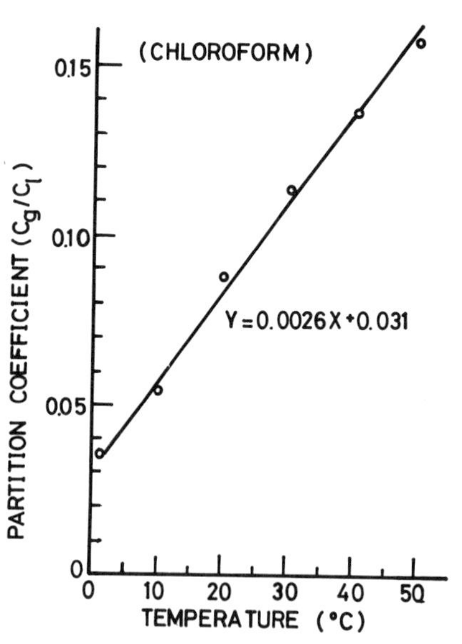

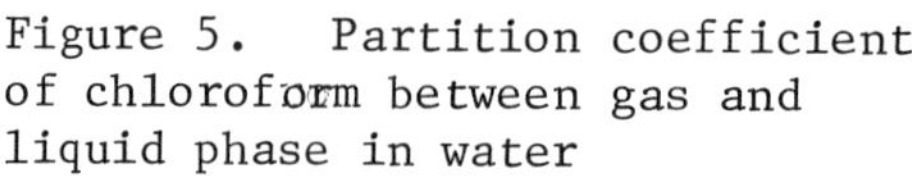
Figure 5. Partition coefficient of chloroform between gas and liquid phase in water

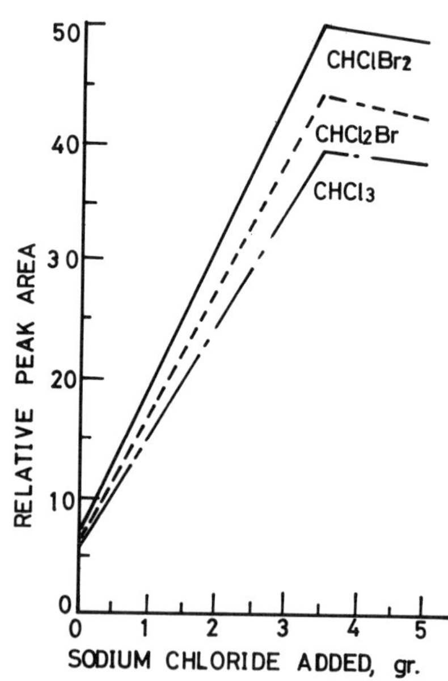

Figure 6. Sensitivity improvement by the addition of salt

SALTING-OUT EFFECT IN HEADSPACE ANALYSIS

The headspace analysis was applied to the raw water, but its reproducibility was not good for the determination of chloroform and carbon tetrachloride, because their concentrations were too low. Some sodium chloride was added to the sample water in order to improve the sensitivity of the method by making the partition coefficient of the compounds favourable to the gas phase. Figure 6 shows the relationship between the peak area and the amount of sodium chloride added.

The peak area increased proportionately with the amount of sodium chloride added up to the saturation point, that is about 3.5 gr, beyond which no change was observed. By adding 3.5 gr of sodium chloride to water, bromoform could also be determined at parts per billion level.

REACTION BETWEEN CHLORINE AND HUMIC ACID

From the results of the measurements of chloroform, it was noticed that the amounts of produced chloroform was influenced by the color of the raw water, which is ordinarily caused by the presence of humic substances. From this observation, chlorination of humic acids was investigated. Figure 7 shows the effect of humic acid concentration on the formation of chloroform. Chlorination of humic acid was carried out at a chlorine dosage of 10 mg/l and at pH 7.0 $\pm$ 0.2. The amount of produced chloroform increased proportionately with humic acid concentration.

Figure 8 shows the effect of pH on the formation of chloroform. Experiment conditions were as follows: humic acid concentration was 10 mg/l, chlorine dosage was 10 mg/l. Chloroform determination was carried out after four hours reaction time. The production of chloroform increased in proportion to pH up to 10. But, below 2.5, chloroform was not produced.

EFFECT OF THE FORM OF CHLORINE ON CHLOROFORM FORMATION

Figure 9 shows the effect of free chlorine dosage on chloroform formation and color reduction in humic acid solution. The amount of chloroform produced increased with chlorine dosage up to 4 ppm, beyond which it was constant. Color decreased linearly in this range. Free residual chlorine began to appear near the dosage where chloroform formation was terminated. These facts indicated that a certain amount of free chlorine, enough to decompose humic acids, was consumed to produce chloroform.

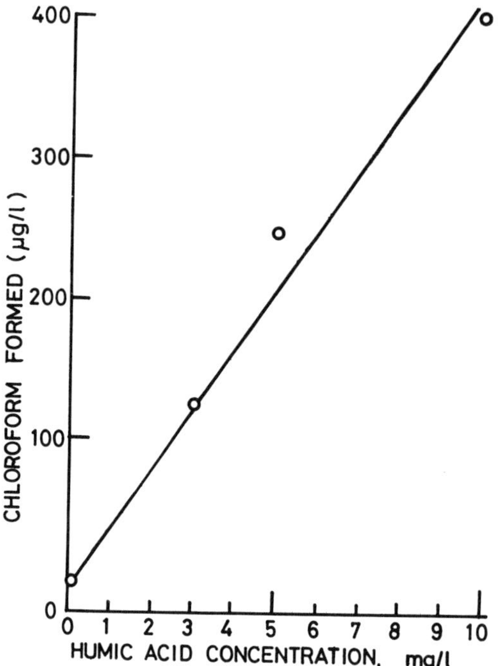

Figure 7. Relationship between chloroform formed and humic acid concentration.

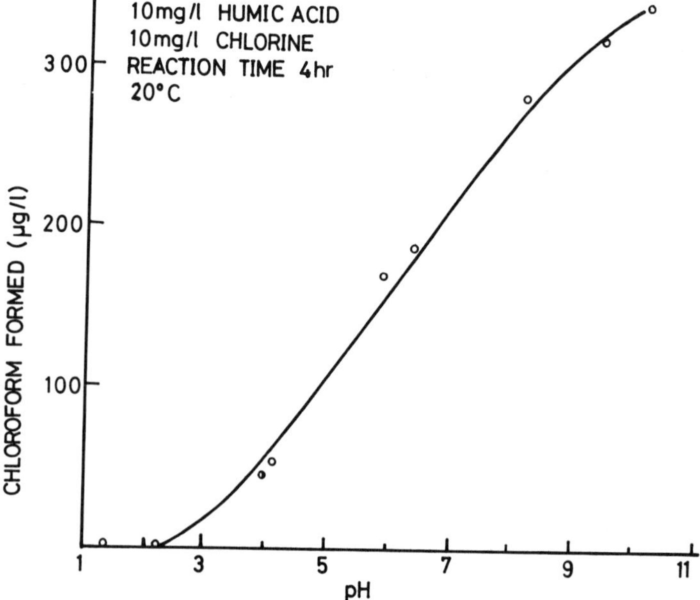

Figure 8. Effect of pH on the formation of chloroform.

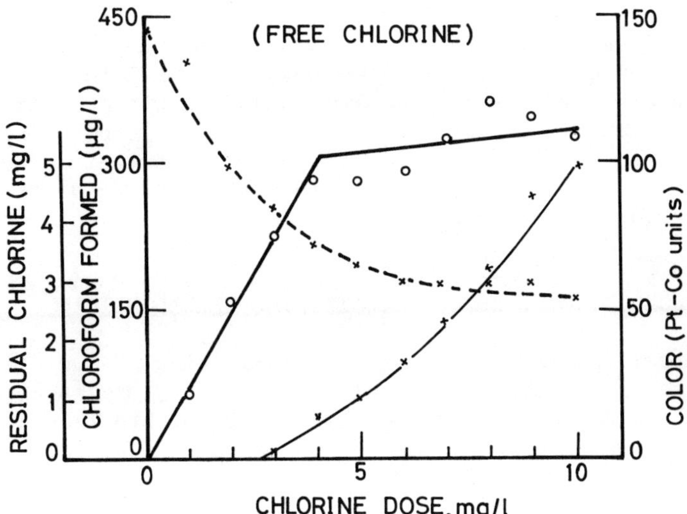

Figure 9. Formation of chloroform by chlorination of humic acids with free available chlorine. Relationship between chlorine dose and the amount of chloroform formed. x-----x: color; o———o: chloroform; x———x : residual chlorine. Humic acid, 10 mg/1. Reaction time, 8 hours.

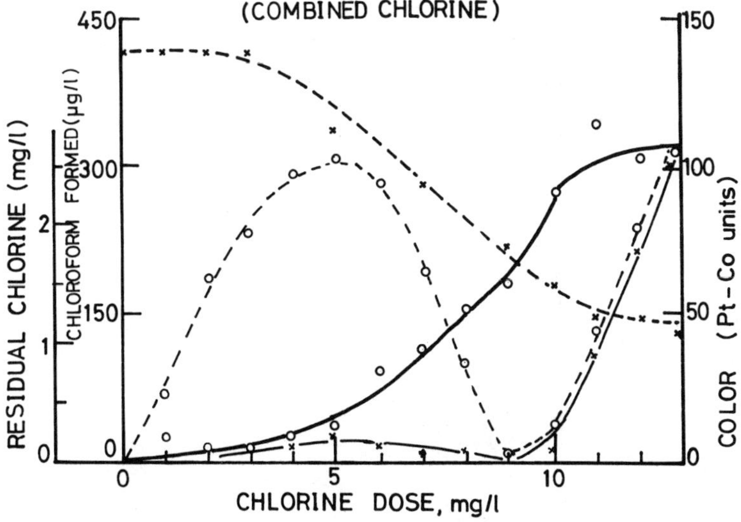

Figure 10. Formation of chloroform by chlorination of humic acids with combined chlorine. Relationship between chlorine dose and the amount of chloroform formed. x-----x: color; o———o: chloroform; o-----o: residual chlorine; x———x: free chlorine. Humic acid, 10 mg/1 Reaction time, 9 hours. Ammonia nitrogen, 0.9 mg/1.

Figure 10 shows the effect of combined chlorine dosage on chloroform formation. Chloroform was not produced where chlorine was present as monochloramine (0-5 ppm). From 5 to 9 ppm in chlorine dosage was the region where chlorine was present in the form of dichloramine. In this region, both chloroform formation and color reduction were observed, not to the extent as much in free chlorine. Dichloramine reacted with humic acids to produce chloroform, while monochloramine did not.

FORMATION OF BROMINE CONTAINING TRIHALOMETHANES

Figure 11 shows the result of chlorination of humic acid solution which contains 5 mg/l of bromide ion. As had been expected, humic acid was found to be a precursor of bromine-containing trihalomethanes. These trihalomethanes were found to be produced by oxidation of bromide ion by chlorine.

HALOGENATED HYDROCARBONS IN OSAKA CITY TAP WATERS

Osaka city supplies water to about three million people. The maximum supply rate is about two million cubic meters per day. Figure 12 shows the treatment stages in the Kunijima water treatment

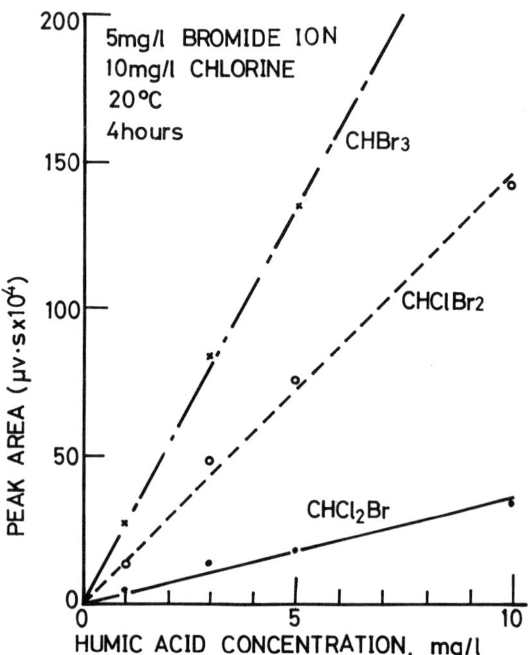

Figure 11. Amounts of bromine-containing trihalomethanes formed by the chlorination of humic acid.

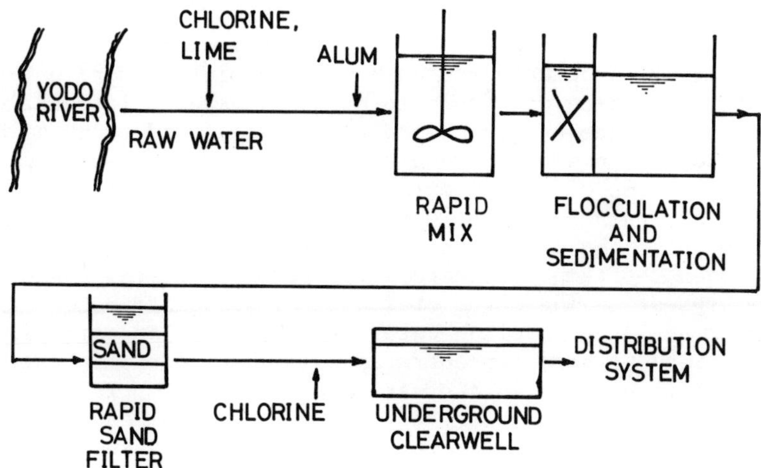

Figure 12. Schematic of the Kunijima water treatment plant

plant, which is one of the city's three plants. Raw water is taken from the Yodo river, which flows down from Lake Biwa, the largest lake in Japan. Before dosing alum, we add lime and chlorine. After flocculation, sedimentation and filtration, water is post chlorinated and distributed to consumers.

Table 1 Concentration of Halogenated Hydrocarbons in Osaka City Tap Water

(A) Area with no pH value adjustment — Jan., 1977 - Mar., 1978

Water temperature range °C	mean °C	pH unit	$CHCl_3$ µg/l	CCl_3CH_3 µg/l	CCl_4 µg/l	$CHCl_2Br$ µg/l	$CHClBr_2$ µg/l	C_2Cl_4 µg/l
6.3-8.9	7.4	6.78	12.5	0.204	0.160	5.8	1.6	0.696
11.3-19.5	15.8	6.81	24.2	0.125	0.118	7.5	2.6	0.528
21.7-28.7	25.4	6.91	36.2	0.085	0.111	14.0	4.3	0.599

(B) Corrosion control area

Water temperature range °C	mean °C	pH unit	$CHCl_3$ µg/l	CCl_3CH_3 µg/l	CCl_4 µg/l	$CHCl_2Br$ µg/l	$CHClBr_2$ µg/l	C_2Cl_4 µg/l
5.4-9.4	7.2	7.22	15.0	0.212	0.163	7.8	2.4	0.808
11.2-19.4	15.6	7.40	29.6	0.126	0.116	9.6	3.4	0.656
21.6-28.6	25.2	7.30	37.0	0.078	0.091	14.6	4.9	0.640

Table 1 shows the concentrations of halogenated hydrocarbons in tap waters in Osaka city. Besides trihalomethanes, tetrachloroethylene, trichloroethane and carbon tetrachloride were detected. Carbon tetrachloride was found to be a contaminant in chlorine. Tetrachloroethylene and trichloroethane were found to be present in raw water. In winter, when water temperature was below 10°C, the concentration of trihalomethanes was lower than in other seasons. In spring and autumn, when the temperature was in the range of 10° to 20°C, the concentration was in the middle. And in summer, when the temperature rose to nearly 30°C, the concentration was the highest. The water of Table (A) was distributed at about 0.5 higher pH value than the water of Table (B) for corrosion control. Trihalomethane concentrations in the waters of Table (A) were higher than those in the water of Table (B). This was because the trihalomethane formation reaction was dependent on pH value and as mentioned before, the higher the pH value, the faster the reaction rate.

CONCLUSION

1. Headspace gas chromatographic analysis is a simple and reliable method for the determination of chloroform and other volatile halogenated hydrocarbons in drinking water.

2. Sensitivity of the headspace analysis was improved by adding sodium chloride in sample water. This modified headspace analysis can be used for the determination of bromoform at μg/l level.

3. Free chlorine and dichloramine reacted with humic acids to form chloroform, while monochloramine did not.

4. Trihalomethanes concentration in Osaka city tap waters had a seasonal change. Chloroform was detected at the concentration range between 12.5 and 37.0 μg/l in 1977.

5. Trihalomethane formation reaction was dependent on pH value. Water distributed at higher pH values for corrosion control of the pipes were found to be richer in trihalomethane.

REFERENCES

Dow Chemical, U.S.A. Jan 7, 1972. "Chlorinated Organics and Hydrocarbons in Water by Vapor Phase Partitioning and Gas Chromatographic Analysis" Method No. QA-466.

Kajino, Masashi. July, 1977. "Formation of Trihalomethanes During Chlorination" J.JWWA, No. 514, 17-36.

Nicholson, Arnold A., and Otto Meresz. 1975. "Analysis of Volatile, Halogenated Organics in Water by Direct Aqueous Injection-Gas Chromatography" Bull. Environ. Contam. Toxicol., Vol. 14, No. 4, 435-456.

Richard, J.J., and G.A. Junk. Jan, 1977. "Liquid Extraction for the Rapid Determination of Halomethanes in Water". J.AWWA, Vol. 69, 62-64.

Rook, J.J. June 1974. "Formation of Haloforms During Chlorination of Natural Waters". Water Treatment and Examination, 23, Part 2, 234-243.

TRIHALOMETHANE LEVELS IN CANADIAN DRINKING WATER

David T. Williams, Rein Otson and Peter D. Bothwell

Environmental Health Directorate

Health and Welfare Canada

Tunney's Pasture, Ottawa, Ontario, K1A 0L2

and

Keith L. Murphy and John L. Robertson

IEC International Environmental Consultants Limited

5233 Dundas Street West, Islington, Ontario, M9B 1A6

The presence of the trihalomethanes chloroform, bromodichloromethane, chlorodibromomethane and bromoform in drinking water which had been treated with chlorine was reported by workers in Holland (Rook, 1974) and the United States (Bellar et al, 1974). It was postulated that the chloroform was formed by the reaction of the chlorine, via the haloform reaction, with the natural organics present in the raw water supply (Rook, 1974; Bellar et al, 1974). The brominated trihalomethanes were thought to be produced by the reaction of chlorine with bromide ion present in the raw water (Rook, 1974; Bunn et al, 1975). Bunn and co-workers (1975) have shown, under laboratory conditions, that this is true with both bromide and iodide ions and have obtained all ten trihalomethanes containing chlorine, bromine and iodine atoms. However, most methods of analysis have been optimised to detect only the four trihalomethanes containing chlorine and bromine atoms. The classical haloform reaction occurs with compounds containing the methyl ketone structure ($-CO.CH_3$) but Rook (1974) demonstrated that compounds containing 1,3-dione or resorcinol type structures also underwent the haloform reaction. It has been generally accepted that it is the resorcinol type of structure, present as part of the

Table 1 USEPA National Organics Reconnaissance Survey Trihalomethane Concentrations

Trihalomethane	Concentration (ug/l) Range	Median
$CHCl_3$	<0.1 to 311	21
$CHCl_2Br$	0.3 to 116	6
$CHClBr_2$	<0.4 to 110	1.2
$CHBr_3$	<0.8 to 92	-

complex structures of humic and fulvic acids, which is the main source of trihalomethanes in natural waters treated with chlorine (Rook, 1974).

In order to assess the extent of the trihalomethane problem, the United States Environmental Protection Agency (USEPA) carried out a National Organics Reconnaissance Survey (Symons et al, 1975) during February to April 1975 (Table 1). The survey covered 80 cities and chloroform levels were found as high as 311 μg/l, although the median value was only 21 μg/l. In most cities chloroform was the major contaminant but in some cities high levels of the brominated trihalomethanes were found.

Similar results were found in a survey of 83 utilities in Region V of EPA (December 1975) which covers the area just south of the Great Lakes. In both surveys the method of analysis was the gas sparge method developed by Bellar and Lichtenberg (1974) and the samples were stored at 2 to 8°C until analysed. However, no thiosulfate was added to destroy residual free chlorine and trihalomethane formation could have continued during storage of the samples.

Starting in December 1974, the Province of Ontario in Canada began a provincial survey of water treatment plants to determine the levels of organic contaminants in Ontario drinking water (Table 2). Chloroform levels as high as 258 μg/l were reported with a median value of 33 μg/l (Smillie et al, 1977; Smillie, 1977). The method of analysis used was the direct aqueous injection method which measures the total potential chloroform levels. They found that under their gas-chromatographic conditions the total potential chloroform levels were 1.7 times higher than chloroform levels found using the gas sparge method. Using this factor the Ontario and USEPA survey results are obviously very similar when compared.

To determine the extent of the occurrence of THM in drinking water across Canada, a national survey of Canadian drinking water was carried out by International Environmental Consultants Ltd.,on

Table 2 Total Potential Chloroform Levels in Ontario Drinking Water

Trihalomethane	Concentration (ug/l) Range	Median
$CHCl_3$	<1.0 to 258	33

behalf of the Department of National Health and Welfare, Canada (1977). This survey covered 70 municipalities with the number of municipalities sampled per province being roughly proportional to population. Because of the provincial survey only a few municipalities were sampled in Ontario. Approximately 38% of the Canadian population was covered by this survey. Samples of water were collected before and after treatment at the water treatment plant and from two points in the distribution system at about 0.5 and 1.0 mile from the treatment plant. The collected samples were treated with sodium thiosulfate and stored at 4°C until analysed. The temperature and pH of the water were measured at each sampling location at the time of sampling, and questionnaires on water treatment practices and chlorine and other chemical dosages were completed for each municipality.

Summarized data from this survey on trihalomethane levels for samples from the treatment plant and distribution systems are given in Table 3.

The levels of trihalomethanes in the raw water were either not detectable or were trace levels. Chloroform levels in distribution system samples ranged as high as 121 μg/l with a median level of 13 μg/l and a mean of 22.7 μg/l.

Chloroform levels in samples of treated water as they left the plant were significantly lower with a median level of 8 μg/l and a mean of 18.4 μg/l.

Total trihalomethane levels in the distribution system samples were significantly higher than in the plant samples but no difference was detectable between the two distribution system samples at the 95% confidence levels.

Trihalomethane levels were determined by two methods in one laboratory; the gas sparging method (Bellar and Lichtenberg, 1974) which gives the actual level of trihalomethanes in the sample, and the direct aqueous injection method (Nicholson et al, 1977) which estimates the "total potential" trihalomethane levels. Both methods were carried out using an electron-capture detector and a comparison

Table 3 Trihalomethanes in Canadian Drinking Water (Concentrations in ug.1).
a Results Reported by IEC International Environmental Consultants Ltd.
b Results from Analyses by Department of National Health and Welfare, Canada.
c Direct Injection

Compound	Method of Analysis	Treated Water[a]			Distrib. System 1[a]			Distrib. System 2[a]			Distrib. System 1 (HWC)[b]		
		Range	Mean	Median	Range	Mean	Median	Range	Mean	Median	Range	Mean	Median
$CHCl_3$	Gas Sparge	0-77	18.4	8.0	0-121	22.7	13.0	0-85	24.5	13.5	0-83	21.6	9.0
	Dir. Inj.[c]	0-144	24.0	9.9	0-150	30.8	16.0	0-122	32.1	18.5	-	-	-
$CHBrCl_2$	Gas Sparge	0-23	2.0	0.8	0-33	2.9	1.4	0-27	2.6	1.3	0-15	1.9	1.0
	Dir. Inj.	0-29	2.2	0.7	0-44	3.1	1.3	0-22	2.7	1.6	-	-	-
$CHBr_2Cl$	Gas Sparge	0-4.6	0.3	0.1	0-6.2	0.4	0.1	0-4.9	0.4	0.1	0-6.5	0.3	-
	Dir. Inj.	0-5.8	0.3	-	0-7.0	0.4	-	0-4.5	0.4	-	-	-	-
$CHBr_3$	Gas Sparge	0-1.5	0.1	-	0-1.0	0.1	-	0-2.1	0.1	-	-	-	-

of the results obtained by the two methods showed that on average the total trihalomethane (TTHM) levels obtained by the sparging method were 70 - 75% of the TTHM levels obtained by the direct injection method. However, on a sample to sample basis there was a significant variation in the ratio of gas sparging to direct injection results. The second laboratory analysed one set of distribution samples (station 1 at 0.5 mile) using the gas sparging method and a Hall electrolytic conductivity detector. The values reported by the laboratories for levels of chloroform and bromodichloromethane in corresponding samples were in good agreement although the values obtained using the conductivity detector were, in general, slightly lower than those obtained using the electron capture detector with the relative difference increasing at low concentration.

At the time of sample collection, data were collected by means of a questionnaire, on details of water source, water quality, and water treatment procedures. Analysis of these data and the survey results showed that average total trihalomethane levels were similar when the raw water source was a lake or river, but much lower for ground water sources. Rivers (39), lakes (20), and wells (11), were sampled in this survey (Table 4).

Correlations were also attempted for total trihalomethane levels, with water quality and water treatment parameters. A correlation between raw water total organic carbon and treated water total trihalomethane levels was not obvious from the raw date. However, if the data were grouped into cells as suggested by Symons and co-workers (1975) a weak correlation could be detected (Figure 1). The median value for total organic carbon was 4.0 mg/l in this survey. Similarly, a correlation between raw water chlorine demand and treated water total trihalomethane levels could be detected by grouping the data into cells (Figure 2). Similar correlations could be obtained for TOC and raw water chlorine demand when compared to total trihalomethane levels in the distribution system samples.

A multivariate statistical analysis showed the chlorine demand was the dominant independent variable and that it could explain 34% of the variation in total trihalomethane levels determined by the

Table 4 Comparison of Trihalomethane Levels for Different Raw Water Sources

	Rivers	Lakes	Wells
Median (umoles/l)	0.10	0.09	0.02
Mean (umoles/l)	0.20	0.21	0.02

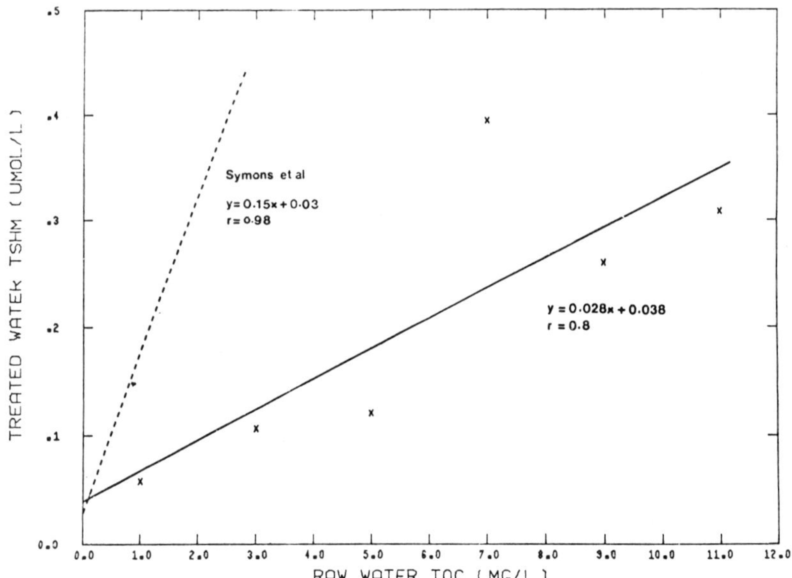

Figure 1. Correlation between treated water total sparged halomethane and raw water TOC - Data in cells.

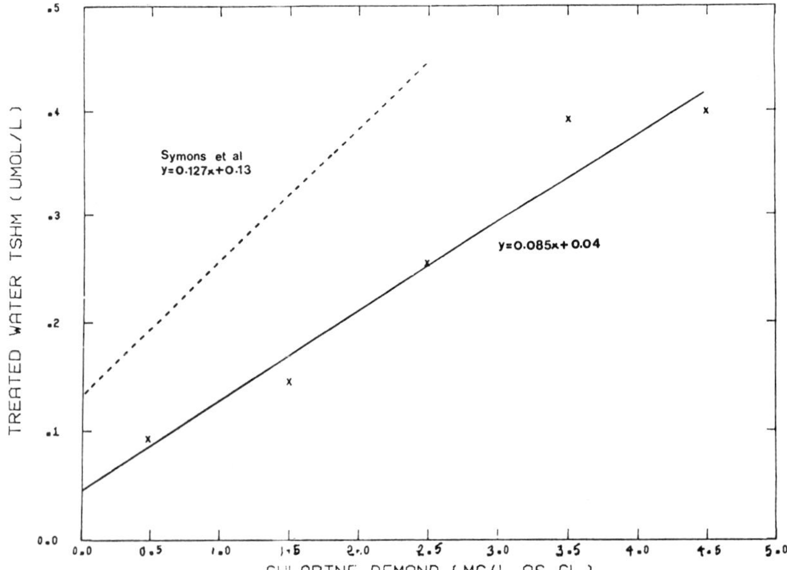

Figure 2. Correlation between treated water total sparged halomethane and chlorine demand - Data in cells.

gas sparge method and 43% determined by the direct aqueous injection method. Approximately 47% of the variation for the gas sparge method and 57% for the direct aqueous injection method could be accounted for by also including pH, temperature and hardness in the regression equations (Tables 5 and 6).

Only chlorine demand and pH of the raw water show a significant correlation with TTHM levels in this multivariate analysis. The correlation of TOC with TTHM (Figure 1) is masked in the multivariate analysis because of the high correlation between TOC and chlorine demand.

During preparations for the national survey the analysis of tap water in our Ottawa laboratory indicated significant variations in chloroform levels through the year (Figure 3). The highest values, 120 μg/l, were found during September 1976 and the lowest values, 10 to 20 μg/l, during the winter months January to March 1977. A similar variation occurred through the following summer and winter.

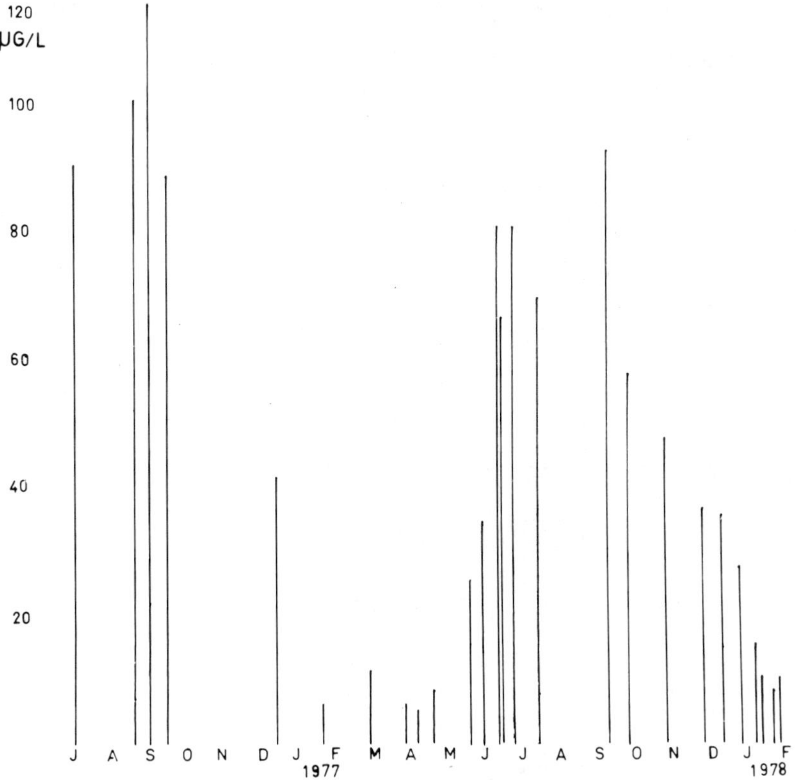

Figure 3. $CHCl_3$ levels - Room 201 tap water.

Table 5 Regression Coefficients and Explained Variation of Gas Sparge TTHM Values for Distribution System Water Samples (a Significant at $p \geq 0.05$)

Independent Variable	Regression Coefficient	Total Explained TTHM Variation (%)	Additional TTHM Explained Variation(%)
Chlorine Demand	0.056	34.4	-
pH (Raw Water)	-0.094	41.7	7.3[a]
TOC (Raw Water)	0.008	44.0	2.3
Temperature (Raw Water)	-0.017	44.4	0.4
Temperature (Distribution Water)	0.021	46.7	2.3
Hardness	0.00	47.5	0.8

Table 6 Regression Coefficients and Explained Variation of Direct Injection TTHM Values for Distribution System Water Samples
(a Significant at $p \geq 0.05$)

Independent Variable	Regression Coefficient	Total Explained TTHM Variation(%)	Additional TTHM Explained Variation(%)
Chlorine Demand	0.086	42.9	-
pH (Raw Water)	-0.130	52.4	9.5[a]
TOC (Raw Water)	-0.022	54.7	2.3
Temperature (Raw Water)	-0.022	56.3	1.6
Temperature (Distribution Water)	-0.00	57.0	0.8
Hardness	-0.008	57.6	0.5

Therefore, we began a one year study in August 1977 to determine seasonal variations in trihalomethane levels in Ottawa-Hull drinking water. In co-operation with the local water treatment plants we intend to try to correlate the trihalomethane variation with changes in water quality parameters or water treatment practices. There are three water treatment plants in Ottawa-Hull and samples are being collected and analysed from each plant and from the distribution system of each plant. Data on TOC, chlorine dosage, chlorine demand and other water quality parameters are being gathered in co-operation with treatment plant personnel.

The variations in chloroform levels on samples from the distribution systems of each plant are shown in Figure 4. Chloroform levels were high, 120 to 160 μg/l, in the summer at the beginning of the survey and fell to their lowest value, 13 μg/l, in January and stayed low during the spring months. Levels of bromodichloromethane in Ottawa-Hull drinking water are very low in summer, 2 to 3 μg/l, and essentially non-detectable in the winter.

The three treatment plants obtain their raw water supply from the same river and, consequently, have similar variations in chloroform levels. However, it can be seen (Figure 4) that in absolute terms the differences in water treatment practices at each plant give rise to different levels of chloroform. Correlations of these factors and all of the data gathered during the study will be made at the completion of the study.

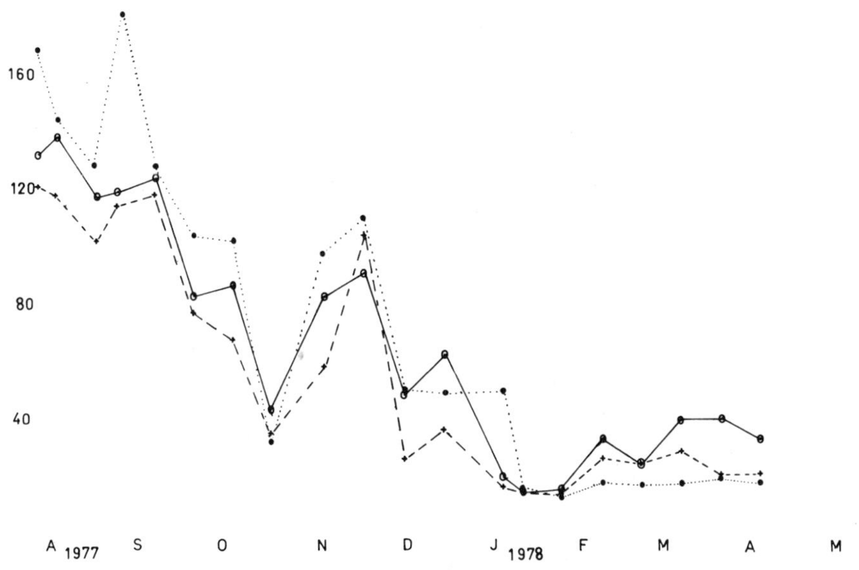

Figure 4. $CHCl_3$ levels in Ottawa-Hull distribution systems.

REFERENCES

Bellar, T.A., and J.J. Lichtenberg, 1974. Determining volatile organics at microgram-per-litre levels by gas chromatography. J. Am. Water Works Assoc. 66: 739-44.

Bellar, T.A., J.J. Lichtenberg, and R.C. Kroner, 1974. The occurrence of organohalides in chlorinated drinking water. J. Am. Water Works Assoc. 66: 703-6.

Bunn, W.W., B.B. Haas, E.R. Deane, and R.C. Kleopfer, 1975. Formation of trihalomethanes by chlorination of surface water. Env. Letters 10: 205-13.

Health and Welfare Canada. National survey for halomethanes in drinking water. Environmental Health Directorate report (77-EHD-9), Ottawa, Ont. (1977).

Nicholson, A.A., O. Meresz, and B. Lemyk, 1977. Determination of free and total potential haloforms in drinking water. Anal Chem. 49: 814-19.

Rook, J.J., 1974. Formation of haloforms during chlorination of natural waters. Water Treat. Exam. 23: 234-43.

Smillic, R.D., A.A. Nicholson, O. Meresz, W.K. Duholke, G.A.V. Rees, K. Roberts, and C. Fung. Organics in Ontario drinking waters. Part II. A survey of selected water treatment plants. Ontario Ministry of the Environment, Rexdale, Ont. (April 1977).

Smillie, R.D. 1977. Chloroform levels in Ontario drinking water. Water Poll. Contr. 115: 8-9.

Symons, J.M., T.A. Bellar, J.K. Carswell, J. DeMarco, K.L. Kropp, G.G. Robeck, D.R. Seeger, C.J. Slocum, B.L. Smith, and A.A. Stevens, 1975. National organics reconnaissance survey for halogenated organics. J. Am. Water Works Assoc. 67: 634-47.

United States Environmental Protection Agency Office of Toxic Substances. Preliminary assessment of suspected carcinogens in drinking water - Report to Congress. Washington, D.C. (December 1975).

AQUATIC ORGANISM PATHOBIOLOGY AS A SENTINEL SYSTEM TO MONITOR ENVIRONMENTAL CARCINOGENS

R.A. Sonstegard[1*] and J.F. Leatherland[2]

[1]Department of Biology, McMaster Univ. Hamilton, Ont.

[2]Department of Zoology, Univ. of Guelph, Guelph, Ont.

INTRODUCTION

Although the etiology of neoplasia in man, with rare exceptions is unknown, it is estimated that 60-90% of the cancers in man are environmentally induced (Epstein, 1974; Hammond, 1975). That environmental factor(s) are the major determinants in human cancer(s) is based on the uneven geographical distribution of the disease (Draser and Irving, 1973; Armstrong and Doll, 1975) and studies of migrant populations which indicate that the prevailing cancer rate is determined by the place of residence rather than origin (Smith, 1956; Haenszel and Kurihara, 1968; Staszewski and Haenszel, 1965; Haenszel et al., 1973). Most of the known carcinogens are considered to be the product of increasing agricultural and technological sophistication.

Carcinogen Burden of Aquatic Environments

The aquatic environment is one of the ultimate recipients of man's chemical waste products. The occurrence and biomedical significance of biorefractories in water, with particular reference to carcinogens, has recently been excellently reviewed by Kraybill (1977, 1978). At present, some 1,300 organic compounds have been identified in drinking waters (Garrison, 1977) but this is undoubtedly only a fraction of those which exist.

At the present time, nationally and internationally there are water shortages and ever increasing larger populations are having to recycle water for drinking purposes. This practice has inherent dangers inasmuch as most water treatment facilities do

* Career awardee of the National Cancer Institute of Canada.

not remove biorefractory compounds, and recently the time-honored chlorination treatment processes have themselves been implicated in the production of carcinogens (Jolly, 1974; Bellar et al, 1974). The reuse of water and the possible associated health effects has recently been reviewed by Shuval (1978). Of increasing concern is, not so much with the acute short term effects of this practice, but rather with the effects of long term chronic doses on the health of man (carcinogenesis).

Human Epidemiology - A Link Between Cancer Frequency and Water Quality

Page and co-workers (1975) compared rates of cancer among populations in the New Orleans region which consumed Mississippi River water for drinking water with those consuming high quality ground water. These studies suggest a significant relationship between cancer mortality and the use of Mississippi River water as a drinking water source.

Reiches (1977) reported that Ohio residents who consume Ohio River and Lake Erie water had a cancer mortality higher than those drinking Scioto River water. Similarly, Buncher (1976) reported that cancer frequency in the vicinity of Cincinnati of a population drinking ground water is lower than in one consuming surface water.

While the above reports cannot be considered conclusive evidence that cancer is, in fact, being caused by consuming contaminated water(s), these very suggesting findings must be fully taken into consideration.

Bioaccumulation

The ability of aquatic organisms to bioaccumulate environmental xenobiotics is well established. For example, the concentration gradient or organochlorines in water compared with those found in Great Lakes salmon is in excess of a million (Reinert, 1970; Reinert and Bergman, 1974; Veith et al, 1977; Norstrom et al, 1978; Sonstegard and Leatherland, 1979a). The bioconcentration gradient has facilitated biochemical quantitation and qualitation of a variety of aquatic xenobiotics, and has led to the detection and identification of compounds whose existence in the environment was previously unsuspected (e.g. Mirex) (Kaiser, 1974). This trait has led to surveillance programs using indicator species (Norstrom et al, 1978) whereby changes in the environment might be accurately monitored, and point sources of pollutants identified. It has also enabled the various Food and Drug directorates to identify potentially dangerous food sources, which has led to legislation prohibiting the sale of certain species.

Dietary consumption of fishes by sportsfishermen represents an uncontrolled source of human exposure and has focused concern as to the possible health effects of this practice (Sonstegard and

Leatherland, 1979a,b,c). For example, Great Lakes salmonids have xenobiotic levels which prohibit their commercial sale. Despite warning from government agencies, sportsfishermen consume these animals in quantity (e.g. 14 million pounds by resident anglers of Michigan, annually) (Humphrey, 1975). Earlier, we eluded to possible health effects of consuming contaminated drinking water. However, when compared to dietary consumption of fish, it would appear that the latter practice far outranks water as an immediate source of environmental xenobiotics. For example, based on organochlorine levels in water compared with salmonid tissue levels, one would have to consume over a million liters of Great Lakes water to match the exposure of a single meal of Great Lakes salmon. Clearly, studies must be undertaken to document the danger of this practice.

The ability of aquatic organisms to bioaccumulate xenobiotics provides a unique probe of assessing carcinogen burden by in vitro assays (e.g. Ames testing) performed on extracts of aquatic organisms (Sonstegard, 1978) and in vivo assays (e.g. fish to laboratory rat feeding trials) (Sonstegard and Leatherland, 1979a,b, c). The latter has particular relevance in that it addresses health hazards of the "environmental cocktail" to which man is exposed.

Environmental Surveillance

One of the most productive lines of investigation in human cancer research have been epidemiological studies which have linked environmental factors as being the major cause of cancer in man. The standard method in cancer epidemiology is to tabulate incidence data for each histological type of neoplasm in man and to compare these amongst various populations nationally and internationally. These types of studies are confounded, however, in that there is a 20-30 year latent period in man before cancers become evident in a given population. Therefore, the environmental problems we encounter today, won't be detectable in the human population as a clinical disease for another two or three decades; at which time they may be widespread.

In recent years there have been rapid developments in a number of areas in biology which have led to the development of a number of rapid, economical, and sensitive in vitro tests for carcinogens (Ames et al, 1975; Hart et al, 1977). These tests are providing an invaluable role in screening compounds for carcinogenic potential, but are designed for evaluation of individual "purified" test chemicals. It is increasingly clear that environmental carcinogenesis is an extremely complex phenomenon, and is likely not a problem caused by an individual chemical, but rather by batteries of chemicals "environmental cocktail" as are found in recycled drinking water or a meal of fish. The interaction of various chemicals (carcinogens and non-carcinogens) is largely unknown, however, there is experimental evidence of synergism, co-carcinogenesis, and tumor promotion by various chemicals.

Aquatic Organisms as Indicators of Environmental Carcinogens

Against this background, it is apparent that there exists a need for new avenues by which to evaluate environmental carcinogens. As an approach to this situation, we have been exploring the potential utility of monitoring tumor frequencies in feral aquatic organisms as "sentinels" of environmental carcinogens. These studies are patterned after human epidemiological investigations in which tumor frequencies of each pathological type are tabulated and compared to various geographical realms.

During the course of these studies, we made field epizootiological studies of tumor frequencies in fishes inhabiting polluted and non-polluted waters of the Great Lakes (Superior, Michigan, Huron, Erie, and Ontario). Over 50,000 fish were captured and necropsied and epizootics of neoplasia in the following species were detected which appear to have an environmental etiology.

Gonadal tumors in cyprinid fishes (carp, *Cyprinus carpio*; goldfish, *Carassius auratus*; and carp x goldfish hybrids) were found throughout the Great Lakes with tumor frequencies approaching 100% in specific age-sex groups. Field epizootiological and laboratory studies together with reviews of museum collections (made prior to 1952) suggests that the tumor is being induced by an environmental factor(s) being discharged into the lake system (Sonstegard, 1974; Sonstegard, 1976; Sonstegard et al, 1976; Leatherland and Sonstegard, 1977; Sonstegard, 1978; Sonstegard and Leatherland, 1978).

Epizootics of benign skin tumors (papillomas) were found in white sucker (*Catastomus commersoni*) populations throughout the Great Lakes basin. However, elevated tumor frequencies (e.g. 50.8%) were found clustered around an industrial complex on Lake Ontario. In addition to the clustering recorded, there was an anatomical shift in the location of the tumor to the lips, an anatomical site which has near constant contact with bottom sediments (the white sucker is a bottom-dwelling species). The anatomical shift in location of the tumor (compared to other sites on the Great Lakes) together with the clustering recorded, strongly suggests that factor(s) associated with bottom sediments is enhancing the expression of a C-type (oncogenic) virus associated with the tumor (Sonstegard, 1973; Sonstegard, 1976; Sonstegard, 1977; Sonstegard, 1978).

Epizootics of thyroid hyperplasia (goiter) were found in Great Lakes coho salmon (*Oncorhynchus kisutch*) with frequencies of occurrence as high as 82% recorded. Interlake epizootiological studies of the occurrence of the goiter in coho salmon (between 1972 and 1978) indicate that the frequency of occurrence of the goiters is increasing, and that environmental goitrogens (possibly pollutants) may be involved in the etiology of the thyroid dysfunction (Sonstegard and Leatherland,1976; Moccia et al, 1977; Leatherland et al, 1978; Moccia et al, 1978; Sonstegard,1978).

Similarly, a number of laboratories have recently reported the occurrence of proliferative and/or neoplastic diseases in a variety of aquatic organisms which appear to have an environmental etiology (Sonstegard,1978). A comprehensive symposium covering this subject has been published by the New York Academy of Sciences entitled "Aquatic Pollutants and Biological Effects with Emphasis on Neoplasia".

Although the prospect of monitoring tumor frequencies in aquatic organisms is attractive, there exists a great deal of work to be done to establish the relevance of such an approach. For example, there exists the need to generate parallel field and laboratory data (e.g. cause-effect studies) to elucidate the natural biology of the animal in relation to the tumor being studied (e.g. the role of viruses, age, sex, etc.). Well defined field models such as feral fishes may, however, be an invaluable asset inasmuch as they cohabit our biosphere and may be particularly sensitive to environmental carcinogens. As such, they may provide an early warning system as to what the future holds for man so that programs can be undertaken to identify and eliminate cancer causing pollution before they have effect(s) on the human population.

Pragmatic View - Human Experimentation is in Progress

Populations drinking recycled water and those consuming contaminated fish are currently engaged in a large scale program of environmental toxicology and carcinogenesis that no scientist could ethically conduct. An increase in cancer frequency and/or health effects in this population in 20-30 years would strongly implicate the danger of this practice and would catalyze research and educational programs designed to limit and/or eliminate carcinogen exposure by this route.

In light of the above situation, we believe that field epizootiological studies monitoring for health effects in well characterized aquatic organism "sentinels" will provide valuable epidemiological pointers as to what the future holds for man. As an adduct to this program, laboratory cause-effect studies should be used to derive parallel etiological and epizootiological data linking feral aquatic animal pathobiology to laboratory systems which ultimately could be extrapolated to human health. Such an approach might lead to the early detection and elimination of environmental health hazards to man and may eventually yield an extraordinary impact in diminishing the incidence of environmentally related diseases.

REFERENCES

Ames, B.N., McCann, J., and Yamasaki, E. 1975. Methods for detecting carcinogens and mutagens with the Salmonella mammalian-microsome mutagenicity test. Mut. Res. 31: 347-364.

Armstrong, B., and Doll, R. 1975. Environmental factors and cancer incidence and mortality in different countries with specific reference to dietary factors. Int. J. Cancer 15: 617-631.

Bellar, T.A., Lichenberg, J.J., and Kroner, R.C. 1974. The occurrence of organohalides in chlorinated drinking waters. J. Amer. Water Works Assoc. 66: 703-706.

Buncher, C.R. 1976. Ohio study links cancer to contaminated water. Toxic Material News. August 18, pp. 133.

Drasar, B.S., and Irving, D. 1973. Environmental factors and cancer of the colon and breast. Br. J. Cancer 27: 167-172.

Epstein, S.S. 1974. Environmental determinants of human cancer. Cancer Res. 34: 2425-2435.

Garrison, A.W. 1977. Analysis of organic compounds in water to support health effect studies. Ann. New York Acad. Sci. 298: 2-19.

Haenszel, W., Berg, J.W., Sezi, M., Kurihara, M., and Locke, P.B. 1973. Large bowel cancer in Hawaiian Japanese. J. Natl. Cancer Inst. 51: 1765-1779.

Haenszel, W., and Kurihara, M. 1968. Studies on Japanese immigrants. J. Natl. Cancer Inst. 40: 43-68.

Hammond, E.C. 1975. The epidemiological approach to the etiology of cancer. Cancer 35: 652-654.

Hart, R.W., Hays, S., Brash, D., Daniel, F.B., Davis, M.T., and Lewis, N.J. 1977. In vitro assessment and mechanism of action of environmental pollutants. Ann. New York Acad. Sci. 298: 141-158.

Humphrey, H.E.B. 1975. Final Rep. on FDA Contract 223-73-3309. Washington, D.C., United States Department of Health, Education and Welfare, Food and Drug Administration.

Jolly, R.L. 1973. Chlorination effects on organic constituents in effluents from domestic sanitary sewage treatment plants. Publ. No. 55. Environ. Sci. Div. Oak Ridge Natl. Lab., Oak Ridge, Tennessee.

Kaiser, K. 1974. Mirex an unrecognized contaminant of fishes from Lake Ontario. Science 185: 523-525.

Kraybill, H.F. 1977. Global distribution of carcinogenic pollutants in water. Ann. New York Acad. Sci. 298: 80-89.

Kraybill, H.F., Helmes, C.T., and Sigman, C.C. 1978. Biomedical aspects of biorefractories in water. IN: Aquatic Pollutants; Transformation and Biological Effects. O. Hutzinger, L.H. van Lelyveld, and B.C.J. Zoeteman, eds. Pergamon Press, New York, pp. 395-403.

Leatherland, J.F., Moccia, R.D., and Sonstegard, R.A. 1978. Ultrastructure of the thyroid gland in goitered coho salmon (Oncorhynchus kisutch). Cancer Res. 38: 149-158.

Leatherland, J.F., and Sonstegard, R.A. 1977. Structure and function of the pituitary gland in gonadal tumour-bearing and normal cyprinid fish. Cancer Res. 37: 3151-3168.

Leatherland, J.F., and Sonstegard, R.A. 1978. Structure of normal testis and testicular tumors in cyprinids from Lake Ontario. Cancer Res. 38: 3164-3173.

Moccia, R.D., Leatherland, J.F., and Sonstegard, R.A. 1977. Increasing frequency of thyroid goiters in coho salmon (Oncorhynchus kisutch) in the Great Lakes. Science 198: 425-426.

Moccia, R.D., Leatherland, J.F., Sonstegard, R.A., and Holdrinet, M.V.H. 1978. Are goiter frequencies in Great Lakes salmon correlated with organochlorine residues. Chemosphere 7: 649-652.

Norstrom, R.J., Hallett, D.J., and Sonstegard, R.A. 1978. Coho salmon (Oncorhynchus kisutch) and herring gulls (Laraus argentatus) as indicators of organochlorine contamination in Lake Ontario. J. Fish. Res. Board Can. 35: 1401-1409.

Page, T., Harris, R., and Epstein, S.J. 1975. Drinking water and cancer mortality in Louisiana. Science 143: 55-57.

Reiches, N. 1977. Studies on cancer mortalities in populations exposed to water contaminants in Lake Erie, Scioto, and Ohio Rivers. Toxic Materials News. March 11, pp. 99.

Reinert, R.E. 1970. Pesticide concentrations in Great Lakes fish. Pestic. Monit. J. 3: 233-240.

Reinert, R.E., and Bergman, H.L. 1974. Residues of DDT in lake trout (Salvelinus namaychus) and coho salmon (Oncorhynchus kisutch) from the Great Lakes. J. Fish. Res. Board Can. 31: 191-199.

Shuval, H.I. 1978. Health aspects of water recycling practices. IN: Aquatic Pollutants: Transformation and Biological Effects. O. Hutzinger, L.H. van Lelyveld, and B.C.J. Zoeteman, eds. Pergamon Press, New York, pp. 345-403.

Smith, R.L. 1956. Recorded and expected mortality among Japanese of the United States and Hawaii with specific relevance to cancer. J. Natl. Cancer Inst. 17: 459-473.

Sonstegard, R.A. 1973. Relationship between environmental factors and viruses in the induction of fish tumors. Symp. Proc.: Viruses in the Environment and Their Potential Hazards. M.S. Mahdy and B.J. Dutka , eds. pp. 119-129.

Sonstegard, R.A. 1974. Neoplasia incidence studies in fishes inhabiting polluted and non-polluted waters in the Great Lakes of North America. XI International Cancer Congress. Panel 17: 172-173.

Sonstegard, R.A. 1976. The potential utility of fishes as indicator organisms for environmental carcinogens. IN: Wastewater Renovation and Reuse. F.M. D'Itri, ed. Marcel Dekker, Inc., New York, pp. 561-577.

Sonstegard, R.A. 1977. Environmental carcinogenesis studies in fishes of the Great Lakes of North America. Ann. New York Acad. Sci. 298: 261-269.

Sonstegard, R.A. 1978. Feral aquatic organisms as indicators of waterborne carcinogens. IN: Aquatic Pollutants: Transformation and Biological Effects. O. Hutzinger, L.H. van Lelyveld, and B.C.J. Zoeteman, eds. Pergamon Press, New York, pp. 349-358.

Sonstegard, R.A., and Leatherland, J.F. 1976. Studies of the epizootiology and pathogenesis of thyroid hyperplasia in coho salmon (Oncorhynchus kisutch) in Lake Ontario. Cancer Res. 35: 4467-4475.

Sonstegard, R.A., and Leatherland, J.F. 1979a. Growth retardation in rats fed coho salmon collected from the Great Lakes of North America. Chemosphere 7: 903-910.

Sonstegard, R.A., and Leatherland, J.F. 1979b. Hypothyroidism in rats fed Great Lakes coho salmon. Bull. Environ. Contam. Toxicol. In Press.

Sonstegard, R.A., and Leatherland, J.F. 1979c. Effects of diets of Great Lakes coho salmon on laboratory rats. XXII Conference of the International Association for Great Lakes Research. In Press.

Sonstegard, R.A., Leatherland, J.F., and Dawe, C.J. 1976. Effects of gonadal tumors on the pituitary-gonadal axis in cyprinids from the Great Lakes. J. Gen. Comp. Endocrin. 29: 269.

Staszewski, J., and Haenszel, W. 1965. Cancer mortality among Polish born in the U.S. J. Natl. Cancer Inst. 35: 291-297.

Veith, G.D., Kuehl, D.W., Puglisi, F.A., Glass, G.E., and Eaton, J.G. 1977. Residues of PCB and DDT in Western Lake Superior ecosystem. Arch. Environ. Contam. Toxicol. 5: 487-494.

BACTERIAL DEGRADATION OF CRUDE OIL AND OIL COMPONENTS IN PURE AND MIXED CULTURE SYSTEMS

A.M. Bobra, A.G. Clark, and the late J.R. Brown

Institute of Environmental Studies

University of Toronto

Since the eventual removal from land and water of spilled hydrocarbons depends upon microbial oxidation, any situation in which there is no oil degrading bacteria, no oxygen available for that oxidation, or too low a temperature for bacterial growth will result in the oil remaining virtually undegraded. It is already known that oil-degrading bacteria exist at low levels in most ecological habitats and that even in Arctic conditions microbial degradation occurs. The only possible limitation, other than time, could be available oxygen, that is, if we assume that the spilt oil does not contain or release compounds which are toxic to the bacterial population.

In 1971, sixty gallons of kerosene had been spilt on a pond at the University of Toronto's Baie du Doré research station. Bacteria taken from the water and water/sediment interfaces during the next five years were predominantly species of *Achromobacter*, *Acinetobacter* and *Bacillus*, along with a range of organisms which were facultative and microaerophilic forms. Approximately 20% of the isolates could aerobically degrade a test C_{12-18} oil fraction as a sole carbon and energy source, although of the dominant genera, neither the *Achromobacter* nor *Bacillus* species could grow in C_{12-18}. However, this did not preclude other hydrocarbons from being possible carbon and energy sources. For example, *Sarcina* isolates which failed to grow on C_{12-18}, grew abundantly on Norman Wells crude oil. What was significant was that this latter growth only occurred anaerobically.

The experimental pond at Baie du Doré was eight feet deep along the centre with a marked temperature and oxygen gradient. The kerosene had sunk to the bottom and this oil rose to the surface

Table 1 Bacteria Selected

Bacteria	Isolated from	Ability to grow on C_{12-18} as sole carbon and energy source
Acinetobacter (A_{10})	Water (Kerosene Pond)	+
Acinetobacter (C_8)	Floating Mass (Kerosene Pond)	-
Acinetobacter (C_6)	Floating Mass (Kerosene Pond)	+
Bacillus sp. (A_5)	Water (Kerosene Pond)	+
Flavobacterium sp. (C_4)	Floating Mass (Kerosene Pond)	-
Sarcina lutea (A_8)	Water (Kerosene Pond)	-
Pseudomonas aeruginosa (E_4)	Oil Mass (Crude Oil Pond)	+

whenever the sediment was disturbed. During the summers of 1973 to 1976, cycles of algal blooms appeared at various intervals. The algae matts on the pond floor seasonally rose to the surface due to entrapped gas. The dominant algae in the floating matts were *Oscillatoria*, a blue-green filamentous alga, commonly found near the effluents of oil refineries, and *Chlorella*, a green unicellular biflagellated alga, which had been found to be tolerant of many hydrocarbons. It was thought possible that the algae were responsible for the rate of oil degradation and the recovery of the damaged ecosystem. The degradation in the relatively anaerobic depths was being controlled by the rate of photosynthesis.

From the range of bacteria isolated during this period, seven distinct bacterial types were chosen for further study (Table 1). These were three *Acinetobacters* (C_8, C_6 and A_{10}), an asporogenous *Bacillus* sp., a *Flavobacterium* sp., and *Sarcina lutea*, together with a strain of *Pseudomonas aeruginosa* which was isolated from a mixed sour blend Alberta crude oil spill in the same area. They were chosen because of their easily differentiated colony morphology, repeated presence in oil ponds and their ability or inability to degrade a test C_{12-18} oil fraction. The ability of the seven bacteria to degrade Norman Wells crude oil aerobically or anaerobically, both in pure and mixed cultures, was examined in the presence or absence of algae.

Table 2 Summary of % of Total Changes in Oil Following Bacterial or Bacterial and Algal Degradation

Microorganism	Bacteria-alone		Bacteria & *Chlorella*		Bacteria & *Gonium**	
	aerobic	anaerobic	light	dark	light	dark
Flavobacterium (C_4)	30.054	1.761	22.569	1.048	0	0.504
Acinetobacter (C_8)	14.968	4.112	18.069	3.998	2.799	3.602
Acinetobacter (C_6)	18.721	0	27.190	4.654	5.823	6.599
Bacillus (A_5)	29.931	3.862	16.829	1.048	0	0.080
Sarcina (A_8)	0	26.156	4.920	18.591	14.770	13.219
Acinetobacter (A_{10})	4.655	6.291	3.000	5.700	2.988	3.838
Pseudomonas aeruginosa (E_4)	32.438	8.443	22.369	6.265	9.895	9.342

* *Gonium* was killed by 1% (v/v) Norman Wells crude oil under light and dark conditions.

During four weeks at 22°C, under aerobic conditions, *Acinetobacter* (A_{10}) and *Sarcina lutea* failed to grow on 1% (v/v) Norman Wells crude oil in a modified Bushnell-Haas salts medium with added trace elements but without other carbon sources. *Acinetobacters* (C_6 and C_8), both of which exhibited some turbidity in this medium, failed to emulsify the oil. However, *Pseudomonas aeruginosa*, the *Flavobacterium*, and *Bacillus* grew well and emulsified Norman Wells crude. The activity of the *Pseudomonas* sp. was the most extensive.

The oil during degradation was analyzed by gas chromatography with Hewlett Packard Gas Chromatograph series 5830A in conjunction with a flame ionization detector at 300°C. Dual ten-foot 1/8" I.D. stainless steel columns contained 60-80 mesh chromosorb P pre-coated with 10% SE 30 ultraphase. It was temperature programmed from 40°C to 280°C at 15°C/minute. The temperature of the injection port was 300°C.

Norman Wells crude had no fraction less than C_5, and 48% of the total crude oil was initially less than C_{12}. The aerobic degradation was mainly in the C_{5-12} region (Table 2). *Flavobacterium*, *Bacillus* and *Pseudomonas* sp. showed the most extensive degradation pattern of the oil under aerobic conditions (Figure 1). For example, *Pseudomonas* caused the disappearance of approximately 32% of the total chromatogram area when compared to the control. All the peaks, up to and including C_9, disappeared, except for the peak represented by toluene which was present but considerably

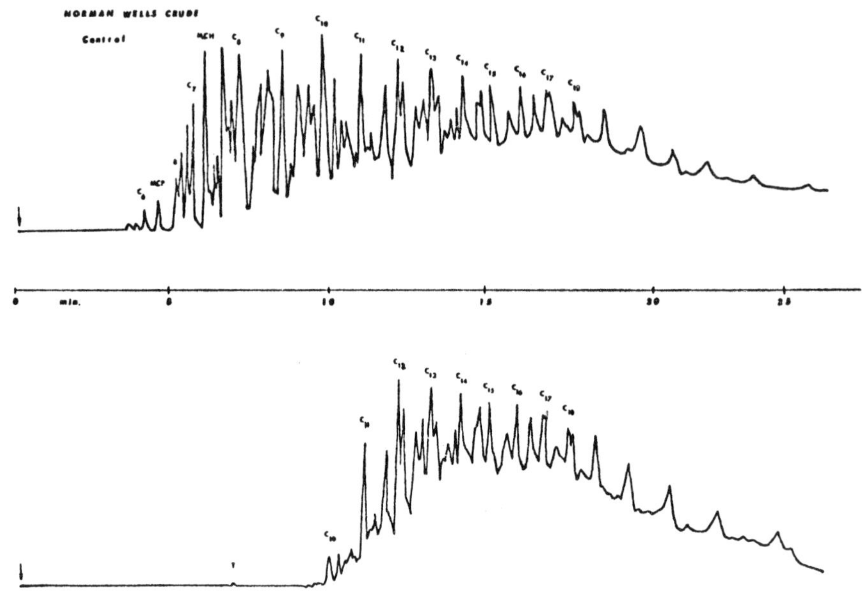

Figure 1. Gas chromatograms of Norman Wells crude oil before (upper) and after (lower) aerobic degradation by Pseudomonas aeruginosa.

reduced. The decane peak was also reduced when compared to the control, indicating that there was some degradation of this peak.

Under anaerobic conditions, the initial dissolved oxygen concentration in sealed and carbon dioxide gassed flasks was 4 ppm. After four weeks of shaking at 22°C, the emulsification of the oil and the increasing turbidity of the mineral salts medium indicated microbial activity in three of the seven bacterial types (Table 2). Sacrina was the most active of the three bacterial types in degrading oil under low oxygen concentration. All the peaks from isopentane, to and including toluene, disappeared. This represented 26% of the total chromatogram. Pseudomonas aeruginosa and Acinetobacter (A_{10}) grew only slightly under the anaerobic conditions with no visible emulsification of the oil. The changes in the oil analysis with these two bacterial types were only slight (6 - 8%).

All of the anaerobic and aerobic flasks were duplicated with one set being incubated in the light, the other in the dark, to check for photo-oxidation effects. After four weeks, a comparison of the chromatograms showed no difference in microbial attack.

The seven bacteria were next grown singly in the presence of an actively growing algal culture of Chlorella luteoviridis in the light in sealed flasks. Flavobacterium, Acinetobacter (C_8), Acinetobacter (C_6), Bacillus, and Pseudomonas aeruginosa grew and emulsified the oil, whereas there was no growth of Sarcina and Acinetobacter (A_{10}) and no change in the physical appearance of the oil. These two bacteria had grown with Norman Wells crude oil under low oxygen concentration but were the only two types which did not grow under aerobic conditions. Pseudomonas aeruginosa grew slightly under anaerobic conditions but grew abundantly and emulsified the oil under aerobic conditions. Twenty-seven percent of the oil was utilized by Acinetobacter (C_6) with algae with light, compared to 18% when grown alone aerobically.

When the same experiment was repeated in the dark, the algae died and the results were similar to the results when bacteria-alone were degrading oil anaerobically. Acinetobacter (C_6) exhibited a slight growth in this experiment. This did not occur in the bacteria-only experiment and could have been due to growth on nutrients liberated by the dying algae.

The flask experiment was repeated with an actively growing culture of Gonium sociale mixed with the individual seven bacteria in a mineral salts medium. However, 1% (v/v) of Norman Wells crude oil killed the Gonium sp. The results were similar to the bacteria-only results under anaerobic conditions.

Thus, in the light where the algae and bacteria could exist together, the actively growing algae produced oxygen which changed the condition of the flasks from anaerobic to aerobic. Otherwise, only species capable of anaerobic growth or species capable of anaerobic respiration with nitrate replacing oxygen as a terminal electron acceptor (e.g. Pseudomonas sp.), could degrade oil anaerobically.

A further investigation of the dynamics of these algal and bacterial mixtures was conducted under dialysis conditions (Figure 2). A pair of one litre spinner flasks with a double nuclepore membrane (0.22 μ-pore size) was used to separate the cultures. One side of the dialysis flask received a known mixture of the seven bacteria, the other side, a bacteria-free culture of Chlorella luteoviridis or Gonium sociale, in the same modified Bushnell-Haas medium. Carbon dioxide was bubbled through the entire system and a bladder of CO_2 was placed on the algal side to maintain a supply of CO_2. Sterile Norman Wells crude oil was added to the bacterial side at 1% (v/v).

The apparatus was placed on stirrers under fluorescent lights on a 12 hour dark and 12 hour light cycle. The temperature was maintained at 22°C.

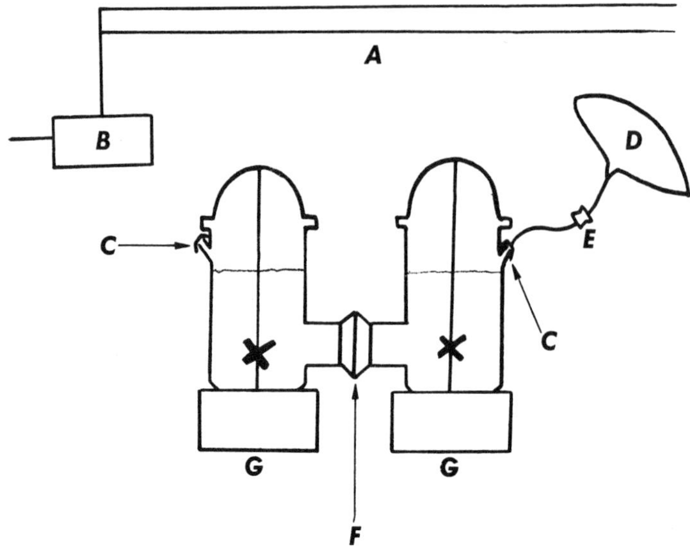

Figure 2. Schematic diagram of the dialysis culture system.
A Fluorescent lights
B Timer: 12 hr on; 12 hr off
C Rubber seals used as injection ports. The bacteria and oil were introduced into the left hand vessel; the pure culture of algae into the right hand vessel.
D Gas bladder of CO_2 on algal side
E Filter to avoid contamination
F Nuclepore (0.22μ) filter separating cultures
G Stirrer blocks

An identical second Bellco dialysis flask was set up in the dark as a control.

During the 28 days incubation period, samples of the aqueous solution from both sides were withdrawn by syringe for microbial and algal counting. A small sample of the floating oil was also removed from the bacterial side. The oil sample was analyzed by gas chromatography, as previously described.

Only in the light was the crude oil extensively attacked by the mixed bacterial culture. Initially, the oil tended to become "stringy" and disintegrated into small droplets which were emulsified throughout the media by the third week. The oil persisted as an unchanged surface layer under dark conditions.

Table 3 Growth of the Seven Bacteria as a Mixture in a Dialysis Culture with Chlorella luteoviridis in the light

Bacterial type	Growth on test C_{12-18} oil fraction	Bacterial counts for week no.				
		0	1	2	3	4
Acinetobacter (C_8)	-	5.2×10^6	2.0×10^5	3.0×10^5	7.0×10^4	$< 10^4$
Acinetobacter (C_6)	+	8.0×10^5	9.9×10^5	9.7×10^7	2.1×10^7	2.8×10^7
Acinetobacter (A_{10})	+	1.1×10^7	2.6×10^8	1.8×10^{10}	1.0×10^9	5.6×10^8
Bacillus (A_5)	+	8.2×10^6	1.0×10^7	7.2×10^{10}	9.9×10^{10}	2.0×10^{11}
Flavobacterium (C_4)	-	5.8×10^6	$< 10^4$	$< 10^4$	$< 10^4$	$< 10^4$
Sarcina lutea (A_8)	-	1.9×10^7	5.0×10^7	3.0×10^8	2.7×10^9	2.6×10^9
Ps. aeruginosa (E_4)	+	2.7×10^6	3.1×10^8	1.4×10^{10}	1.6×10^{10}	9.6×10^9
Total Count		5.3×10^7	6.3×10^8	1.0×10^{11}	1.2×10^{11}	2.1×10^{11}

The initial total bacterial inoculum was 5.3×10^7 per ml (Table 3). Under light conditions, all the bacteria were stimulated, except Flavobacterium and one of the Acinetobacter types The marked response of the Bacillus sp., one of the Acinetobacter (A_{10}) and Pseudomonas aeruginosa accounted for the maximum bacterial population on day 14 (approximately a 3.0 log increase) which remained at this high level until the termination of the experiment. Under dark conditions, the bacterial population only increased marginally (approximately a 0.5 log increase) during the incubation period.

Under light conditions, the algal population increased approximately 2.5 logs, which peaked on day 14 and remained at this new level until the termination of the experiment (Figure 3). The oxygen concentration in the flask was directly proportional to the increase in the algal population. It would appear that in the light, the oxygen generating time of the alga was greater than the oxygen needs of the bacterial population. In the dark, the alga remained constant for seven days and then started to die off. At the termination of the experiment, there were no visible algal cells and the oxygen concentration was low.

When Norman Wells crude oil was analyzed by gas chromatography, a preferential degradation of the C_5 to C_{12} portion was noted which occurred only in the light. After 4 weeks, 21% of the total amount

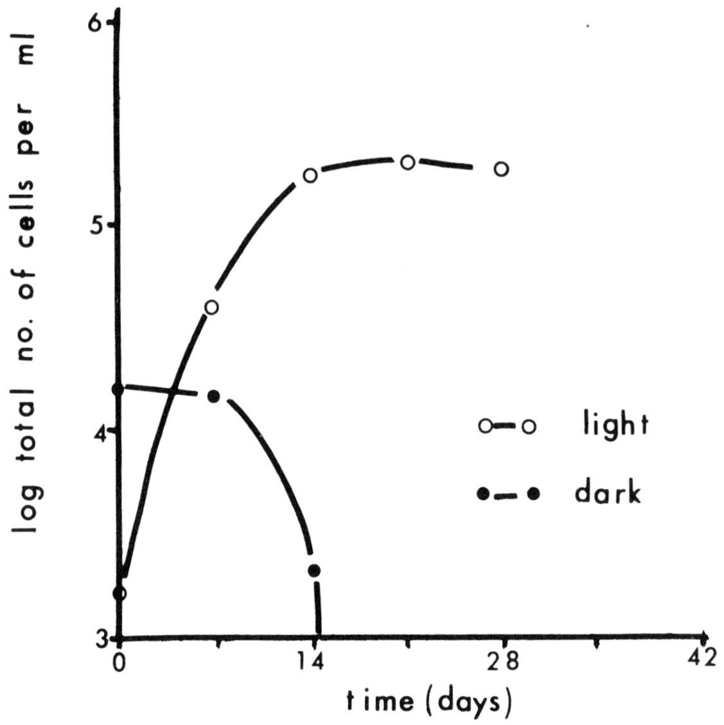

Figure 3. Dialysis flask experiment with Chlorella luteoviridis grown in the light and in the dark. Graph shows the proportion of algal growth

of oil had been degraded in the illuminated bacteria-Chlorella system and 26% in the bacteria-Gonium system. Only 4.9% and 7% were degraded by the same bacteria-alga systems in the dark.

Thus, provided the overlying water in a pond is sufficiently clear and there is a minimum of algicidal and bacteriocidal components liberated by spilt oil, oil sunk to anaerobic depths can be metabolized by bacteria, using oxygen liberated by algal photosynthesis.

The same seven bacterial types were used in a series of experiments involving large laboratory aquatic ecosystems containing sediment, algae and floating plants. Two commercial plastic swimming pools with polyethylene liners were filled to a volume of approximately 10 m^3. One pond served as a control, the other as an experimental.

Table 4 Ecosystem Conditions

Experiment	Hydrocarbons	Method of introducing into system	Amount added	Amount detected in water column
1	Toluene	ring (float)	3.0 liters (2.6 x 10^3 gm)	2 mg/l
2	Naphthalene	ring (float)	200 gm 0.8 liters	.02 mg/l 2 mg/l
3	Naphthalene & Pentane	pump	100 gm 0.8 liters	2.7 mg/l 22 ppm
4	Pentane	pump	0.8 liters	22 ppm
5	Benzene	pump	650 gm	80 ppm
6	Benzene & Phenanthrene	pump	650 gm 2.4 liters	50 ppm 0.7 ppm
7	Cyclohexane	pump	500 gm	24 ppm
8	Mixture of above	ring (float)	mixture of all above single cpds.	*
9	Mixture of above	pump		*

Pure hydrocarbons were chosen to represent the major chemical types present in any one crude oil (Table4). This table lists the compounds, the volume spilt, the method of the spill and the maximum concentration detectable in the water column of the experimental pond. In the two final experiments, the mixture was made up of 500 g of toluene, benzene, cyclohexane and pentane, 100 g of naphthalene and 24 g of phenanthrene. The method of spilling was either to spill a known volume onto the surface of the experimental pond, containing the spill in an aluminum ring, or to introduce the hydrocarbon via a submersible pump. Since many of the compounds were highly volatile, it was found that in floatation spills, volatilization was sufficiently rapid to cause surface loss before substantial amounts could diffuse into the water column.

This study involved a multi-disciplinary approach to develop a comprehensive understanding of the effects of oil components on an ecosystem. All the chemical measurements were made by the Department of Chemical Engineering and Applied Chemistry, University of Toronto, under the direction of Dr. Mackay. Algal and plant assays were taken by the Department of Botany under the supervision of Dr. Hellebust and Hutchinson. Details of the work are available in a report to Inland Waters Directorate of the Department of Fisheries & Environment who funded the project.

We followed the subsequent changes in the bacterial populations in the water and in the sediment. Parallel studies in flasks examined the response of pure cultures to the same spilled hydrocarbons at the same concentrations as maximally attained in the ecosystem.

From the individual graphs for each organism under each spill condition, certain generalizations are possible when viewed with the physical and algal data. For example, if the bacterial response to pure hydrocarbons was initially adverse, it was normally followed by an eventual recovery to the original population size. This population may have continued to increase either due to the hydrocarbons having evaporated to a non-toxic level and now being available for degradation, or due to nutrients liberated from algae and plants damaged by contact with the hydrocarbon. It was also noted that no matter how severe the initial response, at least one organism always thrived but the species varied with the conditions, and if the hydrocarbon did not enter the water column (i.e., was floated onto the water rather than admixed), no significant bacterial response was seen.

The *in vitro* experiments with pure cultures in flasks have shown that no one organism reacted in a predictable manner to any one hydrocarbon and that the flask results could not foretell the population response in the ecosystems. Each bacterial type was either stimulated or inhibited by each compound in a unique fashion. For example, *Sarcina lutea* showed increased growth with all compounds in the flasks, while *Acinetobacter* (C_6) showed no enchancement of growth with any hydrocarbon tested, yet this latter organism was most responsive in the ecosystem experiments; the *Sarcina* was relatively unresponsive. Thus, the *in vitro* results for pure culture bacterial systems should only be extrapolated to large ecosystems (or probably to natural systems) with extreme caution.

REFERENCES

Mackay, D., M.E. Charles, and C.R. Phillips. 1972. The Physical Aspects of Crude Oil Spills on Northern Terrains. An Interim Report to the Arctic Land Use Research Program of the Dept. of Indian and Northern Affairs, Ottawa, Dec.

Oil and Gas Working Group. 1978. Studies of the Effects of Hydrocarbons on Laboratory Aquatic Ecosystems, Report to Inland Water Directorate of Environment Canada. Institute for Environmental Studies, University of Toronto.

INDUCTION OF MIXED FUNCTION OXIDASE ENZYMES IN FISH BY HALOGENATED AND AROMATIC HYDROCARBONS

D.E. Willis, M.E. Zinck, D.C. Darrow and R.F. Addison

Fisheries and Marine Service, Marine Ecology Laboratory

Bedford Institute of Oceanography, Dartmouth, N.S.

ABSTRACT

Several representative halogenated and aromatic hydrocarbon environmental contaminants were fed to trout and their effects on hepatic mixed function oxidase (MFO) enzyme systems studied. The insecticide p,p'-DDT and its metabolite p,p'-DDE did not induce the trout MFO enzyme system whereas the polycyclic aromatic 3-methylcholanthrene and the polychlorinated biphenyl Aroclor 1254 induced various components of this system, most dramatically ethoxycoumarin O-de-ethylase activity. A PCB replacement composed of chlorinated butylated diphenyl ethers (XFS-4169 L : Dow Chemical Co.) had no effect on trout MFO activity other than a slight increase in cytochrome P-450 concentration.

INTRODUCTION

The mixed function oxidases (MFOs) are a group of relatively non-specific enzymes usually localized in the microsomes of various tissues in both vertebrates and invertebrates. They catalyze the degradation of both endogenous and exogenous compounds generally by oxidative pathways, and amongst these reactions, conversions of lipid-soluble nonpolar compounds to more water-soluble polar compounds are carried out. The mammalian and avian MFO enzyme systems can be induced by exposure of the organism to chlorinated hydrocarbons (DDT, cyclodiene insecticides, and PCBs) and aromatic hydrocarbons; this induction presumably represents a defence mechanism by which the organism can degrade and eventually eliminate potentially toxic substances. In principle, the extent or even the occurrence of MFO induction might indicate the sub-lethal exposure of an organism to these materials. We have examined the suitability of MFOs in fish for such monitoring, and in this paper we describe the effect of feeding p,p'-DDT, p,p'-DDE, Aroclor 1254 (PCBs), 3-methylcholanthrene (3-MC) and a mixture of butylated monochlorodiphenyl ethers (XFS-4169L, Dow Chemical Co.), on some of the elements of the MFO enzyme system in trout (*Salvelinus fontinalis*).

MATERIALS AND METHODS

Aroclor 1254 was obtained from the Monsanto Co; p,p'-DDT and p,p'-DDE from Applied Science Labs, Inc.; 3-methylcholanthrene from Eastman Kodak Co. and XFS-4169L from the Dow Chemical Co. Each compound was fed in a different experiment to brook trout to assess its effect on hepatic MFO systems using the methods described fully by Addison et al, 1977.

Brook trout from the laboratory's stock of fresh water maintained fish weighing 100 to 300 g were divided randomly into groups of controls and experimentals for each experiment. The fish were force-fed every other day gelatin capsules containing a single substrate dissolved in an organochlorine-free marine oil to yield a theoretical tissue concentration of 200 μg/g (body weight) except for 3-MC where a concentration of 20 μg/g (body weight) was desired. Control fish were fed in a similar manner with unspiked oil. Both groups were also fed a commercial fish food (Ewoz-Fish Chow) *ad lib*.

After 2 to 3 weeks the fish were killed, weighed, and the livers assayed individually for components of the MFO system including 7-ethoxycoumarin O-de-ethylase, aniline hydroxylase and aldrin epoxidase activities, cytochrome P-450 and microsomal protein. The expected products of enzyme incubations, cytochrome P-450's and protein levels were quantitated fluorometrically, spectrophotometrically, or by gas chromatography. All the data were analyzed statistically by t-test.

Tissue residues of all the compounds fed, except 3-MC, were quantitated using standard methods (Addison et al, 1977; 1978a).

RESULTS AND DISCUSSION

The results are shown in Table 1. None of the pollutants tested produced statistically significant changes in the fish body weights, liver weights, or liver PMS protein concentrations. The difference between experimental and control microsomal protein concentrations for the PCB and 3-MC fed fish was significantly higher in only the 3-MC group whereas the percentage of PMS protein accounted for by the microsomal protein was significantly higher in

Table 1 Indices of Hepatic MFO Activity in Brook Trout Following Feeding p,p'-DDT, p,p'-DDE, XFS-4169L, Aroclor 1254 and 3-methylcholanthrene (3-MC) (means ±s.d.)†

Compound Fed	Fish		Microsomal Protein (mg/g liver)	Microsomal Protein (% Total Protein)	Cytochrome P-450 (nmol/g liver)
DDT	Control	(n=6)	12.3 ±2.5	23.6 ±4.1	11.9 ±2.6
	Experimental	(n=6)	11.2 ±2.4	22.5 ±3.8	12.9 ±5.9
DDE	Control	(n=10)	7.3 ±1.1	13.1 ±1.6	4.2 ±0.9
	Experimental	(n=10)	8.1 ±2.1	14.6 ±3.8	4.1 ±1.4
XFS-4169L	Control	(n=8)	14.5 ±1.6	17.7 ±1.9	4.84 ±1.12
	Experimental	(n=8)	15.5 ±1.1	18.9 ±1.2	5.84 ±0.89*
Aroclor 1254	Control	(n=6)	13.0 ±1.2	17.7 ±0.8	6.0 ±1.0
	Experimental	(n=5)	14.5 ±2.0	19.9 ±1.6*	6.5 ±1.8
3-MC	Control	(n=6)	15.4 ±2.73	19.8 ±3.7	6.4 ±2.7
	Experimental	(n=5)	18.4 ±2.7*	20.9 ±2.8	7.9 ±2.3*

Compound Fed	Fish		Aniline Hydroxylase (nmol p-aminophenol ·g liver^{-1}·hr^{-1})	7-Ethoxycoumarin O-de-ethylase (nmol umbelliferone ·g liver^{-1}·hr^{-1})	Aldrin Epoxidase (nmol dieldrin ·g liver^{-1}·hr^{-1})
DDT	Control	(n=6)	92.6 ±26.5	109 ±21.8	N.A.§
	Experimental	(n=6)	93.6 ±16.4	84.7 ±15.4	N.A.
DDE	Control	(n=10)	97.1 ±29.5	115 ±22.7	106 ±12.5
	Experimental	(n=10)	75.6 ±23.4	96.7 ±19.5	88.4 ±20.6
XFS-4169L	Control	(n=8)	100 ±36.9	79.8 ±19.3	142 ±21.0
	Experimental	(n=8)	115 ±21.0	83.2 ±18.3	141 ±37.0
Aroclor 1254	Control	(n=6)	141 ±30.4	127 ±15.8	N.A.
	Experimental	(n=5)	182 ±69.4	341 ±120**	N.A.
3-MC	Control	(n=6)	155 ±33.7	104 ±19.0	168 ±38.5
	Experimental	(n=5)	220 ±34.1**	216 ±87.6**	237 ±82.5*

* Significantly different from control ($P < 0.05$).
** Significantly different from control ($P < 0.01$).
† Data taken from Addison *et al.*, 1977; 1978a; 1978b.
§ Not analyzed.

only the PCB-fed group. Treatment with p,p'-DDT, p,p-DDE and XFS-4169L produced no difference between experimentals and controls in this indicator. Cytochrome P-450 concentrations remained unaffected by feeding p,p'-DDT and p,p'-DDE but increased slightly in the XFS-4169L, PCB and 3-MC fed fish. However, when these levels were corrected for their respective microsomal protein concentrations, only the XFS-4169L and 3-MC fed groups showed a significant difference between experimental and control fish.

Feeding p,p'-DDT, p,p'-DDE, and XFS-4169L produced no significant differences between experimental and control fish in the indicators of MFO activity studied with the previously noted exception: treatment with XFS-4169L induced a small but significant increase in the concentration of cytochrome P-450. Analysis of the fish tissue for XFS-4169L showed that only 4-chlorodiphenyl ether (a major component, Willis et al, 1978) and a dibutylated chlorodiphenyl ether (minor component) were retained at a level of 8% of theoretical; this may be one reason for the overall absence of effect. For the p,p'-DDT and p,p'-DDE groups the data suggest that either the fish MFO system is already induced maximally by the residue levels of these compounds normally found in the tissues, or that other mechanisms, unrelated to the MFO system, exist to deal with foreign contaminants of this type. It would appear that the fish MFO enzyme system is relatively insensitive to induction by these types of compounds; in spite of feeding massive doses which gave tissue residue levels 200- to 600-fold (except XFS-4169L) above background.

In contrast to these findings the feeding of PCBs and 3-MC produced significant differences between experimental and control fish. Exposure to these compounds resulted in selected increases in the level and activities of cytochrome P-450, aniline hydroxylase, 7-ethoxycoumarin O-de-ethylase, and aldrin epcxidase. As previously noted, the increase in cytochrome P-450 in the PCB group could be accounted for by the corresponding increase in microsomal protein whereas the increase in this indicator for the 3-MC fed fish was significantly different from control values. Aniline hydroxylase activity was not affected to any significant extent by the PCB tested but was induced in the 3-MC fed fish 1½ times as compared to controls. PCBs however, stimulated a 2½-fold induction in O-de-ethylase activity (our most sensitive indicator) and similarly the 3-MC fed fish showed a 2-fold induction in this enzyme activity. Aldrin epoxidase remained unaffected by treatment with PCBs but was induced 1½ times by feeding the fish 3-MC.

The data indicate that exposure by feeding Aroclor 1254 or 3-methylcholanthrene (a polycyclic aromatic) induced selected components of the trout mixed function oxidase enzyme system. Although the effects differed between these two compounds the differences were largely of degree with 3-MC producing a greater overall induction. However, no induction of these parameters of the MFO system was observed when the fish were fed p,p'-DDT and p,p'-DDE. This is in contrast to mammalian and avian systems where these compounds induce the MFO indicators studied here (Bunyan et al, 1972; Ullrich and Weber, 1972).

Polycyclic aromatics such as 3-MC have been shown to induce a synthesis of cytochrome P-448 in rats (Alvares et al, 1973). Goldstein et al (1977) have also shown that in rats only a small percentage of the components of Aroclor 1254 stimulate the production of cytochrome P-448, and the rest operate on cytochrome P-450; these observations would account for the relatively poor response of fish towards this material. Judging from our 3-MC and PCB data it seems reasonable to conclude that a similar *de novo* synthesis occurs in fish, indicating that the fish MFO system would be sensitive to pollutants which stimulate this type of conversion. DDT and its metabolite DDE induce cytochrome P-450 in mammals, and would have little effect on a cytochrome P-448 based system. The PCB replacement XFS-4169L produced only induction in cytochrome P-450 and had no effect on the other indices of MFO activity studied. This fact, and the limited and selective retention of this material by trout, makes it difficult to classify XFS-4169L as being a cytochrome P-450 or P-448 type of inducer.

In general, our results indicate that widely different responses of the fish hepatic MFO enzyme system are observed when exposed by feeding at a sub-lethal level to several representative types of pollutants. Feeding cytochrome P-450 inducers to fish produced no effects whereas feeding cytochrome P-448 inducers produced measurable changes in the indices of the fish MFO system thereby showing the inducibility of this system, and the magnitude of these changes in MFO activity might be used as an index of sub-lethal exposure to environmental contaminants which stimulate a synthesis of cytochrome P-448.

REFERENCES

Addison, R.F., M.E. Zinck, and D.E. Willis. 1977. *Comp. Biochem. Physiol.* 57C: 39-43.

Addison, R.F., M.E. Zinck, and D.E. Willis. 1978a. *Comp. Biochem. Physiol.* In Press.

Addison, R.F., M.E. Zinck, D.E. Willis, and D.C. Darrow. 1978b (Unpublished results).

Alvares, A.P., D.R. Bickers, and A. Kappas. 1973. Proc. Nat. Acad. Sci. U.S.A. 70: 1321-1325.

Bunyan, P.J., M.G. Townsend, and A. Taylor. 1972. Chem. Biol. Interact. 5: 13-26.

Goldstein, J.A., P. Hickman, H. Bergman, J.D. McKinney, and M.P. Walker. 1977. Chem. Biol. Interact. 17: 69-87.

Willis, D.E., M.E. Zinck, and R.F. Addison. 1978. Chemosphere. Submitted.

Ullrich, V. and P. Weber. 1972. Z. Physiol. Chem. 353: 1171-1177.

HALOGENATED BIPHENYL METABOLISM

S. Safe, C. Wyndham, A. Parkinson, R. Purdy and A. Crawford

Environmental Biochemistry Unit, Guelph Waterloo Center for Graduate Work in Chemistry, Guelph Campus

University of Guelph, Guelph, Ontario, Canada

The metabolism of xenobiotics occurs predominantly in the liver and is generally effected in two related phases. Phase I biotransformation includes oxidation, reduction and hydrolysis which either introduces or exposes a more polar substituent or group such as -OH, -SH, $-NH_2$ and -COOH. Phase II biotransformation includes addition reactions which conjugate the electronegative group introduced or exposed during phase I metabolism with endogenous acids. A feature characteristic of these diverse biotransformations is that the metabolic products are generally more hydrophilic than the parent compound and this expedites their renal and biliary excretion.

Whilst the metabolism of most xenobiotics increases their hydrophilicity and excretion, it is erroneous to consider phase I, and to a lesser extent phase II reactions solely as processes of detoxification. Just as the therapeutic efficacy of some drugs relies on their conversion to pharmacologically active metabolites, so the toxicity of certain xenobiotics demands their metabolic conversion from the inert parent compound to an activated toxic species (Gillette, 1975).

Aryl hydrocarbons and halogenated aryl hydrocarbons, being unsuitable substrates for reductive or hydrolytic phase I biotransformations, are oxidised to phenolic derivatives by aryl hydrocarbon hydroxylase (AHH): a highly inducible, microsomal, haem-containing monooxygenase belonging to the family of cytochromes P-450. The cytochrome P-450-mediated oxidation of many toxic [halogenated] aryl hydrocarbons to phenolic metabolites proceeds via an arene oxide intermediate and this lipophilic, electrophilic metabolite has been implicated as an activated toxic species (Jerina and Daly, 1974; Daly et al, 1972). Arene oxides can covalently bind non-critical cellular nucleophiles, such as glutathione, and critical cellular nucleophiles, such as the protein and nucleic acid macromolecules. The latter event is associated with various chemically induced diseases including cancer (Weinstein, 1978).

The approach undertaken in our laboratory is to assume that the metabolism of [halogenated] aryl hydrocarbons is characterised by three metabolic criteria: the phase I (phenolic) metabolites, the phase II (conjugated) metabolites and the covalent macromolecular adducts. This premise is adopted to evaluate more fully the biological effects of microsomal enzyme inducers. Our research interests have recently focussed on the impact of microsomal enzyme inducers on the _in vitro_ metabolism of 4-chlorobiphenyl to the three metabolic fractions described: namely phenols, low molecular weight conjugates and covalent macromolecular adducts (Crawford and Safe, 1977). In addition our interests have also centred on the effects of halogen substituents on substrate metabolic activity.

Previous studies have shown that the metabolism of commercial PCB mixtures and individual isomers yield more polar phenolic metabolites (Sundström et al, 1976) and this pathway was also noted in the _in vivo_ and _in vitro_ metabolism of 4-chlorobiphenyl (Wyndham and Safe, 1978) (Figure 1).

The spectrum of metabolites formed during the _in vivo_ metabolism of 4-chlorobiphenyl by diverse organisms (microorganisms, plants, fish, amphibians, birds and mammals) is dominated by the monohydric phenol, 4'-chloro-4-biphenylol (Sundström et al, 1976). When this major site of hydroxylation is deuterated, the metabolism of the isotopic derivative, 4'-chloro-4[^{2}H]-biphenyl to 4'-chloro-4-biphenylol is accompanied by a 1,2 migration of deuterium and retention of 79% of the deuterium label (Safe et al, 1975). The migration and retention of deuterium represents a [^{2}H] or NIH shift and indicates the intermediacy of an arene oxide (Figure 2) (Jerina and Daly, 1974).

The formation of covalent macromolecular adducts presumably results from the nucleophilic attack of the macromolecules on the 4'-chloro-3,4-biphenyloxide (Wyndham et al, 1976). The diol metabolite retains approximately half the deuterium (39%) found in the phenol and indicates that it forms by hydroxylation of the phenol (Safe et al, 1975). A number of studies have suggested that other PCB isomers are metabolized via arene oxide intermediates (Sundström et al, 1976). Moreover, recent results have confirmed the _in vivo_ and _in vitro_ metabolism and binding of radiolabelled

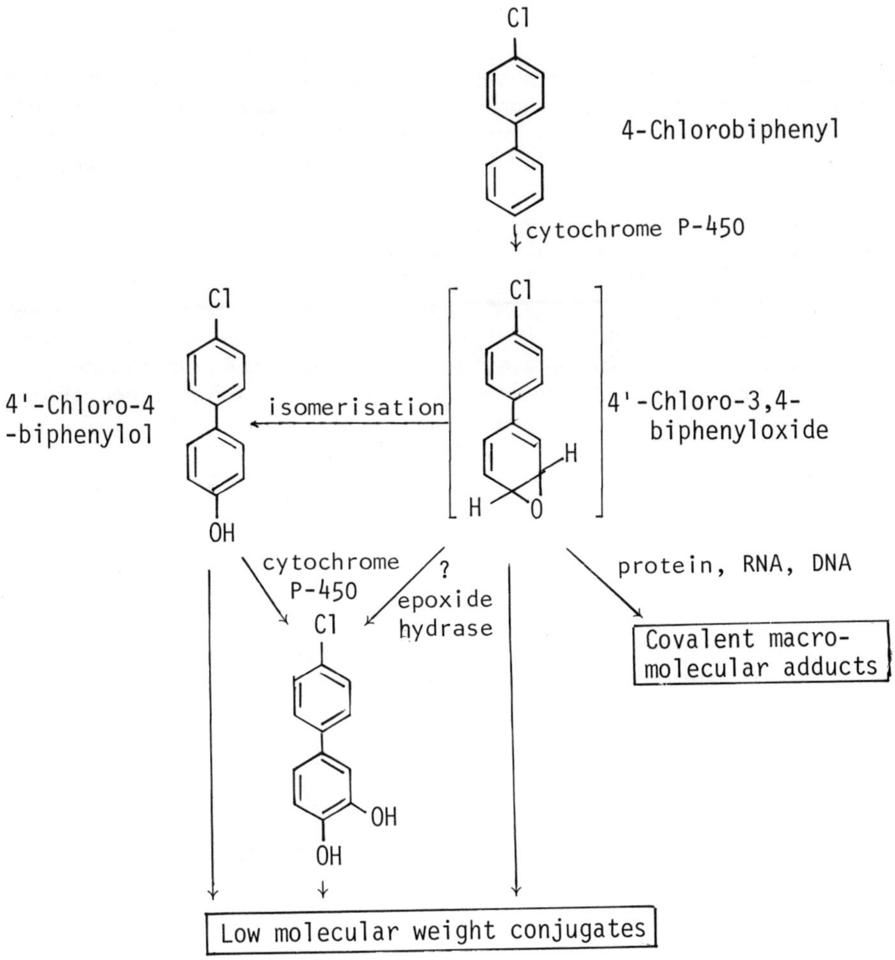

Figure 1. The metabolism of 4-chlorobiphenyl.

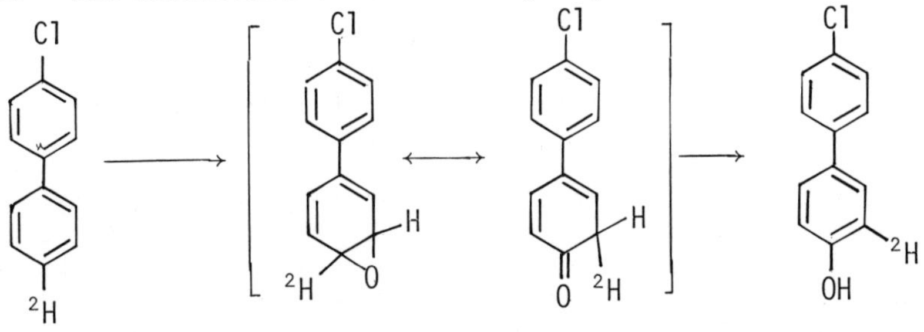

4'-chloro-4[^{2}H]-biphenyl

Figure 2. The NIH shift of deuterium in the cytochrome P-450-mediated hydroxylation of 4'-chloro-4-[^{2}H]-biphenyl to 4'-chloro-4-biphenylol.

KC-300 and KC-500 (commercial PCBs manufactured in Japan) to microsomal macromolecules (Shimada, 1976; Shimada and Sato, 1978). The metabolism of 4-bromobiphenyl was similar to that of the chloro analog and binding to microsomal proteins was also observed (Kohli et al, 1978). Biphenyl is metabolized to give 4-biphenylol as the major product with 2- and 3-biphenylol identified as minor metabolites (Billings and McMahon, 1978). Research in our laboratory has confirmed that biphenyl metabolism is also accompanied by the formation of covalent adducts with microsomal protein (Wyndham and Safe, 1978a).

The major differences in the metabolism of these substrates is observed by comparing the effects of noninduced and induced microsomal enzymes on the metabolic rates. Phenobarbitone (PB) pretreatment is known to enhance the formation of microsomal cytochrome P-450 proteins, whereas, 3-methylcholanthrene (3-MC) pretreatment enhances cytochrome P-448 proteins (Sladek and Mannering, 1966). Both induced microsome preparations have increased monooxygenase enzyme activity and this is evident in the increased rate of substrate hydroxylation. The differences in the induced enzyme activity are generally accompanied by changes in hydroxylation site specifity and a different spectrum of metabolites is formed. Not surprisingly, enzyme inducers can mediate the effects of metabolically related cellular damage by altering the populations and activities of the corresponding activated arene oxide intermediates. The metabolism of biphenyl is enhanced by microsomes isolated from animals pretreated with 3-MC and PB, however, the 3-MC induced microsomal enzymes specifically enhanced biphenyl-2-hydroxylation, whereas the PB induced microsomes enhanced biphenyl-4-hydroxylation (Creaven and Parke, 1966; McPherson et al, 1976).

A comparison of the initial rate of formation of 4-chloro- and 4-bromobiphenyl metabolites with noninduced microsomes (Table 1) clearly illustrates the higher reactivity of the former substrate. Chromatographic analysis of the metabolites formed after incubation of the halobiphenyls with 3-MC and PB induced microsomes indicated that the spectrum of metabolites was not significantly altered with 4'-halo-4'-biphenylol identified as the major product. This contrasted with the changes in site specificity observed using biphenyl as substrate and reflects the directive effects of the halogen substituent. The differences between the 4-bromobiphenyl and 4-chlorobiphenyl substrates are reflected in the effects of the induced microsomal enzyme preparations on metabolic rates. PB and 3-MC induced microsomal enzymes enhanced the overall metabolism of 4-bromobiphenyl whereas only the 3-MC induced microsomes enhanced the metabolism of 4-chlorobiphenyl. The PB induced enzymes did not increase the metabolic rate compared to the noninduced enzyme activity. This unusual enzyme-substrate specificity has been confirmed using rat hepatic microsomes obtained after pretreatment with 3,3',4,4'-tetrachlorobiphenyl, a specific cytochrome P-448

Table 1 A Summary of the Metabolism of Halogenated Biphenyls

Substrate (pretreatment)	Metabolic Fractions - Initial Rate (pmol/mg protein/min.)[a]			
	major p-hydroxy metabolite	total lipophilic metabolites	low molecular wt. conjugates	protein adducts
biphenyl				
(none)	27.1	32.9	5.5	1.6
4-chlorobiphenyl				
(none)	84.8	93.3	11.3	3.14
(PB)	83.5	102	14.4	3.15
(3-MC)	792	962	347	49.2
4-bromobiphenyl				
(none)	57.4	107	36.9	5.0
(PB)	115	175	58.6	11.2
(3-MC)	175	251	97.2	36.2

a the data given are the averages of 3 incubations; the standard deviations were less than 10% for all the initial rate determinations.

inducer and 2,2',4,4'-tetrachlorobiphenyl, a specific cytochrome P-450 inducer (Goldstein et al, 1977). In studies complementary to our own, the degree of binding of radiolabelled commercial PCB mixtures KC 300 (42% chlorine content) and KC 500 (55% chlorine content) to microsomal macromolecules *in vitro* was shown to be dependent on the species, organ and *in vivo* pretreatment with inducers (Shimada and Sato, 1978). The effect of induction on macromolecular binding was more pronounced in rats than mice. The enhancement of macromolecular binding by P-448 inducers (3-MC) but not by P-450 inducers (PB), was observed for rat kidney microsomes but not for liver or lung microsomes. For these latter preparations PB induced microsomes enhanced macromolecular binding more than 3-MC induced microsomal enzymes.

Thus it is clear that the nature and position of the halogen atom are critical factors in determining the metabolic reactivity of halogenated biphenyl substrates. The relative enhancement of the formation of covalent macromolecular-4-chlorobiphenyl adducts by cytochrome P-448 inducers is apparently unrelated to any changes in the site specificity of the hydroxylation reaction since 4'-chloro-4-biphenylol remains the predominant metabolite (>85%) irrespective of the *in vivo* pretreatments. It remains to be shown to what extent the relative enhancement of the *in vitro* formation of covalent macromolecular-4-chlorobiphenyl adducts by AHH inducers can be related to their potentiation of the toxicity of the various [halogenated] aryl hydrocarbons which contaminate our environment. It has been suggested that a correlation exists between high AHH induction, the presence in the environment of AHH inducers (such as polycyclic aryl hydrocarbons) and a high rate of spontaneous cancer in salamanders (*Ambystoma tigrinum*) (Busbee et al, 1978). Toxicity studies using inbred strains of mice in which the phenotypes "aromatic hydrocarbon responsiveness" and "non-responsiveness" have been predetermined suggest that the rate limiting step for the toxicity (mutagenicity or survival time) of benzo[a]pyrene, 7,12-dimethyldibenzo[a]anthracene, 3-methylcholanthrene, σ-aminochrysene, polychlorinated biphenyls, polybrominated biphenyls and 2-acetylaminofluorene is their activation by cytochrome P-448 (Felton and Nebert, 1975; Felton et al, 1976; Thorgeirsson and Nebert, 1977, and Nebert et al, 1977).

ACKNOWLEDGEMENTS

These studies were supported in part by the Research Programs Directorate Health and Welfare Canada, (606-1444-X), the National Cancer Institute (U.S.A.), DHEW, Grant Number 1 B01 CA21814-01, and the National Research Council of Canada.

REFERENCES

Billings, R.E., and R.E. McMahon: Mol. Pharmacol. 14 (1978) 145.

Busbee, D.L., J. Guyden, T. Kingston, F.L. Rose, and E.T. Cantrell: Cancer Letters 4 (1978) 61.

Crawford, A., and S. Safe: Res. Comm. Chem. Pathol. Pharmacol. 18 (1977) 59.

Creaven, P.J., and D.V. Parke: Biochem. Pharmacol. 7 (1966) 16.

Daly, J.W., D.M. Jerina, and B. Witkop: Experimentia 28 (1972) 1129.

Felton, J.S., and D.W. Nebert: J. Biol. Chem. 250 (1975) 6769.

Felton, J.S., D.W. Nebert, and S.S. Thorgeirsson: Mol. Pharmacol. 12 (1976) 225.

Gillette, J.R.: Isr. J. Chem. 14 (1975) 193.

Goldstein, J.A., P. Hickman, H. Bergman, J.D. McKineny, and M.P. Walker: Chem. Biol. Interact. 17 (1977) 69.

Jerina, D.M. and J.W. Daly: Science 185 (1974) 573.

Kohli, J., C. Wyndham, M. Smylie, and S. Safe: Biochem. Pharmacol. 27 (1978) 1245.

McPherson, F.J., J.W. Bridges, and D.V. Parke: Biochem. J. 154 (1976) 773.

Nebert, D.W., R.C. Levitt, M.M. Orlando, and J.S. Felton: Clin. Pharmacol. Therap. 22 (1977) 640.

Safe, S., O. Hutzinger, and D. Jones: J. Agric. Food Chem. 23 (1975) 851.

Shimada, T.: Bull. Environ. Contam. Toxicol. 16 (1976) 25.

Shimada, T. and R. Sato: Biochem. Pharmacol. 27 (1978) 585.

Sladek, N.E., and G.T. Mannering: Biochem. Biophys. Res. Commun. 24 (1976) 668.

Sundström, G., O. Hutzinger, and S. Safe: Chemosphere 5 (1976) 267.

Thorgeirsson, S.S. and D.W. Nebert: Adv. Cancer Res. 25 (1977) 149.

Weinstein, I.B.: Bull. N.Y. Acad. Med. 54 (1978) 366.

Wyndham, C., J. Devenish, and S. Safe: Res. Comm. Chem. Pahtol. Pharmacol. 15 (1976) 563.

Wyndham, C. and S. Safe: Biochem. 17 (1978) 208.

Wyndham, C. and S. Safe: Can. J. Biochem. In press. (1978).

TOXICITY OF HYDROCARBONS AND THEIR HALOGENATED DERIVATIVES IN AN AQUEOUS ENVIRONMENT

D.T. Boyles

Biological Sciences Branch, New Technology Division

BP Research Centre, Chertsey Road

Sunbury-on-Thames, Middlesex, TW16 7LN, England

The interactions between hydrocarbons and halogenated hydrocarbons and biological tissues have frequently been shown to be physical rather than chemical in nature, i.e., the substances cause their effect purely by their molecular presence in some biophase. Even where a breakdown or oxidation product of these substances is known to be the active or main toxicant, the physical properties of the original compound may determine transport from the external environment to the site of toxic action within the biological tissue. Ferguson (1939) showed that equitoxic concentrations of physically acting chemicals had similar thermodynamic activity, which at equilibrium would be uniform throughout the various biophases. In simple terms this means that toxicity of physically acting substances occurs at similar degrees of aqueous saturation - the lower the water solubility the lower the effective concentration. The concept of equitoxicity at constant activity has been confirmed since by investigators working in widely diverse areas of toxicology.

The absolute concentrations of physically acting substances reached in the susceptible biophase when toxicity occurs is difficult to determine experimentally and depends on the chemical composition and physical structure of the biophase. McGowan (1954) calculated the concentrations in the biophase of series of physically acting substances needed to achieve certain biological effects. For any given effect the calculated concentration was constant although external concentrations of active agent varied by many orders of magnitude. Seeman et al (1971) measured the concentration of aliphatic alcohol anaesthetics in erythrocyte membranes as 0.01 to 0.04 molar which was said to agree with partition theories for anaesthesia. Equilibrium conditions are presumed in these studies. However for substances of low water solubility and which are tested at concentrations comparable with aqueous saturation, it is difficult to establish that equilibrium exists and it may be rarely achieved in growing tissues. Diffusion, biodegradation and adsorption processes combined with low solubility may very much reduce the toxicity which many lipophilic compounds are capable of exerting.

Although chemical activity ($\equiv$ % saturation concentration) explains gross differences between absolute equitoxic concentrations of physically acting toxicants, certain factors may influence the measured potency of any substance. Potency has been defined by Crisp et al (1967) as the reciprocal of the thermodynamic activity of equitoxic solutions, and was said by these authors not to depend on polarity or water solubility but on volume and shape of toxic molecules. Small molecules with a high degree of symmetry were said to be the most potent, due to minimum work necessary to be done against the cohesive forces of the biophase molecules. However this effect is only apparent since the actual concentrations of η-aliphatic alcohols measured by Seeman in erythrocyte membranes for anaesthesia decreased with increasing chain length - molecular volume and free energy of binding were here said to be the most general indices of potency. These considerations became especially important when substances are applied to tissues at concentrations in water greater than saturation; in this case Ferguson's theory would predict equitoxicity which in many instances is not observed.

The possibility of devising general toxicity bioassays for use in environmental monitoring, process effluent control and in fundamental studies on toxicity with lipophilic substances and which employ lower forms of life gains strength from the above ideas. It is not surprising that the relative toxic activity of such compounds is similar amongst many forms of life which all possess lipophilic cell membranes and certain other apolar or amphiphilic biophases. Assays for a wide range of biological effects, e.g., acute toxic (respiration, photosynthesis inhibition, permeability changes), chronic toxic and mutagenic (both gene and chromosome), which hydrocarbons and halogenated hydrocarbons are known to cause, can employ lower forms of life, e.g. bacteria, yeasts, algae, green plants, and still be relevant to effects on e.g. fish, birds and mammals. Two such bioassay toxicity tests are presented in this paper together with data on their response to a number of hydrocarbons and derivatives. Both tests are relatively simple, reproducible, cheap and suitable for on line toxicity monitoring of effluents and other aqueous streams. In the first test, which is based on the loss of selective permeability of cell membranes to electrolytes, some physical relationships between the effects of different hydrocarbons applied to higher plants and microorganisms as liquids are demonstrated. The second test is a novel toxicity assay based on the reduction of the specific growth rate of fast growing bacteria. In this test the totality of cell biochemical processes is measured.

MECHANISM OF TOXICITY

The introductory comments regard relative toxicities of hydrocarbons and halogenated hydrocarbons and only suggest the phase in the cell which is the site of toxic action. No information is gained on the mechanism of toxicity. The physiological expression of the interaction between hydrocarbons, etc., and cells is very diverse and depends on the type and complexity of the organism. No unifying mechanism for all the effects has been proposed. At a cellular level effects on respiration, transpiration, photosynthesis, gene and chromosome mutation, membrane permeability and specific growth rates are known. With higher organisms a range of effects is known which include those of anaesthesia, carcinogenicity, synergist to other carcinogens and purely physical disruption of gaseous exchange processes.

The cell site susceptible to the effects of hydrocarbons, etc., is likely to be lipophilic, and in this respect the lipoprotein cell membranes have received most attention. Many physiological processes depend on the integrity of these membranes. Lipophilic substances are thought to either partition into or adsorb on membranes, causing distortion of their physical structure and changes in their properties of selective permeability. Many of the acutely toxic effects of hydrocarbons and their derivatives can be related to disruption of one or more of the cell membrane systems, e.g. plasmalemma and tonoplast membranes in effects on transpiration and cell electrolyte leakage, mitochondrial membranes in respiratory effects, grana membranes in photosynthesis inhibition. Although physico-chemical probes are now available for elucidating membrane structure, little has emerged on the precise changes caused to cell membranes by hydrocarbons and their halogenated derivatives.'

The mutagenic effects of hydrocarbons, etc., towards microorganisms and animals have not been explained by effects on membranes, although lipophilic sites of action are implied by Hansch and Fujita (1964) in an explanation of hydrocarbon carcinogenicity which again involved partitioning of hydrocarbon into the susceptible biophase. The curious contracting effect of lipophilic substances including hydrocarbons on plant chromosomes and inhibition of mitosis, reported by Ostergren (1944) presumably involve apolar reactions with subcellular moieties. Extensive studies have been done on the binding of hydrocarbons to certain proteins (e.g. Wishnia, 1969) but not in the context of hydrocarbon toxicity.

The anaesthetic effects of certain hydrocarbons and chlorinated hydrocarbons have been explained by the formation of microcrystalline hydrates which disrupt nerve responses (Pauling, 1962).

The physico-chemical relationships which determine toxicity of hydrocarbons, etc., will however be similar whichever mechanism is postulated. Clarification of the mechanism of the various hydrocarbon cell interactions awaits the introduction of new and more powerful probes of subcellular structure and composition.

CONDUCTANCE BIOASSAY

Hydrocarbons have been shown to affect the selective permeability of plant cell membranes (Currier, 1951; Van Overbeek and Blondeau, 1954). Electrolyte loss from plant tissues, caused by a variety of lipophilic substances in aqueous solution was studied by Stiles and Jorgensen (1917) who suggested that such substances caused membranes to lose selectively permeability according to first order kinetics. Measurement of the electrical conductance of low ionic media containing suspended tissues has been found to provide a simple, sensitive and reproducible test for the effects of hydrocarbons on cells. In the following experiments the effects of hydrocarbons at a thermodynamic activity of ~ 1 (standard state) are studied. Since an excess of hydrocarbon is used, the problems of low solubility, diffusion, adsorption, etc., already discussed are minimised. The acute or immediate effects on cell permeability are measured - incubation of tissues with these substances over extended periods may of course show a very different pattern of results.

Tests with Higher Plants

Plants were grown to a similar age (3 - 5 weeks) from seed in nutrient solution (Hewitt, 1966) under conditions of controlled temperature, light and relative humidity. 2 ml of test compound were pipetted on the abaxial surfaces of 5 g of plant leaves so as to cover them as evenly as possible. The leaves were immediately immersed in deionised water and the leakage of salts from the leaf cells inferred from changes in conductance across dip-type bright platinum electrodes inserted into the water. The suspension of leaves was maintained at a constant temperature of 25°C and conductance was monitored up to 6 - 7 hours after treatment. A final conductance value which was taken to represent the total electrolytes present in the leaves was taken after boiling the leaf mixtures. None of the compounds tested had an appreciable effect on the final conductance of the leaf suspensions. Where a close comparison between the effects of different compounds on one plant species was required the measurements were performed on the same batch of plants which included either untreated samples or samples treated with the relatively non-toxic n-hexadecane. Approximate proportionality between specific conductance and electrolytes present was assumed since the former was $\lesssim$ 1200 μ Siemens cm^{-1}.

The rate of electrolyte loss from cells of Helianthus annuus (sunflower) was found to depend on carbon chain-length of the applied liquid hydrocarbons (Figure 1). Although the change of conductance with time resembles model curves described by Stiles and Jorgensen from which a coefficient of membrane breakdown could be derived, their function did not adequately describe results obtained with the more toxic hydrocarbons (e.g. nC_6-nC_8). It was decided to express the results as a reciprocal time to lose half the electrolytes present in the cells, which is demonstrated in Figure 2 on a logarithmic scale for the effects of the n-alkanes on H. annuus and Daucus carota (carrot). The latter species, in accord with the known tolerance of members of the family Umbelliferae towards hydrocarbons, is seen to lose internal electrolytes at a much slower rate than the hydrocarbon susceptible species H. annuus. The logarithmic relationship between rate of salt loss from cells and carbon chain length of applied hydrocarbons is suggested (Boyles, 1976) to be related to the equilibrium concentration of hydrocarbon in the susceptible biophase, i.e., plasmalemma and tonoplast membranes. This concentration was thought to diminish logarithmically with

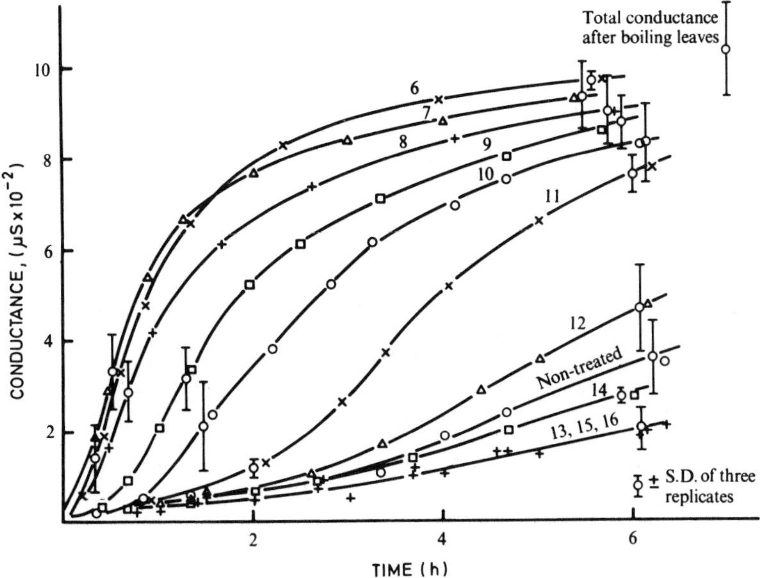

Figure 1. Conductance changes in suspensions of H. annuus leaves treated with n-alkanes C_6-C_{16}.

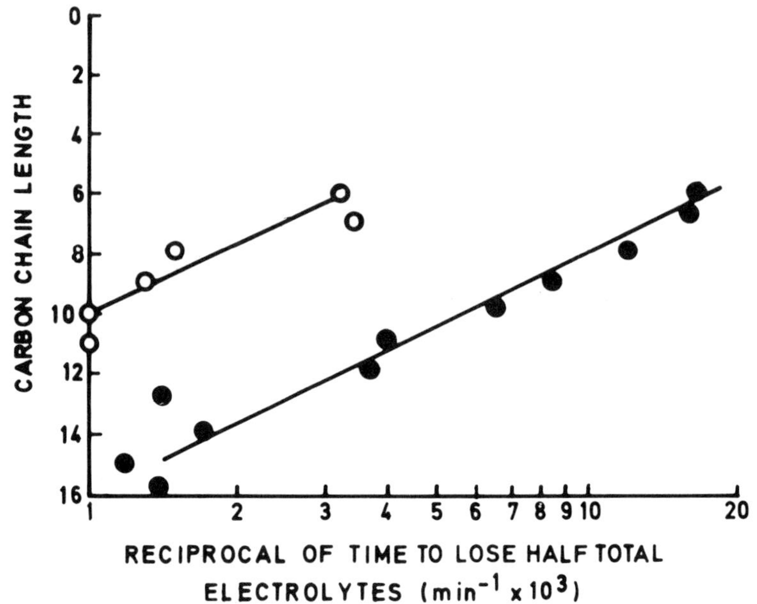

Figure 2. Relation between chain length and rate of loss of electrolytes from (•) H. annuus (o) D. carota leaves treated with n-alkanes.

chain lengths as do hydrocarbon solubilities in hydrophilic solvents. The difference between hydrocarbon susceptible species and hydrocarbon resistant species may be explained in the same way, i.e., partitioning of hydrocarbons into cell membranes is less favourable for oil resistant species. However the alternative explanation, that resistant plant membranes can better tolerate the presence of physically acting substances may not be ruled out.

A further demonstration of the physical nature of hydrocarbon toxicity is the effect of non-toxic diluents. In Figure 3 the rate of loss of electrolytes from leaves of H. annuus and D. carota treated with benzene/n-hexadecane mixtures is shown. Rate of salt loss is proportional to mole fraction of benzene in the mixture - presumably since mole fraction in an external phase would be paralleled by similar changes in the activity of benzene in the susceptible biophase.

The relative effects of hydrocarbons containing 10 carbon atoms but of different chemical type applied to tissues at concentrations > saturation, are shown in Table 1. The results illustrate a general phenomenon for hydrocarbons, that at any given activity in the susceptible biophase (here 1 or the standard state), the order of effect is n-alkane $\simeq$ iso-alkane < olefin < naphthene

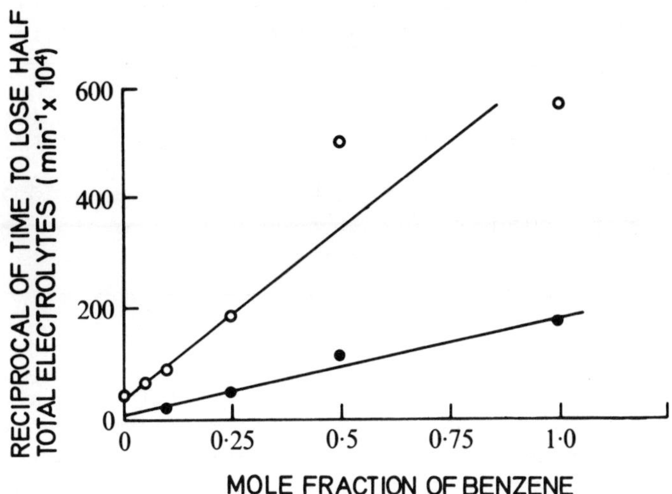

Figure 3. Rate of loss of electrolytes from (o) H. annuus and (•) D. carota leaves treated with benzene/n-hexadecane mixtures.

< aromatic. Substitution of polar moieties, e.g., -Cℓ, -COOH, causes significant increases in potency. These results confirm those of Crisp et al who found much lower thermodynamic activities at equitoxicity for naphthenic, chlorinated or aromatic hydrocarbons than for n-paraffins.

To demonstrate that the effects on electrolyte loss were not due to different rates of bulk movement of hydrocarbons to the susceptible biophase, polyisobutylene (non-toxic) was added (5% w/v) to n-decane to increase the viscosity from 1.17 centistokes to 5.47 (cf n-hexadecane = 3.09 cSt.) No difference was seen between the effects of "normal" and "viscous" n-decane on H. annuus.

Tests with Micro-organisms

With slight modification the conductance technique was also found applicable to micro-organisms. Fresh biomass from chemostat culture was suspended in deionised water (1 g/100 ml). Excess hydrocarbon (1 ml) was shaken vigorously into this suspension and the conductance monitored as with the leaves. In Figure 4 the effects of the n-alkanes and the n-alkylbenzenes on a hydrocarbon utilising yeast, Candida tropicalis, are shown. Besides the chain-length

Table 1 Rates at which Electrolytes were Lost from H. annuus Leaves Treated with C_{10} Hydrocarbons

Hydrocarbon	Boiling Point (°C)	Rate (min^{-1} x 10^4)
Isobutylbenzene	173	140
tert-Butylcyclohexane	171	87
n-Dec-1-ene	171	71
n-Decane	174	58
2,Methyl-3,ethylheptane	161	50
Untreated	-	16

effect, aromaticity was found to increase toxicity for a given molecular weight. C. tropicalis is able to assimilate n-alkanes $> \sim C_{10}$ as carbon and energy growth substrates. Lower boiling alkanes and unsaturated hydrocarbons can inhibit growth.

It was shown with the conductance method using micro-organisms that solid hydrocarbons diffuse into cells only very slowly compared with liquid hydrocarbons since 1,2-dichlorobenzene (a liquid) was

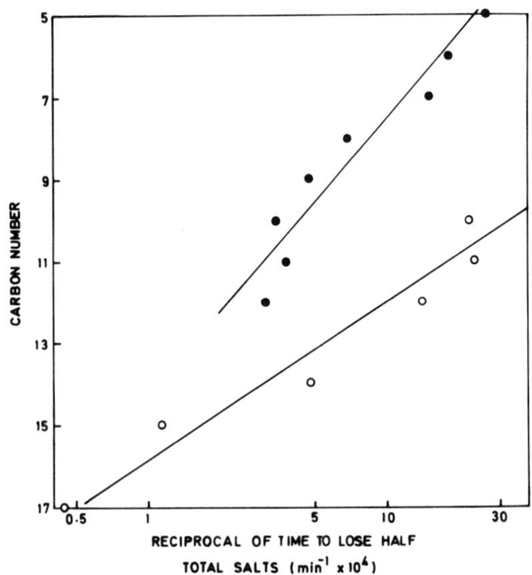

Figure 4. Relation between carbon number and rate of loss of electrolytes from C. tropicalis after treatment with (•) n-alkanes (o) n-alkylbenzenes.

one of the most active compounds tested with the technique (62% of internal electrolytes lost in 6 hours), whereas 1,4-dichlorobenzene (a solid) had only a very small effect (7.3% lost in 6 hours).

GROWTH RATE BIOASSAY

It had been found in work to find a general bioassay of effluent toxicity that microbial metabolic rates of respiration and photosynthesis were less sensitive to common pollutants (e.g., hydrocarbons, heavy metal ions, phenol, etc.) than the specific growth rate (μ) of micro-organisms. Changes in μ in the presence of toxicants (e.g. Bringmann, 1959) involves several days growth of organisms in shake flasks and calculation of growth by turbidity, etc. A simple but sensitive on-line assay of toxicity which would complement chemical analysis of industrial effluents and give warning of toxic build-up, sudden toxic changes, etc., was felt needed especially to protect microbial biofilters which are increasingly used to 'clean up' effluents. A novel rapid method, which employs the oxygen electrode chamber to measure the specific growth rates of fast growing bacteria in the presence of toxicants, was developed.

For an organism growing exponentially:

$$x = x_o e^{\mu t}$$

where x is cell weight or number, x_o cell weight or number at t = o, t is growth time and μ is the specific growth rate. During short periods of steady growth and where microbial yield factor on oxygen is constant:

$x \propto RO_2$, the respiration rate.

$$RO_2 = RO_2^o \, e^{\mu t}.$$

In the oxygen electrode chamber (Rank Brothers, Bottisham, Cambridge, U.K.), which is sealed against the atmosphere, the respiration rate RO_2 of suspensions of micro-organisms in nutrient is related to the dissolved oxygen DO_2 by:

$$RO_2 = \frac{d(DO_2)}{dt}$$

The growth rate of a fast growing culture is then given by:

$$\frac{1}{RO_2} \frac{d(RO_2)}{dt} = \frac{d^2(DO_2)}{dt^2} \cdot \frac{1}{d(DO_2)/dt} .$$

and may be obtained graphically, or by a regression on three groups of DO_2 v t data:

$$\mu = \frac{1}{n} \ln \frac{\sum_{n+1}^{2n} DO_2 - \sum_{2n+1}^{3n} DO_2}{\sum_{1}^{n} DO_2 - \sum_{n+1}^{2n} DO_2}$$

where n is the number of points in each group. The units of μ are the reciprocal time intervals between points in each group. Alternatively, DO_2, t values are accumulated automatically and the regressions $DO_2 = RO_2t + DO_2^{\circ}$ and $\ln RO_2 = \mu t + \ln RO_2^{\circ}$ performed by a programmable calculator with an interface (e.g. Hewlett Packard HP97S). RO_2 and μ are printed out at suitable intervals.

So that several generations of micro-organisms could be tested in a short time, fast growing organisms (generation times 10 - 20 minutes) were employed; the bacteria *Escherichia coli* for low ionic and *Vibrio natriegens* for saline solutions. Standard inocula of these organisms were taken from stock cultures in nutrient broth (Difco) or nutrient broth + NaCl. After dilution to $\sim 10^5$ organisms/ml, growth rates were determined for controls and in the pre-

Figure 5. Oxygen Electrode Chamber

sence of a range of toxic substances. The organisms were placed in the oxygen electrode chamger (Figure 5) which was maintained at 37°C by circulating water from a water bath. The e.m.f. generated by the electrode system, which was proportional to the percentage saturation dissolved oxygen, was recorded on a pen-recorder or logged with the HP97S calculator and the values RO_2 and μ derived during the growth period. The number of organisms was chosen to be sufficient to give a reasonable growth during a period of 30 - 40 minutes and within the limits imposed by the amount of dissolved oxygen available (7 - 8 ppm). The test is rapid, simple, gives reproducible results (if these are always taken as a ratio to non-treated controls) and may be completely automated. In Figure 6 the effects of toluene and 1,2-dichlorobenzene on the specific growth rate of E. coli and V. natriegens are compared. Again it is emphasised that only the immediate effects are measured. Extended incubation with toxicants may increase or decrease the effects due to diffusion, biodegradation, adaptation and selection, etc. If standardised the test may be used as a simple monitor for the toxicity of certain effluents, and for ranking and comparative studies on known and unknown toxic mixtures.

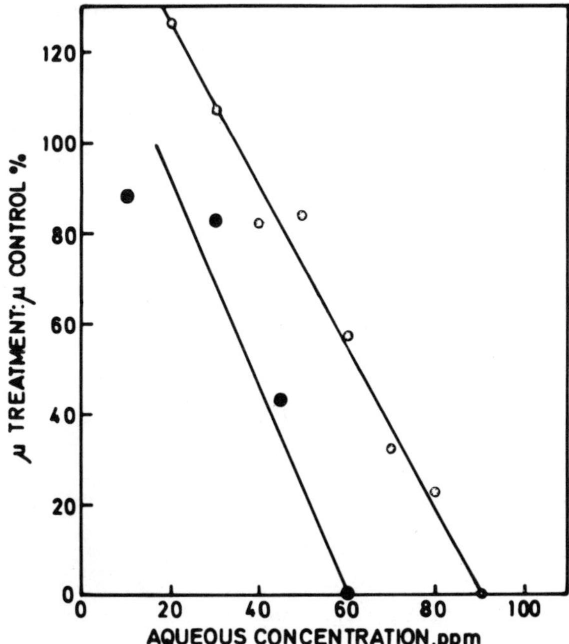

Figure 6. The effect on specific growth rate of toluene on E. coli (o) and 1,2-dichlorobenzene on V. natriegens (•) at 37 C.

CONCLUSIONS

The effects of liquid hydrocarbons and their chlorinated derivatives have been demonstrated in two bioassays, one based on a disruption of cell membranes and leakage of intracellular electrolytes from treated tissues, the other on the specific growth rate of fast growing bacteria. An account of the physical nature of the interaction between cells and hydrocarbons and other lipophilic substances has been illustrated with examples using the bioassays.

In any consideration of hydrocarbon toxicity in the environment which may occur as acute, chronic or mutagenic effects, a key factor is the chemical activity of the hydrocarbon or derivative. This activity is approximated by the percentage of aqueous saturation. Diffusion, biodegradation, adaptation, adsorption on to inert surfaces or non-susceptible biophases may all reduce the inherent toxicity of a compound. Where local concentrations of such substances in the environment exceed aqueous saturation, the substances at equilibrium are present in a susceptible biophase at the standard state - toxicity is then determined by factors such as molecular size and shape of the active agent. For hydrocarbons of similar carbon number the present work indicates an order of potency, aromatic > naphthene > olefin > isoalkane $\simeq$ n-alkane. Substitution of -COOH, -Cℓ increases potency. Differences in susceptibility between different biological species are due to differences in the chemical composition and physical structure of susceptible biophases which may affect either partitioning of hydrocarbon, etc., into the biophase or a differing tolerance of the biophase towards accumulated hydrocarbon or hydrocarbon derivatives.

ACKNOWLEDGEMENTS

Acknowledgement is made to the British Petroleum Company Limited for permission to publish this paper.

REFERENCES

Boyles, D.T. 1976. The loss of electrolytes from leaves treated with hydrocarbons and their derivatives. Ann. App. Biol. 83:113-103.

Bringmann, G. 1959. Vergleichende wasser-toxikologische Untersuchungen an Bakterien, Algen und Kleinkrebsen. Gesundheits-Ingenieur. 80:115-120.

Crisp, D.J., A.O. Christie, and A.F.A. Ghobashy. 1967. Narcotic and toxic action of organic compounds on barnacle larvae. Comp. Biochem. Physiol. 22:629-649.

Currier, H.B. 1951. Herbicidal properties of benzene and certain methyl derivatives. Hilgardia. 20:383-406.

Ferguson, J. 1939. The use of chemical potentials as indices of toxicity. Proc. Roy. Soc. Ser. B. 127:387-404.

Hansch, C., and T. Fujita. 1964. A method for the correlation of biological activity and chemical structure. J. Am. Chem. Soc. 86:1616-1626.

Hewitt, E.J. 1966. Sand and water culture methods used in the study of plant nutrition. 2nd ed. Farnham Royal: Commonwealth Agricultural Bureaux.

McGowan, J.C. 1954. The physical toxicity of chemicals. IV. Solubilities, partition coefficients and physical toxicities. J. App. Chem. 4:41-47.

Ostergren, G. 1944. Colchicine mitosis, chromosome contraction, narcosis and protein chain folding. Hereditas. 30:429-468.

Pauling, L. 1962. Une theorie de l'anesthesie generale. J. Chim. Phys. 59:1-8.

Seeman, P., S. Roth, and H. Scheider. 1971. The membrane concentrations of alcohol anaesthetics. Biochim. Biophys. Acta. 225:171-184.

Stiles, W., and I. Jorgensen. 1917. Studies in permeability. IV. The action of various organic substances on the permeability of the plant cell and its bearing on Czapek's theory of the plasma membrane. Ann. Bot. 3:47-76.

Van Overbeek, J., and R. Blondeau. 1954. Mode of action of phytotoxic oils. Weeds. 3:55-65.

Wishnia, A. 1969. Substrate specificity at the alkane binding sites of haemoglobin and myoglobin. Biochem. 8:5064-5075.

FISH TUMOR PATHOLOGY AND AROMATIC HYDROCARBON POLLUTION IN A GREAT LAKES ESTUARY

John J. Black, Margaret Holmes

Roswell Park Memorial Institute, Buffalo, NY 14263

Paul P. Dymerski, Division of Laboratories and Research

New York State Department of Health, Albany, NY 12201

William F. Zapisek, Department of Chemistry,

Canisius College, Buffalo, NY 14208

The relationship between polyaromatic hydrocarbon (PAH) contamination of the aquatic environment and its potential as a persistent health hazard for the associated biota has been the subject of recent research in this laboratory and others (Payne et al, 1978; Dunn and Stich, 1976). We report here studies of a waterway contaminated with industrial organic compounds, including PAH; these studies present evidence that the pollution is mutagenic and has a point source. In addition, there is a high incidence of fish tumor pathology in the immediate outflow area.

The presence of mutagens, fish tumor pathology and the strong correlation between mutagenicity and carcinogenicity (Commoner et al, 1976; McCann et al, 1975) is of concern to all who eat these fish and drink this water.

The Buffalo River enters Lake Erie near the head of the Niagara River. It is dredged throughout the lower 5.5 miles and is used for commercial navigation by several industries located on the river. The dredged portion of the river has a low gradient (17 cm/km) and low current velocities, and is occasionally subject to reversals in flow direction as a function of changes in Lake Erie water levels (EPA Report, 1975). In this respect the river has an estuarine character.

The river has a history of chemical and domestic sewage pollution and is considered among the most heavily polluted waters in the United States (Kraybill, 1976). The bulk of the flow from the river enters the Black Rock Canal near the beginning of the Niagara River. The canal is used to enable barges to get around the rapids in this part of the Niagara River.

Sediments from the Buffalo River were extracted with methanol and the residue that remained after evaporation was taken up in DMSO and tested for mutagenic potential using the Ames bacterial mutagenesis assay (Ames et al, 1975). Figure 1 shows the results of these assays and indicates a strong source of mutagenic activity in the vicinity of the South Park Avenue bridge (Station 3 in Figure 1). It is also of interest that a sediment sample obtained from the Black Rock Canal also exhibited a moderate degree of mutagenic activity. Subsequent examination of a modest sample of the fish population revealed a high percentage of tissue lesions to be present.

This combination of results prompted us to investigate the aromatic hydrocarbon contamination of this environment. Samples of bottom sediment and samples of fish were obtained from the Buffalo River and a nearby unpolluted stream, the Ishua Creek. The extractions were performed on dry, pulverized samples in a Soxhlet extractor using hexane as a solvent. The primary analytical tool was high performance liquid chromatography (HPLC). A Waters C_{18} reversed phase column was isocratically eluted with an acetonitrile-water (60:40) mixture and detected by an ultra-violet absorbance monitor at 254 nm. (Black et al, in press).

These studies revealed a complex of U.V. absorbing materials to be present in the hexane extracts obtained from the sediment samples (see Figure 2). The properties that these substances exhibit in their absorbance of U.V. light, in reverse phase chromatography and in solvent extraction are consistent with their identification as PAH and their derivatives. Indeed, we have demonstrated that these samples do contain PAH by gas chromatography coupled with mass spectroscopy (GC/MA). Pure samples of PAH known to be present by GC/MS have been co-chromatographed with the sediment samples on HPLC and the retention times of eight (8) PAH have thus been identified including the carcinogens benz(a)anthracene and benzopyrene. To date we have demonstrated the presence of over 20 PAH by GC/MS and confirmed these identifications through analysis of retention times in two different GC systems.

As seen in Figure 2, the concentration of the PAH in the HPLC chromatograms declined downstream from the major source of the Ames test positive materials. Further studies on samples taken every fifty feet in the vicinity of the highest degree of Ames test response (Station 5 in Figure 1) show that the peak of PAH contamination

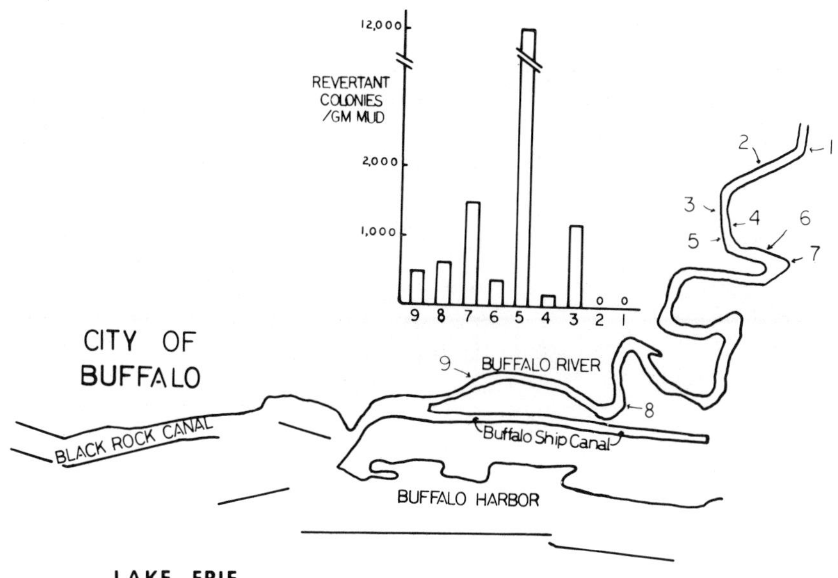

Figure 1. Distribution of Ames test control sediments in the Buffalo River.
1 & 2 above chemical plant
3 adjacent to plant, but upstream from wastewater and flow
4 & 6 opposite shore from chemical plant
5 at wastewater outflow from chemical plant
7 mid-river ½ mile below chemical plant
8 mid-river 2½ " " " "
9 mid river 3½ " " " "

is precisely adjacent to the waste-water outflow of a local dye industry.

When fractions obtained from sediment, fish, and invertebrates from this area were compared, there were obvious similarities, indicating that contaminants found in the sediment were being taken up by the biota (Figure 3).

When the chromatogram of the hexane extract of bottom sediment from the Buffalo River (BRS) shown in Figure 2 was compared to a chromatogram of a hexane extract of Buffalo River carp (BRC), (Figure 3), many points of similarity were noted. It was also obvious that there were many interfering substances that were not common to the two samples. The complexion of local industry led us to suspect that aromatic compounds with basic characteristics (e.g., aniline dyes) were involved, and so we developed a clean-up procedure on this premise. Extraction of BRS from hexane into aqueous acid removes a number of peaks from the BRS chromatogram

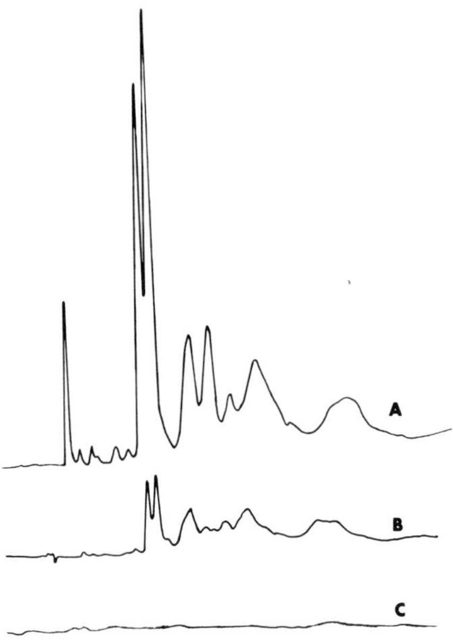

Figure 2. HPLC chromatograms of PAH containing fractions derived from Buffalo River sediment from samples taken at (A), ½ mile downstream (B), and 1 mile upstream (C) from the chemical plant respectively, B and C are concentrated 2x relative to A.

(polar fraction) and leaves the majority of the materials in the hexane (non-polar fraction). The polar fraction can be recovered from the aqueous acid by alkalinizing and reextracting the material into hexane. A chromatogram of such material is shown in Figure 4A.

The same procedure was used to clean up the hexane extracts of Buffalo River carp (BRC), (Figure 4B). A comparison of the chromatograms of the polar fractions of BRS and BRC shows many points of similarity. To test these similarities, we chromatographed a 1:1 mixture of BRS and BRC. The results are seen in Figure 4C. If a peak is unique to one of the samples, it will be diluted to 50% by the other sample. If it is contained in equal quantities in each sample, it will not be diluted, and if it is present in each sample, but in different quantities, it will be averaged and the dilution will be less than 50%. We conclude that the peaks marked with a ▼ in Figure 4 are contained in both fish and sediment based on the criteria of hexane extraction, acid solubility, and co-chromatography. Work is currently in progress to identify the components of the acid-soluble fraction by mass spectrometry and to determine the mutagenicity of each component through the Ames test and by literature search.

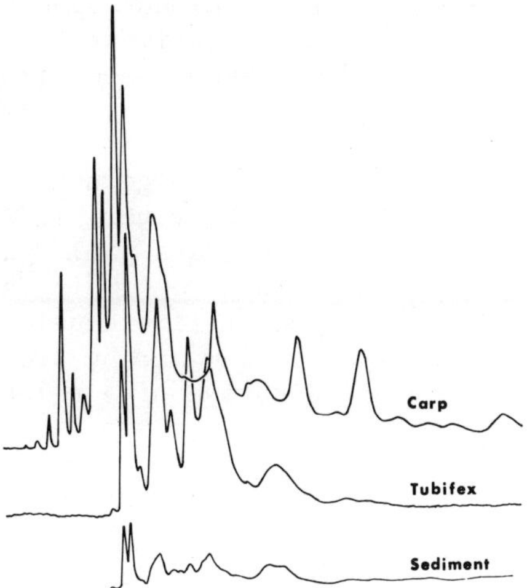

Figure 3. HPLC chromatograms of PAH containing fractions from carp, tubifex and sediment. Relative to the sediment sample,the carp and tubifex samples are concentrated 20 and 6 fold, respectively.

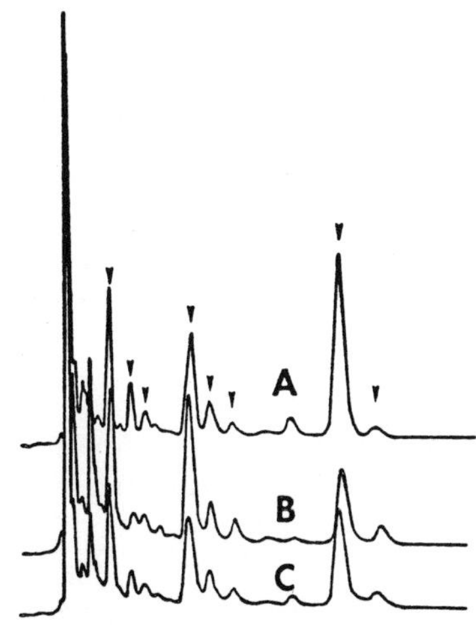

Figure 4. HPLC chromatograms of the polar fraction of (A) Buffalo River sediment from a locale adjacent to chemical industry, (B) Buffalo River carp, (C) a 1:1 mixture of A and B.

This polar fraction from the sediment does not originate from normal bottom sediment or from sewage. Although the polar fraction is strongly present in the bottom sediment opposite a likely industrial source, it is absent from the sediment one mile upstream from that source, and noticeably diminished one-half mile downstream from that source.

While it is difficult to obtain large numbers of fish from the most severely polluted portion of the river, it was possible to collect modest numbers of fish from less polluted portions of the river and from one part of the Black Rock Ship Canal. Because of the shape of the canal at this point (the canal makes approximately a 30 degree turn), there is a shallow marshy area that has a biological community structure indicative of a relatively "clean" environment.

Pathologic examination of a modest number of fish from these two areas revealed a high incidence of proliferative tissue lesions to be present among several bottom-feeding species of fish.

Approximately 35% of goldfish x carp hybrids carried abnormal appearing gonads. Sonstegard has diagnosed a high frequency of gonadal tumors in goldfish hybrids as being of sertoli cell origin (Sonstegard, 1977). Sheepshead (_Aplodinotes grunniens_) are frequently affected (20-25%) with a dermal lesion that appears as a neurolemmoma with H & E stain. Papillomas have been observed in a few white suckers (_Catostomus commensoni_). It may be significant that all of the papillomas were located on the lips. This tumor displays an anatomic shift from a general body distribution to one exclusively on the lips in several polluted areas of the Great Lakes (Sonstegard, ibid). Only a few brown bullheads (_Ictalurus nebulosus_) have been examined but 2 out of 9 displayed dermal tumors. One of those growths was highly malignant and had invaded from the dorsum of the upper jaw through the palate.

While the evidence is circumstantial, it appears that if the pollution is low enough to avoid acute toxicity, chronic exposure to a complex mixture of PAH pollutants will cause several species of bottom-feeding fish to develop neoplasia. This association of environmental pollution and the generation of fish tumors has been previously suggested by Dunn and Stich (1976). That the sediments contain potential carcinogens as evidenced by both HPLC data and Ames test results and that fish can produce other mutagenic metabolites from PAH (Payne et al, 1978) supports this hypothesis.

A variety of sources have been demonstrated for mutagenic and carcinogenic materials in the aquatic environment (Payne et al, 1978). Some of these sources are known to be upstream from the area under study but the distribution of PAH and mutagens in the bottom sediment (Figure 1 and Figure 2) precludes an upstream source, because of the absence of PAH or mutagenesis upstream from the point source. In addition, the maintenance of a navigable ship canal by regular dredging, suggests a very recent origin for the sediment.

Although we have no direct evidence that any of the materials found in the carp are mutagenic, it is tempting to speculate whether a correlation can be made concerning the finding of acid extractable U.V. absorbing materials only in fish from the locale where such material (PAH) is also found in the bottom sediment, the high degree of Ames test revertants from materials extracted from that bottom sediment and the incidence of fish tumor pathology in the Buffalo River and its outflow area.

Our work in progress is directed toward the identification of those materials being accumulated by the biota and toward the development of a quantitative model for the flow of mutagenic materials through this ecosystem and its impact on the biota.

REFERENCES

Ames, B.N., et al. 1975. Mutation Research 31: 347-364.

Black, J., et al. Bull. of Env. Cont. and Tox. (in press).

Commoner, B., et al. 1976. EPA Report 600/1-76-022 Gov. Print. Off., Wash., D.C.

Dunn, B., and H.F. Stich, 1976. J. Fisheries Research Board of Canada, 33: 2040-2046.

Environmental Protection Agency, Report No. 905/9-74-100.

Kraybill, H.F. 1976. Prog. Exp. Tumor Res. 20: 34.

McCann. J., et al. 1975. PNAS 72: 5135.

Payne, J.F., et al. 1978. Science 200: 329-330.

Sonstegard, R.A. 1977. Ann. N.Y. Acad. Sci. 298: 261-269.

BIOACCUMULATION OF ARENE AND ORGANOCHLORINE POLLUTANTS BY *CIPANGOPALUDINA CHINENSIS* (GRAY, 1834) (MOLLUSCA: GASTROPODA) FROM PONDS IN THE ROYAL BOTANICAL GARDENS, HAMILTON, ONTARIO

L. Kalas, A. Mudroch and F.I. Onuska

National Water Research Institute, Canada Centre for Inland Waters, Burlington, Ontario, Canada L7R 4A6

Concentrations of polynuclear aromatic hydrocarbons (PAHs) and polychlorinated biphenyls (PCBs) were determined in cyclohexane and iso-octane-benzene extracts of soft tissues of *Cipangopaludina chinensis* (Gastropoda: Prosobranchia) by gas chromatography and mass spectrometry. Polynuclear aromatic hydrocarbons were also determined in cyclohexane extract of bottom sediment from which molluscs were sampled.

The marshes of the Royal Botanical Gardens, in Hamilton, from which the sediment sample and molluscs were collected, are located on the northern shore of Hamilton Bay of Lake Ontario, in the vicinity of major highways connecting Toronto, Burlington and Hamilton. The area is of a highly populated residential character. Marshes are fed by Hidden Valley Creek, which is of medium size. The concentrations of major elements in water are as follows: Ca^{+2}, 86.0 mg/l; Mg^{+2}, 20.4 mg/l; Na^{+}, 17.0 mg/l; K^{+}, 3.7 mg/l; SO_4^{-2}, 6.30 mg/l; Cl^{-}, 35.0 mg/l; alkalinity 86.7 mg/l. The water is typical of Ca-HCO_3 waters. Marsh substrate is composed of soft, fine silty clay with some sand and constant major elements in the following concentrations: 58.6%, SiO_2; 9.4%, Al_2O_3; 5.7%, Fe_2O_3; 3.0%, MgO; 8.5%, CaO; 1.2%, Na_2O; and 3.0%, K_2O.

Because of the large amount of *Cipangopaludina chinensis* shells in these marshes, it seems that the species population was well established.

Cipangopaludina chinensis (Fig. 1) is a large viviparous gastropod. Adult specimens reach a height of 5 cm and weigh about 45 g (wet weight), which comprises 30 g of aragonitic shell

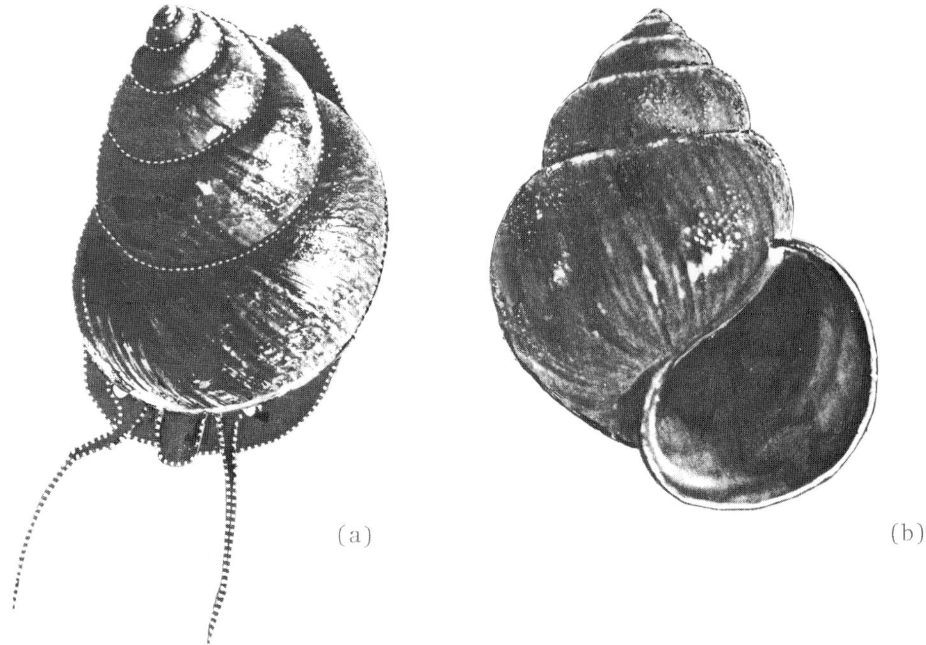

Fig. 1. *Cipangopaludina chinensis*, crawling animal(a) and shell(b).

and 15 g of soft parts. The soft parts contain about 90% water.

Cipangopaludina chinensis is widely spread in Asia, on both the mainland and islands, and in North America (Fig. 2).

In China and Japan, the soft parts of this snail are commonly used as food. During the last century, living specimens of *Cipangopaludina chinensis* were commercially available in most large cities in the USA and Canada, and it is believed that they were unintentionally released into the environment. Today, the species occurs in highly eutrophic freshwater marshes in the vicinity of North American cities such as Vancouver, San Francisco, Philadelphia, Cleveland, New York and Montreal (Pennak, 1953; Perron and Probert, 1973; Plinski, Ławacz, Stańczykowska and Magnin, 1978). Some specimens were also found in rivers (Grand River, southern Ontario, Canada) and in larger quantities in Sandusky Bay of Lake Erie, where they become a nuisance to fishermen (Wolfert and Hiltunen, 1968). Based on the mode of occurrence of the species, it was concluded that it is widely pollution-tolerant.

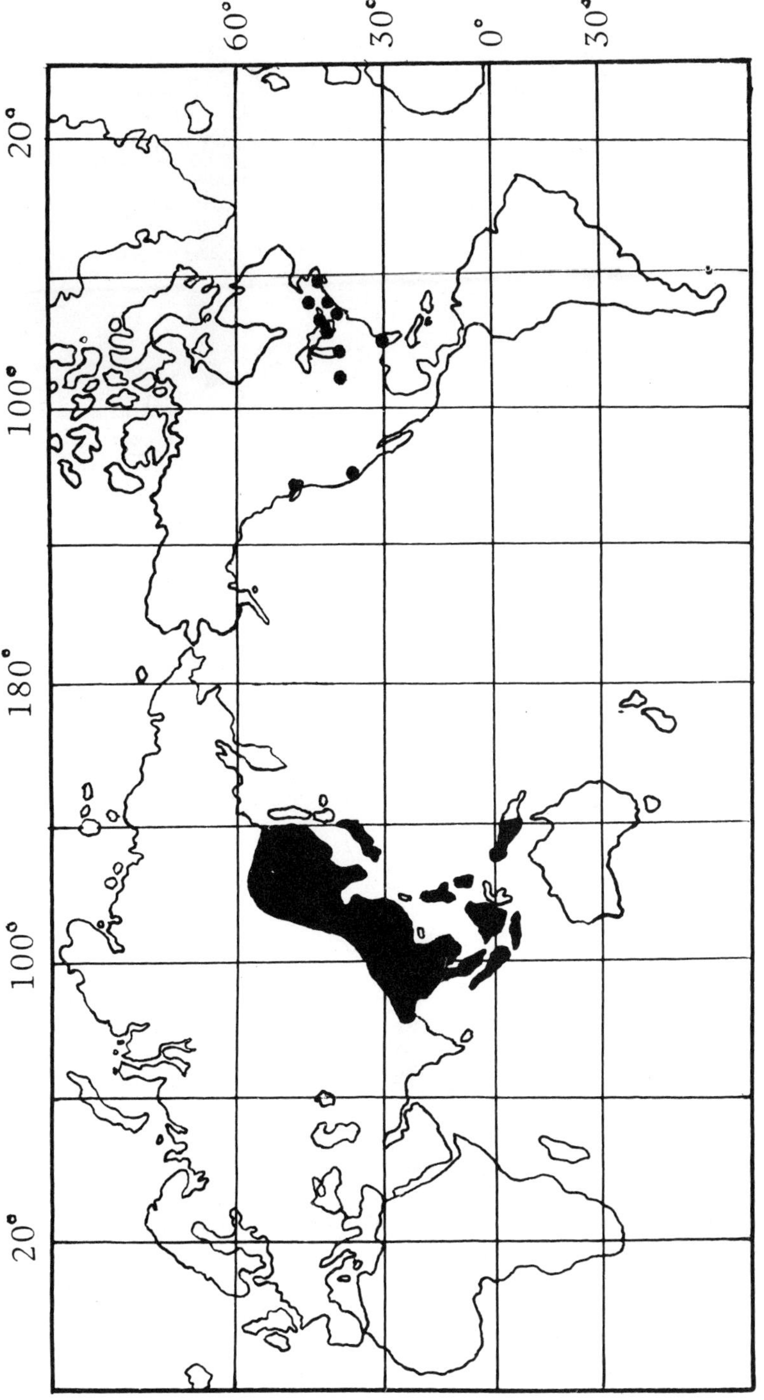

Fig. 2. A map showing geographic distribution of *Cipangopaludina chinensis* in Asia (▬) and North America (●).

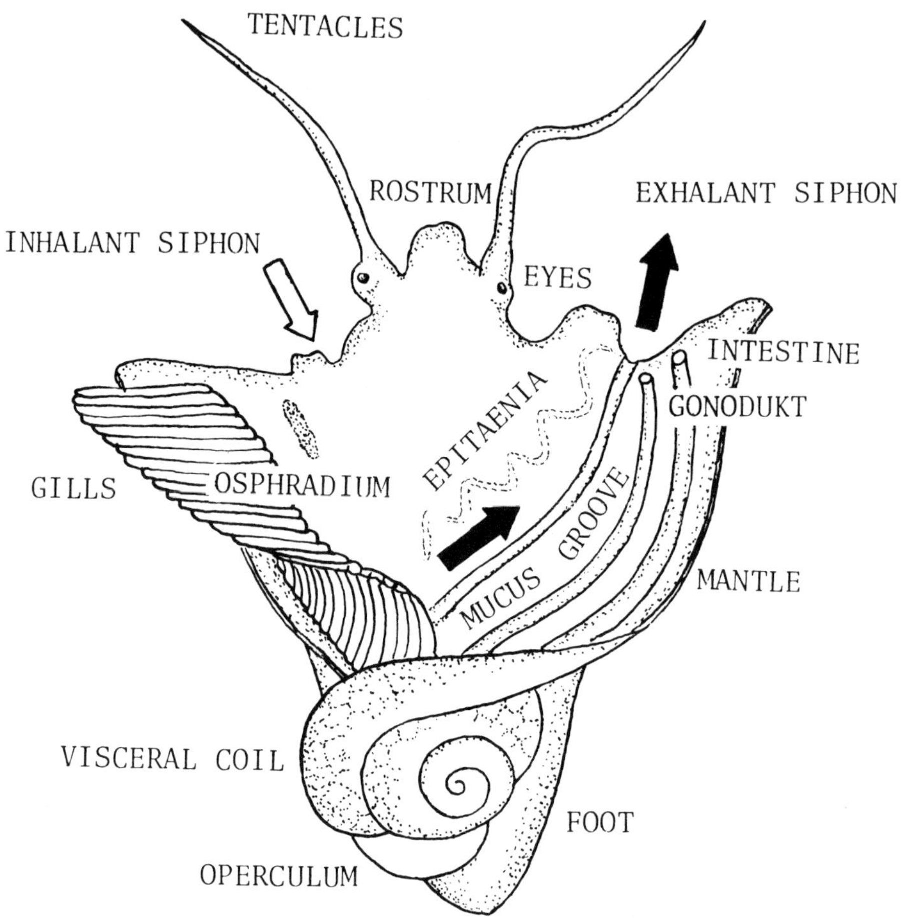

Fig. 3. Cipangopaludina chinensis, a dorsal view of soft parts, showing schematically gills in open mantle cavity and mucus carrying respiratory current (➡).

The mobility of the species in comparison to other freshwater gastropods is rather low. This affects its nutritional and feeding habits. Cipangopaludina chinensis is equipped with gills (Fig. 3) which serve multi-various functions: exchange of gases, cleansing, sorting and collecting of fine particles from the water. These fine particles (seston) are transported with the help of respiratory current and mucus string from the gill cavity out of and then into the animal's mouth, and swallowed. In addition to this mucociliary feeding, other feeding mechanisms were observed, such as scraping and grazing of bottom detritus and scavenging the bottom sediment particles with attached algal growth.

Sexes in Cipangopaludina chinensis are separated. The female produces as many as 40 to 50 offsprings, which are retained in the female's uterus until the crawling stage. The population is subject to predating. Most common predators are racoons, muskrats, waterfowl and, most likely, fish species. Therefore, the species represents an important component of the food chain.

The species has the longest life span (6-15 years, Hyman, 1967) of all the members of freshwater gastropods found in North America. During the winter hibernation, it is buried in pond or river bottom sediments. The period of retardation affects the shell growth and the utilization of fat stored mainly in the mid-gut gland (liver) occupying the greater part of its visceral mass. The fat depleted during starvation may appear in the midgut gland almost immediately after resuming feeding (Rosen, 1932). This fact is regarded as very important in assessing cyclic patterns of uptake, bioaccumulation and release of the pollutants.

The uptake, bioaccumulation and release of the pollutants by molluscs are interrelated processes which also depend on the chemical nature and ambient levels of pollutants in the surrounding environment.

The water sample from the collection site of Cipangopaludina chinensis in the Royal Botanical Gardens, Hamilton was collected into a 1 ℓ glass bottle. The total quantity of PCBs detected in the water was less than 0.02 μg/l (Table 1). A sediment sample was collected at the same locality using a plastic tubing (6.6 cm Ø) which was hand-driven into the sediment. The top 7 cm of this core was separated into a glass jar and freeze-dried. Prior to analysis, the sample was ground to 100 mesh (149 μ) size. The ground sediment was then extracted by cyclohexane and the extract analysed for PAHs. The total amount of PAHs in the sediment was 23.2 μg/kg dry weight. With the exception of chrysene and me-pyrene, the concentations of PAHs with higher molecular weight were under the detection limit (0.01 μg/kg dry weight).

Four adult specimens of Cipangopaludina chinensis were selected for analysis. The soft parts were isolated from the shell after freezing and thawing and the material was freeze-dried and further hand-ground prior to analysis. Part of the sample was extracted by cyclohexane for PAHs' determination and part by iso-octane-benzene for PCBs' determination.

Table 1. Concentrations in Water

Compound	µg/l
Lindane	0.001
Heptachlor	n.d.*
Aldrin	n.d.
Heptachlor epox.	n.d.
p,p -DDE	<0.001
Dieldrin	0.001
p,p -DDD	n.d.
p,p -DDT	<0.001
o,p -DDT	n.d.
Endrin	n.d.
α-Chlordane	<0.001
γ-Chlordane	<0.001
α-Endosulfan	n.d.
β-Endosulfan	n.d.
p,p -Methoxychlor	n.d.
Total PCB	<0.02
α-BHC	n.d.

* n.d. = not detected.

The PAHs' concentration amounted to a total of 1,497.6 µg/kg dry weight of molluscs' soft parts. Concentrations of naphthalene, me-naphthalene, dimethyl naphthalene, and biphenyl were under the detection limit (0.01 µg/kg). These compounds were detected in sediments (Table 2). Concentrations of benzo(a) fluorene, benzo(b) fluorene and chrysene were highest from all of the PAHs determined. However, their concentrations in the sediment were under the detection limit. Generally, with the exception of naphthalene, me-naphthalene, dimethyl-naphthalene, biphenyl and me-pyrene, concentrations of PAHs were from 15 µg to over 200 times higher in the soft parts of *Cipangopaludina chinensis* than those in the sediment. The largest differences were found in the concentrations of benzo(a) fluorene and benzo(b) fluorene (Table 2). The total concentration of PCBs in the soft parts of *Cipangopaludina chinensis* was 600 µg/kg dry weight.

Generally, the aquatic environment receives chronic PAH inputs from various sources. Anthropogenic combustion and natural fires, both forest and prairies, are the most common sources (National Academy of Science, 1972). Blumer *et al.* (1974) found large PAH values outside a town near the highways by studying the concentrations of PAHs in soils in the vicinity of a Swiss

Table 2. Concentrations of Polynuclear Aromatic Hydrocarbons and Polychlorinated Biphenyls in Sediment and *Cipangopaludina* sp.

PAH	Sediment µg/kg dry weight	*Cipangopaludina* µg/kg dry weight
Naphthalene	2.00	<0.01
Me-Naphthalene	0.50	<0.01
Dimethyl Naphthalene	1.00	<0.01
Biphenyl	6.00	<0.01
Me-Biphenyl	0.50	16.00
Anthracene-Phenanthrene	3.00	134.80
Me-Anthracene-Phenanthrene	0.70	60.80
Fluoranthrene	3.00	97.50
Pyrene	2.50	37.00
Azapyrene	<0.01	71.40
Benzo(a)fluorene	<0.01	216.00
Benzo(b)fluorene	<0.01	217.50
Me-Pyrene	1.00	<0.01
Binaphthyl	<0.01	181.90
Chrysene	3.00	302.80
Benzo(a)pyrene	<0.01	25.10
Perylene	<0.01	25.90
Coronene	<0.01	110.90
Total	23.20	1497.60
PCBs (total)	not determined	600.00

mountain town. Blumer *et al.* (1977) and Mair (1964) showed that specific PAHs are produced by geological diagenesis from certain pigments and terpenoids. Recently, it was speculated that some of the PAHs could be synthesized by certain bacteria (Brison, 1969; Niaussat *et al.*, 1970), algae (Borneff *et al.*, 1968) or plants (Hancock *et al.*, 1970). Hase and Hites (1976) found that some bacteria accumulate but do not synthesize PAHs.

Laflamme and Hites (1978) suggested that the distribution of PAHs in soils and sediments could be modified by natural mechanisms, including differential water solubility. The concentrations of some PAHs in sediments and soils vary widely. Laflamme and Hites (1978) found concentrations of pyrene in sediments from Charles River, Mass., and in sediments from the Amazon River, 13,000 µg/kg and 3-6 µg/kg dry weight, respectively. Strosher and Hodgson (1973) found concentrations of PAHs in the Great Lakes' sediments ranging between 0.02-54 µg/kg. Zitko (1965) summarized concentrations of aromatic hydrocarbons in marine fauna and found concentration ranges for clams *Mytilus edulis* and *Mya arenaria* between 6.67 to 39.90 μg/kg dry weight. These values of PAHs may be exceeded many times in clams affected by oil pollution (Onuska, 1975, 1976).

Dunn and Stich (1976) found that benzo(a) pyrene from environmentally contaminated mussels is released on transfer from a polluted area to tanks with clean circulating water. On the other hand, Boehm and Quinn (1977) proved that chronically accumulated hydrocarbons in the clams' lipid pool are strongly retained in the case of the marine clam (*Mercenaria* sp.). Clams were only very slowly and uncompletely depurated.

Concentrations of PAHs and PCBs found in *Cipangopaludina* are only informative. However, they suggest the suitability of this species and other branchiate molluscs for further studies on bioaccumulation of these pollutants from the freshwater environment.

REFERENCES

Blumer, M., W. Blumer and T. Reich, 1977. Polycyclic aromatic hydrocarbons in soils of a mountain valley: Correlation with highway traffic and cancer incidence. Environmental Science & Technology 11: 1082-1084.

Boehm, P.D. and J.G. Quinn, 1977. The persistence of chronically accumulated hydrocarbons in the hard shell clam *Mercenaria mercenaria*. Marine Biology 44: 227-233.

Borneff, J., F. Selenka, H. Kunte and A. Maximos, 1968. Experiental studies on the formation of polycyclic aromatic hydrocarbons in plants. Environ. Res. 2: 22-29.

Brisou, J., 1969. Benzo(a) pyrene biosynthesis and anaerobiosis. C.R. Soc. Biol. 163: 722-724.

Dunn, P.B. and H.F. Stich, 1976. Release of the carcinogen benzo(a) pyrene from environmentally contaminated mussels. Bulletin of Environmental Contamination and Technology 15: 398-401.

Hancock, J.L., H.G. Applegate and J.D. Dodd, 1970. Polynuclear aromatic hydrocarbons on leaves. Atmos. Environ. 4: 363-370.

Hase, A. and R.A. Hites, 1976. On the origin of polycyclic aromatic hydrocarbons in the aqueous environment. In: Identification and Analysis of Organic Pollutants in Water (ed. L.H. Keith).

Hase, A. and R.A. Hites, 1976. On the origin of polycyclic aromatic hydrocarbons in recent sediments: Biosynthesis by anaerobic bacteria. Geochim. Cosmochim. Acta 40: 1142-1143.

Hyman, L.H., 1967. The invertebrates: Mollusca, 792 p.

Laflamme, R.E. and R.H. Hites, 1978. The global distribution of polycyclic aromatic hydrocarbons in recent sediments. Geochim. Cosmochim. Acta 42: 289-303.

Mair, B.J., 1964. Terpenoids, fatty acids and alcohols as source materials for petroleum hydrocarbons. Geochim. Cosmochim. Acta 28: 1301-1321.

National Academy of Sciences, 1972. Particulate polycyclic organic matter.

Niaussat, P., C. Auger and L. Mallet, 1970. Appearance of carcinogenic hydrocarbons in pure *Bacillus badius* cultures relative to the presence of certain compounds in the medium. C.R. Acad. Sci. (Paris) 270D: 1045-1047.

Onuska, F.I., A.W. Wolkoff, M.E. Comba, R.H. Larose, H. Bravo, S. Salazar, L.A.V. Botelo, E. Mandelli, M. Novotny and M.L. Lee, 1975. The determination and identification of polyaromatic hydrocarbons in oysters from the Gulf of Mexico. Restricted report. Canada Centre for Inland Waters, 27 p.

Onuska, R.K., A.W. Wolkoff, M.E. Comba, R.H. Larose, M. Novotny and M.L. Lee, 1976. Gas chromatographic analysis of polynuclear aromatic hydrocarbons in shellfish on short well-coated glass capillary columns. Analytical Letters 9(5): 451-460.

Pannak, R.W. 1953. Freshwater invertebrates of the United States, 769 p.

Perron, F. and T. Probert, 1973. *Viriparus malleatus* in New Hampshire. The Nautilus: 87(3): 90.

Plinski, M., W. Lawacz, A. Stanczykowska and E. Magnin, 1978. Etude quantitative et qualitative de la nourriture des *Viviparus malleatus* (Reeve) (Gastropoda, Prosobranchia) dans deux lacs de la région de Montréal. Can. J. Zool. 56: 272-279.

Rosen, B., 1932. Zur Verdaunungsphysiologie der Gasteropoden. Zool. Bigrad 14.

Wolfert, D.R. and J.K. Hiltunen, 1968. Distribution and abundance of the Japanese snail *Viviparus japonicus* and associated macrobenthos in Sandusky Bay, Ohio. The Ohio Journal of Science, 68(1): 32.

Zitko, V., 1975. Aromatic hydrocarbons in aquatic fauna. Bull. Env. Contam. Toxicol. 14: 621-631.

THE CORRELATION OF THE TOXICITY TO ALGAE OF HYDROCARBONS AND HALOGENATED HYDROCARBONS WITH THEIR PHYSICAL-CHEMICAL PROPERTIES

T.C. Hutchinson, J.A. Hellebust, D. Tam, D. Mackay,

R.A. Mascarenhas, and W.Y. Shiu

Institute for Environmental Studies

University of Toronto, Toronto, Ontario, M5S 1A4, Canada

INTRODUCTION

There are strong incentives to develop methods of predicting the toxicity of environmental contaminants to specific biota. Consideration of the large number of contaminants, the wide variety of biota (many of which may change in susceptibility at different life stages, temperatures and other conditions), the range of possible concentration and exposure times, and the possibility of simultaneous stress from several contaminants suggests that it will be impossible to undertake experimental study of all the combinations of contaminant, biota and conditions which apply in the environment. Economies of effort can be achieved if methods can be developed to calculate the toxic effect from data such as the physical-chemical properties of the contaminant. A suitable approach would be to determine toxicity for some members of a chemically similar series of compounds (such as aromatic hydrocarbons) to a specific organism, correlate the toxicity with physical-chemical properties then infer the toxicity of other untested compounds from a knowledge of their physical-chemical properties.

Fortunately, physical-chemical data are usually more easily and accurately measured than toxicity data, thus it is attractive to assemble the appropriate data, and devise predictive correlations for these data where experimental values are unavailable. Such correlations are invaluable in exposing "bad" data and in treating situations in which the toxicant consists of a large number of individual compounds. Examples are PNAs and PCBs.

In this study, this hypothesis has been tested by examining the effect of 38 hydrocarbons and halogenated hydrocarbons on the photosynthesis of two green algae in aqueous solution under uniform conditions. The first part is a discussion of the selection, measurement and correlation of the physical-chemical properties of these compounds and the second a description of the algal studies and the correlation of toxicity with physical-chemical properties.

Physical-Chemical Properties

Assuming that the mechanism of toxicity is that the toxic molecule diffuses from solution in the water phase into the cell membrane where it has some disruptive effect, it is apparent that the most relevant properties are likely to be aqueous solubility (which controls the amount of toxicant in solution and also the magnitude of diffusion concentration gradients which may be established), lipid-water partition coefficient (which characterises the thermodynamic driving force for transfer from water to lipid phase) and diffusivity (which influences the rate at which the molecule may move in stagnant diffusive layers near the cell wall. Other properties which may be worth considering are vapor pressure (of the liquid, solid, or subcooled liquid, depending on the test temperature and the triple point temperature) and the Henry's Law constant (which is essentially the ratio of vapor pressure to solubility). Unfortunately, it is impractical to measure the actual lipid-water partition coefficient, thus in this work we have used n-octanol and glyceryltrioleate to simulate the lipid.

Aqueous solubilities of some 30 polynuclear aromatic hydrocarbons have been measured using standard techniques (Mackay and Shiu, 1). Similar techniques were used to determine solubilities of other hydrocarbons and halogenated hydrocarbons.

A novel technique (Mascarenhas, 2) was devised to measure octanol-water partition coefficients by headspace analyses. Essentially, it involves equilibrating air and water containing a subsaturated solution of the solute contaminant, sampling a known volume of the air phase and measuring peak height on a GC. This is repeated several times, then a quantity of octanol is introduced, sufficient to reduce the aqueous air concentrations to about half. This quantity is small since the partition coefficient is high. Repeated air samples are then taken following equilibration. The technique is interesting in that comparison of the peak heights from the air samples before and after octanol introduction enables the organic-water partition coefficient to be determined without actual determination of a concentration in either liquid phase, nor is absolute calibration of the GC required, only relative areas are used and the only assurance required is that the concentration-peak area relationship is linear. The system was tested with glyceryltrioleate with less success due to its high viscosity.

No diffusivity or vapor pressure measurements were made.

The correlation of aqueous solubility of PNAs with molecular properties has been fully discussed by Mackay and Shiu (1). The approach is to calculate the aqueous activity coefficient from molecular properties and from it calculate the solubility either directly for liquid solutes or using the solid/subcooled liquid fugacity ratio for solid solutes. A similar approach has been taken by Yalkowsky and Valvani (3). The Hansch-Leo approach (4) was used to estimate octanol-water partition coefficients for some compounds. Interestingly, it can be shown that solubility, partition coefficient, Henry's Law constant and vapor pressure are closely linked thermodynamic properties, suggesting that a fruitful approach is to assemble all these data for specific compounds in some form of data bank system. Diffusivity is best correlated by the Wilke equation.

Ultimately, the aim is to predict these properties from fundamental data such as triple and critical point, vapor pressure, molar volume and shape and functional groups. The methodology for accomplishing such predictions is now emerging.

Toxicological Background

In previous studies of the toxicities of aromatics to algae Kauss et al (5) found a relationship with solubility. Similar results had been obtained by Berry and Brammer (6) and Currier (7). In accord with early work by Ferguson (8), Kauss and Hutchinson (9) found that the activity (product of mole fraction and activity coefficient) of aromatics was a good indicator of toxicity. Since activity coefficient is inversely proportional to solubility this implies that less soluble compounds exert a greater toxic effect at a given concentration. An alternative but similar viewpoint is that the important aqueous concentration toxicologically is not the absolute concentration but rather the fraction which that concentration is of the solubility. For example, benzene (solubility 1780 ppm) and naphthalene (solubility 30 ppm) will exert the same toxic effect at one tenth of saturation, i.e. 178 and 3 ppm respectively, but when both are present at say 3 ppm, the naphthalene will be considerably more toxic. A simple thermodynamic explanation of this effect is that the concentrations achieved in the lipid phases of the organism will depend directly on the activity, with similar hydrocarbons in lipid concentrations being achieved in organisms contacted with say 178 ppm benzene and 3 ppm naphthalene.

This effect can be readily expressed mathematically by equating the fugacities (f) in the lipid (L) and water (W) phases on a Raoult's Law basis as:

$$f_L = x_L\gamma_L f_R = f_W = x_W\gamma_W f_R$$

where γ is the activity coefficient, x is the mole fraction and f_R the reference fugacity (which cancels). Since γ_W is the inverse of the solubility S_W (in mole fraction units) the concentration of toxicant in the lipid phase x_L is given by:

$$x_L = x_W\gamma_W/\gamma_L = x_W/\gamma_L S_W$$

If the toxic effect depends primarily on x_L and not on the nature of the toxicant, and if γ_L is approximately equal for all organic toxicants (which is a reasonable first approximation) then at identically toxic conditions for a series of compounds, x_L will be constant, as will x_W/S_W. In this case solubility can be expressed as mole fraction or mols/litre, these quantities being proportional at these low concentrations.

Octanol is usually selected as the pure chemical closest in properties to a lipid phase; thus if the activity coefficients of a toxicant in octanol and lipid are equal then the octanol water partition coefficient K_{OW} expressed as a mole fraction ratio becomes x_O/x_W or x_L/x_W and equal to γ_W/γ_L or $1/S_W\gamma_L$. This explains the observation that octanol-water partition coefficient is approximately inversely related to solubility S_W (Chiou et al (10), Mackay (11)).

Equal concentrations x_L will thus be obtained for a series of toxicants when $x_W K_{OW}$ is constant. K_{OW} is normally expressed as a concentration (mass/volume) ratio, however, this is proportional to the mole fraction ratio.

If then a series of experiments is designed to determine the different concentrations x_W causing a given toxic effect (such as 50% inhibition of photosynthesis), it would be instructive to examine the constancy of the values of the groups x_W/S_W or $x_W K_{OW}$. Alternatively, plots of log x_W vs log S_W and log K_{OW} can be constructed to elucidate the same relationship.

Materials

The 38 hydrocarbons and halogenated hydrocarbons listed in Table 1 were tested for toxicity and physical-chemical data obtained experimentally or from the literature.

Table 1 Physical-Chemical and Toxicity Data for the Compounds. An Asterisk indicates that the toxicities were extrapolated.

No	Compound	MW	Solubility		Log K_{ow}	Chlamydomonas		Angulosa	Chlorella Vulgaris		
			g/m³	mmol/m³(s)		x	x/s	$xK_{ow} \times 10^{-3}$	x	x/s	$xK_{ow} \times 10^{-3}$
1	hexane	86.2	9.5	110	3.00	94	0.85	94	149	1.35	149
2	3 methylpentane*	86.2	12.8	149	2.80	195	1.31	123	145	0.99	91
3	octane*	114.2	0.66	5.7	4.00	6.5	1.14	65	6.27	1.10	63
4	decane*	142.3	0.052	0.365	5.01	1.15	3.14	118	0.303	0.83	31
5	dodecane*	170.3	0.0038	0.0222	6.10	0.033	1.49	42	0.039	1.74	48
6	tetradecane*	198.4	0.0023	0.0115	7.20	0.056	4.85	884	0.131	11.4	2080
7	cyclopentane	70.1	156	2224.0	2.05	1720.0	0.77	192	1660	0.75	186
8	cyclohexane	84.2	55	654	2.44	455.0	0.70	125	380	0.58	105
9	methylcyclopentane	84.2	42	499	2.35	455.0	0.91	102	315	0.63	71
10	methylcyclohexane	98.2	14	143	2.76	107.0	0.75	62	107	0.75	62
11	cyclooctane	112.2	7.9	70	3.28	26.5	0.38	50	24	0.35	47
12	benzene	78.1	1780	22790	2.13	5900	0.26	796	4000	0.18	540
13	toluene	92.1	515	5589	2.69	1450	0.26	710	2250	0.40	1102
14	1,2,4,5 tetramethyl-benzene*	134.2	9.6	72	2.80	72.4	1.00	46	74	1.03	47
15	ethylbenzene	106.2	152.0	1432	3.15	480	0.335	678	590	0.41	833
16	p-ethyltoluene	120.2	94.85	780	3.63	450	0.58	1920	400	0.51	1706
17	O-ethyltoluene	120.2	93.05	774	3.63	155	0.20	661	340	0.44	1450
18	p xylene	106.2	185.0	1742	3.08	430	0.25	516	990	0.57	1190
19	propyl benzene	120.2	55.0	458	3.60	150	0.33	597	135	0.29	537
20	isopropyl benzene	120.2	50.0	416	3.51	73	0.17	236	177	0.42	572
21	n-butyl benzene	134.2	12.6	93.8	3.86	26	0.28	188	23	0.24	166
22	isobutyl benzene	134.2	10.1	75	4.01	23	0.31	235	26	0.35	266
23	decalin	138.25	6.21	44.8	-	14.25	0.32	-	7.30	0.16	-
24	naphthalene	128.2	31.7	247	3.35	75.0	0.30	168	150	0.61	336
25	1-methyl-naphthalene	142.2	28.5	200	3.86	12.0	0.06	87	36	0.18	261
26	2-methyl-naphthalene	142.2	25.4	178	3.86	31.5	0.18	228	63	0.35	456
27	biphenyl	154.2	7.0	45	4.02	8.3	0.18	87	25	0.56	262
28	phenanthrene	178.2	1.29	7.2	4.63	5.30	0.74	226	6.8	0.94	290
29	anthracene*	178.2	0.042	0.23	4.63	1.34	5.82	57	3.0	13.0	127
30	pyrene*	202.3	0.135	0.67	5.22	1.00	1.50	166	1.64	2.45	272
31	dichloro-methane	84.9	19400.0	228500	-	17400	0.076	-	27000	0.12	-
32	trichloro-methane	119.4	7950.0	66580	-	3200	0.048	-	3400	0.051	-
33	1,1,1-trichloro-ethane	133.4	720.0	5390	-	2100	0.39	-	1150	0.213	-
34	1,1,2,2,2 penta-chloroethane	202.3	480.0	2370	-	120	0.051	-	150	0.063	-
35	1,2,3-tri-chloropropane	147.5	1900.0	12880	-	760	0.059	-	1150	0.089	-
36	chloro-benzene	112.6	471.1	4185	2.83	503	0.12	340	880	0.210	594
37	1,2,3-tri-chlorobenzene	181.5	16.6	91.0	4.27	19	0.21	354	34	0.37	633
38	1,2,3,5 tetra-chlorobenzene	215.9	3.57	16.5	5.05	7.3	0.44	819	11.6	0.70	1301

Toxicity Test System

The two algal cultures were grown in a Bold's Basal medium (BBM) in cotton-plugged 125 ml Erlenmeyer flasks, at a pH of 6.5 under axenic conditions. A 12 hour light-dark cycle was used with a light intensity of 400 foot candles, (f.c.) and a temperature of 19°C. Saturated solutions were prepared in sterile BBM following 16 hours of magnetic stirring for a liquid hydrocarbon and 24 hours for a solid hydrocarbon. Straight and branched chain alkanes, cyclo-alkanes, aromatics, polynuclear aromatics, and chlorinated hydrocarbons were tested. The saturated solution was then decanted or filtered, and dilutions made with BBM to provide 0, 20, 50 and 100 percent of the original saturation level of HC.

$^{14}CO_2$ uptake was used as an indication of photosynthesis using a standard method,( ) to provide an activity of 0.5 μCi/100 ml for *Chlorella vulgaris* cells (#260 Indiana collection) and 1.25 μCi/100 ml for *Chlamydomonas angulosa* cells (#680 Indiana collection). Three-to-four day exponential phase cells were used for experiments, at a cell concentration of 5 x 10^4 cells/ml for *C. angulosa* and 20 x 10^4 for *C. vulgaris*. The isotope was added at time xero and glass stoppered flasks incubated under the described conditions for three hours. A dark set of controls was also run. After filtering, the cells were washed with 0.85 percent saline to remove surface $NaH^{14}CO_3$ and the radioactivity determined on dried Millipore filters with an end window gas flow counter (Nuclear Chicago).

RESULTS

A wide range of toxicities was found among the test compounds when the data were expressed as the percentage of saturation causing a 50 percent reduction in photosynthesis. This was determined by extrapolation from the graph of effects of 0, 25, 50, 75 and 100 percent saturation. The solubility of a number of hydrocarbons was so low that a 50 percent reduction in photosynthesis was not achieved, but had to be extrapolated from reaction curves. Those hydrocarbons with low solubility were more toxic on a molar basis than the high solubility compounds (such as benzene). Table 1 lists the compounds, their physical chemical properties and the toxicity results expressed as x, the concentration necessary to achieve 50% photosynthetic inhibition. The groups x/s and xK_{OW} are shown. In considering these results, it should be noted that the solubilities extend over 8 orders of magnitude and K_{OW} over 6 orders; thus correlation of toxicity, even within 1 order of magnitude, is a considerable accomplishment. The data are also plotted on Figure 1 and 2 which show the striking correlations which indicate that low solubility, high K_{OW} substances are more

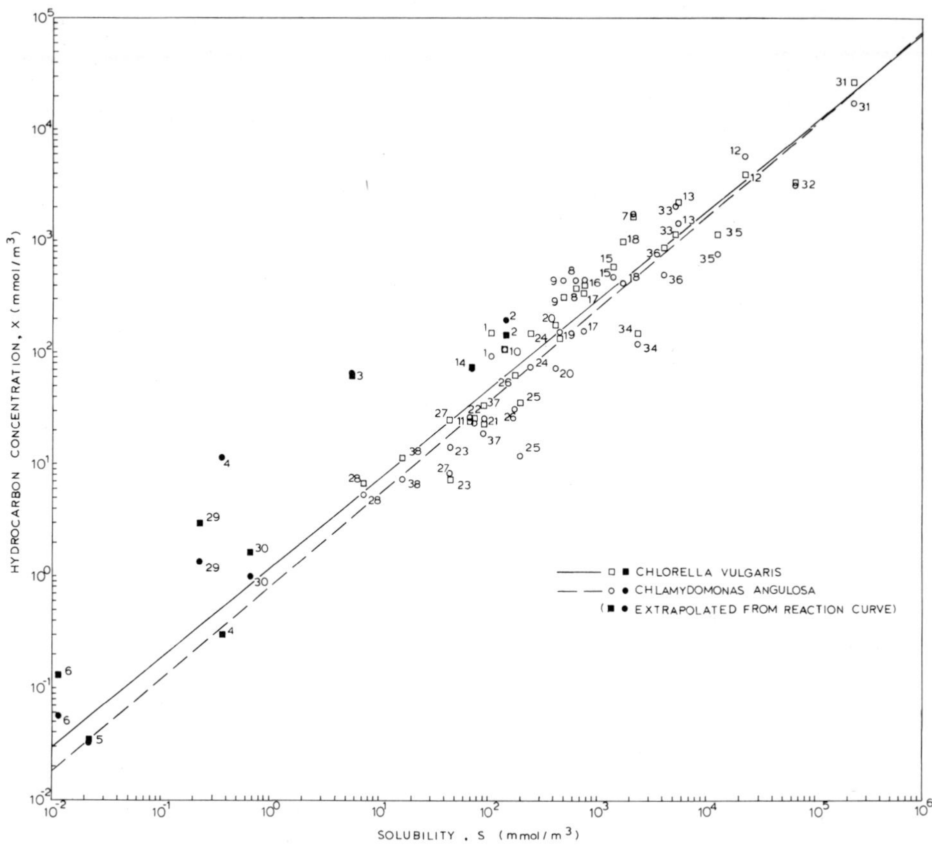

Figure 1. Plot of Hydrocarbon Concentration Necessary to Cause a 50% Reduction in Photosynthesis against Solubility. The lines correspond to the equations in the text.

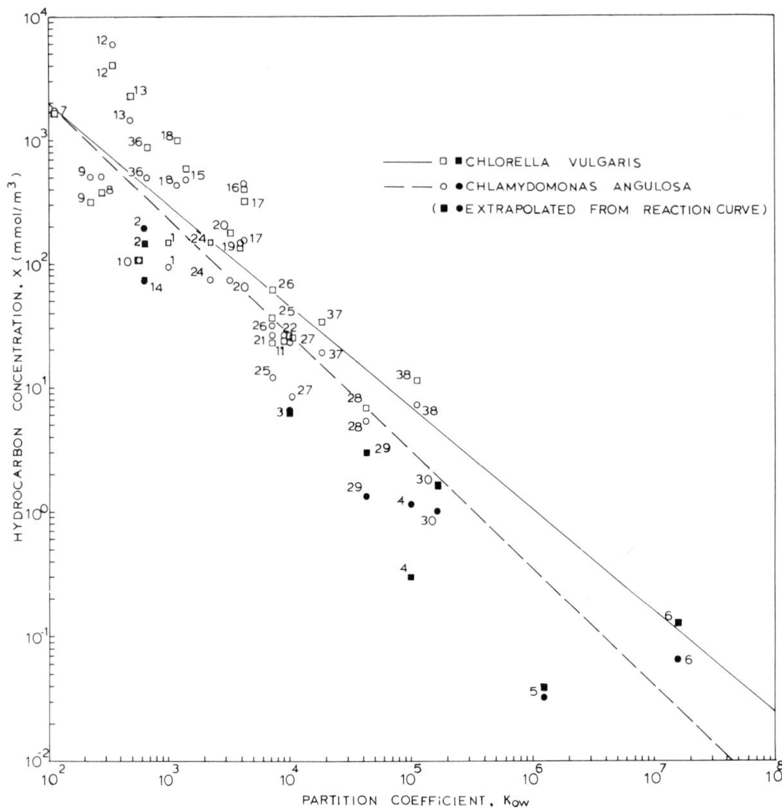

Figure 2. Plot of Hydrocarbon Concentration Necessary to Cause a 50% Reduction in Photosynthesis against Octanol-Water Partition Coefficient. The lines correspond to the equations in the text.

toxic at the same molar concentration. The correlation equations obtained (without including the compounds for which extrapolation is necessary) are as follows.

Chlamydomonas angulosa

$$\log x = -0.101 + 0.826 \log S$$
$$\log x = 5.167 - 0.928 \log K_{ow}$$

Chlorella vulgaris

$$\log x = 0.069 + 0.804 \log S$$
$$\log x = 4.934 - 0.817 \log K_{ow}$$

Here x and S are expressed in $mmol/m^3$. The correlations were done on the logarithmic quantities.

In all cases, the standard error corresponds to approximately a factor of 3 in x which is remarkably small considering the range in concentration from 0.03 to 5900 $mmol/m^3$.

The other test of the hypothesis is to compare the constancy of values of x/S and xK_{ow}. This is equivalent to forcing a regression line of slope +1.0 on the solubility correlation or -1.0 on the K_{ow} correlation. The coefficients are fairly close to unity (0.80 to 0.93); thus only a modest loss in accuracy is obtained. The mean values and standard deviations are as follows.

Chlamydomonas angulosa

x/S	mean 0.350	$SD = 0.255$
xK_{ow}	mean 394×10^3	$SD = 411 \times 10^3$

Chlorella vulgaris

x/S	mean 0.428	$SD = 0.287$
x/K_{ow}	mean 548×10^3	$SD = 477 \times 10^3$

These statistics are distorted to some extent by the presence of a few points which account for most of the variance. For example, the x/S values for both algae average approximately 0.4, and 70% of all the data lie between 0.1 and 0.7. The conclusion is that 50% reduction in photosynthesis is obtained for 70% of the compounds when the concentration lies between 10 and 70% of the solubility, the mean being approximately 40% of solubility. A similar argument can be applied to K_{ow}, the conclusion being that the mean concentration ($mmol/m^3$) necessary to achieve 50% reduction in photosynthesis is ($470000/K_{ow}$) $mmol/m^3$. These approximate characterisations of toxicity may be useful in obtaining order-of-magnitude estimates of toxicity for other hydrocarbons and halogenated hydrocarbons.

CONCLUSIONS

These results are regarded as being encouraging in that there is clearly a commonality of toxic mechanism for the class of compounds studied. Very satisfactory prediction of photosynthetic toxicity can be obtained from either solubility of K_{ow}. This approach can possibly be extended to other organisms, toxic effects and classes of compounds and, if successful, could result in considerable economies of effort and increase predictive capabilities. Although more complex compounds, organisms and effects will probably not yield as simple relationships, it is desirable to test such structure-activity hypotheses to determine the limits of their applicability.

REFERENCES

1. Mackay, D. and Shiu, W.Y., 1977. J. Chem. Eng. Data 22:399

2. Mascarenhas, R., 1978. M.A.Sc. Thesis, Dept. of Chem.Eng. & Applied Chem., University of Toronto.

3. Yalkowsky, S.H. and Valvani, S.C., 1979. J.Chem.Eng. Data 24:127.

4. Leo, A.J., Hansch, C., and Elkin, D., 1971. Chem. Rev. 71:525.

5. Kauss, P.B., Hutchinson, T.C., Soto, C., Hellebust, J.A., and Griffiths, M., 1973, Proc. Conf. on Prevention and Control of Oil Spills, API, Washington, D.C. 703-714.

6. Berry, W.O. and Brammer, J.D., 1977. Environmental Pollution, 13:229.

7. Currier, H.B., 1951. Hilgardia, 20:383.

8. Ferguson, J., 1939. Proc. of the Royal Society, 127:387.

9. Kauss, P.B. and Hutchinson, T.C., 1978. Environmental Pollution, 9:157.

10. Chiou, C.T., Freed, V.H., Schmedding, D.W., Kohnert, R.L., 1977. Envir. Sci. Technol., 11:475.

11. Mackay, D., 1977. Envir. Sci. Technol., 11:1219.

INDEX

The page numbers refer to the title page of each paper